APPLIED MULTIWAY
DATA ANALYSIS

APPLIED MULTIWAY DATA ANALYSIS

Pieter M. Kroonenberg

Leiden University
Department of Education
The Netherlands

A JOHN WILEY & SONS, INC., PUBLICATION

Published by John Wiley & Sons, Inc., Hoboken, New Jersey.
Published simultaneously in Canada.

For general information on our other products and services or for technical support, please contact our
Customer Care Department within the United States at (800) 762-2974, outside the United States at
(317) 572-3993 or fax (317) 572-4002.

Wiley also publishes its books in a variety of electronic formats. Some content that appears in print may
not be available in electronic format. For information about Wiley products, visit our web site at
www.wiley.com.

Library of Congress Cataloging-in-Publication Data:

Kroonenberg, Pieter M.
Applied multi-way data analysis / Pieter M. Kroonenberg.
     p. cm. (Wiley series in probabity and statistics)
  Includes bibliographical references and indexes.
  ISBN 978-0-470-16497-6 (cloth : acid-free paper)
 1. Multivariate analysis. 2. Multiple comparisons (Statistics). 3.
Principal components analysis. I. Title.
  QA278.4.K76 2008
  519.5'35—dc22   2007029069

Printed in the United States of America.

10 9 8 7 6 5 4 3 2 1

To Lipke Bijdeley Holthuis

The Multiway Data Analyst

Escher's "Man with Cuboid" symbolizes a Multiway Data Analyst intrigued by an apparently unsolvable three-mode problem.

CONTENTS

FOREWORD

In 1988, we were involved in an Italian-Dutch-French-English consortium that organized an international meeting on the Analysis of Multiway Data in Rome. This meeting brought together many of the pioneering researchers in the field for the first time. The Rome conference resulted in the book *Multiway Data Analysis*, published in 1989 by North-Holland, which contained (mostly technical) papers of authors from a variety of backgrounds in statistics, psychometrics, linear algebra, and econometrics. Note that this was before chemometrics had joined the three-way community. Although certainly a landmark volume - it widened the scope of the field and deepened our insights into the breathtaking ramifications of the underlying mathematics - the book was almost inaccessible to the uninitiated in the three-way field: no big surprise for a first attempt to bring order into chaos. In addition, the Multiway88 book also suffered from a paucity of substantial applications.

Applied Multiway Data Analysis is a timely successor that marks the increased maturity of the field. The book brings a unified treatment of the major models and approaches, and of the decisions that must be made to get meaningful results. It gives practical advice on preparing the data for analysis (a major issue in three-way analysis), how to treat missing data, how to select a model, how to make revealing plots, and so on. There is a lot of detail in three-way data analysis, and the author had the stamina to cover it all. Based upon an important distinction between the roles played by the different sides of the data box that is the target of three-way analysis, the present book has special chapters on four major data types: three-way or multi-way profile data, three-way rating scales, multivariate longitudinal data, and multi-way contingency tables. Keeping this distinction consequently in focus is a major methodological asset of *Applied Multiway Data Analysis*.

Pieter Kroonenberg is eminently qualified to write about multi-way analysis techniques from the perspective of their application. He belongs to the generation that made the field professional, and knows the literature as no one else. Besides groundbreaking theoretical papers he contributed many applied papers, and has more than thirty years' hands-on experience in software development and data analysis. Moreover, he is a most dedicated teacher, as we can personally testify from our own experience in co-teaching the graduate course "Applied Multivariate Data Analysis" with him for many years.

In 2004 Pieter Kroonenberg was awarded a personal chair in "Multivariate Analysis of Three-Way Data" that was sponsored by the Foundation for the Advancement of Data Theory. This foundation administers a legacy bestowed by John P. van de Geer, who - besides his other major accomplishments - put three-way data analysis on the Dutch statistical research agenda in the early nineteen-seventies. Pieter Kroonenberg presented an insightful genealogy of three-way data analysis in the Netherlands in his 2005 inaugural lecture. The inspiration to write the present book originated from a cooperative project with Jos ten Berge and Henk Kiers at Groningen University that focused on three-way methods for the analysis of longitudinal data.

In the present work, Pieter Kroonenberg aims at winning the hearts and minds of scientists from all empirical disciplines. Applied researchers will appreciate the wealth of in-depth analyses of real-life data sets that convincingly demonstrate the additional benefits of adopting the three-way view of the world. The book contains excellent chapters on data handling, model selection, and interpretation of results, which transcend the specific applications and nicely summarize "good practices" that will be useful in many disciplines. We believe that this book will offer applied researchers a lot of good advice for using three-way techniques. In addition, *Applied Multiway Data Analysis* will turn out to be a valuable source of reference for three-way specialists.

WILLEM J. HEISER & JACQUELINE MEULMAN

Leiden, The Netherlands
28 September 2007

PREFACE

The core of this book is concerned with practical issues in applying multiway component techniques to multiway data with an emphasis on methods for three-way data. In
particular, most of the data sets considered in this book are fully crossed and the data
are generally numerical rather than categorical. Thus, leaving aside missing data, all
subjects or objects have scores on all variables at all occasions under all conditions,
all judges score all concepts in all situations, and each year all crop varieties have
values for all attributes in all environments, to name but a few research designs. Also,
the emphasis is on exploratory component models rather than confirmatory factor
models.

The analyses presented in this book were primarily carried out with my own software, 3WayPack, of which a light version can be downloaded from the website of the
The Three-Mode Company (*http://three-mode.leidenuniv.nl*). In addition, several of
the data sets analyzed in this book can be found there. Furthermore, information is
supplied on multiway software both on stand-alone programs and programs operating
under the MATLAB and R systems.

A brief word about both the title of the book and the use of the words "way" versus
"mode". Later a detailed discussion of this distinction will be presented, but here

I would like to remark that my personal preference is to use the word "mode" for methods and models, and "way" for describing the dimensions of a data array. Thus, I prefer to use terms like "multimode analysis" over "multiway analysis" but I tend to use the term "multiway data". However, the present parlance favors the use of "multiway" throughout, and I have decided to conform to this usage. Internet searches also favor the use of the term multiway (with or without hyphen) as this term is primarily related to statistical techniques, while multimode is primarily related to optical physics where spectra show several modes or peaks. I will make an exception for models and methods in the three-way case as the term "three-mode" is so closely related to the origins of the techniques.

This book was written from the perspective of analyzing data, but the applications are primarily treated as examples of a type of data rather than as examples of subject areas. The fields of application range from the social and behavioral sciences to agriculture, environmental sciences, chemistry, and applications in many other disciplines have appeared as well; a full bibliography of such applications can be found on the website of The Three-Mode Company. The background required is a firm understanding of principal component analysis and related techniques, but in most chapters an attempt has been made to keep the presentation at a conceptual rather than a mathematical level. Readers are expected to have, at least, a basic knowledge of linear algebra and statistics, be it that some applications can be understood without them.

The introductory chapter was written with a more general audience in mind and is aimed at those who want to get a basic acquaintance with what multiway analysis is and what kind of problems can be tackled in this way. It may also serve as a gentle, nontechnical introduction for those who aim to use the techniques in the future.

The book consists of three parts. The first part discusses multiway data, multiway models, and ways to estimate these models. The second part is devoted to practical aspects relevant for carrying out analyses, and the third part is devoted to applications. The latter are arranged around general data types, such as profile data, rating scale data, longitudinal data, and categorical data. In several cases, additional explanation of techniques was necessary to understand the analyses, such as in the chapters on clustering, and the analyses of categorical and binary data.

A general presentation scheme has been devised, especially for the chapters dealing with applications, in order to structure the material as much as possible. First, a brief introduction into the general problems, types of analyses, or data is presented, followed by a more concrete overview of the chapter. Next, an exposition of the chapter's techniques and data provides some insight into the background to the analyses. There will necessarily be some overlap between the earlier and later chapters but this is unavoidable given the aim to make the latter, applied chapters reasonably self-contained. To set the scene for the discussion of the practical aspects of carrying out analyses, a relatively brief application is presented next, which is then used to discuss the logic, practical problems, and tricks that allow the reader to appreciate outcomes of the analyses. These are then illustrated with a fully fledged application. Some chapters are rounded off with a conclusion and advice on how to proceed further.

As much as possible, each application will have a standard format. In principle, the following sections will be included.

- *Stage 1: Objectives.* The purpose of the research from which the data originated is set out, and the questions that are going to be answered will be outlined.

- *Stage 2: Data description and design.* Details and format of the data are presented, and the way they will be used to answer the research questions.

- *Stage 3: Model and dimensionality selection.* Either a model will be selected from several model classes or an appropriate dimensionality will be chosen for a model from a single model class, or both.

- *Stage 4: Results and their interpretation.* The outcomes of the analysis will be presented and interpreted using the aids for interpretation set out in Part II.

- *Stage 5: Validation.* This heading covers both internal validation (checking the results against the data via examination of residuals, etc.) and external validation (using external information to enhance the interpretation).

This format for presenting applications was inspired by, but not completely copied from, a similar setup by Hair, Anderson, Tatham, and Black (2006) in their book on multivariate data analysis. It was necessary to deviate from their format, but its spirit has been maintained.

In the text many references have been included to other sections in the book. When such a section is in a different chapter the page number of the beginning of that section has been given as well. The entries in the author index have the form Author (year), and refer to the page where the article of that first author published in the year indicated is cited. Thus, second (and subsequent) authors do not appear in the author index. The book contains a glossary of multiway-related terms, but the entries in the glossary do not appear in the subject index.

Writing a book involves making choices. Given a life-time involvement with multiway analysis, I started with the intention to write a book that would reflect the state-of-the-art on the subject. The idea was to cover everything that had been written about the subject, both theoretical and applied. Clearly this was a mad notion, and when sanity returned, I conceived a book that emphasized my personal experience and the practical side of performing multiway analyses. One effect of this is that I performed virtually all analyses presented in the book myself, so that many topics fell along the wayside because I did not have the time to acquire experience in analyzing data from a particular field or getting personally acquainted with a particular method. I apologize for this, especially to those authors whose work should have been included but was not. Maybe my next move should be to edit a book which contains contributions of all those colleague researchers, model builders, and program developers who are insufficiently represented in this book.

Topics which at least deserve to be treated in a similar form as those discussed in this book are fMRI, EEG, and similar data, signal processing data, multiset data in which at least one of the modes consists of different entities in each level, stochastic multimode covariance matrices, individual differences multidimensional scaling in its various forms, several multiway clustering techniques, optimal scaling for multiway data, mixed model methods for multivariate longitudinal data, multiway generalizations of the general linear model such as multiblock techniques, N-way partial least squares, and possibly others which I may have overlooked. Fortunately, chemistry applications are already well covered in Smilde, Geladi, and Bro (2004). It would be nice to have chapters showing how multiway methods have contributed to theoretical developments in a variety of scientific disciplines. Furthermore, an overview of mathematical developments in tensor algebra, the rank of multiway arrays, degeneracy of Parafac models, uniqueness of multiway models, and algorithmic improvements should be written. In other words, enough to do for another academic life or two.

Acknowledgments

Many organizations contributed directly and indirectly to the realization of this book, but special thanks go to the following institutions for providing financial support during the writing of the book. The section Algemene en Gezinspedagogiek – Datatheorie, the Department of Education and Child Studies, and the Faculty of Social and Behavioural Sciences, all Leiden University, provided me with the opportunity and time during the entire process of writing this book. The idea of the book was conceived in 2000 within project "Multiway Analysis of Multivariate Longitudinal Data" (NWO 575-30.005 t/m .008) financed by the Netherlands Organisation for Scientific Research (NWO). The majority of the chapters were written or conceived during my one-year stay (2003–2004) at the Netherlands Institute for Advanced Study in the Humanities and Social Sciences (NIAS). Further financial support was provided again by NWO, the Nissan Foundation, and the Ethel Raybould Foundation, which allowed me to consult my Japanese and Australian colleagues about matters concerning several chapters of the book. Special thanks go to the Board of the Stichting ter Bevordering van de Wetenschap der Datatheorie (the Foundation for the Advancement of Data Theory) who established the chair: "Multivariate Analysis of Three-Way Data."

Many researchers contributed by making their data available and by commenting on the analyses that I carried out on their data. Of course, misinterpretations are entirely my own. In particular, I would like to mention Claus Andersson, Carolyn Anderson, Eva Ceulemans, Fernando De La Torre, Paul Gemperline, Hideo Kojima, Nick Martin, Iven van Mechelen, Takashi Murakami, the National Institute for Child Health and

Human Development (NICHD), Irma Röder, Michel Sempé, Ivana Stanimirova, Rien van IJzendoorn and Iven van Mechelen.

A number of data sets were derived from published papers: Multiple personality data (Osgood & Luria, 1954, pp. 582–584, Tables 2–4), Davis's Happiness data (Clogg, 1982, Table 2), Electronics industries data (D'Ambra, 1985) and Chromatography data (Spanjer, 1984, pp. 102, 103).

Thanks to all proofreaders of chapters: Kaye Basford, Rasmus Bro, Stef van Buuren, Eva Ceulemans, Richard Harshman, Mia Hubert, Henk Kiers, Marleen van Koetsveld, Wojtek Krzanowski, Takashi Murakami, Ineke Smit, Ivana Stanimirova, Giorgio Tomasi, and Iven van Mechelen. Ives de Roo (NIAS) developed a new interface for the program suite 3WayPack, and Mariska van Eenennaam assisted in building the bibliographic data base both for the book and the website of The Three-Mode Company.

Many colleagues assisted me over the years in clarifying points of multiway analysis, and they come from all over the world and from many disciplines. Many of them were involved in a series of specialized conferences on three-way and multiway methods, especially the TRICAP meetings and recently the Tensor workshops. These gatherings were and are the basis of the existing worldwide multiway community, which is my habitat. To all these individuals, my thanks.

PIETER M. KROONENBERG

Leiden, The Netherlands
September 2007

PART I

DATA, MODELS, AND ALGORITHMS

In Part I we first present an overview of multiway data and their analysis. Then a detailed overview is presented of the kind of multiway data one may encounter, and finally models, methods, and techniques for the analysis of multiway data are discussed.

CHAPTER 1

OVERTURE

This book is devoted to the analysis of multiway data, which have a richer and more complex structure than just objects or subjects have scores on a number of variables. Such data are measured several times and/or under several conditions, are ubiquitous, and are collected on a regular basis in many disciplines. They can, for instance, be found in large-scale longitudinal studies and in agricultural experiments and are routinely produced during the study of chemical processes, especially in analytical chemistry. Also, signal processing data and assessments of cracks in the surface of paintings lend themselves to multiway analysis. In fact, it will be shown that scientific research cannot do without multiway analysis, even though not everybody knows this yet. Multiway data can be found in every walk of science, if one only looks. Once alerted to their existence, they can be seen everywhere, in much the same way that first-time prospective parents suddenly notice pregnant women everywhere.

In this introduction, multiway data, in particular, three-way data, are presented and a general introduction is given on how they can be tackled. By means of a detailed example, the value of multiway analysis is brought to the fore as a type of analysis that can be fruitfully used in many disciplines. The introduction will close with a

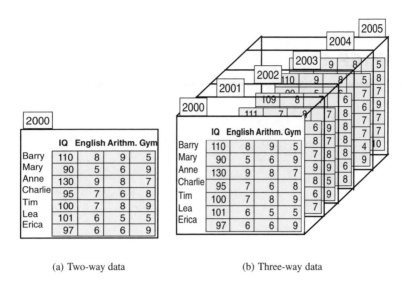

(a) Two-way data   (b) Three-way data

Figure 1.1   Storing several two-way data sets in three-way box.

sketch of the origins of multiway analysis and its explosive development since around 1990. The presentation will use primarily three-way examples because they are more common and easier to understand, and because the general practice in multiway data analysis is to count "one, two, multi".

1.1   THREE-WAY AND MULTIWAY DATA[1]

What are three-way data? To put it at its simplest, three-way data are data that no longer fit onto one index card, but need a box to contain them. Looking, for instance, at an example from child studies we note that many data sets there take the form of the scores of a number of children on a number of variables, such as intelligence, and marks for English, arithmetic, and physical education (gym) (Fig. 1.1(a)).

Such data may be written on one index card, in the form of a table with the children in the rows and the variables in the columns. We now have two "ways": Children and variables. If the data form part of a longitudinal investigation in which the children are measured every year on the same variables, we have three-way data, the "ways" being children, variables, and years or occasions. One index card will now no longer suffice, but we need a separate card for each year, and the set of cards is kept in a box: Three-way data fit in boxes, as shown in Fig. 1.1(b).

---

[1]The text of this and the following section is largely based on Kroonenberg (2005a), which also contains the genealogy of Dutch three-way researchers mentioned in the Foreword. Available from *https://openaccess.leidenuniv.nl/dspace/bitstream/1887/3494/3*. Accessed May 2007.

Once we have realized this, we suddenly notice data in boxes everywhere. A *plant breeder* has planted several varieties of peanuts (first way) in different locations (second way) and measures the characteristics of the harvested plants, such as yield, quality, and the percentage of saturated and of unsaturated oils (third way). A *chemist* has ten containers with different sugar solutions, sends light of varying wavelengths through the containers, and measures the excitation wavelengths emerging at the other side. A *medical pharmacologist* has derived the spatial structures of a number of cocaine variants from crystallographic measurements, and wants to know to what extent their spatial structures are the same, and to what extent they are different. The molecules form the first way, their constituent atoms the second, and the spatial coordinates the third.

It may seem a miracle that a radio mast, which receives signals not only from one's own mobile phone but also from countless other cell phones, knows exactly that the answer to a specific person's outpourings should be sent to her, rather than her neighbor who has nothing to do with the exchange. This process, too, may be analyzed through three-way models.

Extending the concept to multiway data, one might think of the typical plant breed- ing experiments, which are conducted over a number of years so that four-way data are obtain: peanut varieties×attributes×locations×years. Similarly, any research in which subjects are measured on several variables under various conditions at a num- ber of points in time generates four-way data. Moreover, fMRI scans, which are becoming more and more routine in hospitals and research, also provide multiway data, such as voxels×time points×subjects×trials×task conditions. However, ac- tual situations in which higher-way data are collected are still few and far between. Analyses of such data in their full multiway appearance are even rarer, but they are on the increase.

1.2   MULTIWAY DATA ANALYSIS

Three-way analysis is no more than the analysis of data that fit in boxes, and multiway analysis is the analysis of data that fit in more-dimensional boxes or *hyperboxes*. This statement might not be very enlightening to the uninitiated and it is better to rephrase it as "What type of research questions can be tackled via multiway analysis?" Such a question has the advantage that it is formulated in substantive rather than methodological terms, and we will discuss this in the three-way context.

Let us again look at the example of the children tracked over a number of years. What questions would the researchers have had in mind when they started to col- lect data? There are of course many possibilities, but let us confine ourselves to those questions that may be handled via three-way analysis. In this case, the *central questions* might be:

- What are the relations between the variables?

- What trends may be discovered over time?

- Are there different types of children?

These are three questions, one for each way. Although these types of questions are interesting, they only apply to one way at a time. Three-way analysis has been devised especially to deal with more complex questions such as:

- Do the relations between the variables change over time? For instance, it is well-known that in very young children intelligence is still very amorphous, but that as they get older some children develop better on some aspects of intelligence than on others. In other words, time brings a change in the structure, that is, the interrelations between the various parts, of an intelligence test.

- An even more complex question is: Does the structure of the variables change over time in a different way for different groups of children, for instance, for boys and for girls, or for children with different levels of mental handicap?

With such complex questions, involving all three aspects, or ways, of the data, three-way analysis really comes into its own. *Plant breeders*, for instance, are interested in the specific adaptation of crops; in other words, they like to know which varieties of a plant will perform well on specific attributes in locations with specific characteristics. In concrete terms, where should one grow what type of peanut plants, in order to obtain peanuts that are specifically suitable for making peanut butter?

One's first acquaintance with techniques for multiway analysis is often a bit of a shock, because of the complications involved in the correct understanding and interpretation of those techniques. However, this is unavoidable, as we are dealing with complex techniques intended to solve complex questions. Testing differences in average lengths between boys and girls is child's play for the average student, but three-way questions are more intricate. However, it is this complexity that can yield deeper insights. The picture opposite the content page of this book, a woodcut by M.C. Escher[2], shows a multiway data analyst intrigued by an apparently unsolvable three-mode problem.

1.3   BEFORE THE ARRIVAL OF THREE-MODE ANALYSIS

How were three-way data analyzed before the arrival of thee-way data analysis techniques? To put it briefly: by flattening the box, or stringing out its contents. In both cases the idea is to make three-way data into two-way data by eliminating one of the ways. Instead of looking at the interactions between three types of units (or ways), one then only needs to analyze two.

[2]M. C. Escher, *Man with Cuboid*. Copyright © 2007 The M.C. Escher Company B.V., Baarn, Holland. *http://www.mcescher.com/*. All rights reserved.

Flattening consists of making a two-way matrix out of a three-way array by removing one of the ways. It is typically done by taking the averages over all cards in the box, so that one is left with one card containing means. These may be averages over all years, so that one loses sight of trends over time, but one may also take averages over all subjects, so that individual differences disappear below the horizon.

Stringing out is creating a two-way matrix out of a three-way array by either laying out all the cards in one long row, so that the relation between similar variables at different time points is neglected, or laying out all cards in one tall column, so that the connections between people's scores at different moments in time are lost. The technical term for this procedure is matricization. In all these cases, the data and their analysis are shrunk from three-way to two-way.

Sometimes this may do no harm, because it is possible that a three-way analysis leads to the conclusion that no three-way analysis is necessary: for instance, if nothing changes over time, or if all subjects may be viewed as having been randomly drawn from one single population. However, if this is not the case, the flattening or stringing out of three-way data leads to an unnecessary and sometimes unacceptable simplification.

1.4 THREE-MODE DATA-ANALYTIC TECHNIQUES

Since the beginning of the 1960s, a series of techniques have been devised specifically aimed at doing justice to three-way data, and these techniques were later extended to four-way and higher-way data. They bear such intriguing names as three-mode principal component analysis, multilinear component analysis, three-mode factor analysis, three-way cluster analysis, parallel factor analysis, multiway covariance analysis, multidimensional scaling techniques for individual differences, generalized Procrustes analysis, multivariate longitudinal analysis, and many more of this kind.

Most of these methods have a strongly *exploratory character*, which means that one tries to find the patterns among the elements of the three ways, without *a priori* postulating specific configurations and without applying tests to these patterns. This is partly because it is difficult to specify such patterns beforehand, and partly because hypothesis testing supposes that something is known about the distributions of the scores, which for multiway data is only very rarely the case. It is, however, perfectly possible to determine the stability of the estimated values of the parameters via repeated sampling from the sample in question (*bootstrap* method), but these developments are still in their infancy in three-way analysis.

1.5 EXAMPLE: JUDGING CHOPIN'S PRELUDES

In order to give an idea of multiway analysis, we will look, rather superficially, at a three-mode analysis of data produced by Japanese students, in reaction to listening to (parts of) the 24 preludes by the Polish composer Chopin played by a French

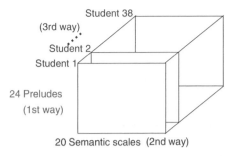

Student 38
(3rd way)
Student 2
Student 1

24 Preludes
(1st way)

20 Semantic scales (2nd way)

Figure 1.2   The Chopin prelude data set: 24 preludes by 20 semantic scales by 38 students.

pianist. The model used for this analysis is the Tucker3 model, which is presented in Section 4.5.3 (p. 54), and a further analysis of these data can be found in Section 8.9 (p. 190). In this example a type of data is used, called "semantic differentials", that actually initiated the development of three-mode analysis, as will be discussed in more detail later (p. 12).

The *research questions* were the following:

1. Classical music may be characterized in *technical* terms such as key, tempo, and mode (i.e., major/minor). To what extent are *nonprofessionals* sensitive to these aspects when we ask them to describe music in everyday adjectives such as loud, gloomy, and tender?

2. Is it possible at the same time to establish, by means of the same adjectives, their preference for specific types of music expressed as key signature, tempo, and mode?

The study was conducted among 38 Japanese students who were familiar with classical music (the first way). The 24 preludes were played to them (the second way), and after every prelude (or the first 80 seconds of it) they were asked to fill out a set of 20 semantic rating scales (the third way); these are given in Table 1.1. Note that these scales consist of two complementary concepts (restless – calm, fast – slow, strong – weak, etc.).

What we are trying to find is the connection between the preludes and the semantic rating scales as related to the individual differences between students. The main result of the investigation was that the students largely agreed on the technical, cognitive aspects of the preludes, but differed as to the affective elements, that is, the sort of music they preferred. This will now be explained in more detail.

1.5.1   Consensus on the musical-technical aspects

Scales.   Let us look at the *Semantic space* (Fig. 1.3), in which we see the relations between the 20 bipolar scales as reflected in the students' assessments. Every scale

Table 1.1   Semantic differentials used in the Chopin prelude study

Calm	-	Restless	Gentle	-	Severe
Quiet	-	Noisy	Lyrical	-	Dramatic
Tranquil	-	Vehement	Weak	-	Strong
Slow	-	Fast	Still	-	Loud
Light	-	Heavy	Cheerful	-	Gloomy
Bright	-	Dark	Soft	-	Hard
Happy	-	Sad	Clear	-	Cloudy
Warm	-	Cold	Small	-	Large
Delicate	-	Coarse	Thin	-	Thick
Unattractive	-	Attractive	Uninteresting	-	Interesting

English translations of the Japanese terms; scores run from 1 to 7.

is represented by an arrow, with one adjective at the tip (*restless*) and its antonym on the other side (*calm*). In the interest of clarity there are only four arrows of which both poles have been drawn. What does it mean that some arrows are close together, and that others are at wide angles to each other? Let us first look at the arrows marked *restless* and *dramatic*. These are close together, because students generally gave preludes similar scores on these two scales. Hence, *restless* preludes are also *dramatic*, and *calm* preludes are also *lyrical*. When the arrows are at right angles to each other, such as, for instance, *fast* and *cold*, this means that according to the students those scales have no relation to each other at all: *fast* preludes may be *hot* as well as *cold*, and the same applies to *slow* preludes.

Preludes.   The *Prelude space* shows the configuration of the preludes as a reflection of the scores the students assigned to them (Fig. 1.4). For a correct interpretation of the prelude space, one should imagine the scale space as an overlay on top of the preludes, with the same orientation and the axes aligned.

Using the information from the scales, we can deduce that Prelude no. 16 in b♭, tempo indication *presto*, was judged especially noisy and fast. Prelude no. 20 in c, *largo*, located in the bottom right-hand corner of the Prelude space, was judged by the students as coarse, heavy, sad, and gloomy. However, Prelude no. 15 in D♭, *sostenuto*, in the bottom left-hand corner of the Prelude space, was rated calm and lyrical.

Technical merits.   If we now judge the configuration of the preludes in the Prelude space on its technical merits, we note that the students have made a clear division in Tempo (fast and slow) and in Mode (major and minor). In other words, on the basis of the semantic scales the students have arranged the preludes in a pattern that is found to correspond to an arrangement based on musical-technical aspects. This

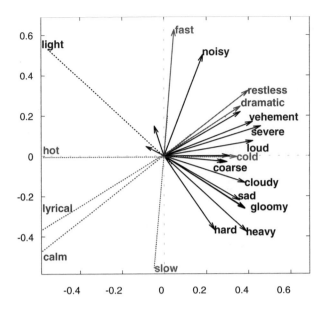

Figure 1.3   Relationships between the semantic differential scales. Narrow angles indicate high correlations between scale; perpendicular scales are unrelated.

is not all, however; closer inspection reveals that the arrangement of key signatures over the area largely corresponds to that of the circle of fifths (see Fig. 1.5). We may now even note two anomalies. These are Preludes nos. 9 and 10: Prelude no. 9 is in a major key and is situated in the "minor" area, and Prelude no. 10 is in a minor key and is found in the "major" part of the Prelude space. The probable cause is that in some longer preludes key changes occur, and the students were offered only the first 80 seconds, which may have given them a distorted impression of these preludes.

1.5.2   Individual differences

Figure 1.4 showed the consensus among the students. What still needs to be discussed is to what extent the students *differed*. There is no need for a figure to illustrate this, because it may easily be said in words. The students especially liked either fast pieces in a major key, or slow pieces in a minor key. Of course, this is as one would expect.

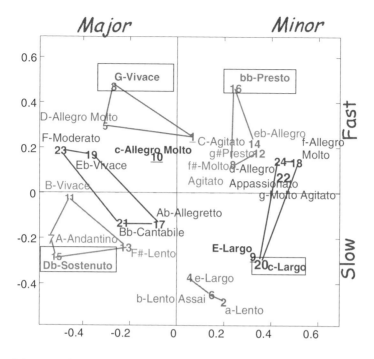

Figure 1.4   Relationships between the preludes. Horizontal axis: Mode — major, minor; Vertical axis: Tempo — fast, slow.

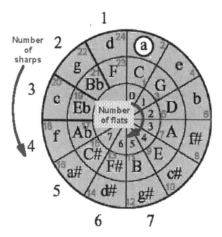

Figure 1.5   Circle of fifths, labeled with the Prelude numbers.   Adapted from *http://www.uncletim.com/answers/circle.htm.*   Accessed May 2007.; reproduced with kind permission from Tim Gillespie.

1.6   BIRTH OF THE TUCKER MODEL

The history of techniques for three-way data analysis starts with the late Ledyard R Tucker (†2004)[3], who, among other places, worked at the University of Illinois and the Educational Testing Service in Princeton. Another prominent godfather is Douglas Carroll, who spent a large part of his academic career at Bell Laboratories and is now on the staff at Rutgers University. A third founding father of three-way techniques is Richard Harshman, based at the University of Waterloo in London, Canada.

In an interview with Neil Dorans (2004), Tucker had the following to say on this topic:[4]

> Three-way factor analysis grew out of this multidimensional scaling work. While I was at ETS[5], I had observed that Charles Osgood of the University of Illinois had collected data from three modes—concepts, scales, and subjects— in his semantic differential research. I thought that the data should be analyzed differently than it was. He collapsed over people and threw away individual differences data [i.e., he flattened his data box]. *So I developed the 3-mode factor analysis approach, a very general model for evaluating individual differences data. It was able to deal with the variety of people* (p. 8).

Tucker illustrated his objection to "flattening" by the example of the car manufacturer Chrysler, who designed a car for the "average customer" without appreciating the importance of individual differences. Unfortunately, the car did not sell, because there was no such thing as the *average* customer, and Chrysler almost went bust on this enterprise. The necessity of paying attention to individual differences is still the guiding principle for almost all work in the area of three-way analysis.

1.7   CURRENT STATUS OF MULTIWAY ANALYSIS

A full and detailed survey of the development of multiway analysis will not be undertaken here; see, for instance, Section 2.4 (p. 20) and Smilde et al. (2004, pp. 57, 58, for a brief historical introduction). Some idea of its fortunes over the years can be seen in Fig. 1.6, which shows the roughly calculated citation curves for the main protagonists, divided into founding fathers (Tucker, Carroll, and Harshman), psychometricians, and chemometricians. What is most striking in this figure is the steep increase in the number of citations during the 1990s, and the fact that this escalation is almost exclusively due to the chemometricians. Moreover, the rising number of citations for the psychometricians is also due to the stormy developments in chemometrics, because it was in psychometrics that the initial developments had taken place. Especially the models developed by the psychometrician Richard Harshman

---

[3]Note that R is Tucker's middle name and not an initial, so that it should be written without a full stop.
[4]Reproduced with kind permission from Neil Dorans.
[5]Educational Testing Service, Princeton.

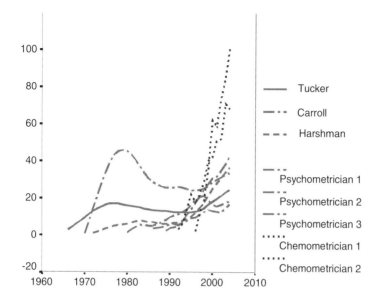

Figure 1.6 Popularity of multiway analysis as expressed in the number of citations per year to work by the founding fathers, three psychometricians, and two chemometricians (1963-2004).

corresponded directly to physical models in use in analytical chemistry. What the chemometricians needed were methods to estimate the parameters of their multiway models and these had been a subject of intensive research in the psychometric field. Moreover, new experiments in chemistry could easily be designed to take advantage of the uniqueness properties of Harshman's Parafac model. Also, important commercial applications immediately offered themselves in the chemistry domain. All these factors led to an increase in publications on multiway analysis.

An interesting aspect of this development is the fact that psychometricians also started to publish in chemometrics journals, because that was where their biggest customers were located. Whereas in the early 1990s applications in the social and behavioral sciences were at a rather low level, applications in applied chemistry were soaring and in other branches of science, such as agriculture, signal processing, medicine, and mathematics, applications can now also be found. This increased interest in applying existing multiway methods also created a considerable drive in further developing technical aspects of the techniques. Foundations were laid and are being laid for mathematical and statistical extensions of what were initially relatively straightforward three-way techniques, to sophisticated multiway methods in various forms.

CHAPTER 2

OVERVIEW

2.1 WHAT ARE MULTIWAY DATA?[1]

Most statistical methods are used to analyze the scores of objects (subjects, groups, etc.) on a number of variables, and the resulting data can be arranged in a two-way *matrix*, that is, a rectangular arrangement of rows (objects) and columns (variables). However, data are often far more complex than this. For example, the data may have been collected under a number of conditions or at several time points. When either time or conditions are considered, the data are no longer two-way, but become *three-way data* (see Fig. 2.1). In such a case, there is a matrix for each condition, and these matrices can be arranged next to each other to form a *wide combination-mode matrix* of subjects by variables×conditions. Alternatively, one may create a *tall combination-mode matrix* of subjects×conditions by variables. The term *combination-mode* is used to indicate that one of the ways of the matrix consists of the combination of two

---

[1]Portions of this chapter have been taken from Kroonenberg (2005b); 2005 ©John Wiley & Sons Limited. Reproduced and adapted with kind permission.

Applied Multiway Data Analysis. By Pieter M. Kroonenberg
Copyright © 2007 John Wiley & Sons, Inc.

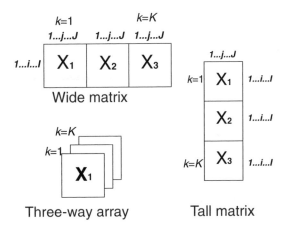

Figure 2.1 Two-way combination-mode matrices and three-way arrays

modes rather than a single one. The third possibility is to arrange the set of matrices in a three-dimensional block or three-way *array*. When both time and conditions play a role, the data set becomes a four-way array, and including even more aspects leads to higher-way data. The collection of techniques designed to analyze multiway data are referred to as *multiway methods*, or in the three-way case mostly three-mode methods. Making sense of multiway data is the art of *multiway data analysis*. The term *multiway array* is now the standard; sometimes the term *multiway tensor* is used instead of array. We will, however, avoid the term tensor because it has a more specific meaning in mathematics and we are primarily concerned with data arrays rather than algebraic structures.

The words *multiway* and *multimode* can be given very precise meanings but in this book the usage is not always very exact. The word *way* is considered more general, referring to the multidimensional arrangement irrespective of the content of the data, while the word *mode* is more specific and refers to the content of each of the ways. Thus, objects, variables, and conditions can be the modes of a three-way data array. When the same entities occur in two different ways, as is the case in a correlation matrix, the data are one-mode two-way data. When correlation matrices for the same variables are available from several different samples, one often speaks of a two-mode three-way data array, where the variables and the samples are the two modes. In this book, we generally refer to multiway and three-way data and multiway and three-way arrays. However, following historical practice we will generally refer to three-mode methods, three-mode analysis, and three-mode models, but to multiway methods, multiway analysis, and multiway models. Terms commonly used in connection with multiway models are *trilinear*, *quadrilinear*, and *multilinear*, referring to linearity of these models in one set of their parameters given the other sets of parameters.

Individual differences is another expression that deserves attention in the context of multiway data analysis. In standard statistical theory it is assumed that subjects are drawn from a population, and thus are exchangeable. Two subjects are exchangeable, if they are drawn from the same population and it is irrelevant to the validity of generalizations made from the sample which of the entities is included in the sample. Entities in random samples are automatically exchangeable. In multiway analysis random samples are rare, but one hopes that the entities included in a sample are at least exchangeable with nonsampled entities from the population. On the other hand, in many multiway analysis there is not necessarily a sampling framework. Each individual or object is treated as a separate entity and its characteristics are relevant to the analysis. In some research designs this aspect of the data is more important than in others. For instance, in an agricultural experiment each variety of a crop is specifically chosen to be planted and compared with other varieties. Thus, in such studies the idea of a sample from a population is less relevant. However, in studies where the emphasis is exclusively on the development of a correlational structure over time, the specific subjects in the study may be of less interest.

2.2   WHY MULTIWAY ANALYSIS?

If there are so many statistical and data-analytic techniques for two-way data, why are these not sufficient for multiway data? The simplest answer is that two-mode methods do not respect the multiway design of the data. In itself this is not unusual since, for instance, time-series data are often analyzed as if the time mode were an unordered mode and the time sequence is only used in interpretation. However, it should be realized that multiway models introduce intricacies that might lead to interpretational difficulties. One of the objects of this book is to assist in using multiway models in practical data analysis. Thus, the book can be seen as an attempt to counter the statement by Gower (2006, p. 128) that "many, but not all, of the multilinear models developed by psychometricians are, in my view, uninterpretable – except perhaps to their originators". Luckily, support for the position taken in this book can be had from various quarters. In analytical chemistry, for instance, researchers adopt an entirely different tone. As an example, we may quote Goicoechea, Yu, Olivieri, and Campiglia (2005, p. 2609), who do *not* belong to the originators: "multidimensional data formats, which combine spectral and lifetime information, offer tremendous potential for chemometric analysis. High-order data arrays are particularly useful for the quantitative analysis of complex multicomponent samples and are gaining widespread analytical acceptance".

Multiway data are supposedly collected because all ways are necessary to answer the pertinent research questions. Such research questions can be summarized in the three-way case as: "Who does what to whom and when?", or more specifically: "Which groups of subjects behave differently on which variables under which conditions?" or "Which plant varieties behave in a specific manner at which locations on which attributes?". Such questions cannot be answered by means of two-mode methods, because these have no separate parameters for each of the three modes. When analyzing three-way data with two-mode methods, one has to rearrange the data as in Fig. 2.1, and this means that either the subjects and conditions are combined into a single mode (*tall combination-mode matrix*) or the variables and conditions are so combined (*wide combination-mode matrix*). Thus, two of the modes are always confounded and no independent parameters for these modes are present in the model itself, except when models are used specifically geared toward this situation.

In general, a multiway model uses fewer parameters for multiway data than an appropriate two-mode model. To what extent this is true depends very much on the specific model used. In some multiway component models low-dimensional representations are defined for all modes, which can lead to enormous reductions in parameters. Unfortunately, that does not means that automatically the results of a multiway analysis are always easier to interpret. Again this depends on the questions asked and the models used.

An important aspect of multiway models, especially in the social and behavioral sciences, is that they allow for the analysis of individual differences in a variety of conditions. The subjects do not disappear in means, (co)variances, or correlations, and possibly higher-order moments such as the kurtosis and skewness, but they are examined in their own right. This implies that often the data set is taken as is, and not necessarily as a random sample from a larger population in which the subjects are in principle exchangeable. Naturally, this affects the generalizability, but that is considered inevitable. At the same time, however, the subjects are recognized as the "data generators" and are awarded a special status, for instance, when statistical stability is determined via bootstrap or jackknife procedures (see Section 9.8.2, p. 233; Section 8.8.1, p. 188; Section 8.8.2, p. 188). Furthermore, it is nearly always the contention of the researcher that similar samples are or may become available, so that at least part of the results are valid outside the context of the specific sample.

2.3 WHAT IS A MODEL?

2.3.1 Models and methods

In this book a *model* is a theoretical, in our case a mathematical, construct or "a simplified description of a system or complex entity, esp. one designed to facilitate descriptions and predictions" (*Collins English Dictionary*). A well-known example is the normal distribution, which often serves as a simplified description of a distribution of an observed variable. The simplification comes from the fact that the normal distribution is determined by its shape (described by a formula) and the values of its two parameters, the mean and the variance. However, to describe the *empirical* distribution of a variable we need to know all data values, however many there are. Thus, if we are prepared to assume that a variable is approximately normally distributed, we can use the properties of that distribution, and we can describe and interpret its shape using the parameters to describe and compare the distributions of different variables. All of which is extremely difficult to do with empirical distributions because there is no real reduction in complexity.

A central question is whether a particular model is commensurate with a particular data set, for instance, whether the empirical distribution of a variable can be adequately represented by a normal distribution with the specific mean and variance. To study this, one often defines *loss functions* (also called discrepancy functions), which quantify the difference between the data and the model. In the case of the normal distribution, this could take the form of the sum of (squared) differences between the observed values and those derived from the normal distribution.

In physical sciences, a model is often determined by physical laws that describe the phenomenon under study, such as the way a pendulum swings. In such a case the interpretations of the parameters are known beforehand, and one only has to estimate from the data the values of the parameters for the case at hand.

In the social and behavioral sciences, similarly defined models are rare. Instead, the idea of a model takes on a different meaning. Models are conceived as descriptive of structural relationships between the entities in the data, and models are not grounded in a particular theory but have a more general character. The principal component analysis (PCA) model is an example of this. It is assumed that the patterns in the data can be described by a bilinear relationship between components for the subjects (rows) and those for the variables (columns), as described in Eq. (2.1):

$$x_{ij} = \hat{x}_{ij} + e_{ij} = \sum_{s=1}^{S} a_{is} f_{js} + e_{ij} \ (i = 1, \ldots, I; j = 1, \ldots, J), \qquad (2.1)$$

where the a_{is} are the coefficients of the subjects on the components of the subjects (often called *scores*), the f_{js} are the coefficients of the variables (often called *loadings*), and S is the number of components used to approximate the data (see Chapter 9, especially Section 9.3.1, p. 215). The relationship is called bilinear because given the coefficients of one mode the equation is linear in the other mode, and vice versa. Given a solution, not only the parameters need to be interpreted but also the structure they describe. Generally, there are no detailed laws that explain why the description of the data by a principal component model is the proper and unique way to represent the relationships in the data. However, it is possible to specify via a loss function whether the PCA model provides an adequate description of the data in terms of fit. In this situation $\hat{x}_{ij}$ will be called the *structural image* of the data. The description of the structure via principal components in Eq. (2.1) is then the *structural representation*. However, it is generally more convenient to simply use the word model in this case, keeping in mind the more formal correct designation.

Some multiway techniques described in this book do not conform to the idea of a model, because there is no single formulation of the relationships between the variables. Consequently, it is not possible to define a loss function that indicates how well the solution fits the data. Such techniques are called *methods*. They generally consist of a series of procedures or steps that have to be followed to obtain a solution, but it is not possible at the end of the computations to say how well the solution fits the data.

In two-mode analysis, many clustering procedures are methods without underlying models. An example in three-mode analysis is the procedure called STATIS (Structuration des tableaux à trois indices de la statistique or Structuring of triple-indexed statistical tables (L'Hermier des Plantes, 1976; Lavit, 1988)), which is much used in France. It consists of three separate but linked steps, but no overall loss function can be specified (see Section 5.8, p. 105).

The disadvantage of not having a loss function is that it is not possible to compare the fit of different solutions because there is no common reference point. Furthermore, solutions cannot be statistically evaluated via significance tests, should one desire to do this. The models discussed in this section are typically data-analytic models, because there is no reference to statistical properties of these models, such as the type of distributions of the variables or of the errors. When a model has an explicit error structure and distributional assumptions are made, we will use the term *statistical models*. Nearly all models for multiway data are data-analytic models. Notable exceptions are the models underlying the three-mode mixture method of clustering (see Chapter 16) and three-mode (common) factor analysis (see Section 4.3.5, p. 47).

2.3.2   Choice of dimensionality

Characteristic of component models is that they in fact form classes of models, because within each class we may specify the dimensionality of the solution. A principal component model as given in Eq. (2.1) fits the data differently for each value of S. Thus, in many cases we have to choose both the class of models we will use to fit the data, and the dimensionality of the model within the class. For some multiway model classes we have to choose several dimensionalities, one for each mode.

2.4   SOME HISTORY

2.4.1   Prehistory

Very early in the twentieth century mathematicians, especially those working in linear algebra, were interested in handling more than one matrix at a time and in examining the properties and eigendecompositions of sets of matrices and multiway arrays. The latter were referred to as tensors or polyadics (analogous to dyadic and triadic). In fact, one can even trace such an interest back as far as Camille Jordan (1874)[2], who was interested in simultaneously diagonalizing two matrices at a time. Models

---

[2]His paper is primarily concerned with one of the first descriptions of the singular value decomposition.

now known as Parafac and Tucker3 model were already known to Frank Hitchcock, who also discussed rank problems related to multiway structures (Hitchcock, 1927a, 1927b). Rufus Oldenburger (1934) also wrote about properties of multiway arrays. However, these exercises in linear algebra were not linked to real data analysis but stayed within the realm of mathematics.

2.4.2 Three-mode component models

Three-mode analysis as an approach toward analyzing three-way data started with Ledyard Tucker's publications (Tucker, 1963, 1964, 1966)[3] about what he called *three-mode factor analysis*, but what is now generally referred to as *three-mode component analysis*[4]. His model is now often referred to as the *Tucker3 model*. As noted above, before Tucker's series of papers, various authors had been concerned about what to do with sets of matrices, especially from a purely linear algebra point of view, but studying three-way *data* really started with Tucker's seminal work. In his earlier papers, Tucker formulated two models (a principal component model and a common factor model) and several computational procedures. Tucker also wrote or collaborated on about ten applications, not all of them published. One of his PhD students, Levin (1965), wrote an expository paper on applying the technique in psychology, and after that the number of applications and theoretical papers gradually, but slowly, increased. In an amazingly concise paper, Appellof and Davidson (1981) independently reinvented the Tucker3 model in passing, while discussing the merits of the Parafac model for chemistry.

A second step was taken when Kroonenberg and De Leeuw (1980) presented an improved, alternating least-squares solution for the original component model as well as a computer program to carry out the analysis. A very similar algorithm called COM-STAT was independently proposed by Röhmel, Streitberg, and Herrmann (1983) (see also Herrmann, Röhmel, Streitberg, & Willmann, 1983) but these authors do not seem to have followed up their initial work on this algorithm. Kroonenberg (1983c) also presented an overview of the then state of the art with respect to Tucker's component model, as well as an annotated bibliography (Kroonenberg, 1983a). Hierarchies of three-mode models from both the French and Anglo-Saxon literature, which include the Tucker3 and Tucker2 models, respectively, have been presented by Kiers (1988, 1991a); see Section 8.7 (p. 186).

In several fields multiway developments have taken place such as unfolding, block models, longitudinal data, clustering trajectories, conjoint analysis, and PLS modeling, but a historical overview of these specialized proposals will not be given here. An extensive bibliography on the website of The Three-Mode Company[5] contains references to most of the papers dealing with multiway issues.

---

[3]By August 2007, his work on three-mode analysis was cited about 550 times in the journal literature.
[4]The expression "three-mode components analysis" also exists, but as the common term is "factor analysis", component analysis without the s seems more appropriate.
[5]*http://three-mode.leidenuniv.nl/*. Accessed May 2007.

2.4.3   Three-mode factor models

A stochastic version of Tucker's three-mode common factor model was first proposed by Bloxom (1968) and further developed by Bentler and co-workers – Bentler and Lee (1978b, 1979), S.-Y. Lee and Fong (1983), and Bentler, Poon, and Lee (1988). Bloxom (1984) discussed Tucker's factor models in term of higher-order composition rules. Much later the model was treated in extenso with many additional features by Oort (1999, 2001), while Kroonenberg and Oort (2003a) discussed the link between stochastic three-mode factor models and three-mode component models for what they called multimode covariance matrices.

2.4.4   Parallel factor model

Parallel to the development of three-mode component analysis, Harshman (1970, 1972a) conceived a three-mode component model that he called the *parallel factor model* (PARAFAC). He conceived this model as an extension of regular component analysis and, using the parallel proportional profiles principle proposed by Cattell and Cattell (1955), he showed that the model solved the rotational indeterminacy of ordinary two-mode principal component analysis.

At the same time, Carroll and Chang (1970) proposed the same model, calling it canonical decomposition — (CANDECOMP). However, their development was primarily related to individual differences scaling, and their main contribution was to algorithmic aspects of the model without further developing the full potential for the analysis of "standard" three-way arrays. This is the main reason why in the present book the model is consistently referred to as the Parafac model. In various other contexts the Parafac model was independently proposed, for instance in EEG research by Möcks (1988a, 1988b), under the name *topographic component model*.

A full-blown exposé of the model and some extensions are contained in Harshman and Lundy (1984c, 1984d), a more applied survey can be found in Harshman and Lundy (1994b), and a tutorial with a chemical slant in Bro (1997). The Parafac model has seen a large upsurge in both theoretical development and applications, when it was realized that it corresponded to common physical models in analytical chemistry; for details, see the book by Smilde et al. (2004). Similarly, in signal processing and communications, many applications have evolved, including blind multiuser detection in direct-sequence code-division multiple-access (DS-CDMA) communications (Sidiropoulos, Giannakis, & Bro, 2000c), multiple-invariance sensor array processing (Sidiropoulos, Bro, & Giannakis, 2000b), blind beamforming in specular multipath (Liu & Sidiropoulos, 2000; Sidiropoulos & Liu, 2001), and, in more general terms, blind diversity-combining in communications (Sidiropoulos & Bro, 2000d).

2.4.5   Individual differences scaling

Carroll and Chang (1970) developed a reliable algorithm for the weighted Euclidean model, which they called the individual differences scaling (INDSCAL) model. This model introduced individual differences into multidimensional scaling and formed a milestone in multiway analysis[6]. The model is an extension to three-way data of existing procedures for two-way similarity data. Carroll and Chang's (1970) work extended and popularized earlier independent work by Horan (1969) and Bloxom (1968). Over the years, various relatives of this model have been developed such as the individual difference in orientation scaling (IDIOSCAL) model (Carroll & Chang, 1972b), a less restricted variant of INDSCAL, the Parafac model, which can be interpreted as an asymmetric INDSCAL model, and the Tucker2 model, which also belongs to the class of individual differences models of which it is the most general representative. The INDSCAL model has also been called a "generalized subjective metrics model" (Schönemann, 1972). Other, similar models have been developed within the context of multidimensional scaling, and general discussions of individual differences models and their interrelationships can, for instance, be found in Arabie, Carroll, and DeSarbo (1987), F. W. Young (1984), F. W. Young and Hamer (1987), and their references.

2.4.6   Three-mode cluster models

Carroll and Arabie (1983) developed a clustering version of the INDSCAL model. This model consists of a set of common clusters for which each sample or subject in the third mode has weights that indicate whether they do or do not belong to these clusters. The procedure was called individual differences clustering (INDCLUS), and is applied to sets of similarity matrices. Within that tradition, several further models were suggested including some (ultrametric) tree models (e.g., see Carroll, Clark, and DeSarbo, 1984; and De Soete and Carroll, 1989).

Sato and Sato (1994) presented a fuzzy clustering method for three-way data by treating the problem as a multicriteria optimization problem and searching for a Pareto efficient solution. Coppi and D'Urso (2003) provided another contribution to this area.

Based on multivariate modeling using maximum likelihood estimation, Basford and McLachlan (1985b) developed a mixture method approach to clustering three-mode continuous data, which has seen considerable application in agriculture. It will be discussed in detail in Chapter 16. Extensions to categorical data can be found in Hunt and Basford (1999, 2001), while further contributions in this vein have been made by Rocci and Vichi (2003a) and Vermunt (2007).

Recently, three-mode partitioning models were proposed independently by several authors (Rocci & Vichi, 2003b; Kiers, 2004b; Schepers & Van Mechelen, 2004). The

---

[6]By August 2007, Carroll and Chang's paper was cited over 1000 times in the journal literature.

models attempt to partition all three modes at the same time by specifying binary component matrices for each of the modes with a real-valued core array to link the three modes. A number of k-means-like algorithms have been proposed for the estimation of the model, and the various proposals have been evaluated in great detail by Schepers, Van Mechelen, and Ceulemans (2006).

2.4.7 Multiway proposals and algorithms

Several extensions now exist generalizing three-mode techniques to multiway data. The earliest reference to the concept of three-way data probably goes back to Cattell (1946), who initially only considered three-way research designs and described these via his three-dimensional *Covariation Chart*; see also Section 19.1, p. 469. Twenty years later, he proposed the probably most comprehensive system for describing relations for multiway research designs (Cattell, 1966). Carroll and Chang (1970) were the first to propose actual algorithms for multiway data analysis and developed a seven-way version of their CANDECOMP program. Lastovicka (1981) proposed a four-way generalization for Tucker's non-least-squares solution, and applied it to an example based upon viewer perceptions of repetitive advertising. Kapteyn, Neudecker, and Wansbeek (1986) generalized the Kroonenberg and De Leeuw (1980) alternating least-squares solution to the Tucker$n$ model. The book by Smilde et al. (2004) contains the references to the developments that have taken place especially in chemometrics with respect to multiway modeling, including multiway block models in which not one but several multiway blocks are present. Such models often have a dependence structure in which the values in one block are used to predict those in another block. Several applications of these models exist in process analytical chemistry, but they will not feature in the present book; details and references can be found in Smilde et al. (2004). Chapter 19 treats multiway data analysis in more detail.

2.5 MULTIWAY MODELS AND METHODS

By way of an overview of the wide array of methods now available for the analysis of multiway data, Tables 2.1–2.3 list a large number of more and less known multiway methods. Given the developments in the field, the lists will necessarily be incomplete by the time they appears in print. The references mentioned are primarily meant for further orientation and are not necessarily the original sources. An attempt has been made to include comprehensive or tutorial papers wherever feasible.

2.6 CONCLUSIONS

As can be seen from Tables 2.1–2.3, many two-mode techniques have their multiway equivalent. In this book the choice was made to concentrate on the most common

Table 2.1 Methods and Models Chart. Part 1 — Profile and rating scale data

Dependence techniques: General linear model methods

multiblock multiple regression	Smilde, Westerhuis, and Boqué (2000)
multiway partial least-squares models	Bro (1996a)
multiway multiblock models	Smilde et al. (2000)
multivariate longitudinal data analysis	Fitzmaurice, Laird, and Ware (2004)
three-mode redundancy analysis	Kroonenberg and Van der Kloot (1987d)
multivariate repeated measures analysis	Hand and Taylor (1987)
ANOVA	

Interdependence techniques: Component methods

multiway component analysis	Tucker (1966); Kroonenberg (1983c)
parallel factor analysis	Harshman and Lundy (1984c, 1984d)
three-way correspondence analysis	Carlier and Kroonenberg (1996)
latent class analysis	Carroll, De Soete, and Kamensky (1992)
spatial evolution analysis	Carlier (1986)

Dependence and interdependence techniques

multiset canonical correlation analysis	Kettenring (1971)
nonlinear canonical correlation analysis	Van der Burg and Dijksterhuis (1989)
generalized Procrustes analysis	Gower and Dijksterhuis (2004, Chap. 9)

Clustering methods

three-mode mixture methods of clustering	Hunt and Basford (1999)
fuzzy three-mode clustering	Sato and Sato (1994)
hierarchical classes clustering	Ceulemans and Van Mechelen (2005)
three-mode partitioning	Schepers et al. (2006)
simultaneous clustering and data reduction	Vichi, Rocci, and Kiers (2007)

multiway methods, which have been applied in various contexts and with which extensive practical experience has been gained. The focus of this book is providing guidelines for applying the methods mentioned in Table 2.1 to real data, with a special emphasis on component models. Some other topics were added partly because of their importance, personal preference, and experience. Several papers have also provided overviews of the field, for instance, Kiers (1998e), Bro (2006), Kroonenberg (2005b), Kroonenberg (2006b), and the book by Smilde et al. (2004).

The stochastic methods mentioned in Table 2.2 are well covered in books on structural equation models, such as Bollen (1989) and the references mentioned in the table. Their exploratory counterparts are relatively new and unexplored and require much more experience before firm practical recommendations can be made. For the methods mentioned in Table 2.3, recent book chapters on the subject can be found in Cox and Cox (2001, Chapters 10–12) and Borg and Groenen (2005, Chapter 22).

Table 2.2 Methods and Models Chart. Part 2 — Covariance matrices

---

Stochastic covariance (or structural equations) models
Repeated measures analysis
    three-mode common factor analysis   Bentler et al. (1988)
    modeling multimode covariance matrices   Browne (1984); Oort (1999)
    invariant factor analysis   McDonald (1984)
Cross-sectional methods
    simultaneous factor analysis   Jöreskog (1971)

Exploratory covariance model methods
Repeated measures methods
    quasi three-mode component analysis   Murakami (1983)
Cross-sectional and multiset methods
    simultaneous component analysis   Timmerman and Kiers (2003)
    indirect fitting with component methods   Harshman and Lundy (1984c)

Table 2.3 Methods and Models Chart. Part 3 — Similarity and preference data

---

Multidimensional scaling models
    individual differences scaling   Carroll and Chang (1970)
    general Euclidean models   F. W. Young (1984)
    three-way multidimensional scaling   Tucker (1972)
Clustering methods
    individual differences clustering   Carroll and Arabie (1983)
    three-way ultrametric tree models   De Soete and Carroll (1989)
Unfolding models
    three-way unfolding   DeSarbo and Carroll (1985)

CHAPTER 3

THREE-WAY AND MULTIWAY DATA

3.1 CHAPTER PREVIEW

In this chapter we will introduce the terminology used in this book to set the scene for a discussion of the various types of multiway data. The terminology and notation is largely in line with the proposals put forward by Henk Kiers in the *Journal of Chemometrics* (Kiers, 2000b). An overview of the notation in this book may be found in Appendix A (p. 490). We will also briefly review the original way of handling three-way data arrays and continue with a discussion of the various principles that can be used to describe different types of data. The major part of the chapter contains an overview of the data types encountered in multiway analysis. The multiway data treated in this book have as a common characteristic that, leaving aside missing data, the variables (or similar entities) are the same across conditions and/or time points. In particular, data in which the variables can be partitioned into different disjunct sets will not be treated in this book. This means that techniques for such data (such as multiset canonical correlation analysis and multiway multiblock data) will only be mentioned in passing. The discussion in this chapter is restricted to three-way

Applied Multiway Data Analysis. By Pieter M. Kroonenberg

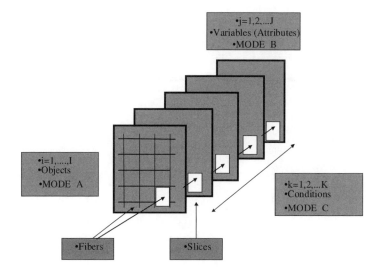

Figure 3.1 Three-way data array with fibers and slices indicated. Reproduced with kind permission of Kemal Büyükkurt.

data with the tacit assumption that the generalization to multiway data is relatively straightforward. A fuller discussion of multiway data is relegated to Chapter 19.

3.2 TERMINOLOGY

A common form of three-way data consists of subjects by variables by conditions (or similar type of entities) where subjects, variables, and conditions should be taken as generic terms that have a different content in different disciplines, say, spectra, plant varieties, or signals. Such data can be arranged into a three-dimensional block (see Fig. 3.1), and the three-way data block is referred to as a *three-way data array*. Multiway data with multi larger than three arise when, for instance, in addition a time dimension is included or there is a fully crossed design on the conditions.

The word *way* is most often used to indicate a collection of indices by which the data can be classified, while the word *mode* mostly indicates the entities that make up one of the ways of the data box. Thus, data boxes always have three ways, but it needs three different types of entities to make them three-mode. Sets of correlation matrices have three ways but only two modes, while the example in Fig. 3.1 has three ways as well as three modes.

The first way (subjects) has index i running along the vertical axis, the second way (variables) has index j running along the horizontal axis, and the third way (conditions) has index k running along the "depth" axis of the box. The number of *levels* in each way is I, J, and K, respectively. The word *levels* is thus used in the

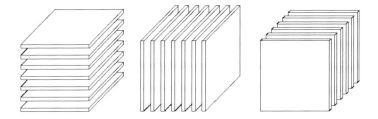

Figure 3.2   Slices of a three-way data array: Horizontal slices, $\mathbf{X}_i$ or $\mathbf{X}_a$; lateral slices, $\mathbf{X}_j$ or $\mathbf{X}_b$; frontal slices, $\mathbf{X}_k$ or $\mathbf{X}_c$.

same manner as in analysis of variance, be it that different variables in the variable mode are also designated as levels. The $I \times J \times K$ three-way data array $\mathcal{X}$ is thus defined as the collection of elements

$$\{x_{ijk} \mid i = 1, \ldots, I; j = 1, \ldots, J; k = 1, \ldots, K\}. \tag{3.1}$$

A three-way array can also be seen as a collection of "normal" (i.e., two-way) matrices, often referred to in this context as *slices* or *slabs*. There are three different arrangements for this, as shown in Fig. 3.2. The different types of slices will be referred to as *frontal* slices, *horizontal* slices, and *lateral* slices. When three-way data can be conceptualized as a collection of two-way matrices, say, $I \times J$ frontal slices, the third way is generally used for the matrix indicator, thus $\mathbf{X}_k$. In addition, one can break up a three-way matrix into one-way submatrices or vectors, in the present context generally called *fibers* (see Fig. 3.3). The different types of fibers will be referred to as *rows*, *columns*, *tubes*, and *pipes* for the fourth way, but the last term is not standard. The primary reference paper for multiway terminology is Kiers (2000b).

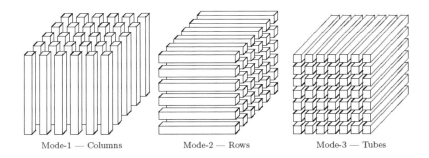

Mode-1 — Columns   Mode-2 — Rows   Mode-3 — Tubes

Figure 3.3   Fibers of a three-way data array; Columns, $\mathbf{x}_{jk}$; rows, $\mathbf{x}_{ki}$; tubes, $\mathbf{x}_{ij}$.

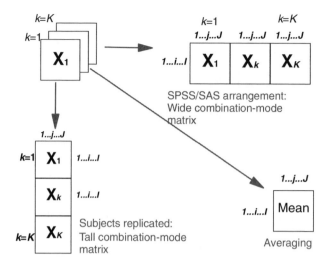

Figure 3.4   ,Fitting a three-way data array into a two-way data matrix

3.3   TWO-WAY SOLUTIONS TO THREE-WAY DATA

Traditionally, when one was confronted with a three-way data array, it was transformed into a two-way data matrix, which was then analyzed using standard techniques for two-way analysis, such as two-mode principal component analysis. Two-way data were constructed from three-way data by averaging, arranging all the scores of a subject on a single record or a sequence of records in order to create a wide combination-mode matrix, the typical arrangement in statistical packages like SPSS and SAS. Alternatively, a subject's measurements were treated as unrelated to the subject's measurements on the same variables under another condition, thus as a tall combination-mode matrix (subjects replicated) (see Fig. 3.4). The disadvantage of these approaches is that if, for instance, the structure of the variables is the same for all conditions, this structure cannot be properly examined. In the case of averaging, one mode entirely disappears from the analysis. This might not be a problem when, for instance, subjects are a random sample from a population, or similarly, if the subjects are exchangeable, that is, each subject can be exchanged for another from the population without affecting the generalization of the results. However, if there is an interest in individual differences, the destruction of the three-way structure by reducing the data to two ways is not a good idea.

3.4 CLASSIFICATION PRINCIPLES

There are at least four principles that can be used to classify three-way data: (1) Raw data and derived measures, (2) Repeated measures versus cross-sectional data, (3) Stochastic and nonstochastic variables, (4) Dependent and interdependent set of variables.

3.4.1 Raw data and derived measures

In many sciences, data in their most basic form are the scores of subjects (objects, plant varieties, entities, etc.) on variables and, in the case of raw three-way data, the variables are measured under several conditions or at several occasions. Depending on the interest in the subjects and the sampling situation, such data are routinely converted into derived measures, such as means and covariance or correlation matrices, or, in the case of categorical data, into frequencies. Especially in classical statistics, with assumptions about the normality of the population distributions, these summary measures contain all the information one needs since they are sufficient for describing the population distributions.

Most of the questions posed in multiway analysis also include questions about individual differences, and in such situations there is a preference for dealing with the raw data rather than derived measures. But in situations with different subjects under different conditions this is not always possible, so that derived measures may have to be used.

3.4.2 Repeated measures versus cross-sectional data

In terms of design, a distinction must be made between repeated measures data and cross-sectional ones. In the former, the objects are the same across conditions and, leaving aside missing data, the design is a fully crossed one. When the focus is primarily on means and the differences between them over conditions, these types of data are generally analyzed using repeated measures analysis of variance. When the emphasis is more on relationships between variables and how they behave under different conditions, different methods are called for, such as those discussed in this book.

Cross-sectional data are different in that, for each occasion or condition, different subjects have been measured. This introduces limitations on the techniques that can be used for analysis. Typically, such data are converted to sets of correlation or covariance matrices, which are then analyzed jointly.

3.4.3 Stochastic and nonstochastic variables

When the main emphasis is on the structure of data rather than on testing specific hypotheses, the data are assumed to represent an entire population rather than a sample from a population. Ultimately, this is not a tenable position, because sampling is always involved, even if it is unknown from which population. A practical approach in such cases is that generalizability should follow from replication and cross-validation, rather than from making assumptions about the sampling process, population distributions, and how the observations have been sampled.

In part, this position has to do with a genuine interest in individual differences, so that means and covariances are not enough. In addition, it is often too difficult to assess the validity of assumptions in complex multivariate data which makes hypothesis testing difficult. Furthermore, in some situations the stochastic aspect of the data is not included, for instance, when analyzing agricultural data with specific varieties planted at specific locations measured on specific attributes. The values to be analyzed are in general plot means calculated over individual plants of a specific variety. In such cases, the stochastics come into play when repeating the experiment, but this is equivalent to replication.

Statistical theory does enter the analysis when comparing and selecting models via an assessment of residuals or residual sums of squares. As we are interested in model fitting, we compare models, and (pseudo-)hypothesis tests are considered. However, such testing will be primarily descriptive. Numbers of parameters will be offset against the gain in fit or reduction of error.

To acquire some control over the stability of the solution, performing bootstraps is the appropriate procedure. At present, however, experience with three-way data has only been limited (see, however, Kiers, 2004a), and it is clearly an area that needs to be explored further (see also Sections 8.8.1 and 9.8.2). Note that not one general bootstrap design will serve all, but that the specific design will depend on the details of the design of the data.

3.4.4 Dependent and interdependent set of variables

In standard multivariate analysis the general linear model is the dominate one. The available variables are grouped in two sets, the criterion (or dependent) variables and the predictor (or independent) variables, and the aim is to use the latter to predict or model scores of the former. Such an analysis was called an *analysis of dependence* by Kendall (1957, pp. 1–4). In the case of one set of variables, Kendall referred to an *analysis of interdependence*. Alternative terms proposed by Harshman (2005b) are *internal relationship analysis* and *external relationship analysis*.

Most three-way data sets consist of one type of variables and are therefore generally analyzed via an analysis of interdependence. However, especially in analytical chemistry, *multiblock* data sets are now starting to be collected, so that the analysis of dependence also enters into the multiway domain (see Smilde et al., 2004).

Table 3.1 Overview of multiway data designs

Designs	Data types
Fully crossed designs	Multiway profile data
	Multiway rating scale data
	Multivariate longitudinal data
	Multiway factorial data; one observation per cell
	Multimode covariances
Nested designs	Multisample data
	Multiset data
	Sets of configurations
Multidimensional scaling designs	Sets of dissimilarity or similarity matrices
	Sets of preference rank orders
	Three-mode distances
Multiway categorical data	Multiway binary data
	Multiway contingency tables
	Multiway ordinal data

3.5 OVERVIEW OF THREE-WAY DATA DESIGNS

In Table 3.1 an overview is given of various types of multiway data. In the following sections we will give a brief characterization of these designs and data types, and in the next chapter we will provide an overview of appropriate models and methods for these data types. However, the analysis methods suggested should not be considered as a closed set.

3.6 FULLY CROSSED DESIGNS

There are three basic types of three-way three-mode data for fully crossed designs: Profile data, rating scale data, and multivariate longitudinal data. *Profile data* are data in which each subject is considered to have scores for a set of variables, and these scores are called the profile of a subject. For *three-way profile data*, the profile is defined either over all conditions together, so that the profile is a slice of the data array, or as a collection of profiles, one for each condition (for analyses of such data, see Chapter 13). There is no explicit assumption of a fixed order in the condition mode. *Rating scale data* are data in which a subject is seen as a judge who awards a score to a variable or concept on a particular scale. In the behavioral sciences often such data consist of stimulus–response data rated by judges. All scales have the same form with generally three, five, seven, or nine scale values. Examples of rating scale data are the graded scale from "not good at all" to "very good" (monopolar scales)

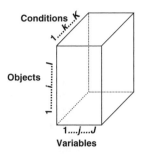

Figure 3.5 Three-way profile data

or bipolar scales with antonyms at the end points of the scales, for instance, "bad" versus "good" with a neutral point in the middle. For *three-way rating scale data*, the subjects are mostly placed in the third mode, as they are the replicates, and produce a complete frontal slice of judgments (for analyses of such data, see Chapter 14). *Multivariate longitudinal data* are like profile data, but it is explicitly assumed that there is a fixed time order in one of the modes, mostly the third (for analyses of such data, see Chapter 15).

3.6.1 Three-way profile data

Profile data are the bread-and-butter of regular data analysis. They generally take the form of objects by variables, where objects may be persons, wines, chemical substances, grain varieties, and so on; similarly, variables come in many forms and definitions (Fig. 3.5). The standard arrangement is that the objects form the rows and the variables the columns. The term *profile data* derives from the idea that each object has a profile of values across all variables. Thus, the rows in the data matrix are these profiles, but in the three-way case slice profiles can be defined as well. In most statistical designs, variables are random variables and the objects are random samples from some hopefully well-defined population. In the nonstochastic approach taken here, there is a much stronger emphasis on individual differences, so that the object mode often comes under close scrutiny. As in statistical designs, the status of the objects is different from the entities of the other modes, because they are the "data generators"; they are the objects whose attributes are measured. However, we will also come across situations where one cannot speak of objects when their mode consists of aggregates. This occurs when the measurements are means across the objects as is often the case in agricultural applications where the individual plants are averaged over plots.

Three-way profile data typically consist of scores of objects on a set of variables measured under a number of conditions (or time points, occasions, environments, chemical mixtures, etc.); this setup is referred to as the *profile arrangement* of three-way data. The names for the modes are again generic. The standard arrangement is that objects form the first mode (or Mode A), the variables the second mode (Mode B), and the conditions the third (Mode C). Generally, but not always, the variables have different measurement scales and the values are not necessarily comparable. Objects have been assigned (or assigned themselves) scores on the variables and have been measured more than once. In the basic form treated here, we do not assume a specific time structure or order relationship between the measurement conditions. The implication of this is that permuting the conditions has no effect on the analysis. A four-way extension would typically include both an occasion mode and a time mode.

- *Primary analysis methods*: *Ordination*: Tucker and Parafac analyses; *Clustering*: Three-mode mixture method of clustering, three-mode partitioning.
- *Secondary analysis methods*: Simultaneous component analysis, STATIS.

3.6.2   Three-way rating scale data

Three-way rating scale data typically consist of scores of subjects on a set of rating scales, which are used to evaluate a number of concepts. The names for the modes are obviously generic, but they share the characteristics that all scales have the same range and the values are considered to be comparable. Subjects are assumed to assign a rating to a concept by evaluating how strongly the scale applies to the concept, so that these ratings function as a measure of the strength of the relationship (or similarity) between scale and concept. To emphasize this, the subjects will be placed in the third mode and be considered judges, each producing a (similarity) matrix of concept by scales (*scaling arrangement*; see Fig. 3.6). Very often in psychology such data can be characterized as stimulus–response data of a set of individuals, where the concepts or situations are the stimuli and the answers to the rating scales the responses. Three-way rating scale data can also be analyzed as if they are three-way profile data (*profile arrangement*; see Fig. 3.6). Which of the two arrangements is used depends mainly on the research questions and the type of rating scales used.

- *Primary analysis methods for scaling arrangement*: Tucker analysis.

- *Secondary analysis methods for scaling arrangement*: Parafac analysis, simultaneous component analysis, STATIS.
- *Profile arrangement*: See Profile data.

3.6.3   Multiway factorial data

Another type of fully crossed designs yielding three-mode three-way data consists of means taken over individuals, where each of the modes can be conceived, as a factor with a number of levels. For instance, three-way rating scale data can sometimes be conceived of in this way when the scales measure a different aspect of a more

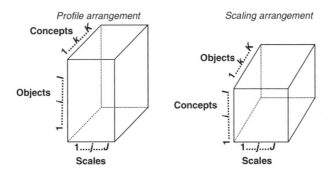

Figure 3.6   Three-way rating scale data

general variable and thus can also be seen as levels rather than separate variables. The data of the perceived reality study (Kroonenberg & Van der Voort, 1987e) may serve as an example. Primary school children rated 8 film types on 11 three-point scales, each of which attempted to measure some aspect of the perceived reality of these film types. The scores were aggregated over grades so that the entries of the array were grade means, and the individuals mode was transformed into a grades mode. The scales could be conceived of as individual scales ("Could the events in this film occur in reality?") or as a level of the variable Perceived reality. Similarly, many agricultural studies produce three-way factorial data with many levels per factor (dependent variable: mean yield; factors: varieties, environments, years). Such data can be analyzed in several ways; often several main effects are removed before a three-mode analysis is attempted (see also Section 6.8, p. 137).

- *Primary analysis methods*: Univariate three-way analysis of variance with one observation per cell, Tucker and Parafac analyses.

3.6.4   Multivariate longitudinal data

The multivariate longitudinal data considered in this book are primarily treated as three-way profile data without attention to order restrictions on the time or occasion mode. Ideally, time order should be prominent in the methods to analyze such data, and a large literature exists on the statistical analysis of multivariate longitudinal data (e.g., see Fitzmaurice et al., 2004). However, we will not touch upon these methods here. Given the exploratory slant of this book, we will use standard three-mode profile methods and use time as an interpretational rather than a modeling device. Progress is being made to include order restriction into three-mode analysis (e.g., see Timmerman & Kiers, 2000), but not much of this work is part of standard analytical practice. An extensive exposé of stochastic three-mode modeling for longitudinal data can be found in Oort (2001).

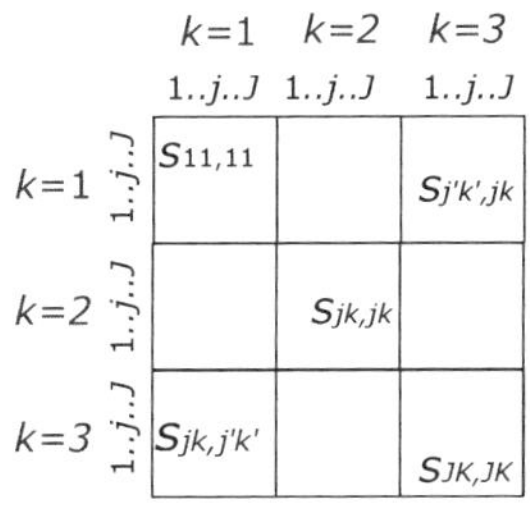

Figure 3.7   Multimode covariance matrix. $s_{jk,jk}$ = variance of variable j at time k; $s_{jk,j'k'}$ = covariance of variable j at time k with variable j' at time k'.

- *Primary analysis methods*: Repeated measures analysis of variance, multivariate longitudinal analyses techniques, Tucker and Parafac analyses, possibly using restrictions on the time components.

3.6.5 Multimode covariances

When there are many subjects and individual differences are not the main concern, an option is to transform the three-way data array into an $JK \times JK$ multivariable–multioccasion covariance matrix, also referred to as a multimode covariance matrix (see Fig. 3.7).

The general consensus is that it is better to analyze multimode covariance matrices rather than multimode correlation matrices, although in reporting it might be easier to talk about correlations because of the standardization involved. The case for the covariances rests on the one-to-one relation between a standard deviation of a variable and the underlying scale. If all variables were scaled within conditions, the relationship between the standard deviation and the number of units of the underlying scale would be different in each case. Of course, when such considerations are not important, multimode correlation matrices can be used. One proposed alternative is to use the standard deviation of one condition for all other conditions, or to use some average standard deviation for all conditions — equal average diagonal standardization (Harshman & Lundy, 1984c, p. 141). Examples of the analysis of multimode covariance matrix with nonstochastic three-mode techniques can be found in Murakami (1983) and Kroonenberg and Oort (2003a).

- *Primary analysis methods*: Stochastic methods — Structural equation modeling, three-mode common factor analysis; Nonstochastic methods — Tucker3 for multimode covariance matrices, Tucker2 for multimode covariance matrices (= quasi three-mode component analysis).

- *Secondary analysis methods*: Nonstochastic: Simultaneous component analysis, STATIS.

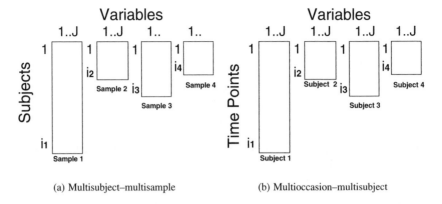

(a) Multisubject–multisample   (b) Multioccasion–multisubject

Figure 3.8   Different types of multisample data.

3.7   NESTED DESIGNS

When the data come from different samples but the subjects in these samples have scores on the same variables, the term three-way data is not really appropriate because subjects are nested in samples and are different across samples. We will refer to such data as *multisample data*. One way to handle nested three-way data is to convert them to covariances or correlations, so that they become three-way two-mode data. In such *multiset data*, sets of covariance matrices are analyzed.

3.7.1   Multisample data

Multisample data do not have a true three-mode structure because the levels of the modes are not fully crossed, but the levels of the row mode are nested within the levels of the third mode. Even though the third mode often consists of samples, it may also consist of conditions, groups, or subjects, while the row mode may consist of persons, objects, or time points. For example, in Fig. 3.8(a), there are four different samples of subjects who have scores on the same set of variables, and their data are referred to as *multisubject–multisample data*. The subjects are thus nested in samples, and no common components can be defined for the subjects. Because one might argue that it is not proper to use the words "first mode" when the entities (subjects) are different in each sample, one may refer to the *rows of a sample* instead.

Another type of multisample data is multioccasion–multisubject data, which can be arranged so that they can be handled as multisubject–multisample data (see Fig. 3.8(b)). In this case, the time series form a set of levels that are nested within subjects. As an example, assume a number of persons have been measured regularly over time on a set of variables, but the times of measurement are different for each person, and the number of observations per person vary as well. If we arrange the data as measurement moments by variables by persons, the data set is a multisample one. The order in the rows for each subject can be subjected to smoothness restrictions or a specific function to do justice to the time-series design. Of course, the process of interpretation will be radically different in both of the mentioned cases of multisample data. For a detailed treatment of multisample data, see Timmerman (2001).

- *Primary analysis methods*: Nonstochastic — Simultaneous component analysis, STATIS.

3.7.2 Multiset data

When it is not necessary to model the rows in nested three-way designs, one generally converts the raw data to covariance or correlation matrices per sample or condition, so that the data become three-way two-mode data of variables by variables by samples or conditions. In such *multiset data*, typically sets of covariance matrices rather than correlation matrices form the basis of the analysis on the same grounds as mentioned in Section 3.6.5. Multiset data may be matrix-conditional, because values across (covariance) matrices are not necessarily comparable (e.g., see Gifi, 1990, for a discussion of matrix conditionality) .

Even though multiset data usually originate from multisample data, not uncommonly two-mode data of objects by variables have a design on the subjects (between–subject design). In this case one may want to check that each subgroup has the same correlational structure for the variables. This can be done by creating a multisample or a multiset data set.

- *Primary analysis methods*: Stochastic — Structural equation modeling, simultaneous factor analysis; Nonstochastic — Simultaneous component analysis.
- *Secondary analysis methods*: Tucker2, Parafac and Parafac2 analyses — Indirect fitting.

3.7.3 Sets of configurations

In several disciplines, such as sensory perception, separate stimulus configurations are derived using multidimensional scaling from stimulus by stimulus comparisons. The configurations of the subjects form a type of three-way three-mode data with special characteristics. For each configuration, the distances between the stimuli with respect to each other are known. However, the coordinate axes of the configurations with respect to each other are not known and may be reflected, translated, and/or rotated versions of each other. Because of these characteristics special methods of analysis are required, such as generalized Procrustes analysis (see Gower & Dijksterhuis, 2004).

In medicinal pharmacology it is often necessary to align molecules to examine their properties. Their three-dimensional coordinates are determined via crystallographic measurement without anchoring the coordinate system. A set of such molecules can be aligned using generalized Procrustes analysis as was shown for the conformal alignment of cocaine molecules by Kroonenberg, Dunn III, and Commandeur (2003b) and Commandeur, Kroonenberg, and Dunn III (2004).

- *Primary analysis method*: generalized Procrustes analysis.
- *Secondary analysis methods*: Tucker and Parafac analyses.

3.8   SCALING DESIGNS

A distinction can be made between regular three-way data and three-way data such as individual differences dissimilarity or similarity data and preference rank order data which are commonly treated within the multidimensional scaling framework. There is an extensive literature on treating such data, and a large number of books on multidimensional scaling include chapters on three-way multidimensional scaling (Arabie et al., 1987; F. W. Young & Hamer, 1987; Borg & Groenen, 2005).

3.8.1   Dissimilarities and similarities

A common research design, especially in psychology, marketing, and food research, is to ask subjects to judge the similarity between two stimuli, such as the taste of two foodstuffs, the visual similarity between two posters, or the relative preference for one of two perfumes. Alternatively, a subject has to choose one of two objects, leading to binary data. Given that such judgments are made by a number of subjects, we have three-way two-mode data of stimulus by stimulus by subjects (see Fig. 3.9). Because dissimilarities and similarities can easily be expressed in terms of each other, the choice of which type to use in an analysis depends on the analysis technique. In general, scaling techniques work with dissimilarities because they are comparable to distances, large dissimilarities representing large distances, while in component analyses similarities are more usual because large similarities correspond to large inner products.

- *Primary analysis methods*: INDSCAL, IDIOSCAL, and INDCLUS.
- *Secondary analysis methods*: Tucker2, Parafac, and Parafac2 analyses.

3.8.2   Preference rank orders

Another design often encountered in psychology, marketing, and sensory perception is the rank ordering of stimuli. A judge (subject) has to rank order a set of products with respect to a set of criteria, say, likelihood of purchase, desirability, and estimated price. Such rank-order preference data become three-mode when the task is performed by several judges. These data have special characteristics because the rank order of one

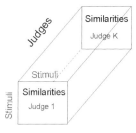

Figure 3.9   Three-way two-mode similarity data.

judge cannot be directly compared to the rank order of another judge (see Fig. 3.10). Special techniques are necessary to handle these data, such as three-mode unfolding (e.g., DeSarbo & Carroll, 1985), but they fall outside the scope of this book.

- *Primary analysis methods*: Three-mode unfolding.

3.9 CATEGORICAL DATA

Most categorical data are profile data with subjects having scores on categorical variables. Several categorical variables are often arranged into multiway contingency tables so that multiway data-analytic techniques can be used for their analysis. Binary variables take a separate place due to their 0-1 character, which makes special types of analysis techniques possible.

3.9.1 Binary data

If the data are basic three-way data, for instance, of objects by variables by conditions, and the variables are binary, they may be treated as ordinary continuous data, although

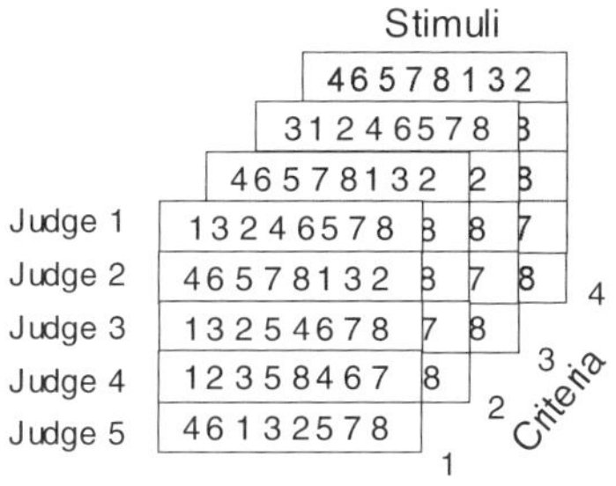

Stimuli

	4 6 5 7 8 1 3 2							
	3 1 2 4 6 5 7 8	3						
	4 6 5 7 8 1 3 2	2	3					
Judge 1	1 3 2 4 6 5 7 8	3	8	7				
Judge 2	4 6 5 7 8 1 3 2	3	7	8	4			
Judge 3	1 3 2 5 4 6 7 8	7	8					
Judge 4	1 2 3 5 8 4 6 7	8		2	3			
Judge 5	4 6 1 3 2 5 7 8		1		Criteria			

Figure 3.10   Three-mode rank-order data.

very skewed distributions may be troublesome. Pure binary data or dichotomized data can be handled with Boolean algebra, which defines multiplication and addition using propositional logic. Chapter 18 contains a brief overview of the work on *hierarchical classes* (HICLAS) *models* for binary data, as well as an example.

- *Primary analysis methods*: HICLAS analysis.
- *Secondary analysis methods*: Tucker and Parafac analyses.

3.9.2 Multiway contingency tables

When we have several categorical variables, the frequencies of their joint occurrences can be placed into a multiway contingency table. Handling these tables with multiway methods is discussed in Chapter 17, in particular three-way correspondence analysis.

Multiway contingency tables come in two forms. They are either matrix-conditional, that is, the data are repeated measures of several categorical variables, or they come from a single sample measured on several variables. The first case violates assumptions about independence of observations often made in loglinear models and association models (Anderson, 1996), because the same subjects provide data at each measurement time. However, for standard multiway models this is not necessarily a problem. In the second case, this problem does not exist. Multiway models are typically not sensitive to the differences between these two types and the analysis proceeds in the same way.

- *Primary analysis methods*: Multiway loglinear modeling, RC(M)-association modeling, multiway correspondence analysis.

3.9.3 Multiway ordinal data

Ordinal categorical multiway data cannot be handled by standard multiway methods without special provisions. Especially effective is to include optimal scaling in multiway techniques (see Gifi, 1990), but this has only been done for the Parafac model under the name of ALSCOMP3 by Sands and Young (1980) (see Section 4.6.1, p. 57). For some comments on this approach, see Harshman and Lundy (1984c, pp. 188–191). Once optimal scaling is included any mix of measurement levels can be dealt with. Another way to analyze multiway ordinal data is to first quantify nonmetric variables with optimal scaling techniques using programs such as the SPSS module *Categories* (Meulman, Heiser, & SPSS Inc., 2000), and analyze the quantified data using regular three-way techniques.

- *Primary analysis methods*: Multiway loglinear modeling and RC-association modeling with ordinal restrictions, optimal scaling multiway models, such as ALSCOMP3.
- *Secondary analysis methods*: Optimal scaling followed by standard multiway techniques, three-way correspondence analysis.

CHAPTER 4

COMPONENT MODELS FOR FULLY-CROSSED DESIGNS

4.1 INTRODUCTION

4.1.1 Three-way data and multiset data

Of the designs for three-way data, the distinction between fully crossed designs and multiset designs is particularly relevant (see also Kiers, 1988, 1991a). In multiset designs, the entities making up at least one of the modes are not the same, for example, there are different individuals in the samples which constitute the third mode (see Section 3.7.2, p. 39). If the individuals have scores on the same variables, a common variable structure can be investigated, but a common structure for the individuals does not exist. In fully crossed three-way data, observations exist for one sample of individuals on all variables at all occasions, and therefore relationships between entities in all modes can be examined. Multiset data can often be converted to two-mode three-way data by calculating correlations or similarities between the variables, so that the mode containing different entities (individuals) is removed. In this chapter, we will primarily deal with fully crossed designs, and only mention component models for

Applied Multiway Data Analysis. By Pieter M. Kroonenberg
Copyright © 2007 John Wiley & Sons, Inc.

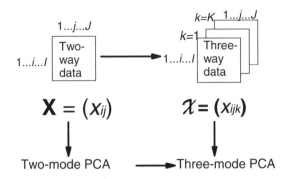

Figure 4.1   Extending two-mode PCA to three-mode PCA.

multiset and two-mode three-way designs in passing. The discussion will concentrate on the three-mode case, because all the basic principles of multiway analysis can be demonstrated on methods for three-way data. In Chapter 19 the focus is on the case where "multi" is larger than three.

4.1.2   Two-mode and three-mode models

A standard problem in multivariate data analysis is the assessment of the structure of a two-mode data set consisting of individuals by variables. The variables could be the items of a questionnaire constructed to cover a number different conceptual domains, or they cover many overlapping concepts. Many other situations may arise in which a reduction of the original variables to a limited number of composite ones is desired. Alternatively, one may want to look at major relationships between the variables in a lower-dimensional space to understand their structure. For all these analysis purposes, (principal) component analysis has been the major vehicle in two-mode analysis. For three-way data, it seems natural to use generalizations of two-mode component analysis to tackle the investigation into their structure (see Fig. 4.1). In this chapter we will use the example of scores of individuals on a number of variables measured under several conditions, but these terms should be taken as generic labels without assuming that the exposition is limited to such entities.

The extension of the basic two-mode component model to three-mode models is based on Hotelling's view of principal component analysis. He describes principal component analysis as a technique to decompose a (data) matrix in two parts, the subject space and the variable space. Components are the directions in these spaces which explain successively the most variance. The other, Pearson's, view of principal component analysis is that components are linear combinations of the original variables, so that components are seen as a set of new uncorrelated (random) variables of which the first explains as much variance as possible, and the others successively explain as much of the variance that is left. One might say that Hotelling took a geometric view and Pearson a statistical view of component analysis (see for a full treatment of the distinction and references Ten Berge & Kiers, 1996a).

In this chapter we assume that the original data have already been preprocessed, that is, that the appropriate means and scale factors have been removed, so that the component models are applied to the preprocessed data. Details about preprocessing can be found in Chapter 6.

4.2 CHAPTER PREVIEW

The most common three-mode component models for fully crossed designs in particular, the Tucker2, Tucker3, and Parafac models, are introduced as generalizations of the two-mode principal component model or more specifically as generalizations of the singular value decomposition. Core arrays will also be discussed in some detail, but the interpretation of the models and their core arrays will be discussed in Chapter 9, and in the chapters devoted to applications.

4.3 TWO-MODE MODELING OF THREE-WAY DATA

Before discussing the various three-mode component models, we will look at the ways three-way data can be transformed into two-way data, and why this might not always be a good idea. There are several ways to transform three-way data so that they can be analyzed with two-mode component models. Fig. 4.2 shows several of these options, these will be briefly discussed in turn.

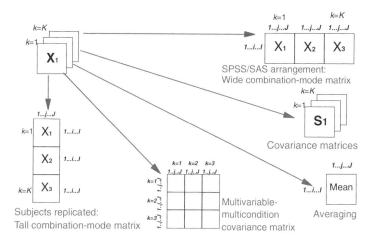

Figure 4.2   Transforming three-way data to two-way data.

4.3.1   Averaging

When the individuals are representative samples from the same population, and it is known or reasonable that there are no (relevant) subpopulations, or, more or less equivalently, that the individuals in the sample are exchangeable with respect to any other member of the population, there is little objection to reduce the three-way data set to a two-way one of averages. Once this is done, standard two-mode data analysis (see Section 4.4.1) can be applied to the resulting average matrix.

Notwithstanding the removal of the three-mode character of a three-way array by averaging, a basic component analysis of the average matrix can serve as a kind of baseline for more complex three-mode models. Such models should uncover patterns in the data that were not shown by the analysis of the average matrix.

4.3.2   Row-wise arrangement of frontal slices

The standard manner in which three-way data are stored and treated in the main statistical packages, such as SPSS, SAS, etc., is as a two-way matrix of individuals by variables×occasions (*wide combination-mode matrix*), in other words, the frontal slices of the three-way array are arranged in a row. Such stringing out of a three-way array into a two-way matrix is now commonly called *matricization*, but was often referred to in the chemical literature as *unfolding*; this use is now discouraged. Tucker (1966, p. 281) introduced the term *combination-mode matrix* for the resulting matrix. The most common analysis methods which respect the three-way character are repeated measures analysis.

The disadvantage of this arrangement in component analyses is that the analysis does not do justice to the multiple measurements on the same variables. In the analysis the variability among the variables is confounded with the variability between the conditions, and the "main effects" of variables and of occasions are not explicitly modeled. The independence of individuals is, however, maintained.

4.3.3   Column-wise arrangement of frontal slices

An alternative to the above arrangement is to treat the three-way data as a *tall combination-mode matrix* of individuals×occasions by variables, in other words, the frontal slices are arranged in a column. No standard two-way techniques exist which explicitly respect such an arrangement, primarily because the observations (individuals×occasions) are no longer independent. Each individual occurs K times in the analysis which goes against the assumptions of most statistical analyses. In a nonstochastic context, such as is often the case in component analyses, the structure in the row mode can be used in interpretation, for instance, by connecting the points of an individual across occasions in a graph (so-called *trajectories*); see, e.g., Lavit, Escoufier, Sabatier, and Traissac (1994), and Fig. 15.11, p. 399. Note that again there is a confounding, this time of the variability between the individuals and the variability among the conditions.

4.3.4 Sets of covariance or correlation matrices

Another way of handling three-way data is to calculate for each sample a covariance or correlation matrix, and analyze the set of covariance matrices. A clear disadvantage is that the subjects disappear from the data and cannot be recovered. This is, of course, not a problem if we deal with truly random samples from populations. However, the data are treated as if they are cross-sectional, and longitudinal information such as trends in the subject scores are lost.

4.3.5 Multimode covariance matrices

To do justice to the design on the variables, one can convert the wide combination-mode matrix into a covariance matrix of variable×occasions by variable×occasions, which in this book is referred to as a multimode covariance matrix, but may also be called a multivariable–multioccasion matrix. The standard analysis is via structural equation modeling, in which case, models can be built which separate the variability between the variables from the variability among conditions, but the disadvantage is that no attention is paid to differences between individuals (e.g., see Bentler & Lee, 1978b; Oort, 1999). Moreover, problems occur when there are too few subjects or too many variables.

4.3.6 Conclusions

Looking at the four options to reduce a three-way array to a two-way matrix before analysis, it should be obvious that none of the methods is ideal, because either individual differences are neglected, or in the analysis the within-mode variabilities of two modes are confounded. Both conceptually and practically, it is more advantageous to analyze the within-mode variability for each mode separately and treat the interaction of these variabilities as a separate, possibly second-order phenomenon.

4.4 EXTENDING TWO-MODE COMPONENT MODELS TO THREE-MODE MODELS

4.4.1 Singular value decomposition

The basic form of two-mode principal component model for two-way data matrix $\mathbf{X} = (x_{ij})$ is the decomposition of $\mathbf{X}$ into scores for the individuals and loadings for the variables, that is,

$$x_{ij} = \hat{x}_{ij} + e_{ij} = \sum_{s=1}^{S} a_{is} f_{js} + e_{ij}, \tag{4.1}$$

or in matrix notation,

$$\mathbf{X} = \hat{\mathbf{X}} + \mathbf{E} = \mathbf{A}\mathbf{F}' + \mathbf{E}, \tag{4.2}$$

where $\mathbf{A} = (a_{is})$ is an $I \times S$ normalized matrix of subject coefficients (i.e., of length 1), $\mathbf{F} = (f_{js})$ is a $J \times S$ matrix of variable coefficients, $\mathbf{E}$ is the $I \times J$ matrix of residuals, and the columns of $\mathbf{A}$ and those of $\mathbf{F}$ refer the same components. It is important to note that, similar to main effects in analysis of variance, the coefficients for the subjects and variables are independent of each other and depend only on either the index of the subjects or that of the variables. The coefficients of subjects and variables combine multiplicatively to model the data, whereas in analysis of variance the main effects combine additively. Principal component analysis is a multiplicative model containing sums of multiplicative terms, which are only dependent on one of the modes at a time. The matrix $\hat{\mathbf{X}}$ is called the *structural base* of the data matrix (Bro, 1998c), the *model matrix* by Van Mechelen and colleagues (e.g., in Ceulemans, Van Mechelen, & Leenen, 2003), and, in this book, $\hat{\mathbf{X}}$ is called the *structural image* of the data independent of the model used.

As illustrated in Fig. 4.4, this component model can be further decomposed into the *basic structure* of the matrix $\mathbf{X}$ by writing the variable coefficients as the product of the normalized coefficients of the variables, and the variability of the components, that is,

$$x_{ij} = \sum_{s=1}^{S} a_{is} f_{js} + e_{ij} = \sum_{s=1}^{S} a_{is} [g_{ss} b_{js}] + e_{ij}, \tag{4.3}$$

or in matrix notation,

$$\mathbf{X} = \mathbf{A}\mathbf{G}\mathbf{B}' + \mathbf{E}, \tag{4.4}$$

where $\mathbf{B} = (b_{jp})$ is a $J \times P$ matrix of variable scores and $\mathbf{G} = (g_{pp})$ is a square $P \times P$ diagonal matrix. This decomposition is the *singular value decomposition* (see Eckart & Young, 1936) and in order for the matrix $\mathbf{G}$ to be diagonal, both $\mathbf{A}$ and $\mathbf{B}$ must be orthogonal. When their columns also have lengths 1, that is, if they are orthonormal, the entries on the diagonal of $\mathbf{G}$ are the *singular values* which are the square roots of the eigenvalues. $\mathbf{A}$ and $\mathbf{B}$ are called the *left singular vectors* and *right singular vectors*, respectively. Even though not always correct, we will often refer to the elements in $\mathbf{B}$ as *loadings* to distinguish them clearly from the elements of $\mathbf{A}$, which will be called *scores*. Strictly speaking, in most contexts the elements of $\mathbf{F}$ should be called loadings, as, given proper centering and normalization, they are correlations between the components and the variables. According to standard practice, singular values in this book will always be arranged in such a way that their sizes are in descending order.

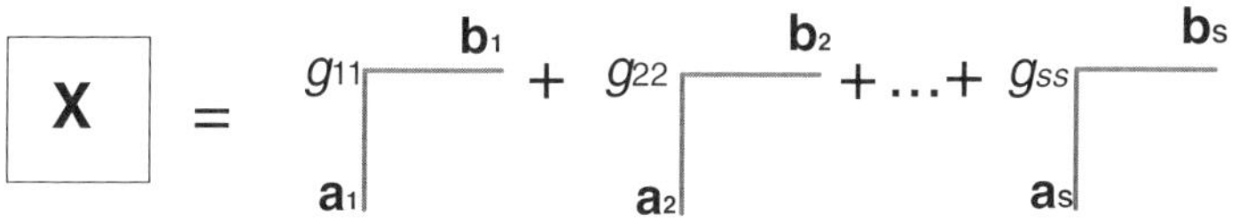

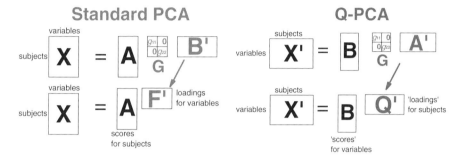

Figure 4.3   Principal component analysis as a sum of rank-one products.

Figure 4.4   Singular value decomposition and its relation with two-mode PCA and Q-PCA.

A final way of writing the singular value decomposition is as the sum of *rank-one matrices* (see Fig. 4.3).

$$\mathbf{X} = \sum_s g_{ss}(\mathbf{a}_s \otimes \mathbf{b}_s) + \mathbf{E}$$

$$\mathbf{X} = g_{11}(\mathbf{a}_1 \otimes \mathbf{b}_1) + g_{22}(\mathbf{a}_2 \otimes \mathbf{b}_2) + \cdots + g_{SS}(\mathbf{a}_S \otimes \mathbf{b}_S) + \mathbf{E}. \quad (4.5)$$

The idea behind writing the singular value decomposition in this way is that each *outer-product* $(\mathbf{a}_s \otimes \mathbf{b}_s)$ produces a matrix of the same size as the data but only has a rank equal to 1 as it is the product of two vectors or rank-one submatrices. This matrix is an approximation of the data matrix when given its proper weight g_{ss}. One could compare this with the contribution of a predictor $\hat{y}_s = b_s X_s$ in regression analysis, where $\hat{y}_s$ is the contribution of X_s to modeling the criterion variable y. Generally, one such approximation of the data matrix is not enough, so that one needs S approximations each weighted by its own g_{ss}. Representing component models with rank-one matrices is especially useful when we want to generalize to multiway component models.

For the sequel it is important to observe that the two-way matrix with singular values or *two-way core array* $\mathbf{G}$ is diagonal, so that the elements $g_{ss'} = 0$ if $s \neq s'$. Thus, the combination $a_{is}b_{js'}$ has a weight 0, so that this product or combination of components is not contributing to the structural image. In the singular value decomposition, $\mathbf{a}_s$, the sth column of $\mathbf{A}$ is exclusively linked to $\mathbf{b}_s$, the sth column of

B, as outer products of $\mathbf{a}_s$ and $\mathbf{b}_{s'}$ have a zero weight if $s \neq s'$. Because the exclusive links between the columns for **A** and **B** with the same index, it is natural to say that $\mathbf{a}_s$ and $\mathbf{b}_s$ refer to the *same* component with a single interpretation. In addition, the number of columns in **A** and **B** is necessarily the same, because any $g_{ss'} = 0$. In some but not all three-mode models, both the one-to-one relationships between columns of component matrices and the necessity of equal numbers of columns in all component matrices are lost.

Another important property of the singular value decomposition is the *uniqueness* of the solution for each number of components. As mentioned earlier, the singular value decomposition is also known as the basic structure of a matrix, which consists of the orthonormal matrices **A** and **B** with singular vectors, and the diagonal matrix **G** with singular values. If only $S < \min(I, J)$ components are used, **A** and **B** are the unique, maximum-variance S-dimensional subspaces. Thus, there is only one set of $(\mathbf{A}, \mathbf{B}, \mathbf{G})$ that fits Eq. (4.4) and has the largest amount of variance explained. Note, however, that the orthogonality is essential. The uniqueness breaks down is when several of the g_{ss} are equal and some of their associated singular vectors lie within the chosen S-dimensional subspace and some others outside it. However, with real data this hardly ever occurs and we will here disregard this possibility. Solutions, however, may become unstable when two g_{ss} have values that are very close together and only one of them is included in the S-dimensional subspace.

If we drop the orthogonality and diagonality restrictions, we can find any number of both orthogonal and oblique rotations for both **A** and **B**, which do not affect the fit of the model to the data. However, after rotation the core matrix **G** is no longer diagonal. In other words, the subspaces are unique but not their coordinate axes. The Parafac model to be discussed in Section 4.6.1 is unique in a different sense. In that model the components themselves (i.e., the axes of the subspaces) are unique, but such axes are not necessarily orthogonal. By uniqueness in the context of three-mode models, usually the uniqueness of the coordinate axes of the space is meant, and we will almost always use the term in this sense. A possible term for this is *component uniqueness* to contrast it with the *subspace uniqueness* of the singular value decomposition.

The singular value decomposition is the foundation for many multivariate techniques and the three-mode component models have as their basis generalizations of this decomposition. However, there are several generalizations that have different properties, and no generalization has all the properties of the two-mode singular value decomposition at the same time.

4.4.2   Separate singular value decompositions

As soon as we have measured our individuals on more than one occasion, we have a three-way array $\mathcal{X}$ of subjects by variables by conditions. The most straightforward way to model this three-way array is treat it as a set of matrices and to apply the

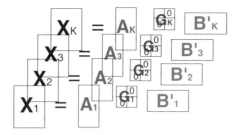

Figure 4.5   Separate singular value decompositions on the frontal slices of a three-way array.

(two-mode) singular value decomposition to each frontal slice of the three-way array (see Fig. 4.5),

$$\mathbf{X}_k = \mathbf{A}_k \mathbf{G}_k \mathbf{B}'_k + \mathbf{E}_k, \tag{4.6}$$

where the $\mathbf{A}_k$, $\mathbf{B}_k$, $\mathbf{G}_k$, and $\mathbf{E}_k$ are defined as before, but this time for each frontal slice k (see Fig. 4.5).

This approach to investigating the basic structure of a three-way array has little to recommend itself from the point of view of parsimony. All analyses are independent and in no way is the singular value decomposition of one frontal slice related to that of another frontal slice. Such analyses have often been carried out in the past, but in order to get some notion of what the solutions have in common, the separate solutions have to be subjected to transformation procedures to a common target (e.g., derived from the singular value decomposition of their average). Alternatively, the collections of the $\mathbf{A}_k$ or the $\mathbf{B}_k$ have to be subjected to a three-mode technique to find a common orientation, so that information becomes available about how much the solutions resemble one another. The technique mostly employed for this is generalized Procrustes analysis; see Borg and Groenen (2005, Chapter 21) or Gower and Dijksterhuis (2004).

There are situations in which this roundabout way to analyzing a three-way array is advantageous, such as when there are no individual differences or when there is no interest in modeling them. However, in terms of directly modeling three-way data, typically one should search for more parsimonious approaches with built-in restrictions for the components.

4.5 TUCKER MODELS

4.5.1 Tucker1 model

Description.   A natural restriction is to demand that the component matrices for one of the modes are the same, for example, that in Eq. (4.6), $\mathbf{A}_k = \mathbf{A}$ or $\mathbf{B}_k = \mathbf{B}$. In the

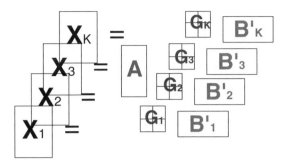

Figure 4.6   Tucker1 model with $\mathbf{A}_k = \mathbf{A}$, but different $\mathbf{B}_k$ — Eq. (4.7).

first case, the individual differences are constant across all k, so that when we replace $\mathbf{A}_k$ in Eq. (4.6) by $\mathbf{A}$, we assume that the common score array is the same across all conditions. This leads to the model

$$\mathbf{X}_k = \mathbf{A}\mathbf{G}_k\mathbf{B}'_k + \mathbf{E}_k \tag{4.7}$$

also shown in Fig. 4.6. In the second case, there are few differences in structure in the variable mode across occasions. Thus, when we replace $\mathbf{B}_k$ in Eq. (4.6) by $\mathbf{B}$, we assume that the common loading matrix is applicable to all occasions. This leads to the model

$$\mathbf{X}_k = \mathbf{A}_k\mathbf{G}_k\mathbf{B}' + \mathbf{E}_k \tag{4.8}$$

Both Eqs. (4.7) and (4.8) may be referred to as Tucker1 models (Kroonenberg & De Leeuw, 1980), where 1 refers to the number of ways for which components are independently determined. These models can be solved by ordinary PCA on the matricized three-way array or the *combination-mode matrix*. Matricizing three-way arrays and subsequently analyzing the resulting two-way matrix has had a long history and has been reinvented many times. For instance, Bentler (1978a) used the Tucker1 model for longitudinal data, using a tall combination-mode matrix. More recently, Nomikos and MacGregor (1994) developed a similar approach for monitoring batch processes, giving their procedure the (in hindsight) unfortunate name "multiway principal component analysis". This term had already been used before for principal component analysis on multiway arrays without matricizing. Several later papers provide examples of Tucker1 model analysis of multiway arrays, also calling it multiway PCA, such as Gallagher, Wise, and Stewart (1996).

Of course, we can also conceive of a model with common subject scores *and* common scale loadings, which is the subject of Section 4.5.2. If we take the restrictions even one step further and also assume that the relative weights of the components are identical (i.e., $\mathbf{G}_k = \mathbf{G}$), we are back at the standard singular value decomposition but this time with the same decomposition for each of the K frontal slices. If we first average across the third mode, this leads to the following results,

$$\mathbf{Z}_k = \mathbf{AGB}' + \mathbf{E}_k = \mathbf{AF}' + \mathbf{E}_k. \tag{4.9}$$

This model might, somewhat perversely, be called the Tucker0 model, but was called SumPCA by Kiers (1988).

Properties.   The Tucker1 models have the same technical properties as the singular value decomposition, but what is really different is that there is not just one but K error terms $\mathbf{E}_k$. Which of the three possible models fits the data best may be evaluated via comparison of model fits (see Section 4.9).

4.5.2   Tucker2 model

Description.   The simultaneous restriction of a common set of scores across all occasions $\mathbf{A}_k = \mathbf{A}$ and a common set of loadings across occasions $\mathbf{B}_k = \mathbf{B}$, while maintaining the orthonormality of the components, causes a breakdown of one of the properties of the singular value decomposition, in particular, the unique linking of the columns of $\mathbf{A}$ and $\mathbf{B}$. This means that the $\mathbf{G}_k$ are no longer necessarily diagonal but the all elements g_{pqk} can be nonzero. To emphasize this, we will write $\mathbf{H}_k$ instead of $\mathbf{G}_k$. The mathematics show that the numbers of columns of $\mathbf{A}$ and $\mathbf{B}$ are no longer necessarily equal. The number of components of $\mathbf{A}$ will be designated by P and that of $\mathbf{B}$ by Q. The $\mathbf{H}_k$ are consequently $P \times Q$ matrices. In summary, we get

$$\mathbf{X}_k = \mathbf{AH}_k\mathbf{B}' + \mathbf{E}_k, \tag{4.10}$$

where $\mathbf{H}_k$ is the matrix representing the relative strength of the links between columns of $\mathbf{A}$ and $\mathbf{B}$. For example, if the data matrix of occasion k, $\mathbf{X}_k$, is approximately recovered by the sum of the products of the 2nd column of $\mathbf{A}$ and the 5th column of $\mathbf{B}$, and the 3rd column of $\mathbf{A}$ and the 1st column of $\mathbf{B}$, then h_{25k} and h_{31k} are nonzero and the remaining elements of $\mathbf{H}_k$ are zero. In sum notation, Eq. (4.10) becomes

$$x_{ijk} = \sum_{p=1}^{P} \sum_{q=1}^{Q} h_{pqk}(a_{ip}b_{jq}) + e_{ijk}. \tag{4.11}$$

The model in Eq. (4.10) is known as the Tucker2 model (Kroonenberg & De Leeuw, 1980). The 2 indicates that components are present for two of the three ways of the data array.

In its basic form the Tucker2 model, $\mathbf{A}$ is an $I \times P$ column-wise orthonormal matrix, $\mathbf{B}$ is a $J \times Q$ column-wise orthonormal matrix (P not necessarily equal to Q), and the P by Q matrix $\mathbf{H}_k$, is called a frontal slice of the *extended core array* $\mathcal{H}$, which is a three-way array of order $P \times Q \times K$. In the Tucker2 model the elements of the core slices express the linkage of the columns of $\mathbf{A}$ and $\mathbf{B}$. Given $\mathbf{A}$ and $\mathbf{B}$ are orthonormal, it can be proved that the square of each element h_{pqk} expresses the size of variability, which is explained by the combination of pth column of $\mathbf{A}$ and the qth

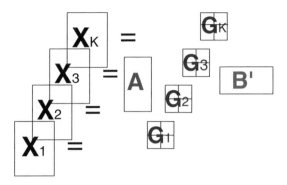

Figure 4.7   The Tucker2 model.

column of **B**, given the parameters have been estimated by a least-squares algorithm (Kroonenberg & De Leeuw, 1980).

Various kinds of differences between occasions can be modeled via special structures in the extended core array. Detailed structures in the extended core array are discussed in Sections 4.8 and 14.3 (p. 346). In Section 4.6.1 we will discuss the Parafac model, which can also be seen as a model with restrictions on the extended core array. Often, in the analysis of real data, there are complex combinations of components, and patterns emerging in the extended core array may be rather messy and sometimes not easy to explain.

Properties.   The Tucker2 model possesses some of the properties of the singular value decomposition. In particular, it allows for a complete decomposition of the fallible three-way array, given enough components. It is also possible, given orthonormal components, to make a complete partitioning of the variability due to components. This property extends to the partitioning of the variability per level of each mode, and per element of the core array; for details, see Ten Berge, De Leeuw, and Kroonenberg (1987), Kroonenberg (1983c, Chapter 7), and Section 9.5, p. 225. Properties this model lacks are *nesting of solutions*, that is components from a solution with a smaller number of components are not necessarily the same as the components from a solution with a larger number of components, and *component uniqueness*, that is, any transformation of the components and/or core array with the inverse transformations of the appropriate other matrices will not affect the fit. In general, it does, however, have *subspace uniqueness*. For a discussion of the Tucker2 model as a mixture of first-order and second-order components for loadings and scores, see Section 9.4.5.

4.5.3   Tucker3 model

Description.   The Tucker3 model (Tucker, 1966) has components for all three modes, thus including the remaining mode of occasions (Fig. 4.8). Moreover, it contains a

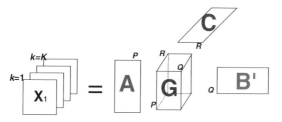

Figure 4.8 The Tucker3 model.

full core array $\mathcal{G}$, which contains the strength (or weight) of the linkage between the components of the different modes. The Tucker3 model can be conceived of as a model in which the extended core array is modeled as a linear combination of several more basic core slices, that is,

$$\mathbf{X}_k = \mathbf{AH}_k\mathbf{B}' + \mathbf{E}_k = \mathbf{A}\left(\sum_r c_{kr}\mathbf{G}_r\right)\mathbf{B}' + \mathbf{E}_k, \tag{4.12}$$

where $\mathbf{C} = (c_{kr})$ is the component matrix for occasions. From $\mathbf{H}_k = \sum_r c_{kr}\mathbf{G}_r$, we may deduce that the occasion components are some kind of second-order components of the extended core array, a view that was already implicit in Tucker's 1966 paper, and further elaborated by Van de Geer (1974); see also Kroonenberg (1983c, p. 18). Equation (4.12) may be written in more compact form with Kronecker products:

$$\mathbf{X} = \mathbf{AG}\left(\mathbf{C}' \otimes \mathbf{B}'\right) + \mathbf{E} \tag{4.13}$$

where $\mathbf{X}$ is a $I \times JK$ wide combination-mode matrix obtained by matricizing $\mathcal{X}$, that is, juxtaposing the $\mathbf{X}_k$. $\mathbf{G}$ is the $P \times QR$ matricized core array constructed in a similar fashion, $\mathbf{C}$ is a $K \times R$ column-wise orthonormal matrix consisting of occasion components, and $\mathbf{E}$ is the $I \times JK$ matrix of residuals. In sum notation this equation becomes

$$x_{ijk} = \sum_{p=1}^{P}\sum_{q=1}^{Q}\sum_{r=1}^{R} g_{pqr}(a_{ip}b_{jq}c_{kr}) + e_{ijk} \tag{4.14}$$

which shows that the g_{pqr} function as weights for the combinations of components. If a particular $g_{pqr} = 0$, then the pth, qth, and rth combination of components does not contribute toward the structural image. By properly normalizing the data and the solution, the scale components become loadings (variable–component correlations; principal coordinates) and the individual and concept components become scores with normalized or standard coordinates.

Taking up the rank-one notation from Section 4.4.1 on the singular value decomposition and extending it to three modes, the Tucker3 model takes the form

$$\mathcal{X} = \sum_{p}^{P} \sum_{q}^{Q} \sum_{r}^{R} g_{pqr}(\mathbf{a}_p \otimes \mathbf{b}_q \otimes \mathbf{c}_r) + \mathcal{E}, \tag{4.15}$$

where the term $(\mathbf{a}_p \otimes \mathbf{b}_q \otimes \mathbf{c}_r)$ is called a *rank-one array*. Advanced treatments of multiway models in terms of rank-one arrays or tensors can be found in Franc (1989), De Lathauwer, De Moor, and Vandewalle (2000), Comon (2001), Kolda (2001), and their references, but its use dates back at least as far as Hitchcock (1927a, 1927b). Within this context, models such as the Tucker3 model are also referred to as *higher-order singular value decomposition* or HOSVD. Each term in Eq. (4.15) produces an $I \times J \times K$ three-way array, which is part of the structural image. The Tucker3 model is complex because we need $P \times Q \times R$ terms of the form $(\mathbf{a}_p \otimes \mathbf{b}_q \otimes \mathbf{c}_r)$ to construct the structural image, each weighted by its own g_{pqr}. If several of these weights are zero, or close to zero the model is much simpler, both to look at and to interpret.

Properties

Decomposition. The Tucker3 model possesses some of the properties of the singular value decomposition; in particular, it allows for a complete decomposition of the three-way array. It is also possible, given orthonormal components, to make a complete component-wise partitioning of the variability. This property extends to the partitioning per level of a mode and per element of the core array; for details, see Ten Berge et al. (1987), Kroonenberg (1983c, Chapter 7), and Section 9.5, p. 225. The model has subspace uniqueness. Properties this model lacks are component uniqueness (any transformation of the components and/or core array with the inverse transformation of the appropriate other matrices will not affect the fit) and nesting of solutions.

Restrictions on the number of components. Even though the numbers of components for the three modes may differ, there are some built-in restrictions in the model. In particular, the *minimum-product rule* says that the product of the numbers of components of two modes must always be equal or larger than that of the third mode, so that $P \times Q \geq R$, $P \times R \geq Q$, and $Q \times R \geq P$. The restriction is a consequence of the three-way orthogonality of the core array (see Section 4.8).

A single component in one mode. A special situation arises when there is only one component in one of the modes. As has been shown by Ten Berge et al. (1987), in such a situation the single remaining core slice corresponding to the single component is necessarily diagonal, given that $\mathbf{A}$ and $\mathbf{B}$ are orthonormal. This model which has been called *replicated* PCA by Van IJzendoorn and Kroonenberg (1990) and *weighted*

PCA by Krijnen and Kiers (1995), has the form

$$x_{ijk} = \sum_{p=1}^{P}\sum_{q=1}^{Q} g_{pq1}(c_{k1}a_{ip}b_{jq}) + e_{ijk}$$

$$= c_{k1}\sum_{p=1}^{P}\sum_{p=1}^{P} g_{pp1}(a_{ip}b_{jp}) + e_{ijk}, \quad (4.16)$$

or in matrix notation,

$$\mathbf{X}_k = c_{k1}\mathbf{A}\mathbf{G}_1\mathbf{B}' + \mathbf{E}_k. \quad (4.17)$$

This model is thus a very simple generalization of two-mode PCA in which there is a single diagonal core slice $\mathbf{G}_1$. Each level k of the third mode weights $\mathbf{A}\mathbf{G}_1\mathbf{B}'$ with a coefficient c_{k1}. If we define the matrix $\mathbf{C}^* = c_{k1}\mathbf{G}_1$, then $\mathbf{C}^*$ can be seen as a component matrix with P columns that are proportional with proportionality constants equal to $g_{pp}/g_{p'p'}$.

Looking ahead to the Parafac model treated next, it looks superficially like the weighted PCA model with orthonormal component matrices. However, the perfect proportionalities of the columns of $\mathbf{C}^*$ prohibit it being a Parafac one (see below). One of the consequences of the perfect proportionality is that the components in the model are not unique because they can be rotated or nonsingularly transformed in the same way that they can be transformed in two-mode PCA. This is in contrast with the Parafac model in which there is component uniqueness (see Section 4.6.1).

4.6   PARAFAC MODELS

4.6.1   Parafac model

The other major component model for three-way data is the Parafac model (Harshman, 1970; Harshman & Lundy, 1984c, 1984d). The same model was independently proposed in a different context by Carroll and Chang (1970) under the name CANDE-COMP (canonical decomposition), but we will use the name Parafac model, because it was Harshman who developed its full potential for handling fully crossed three-way data. Before entering into the description of the model, it is necessary to look at the principles that underlie the Parafac model and that motivated its development and extension by Harshman (1970).

Parallel proportional profiles.   The fundamental idea underlying the Parafac model was first formulated by Cattell (1944) in the form of the principle of *parallel proportional profiles* which he explained in Cattell and Cattell (1955, p. 84; their italics):

> The basic assumption is that, *if a factor corresponds to some real organic unity, then from one study to another it will retain its pattern, simultaneously raising*

or lowering all its loadings according to the magnitude of the role of that factor under the different experimental conditions of the second study. No inorganic factor a mere mathematical abstraction, would behave this way. [...] This principle suggests that every factor analytic investigation should be carried out on at least two samples, under conditions differing in the *extent* to which the same psychological factors [...] might be expected to be involved. We could then anticipate finding "true" factors by locating the unique rotational position (simultaneously in both studies) in which each factor in the first study is found to have loadings which are proportional to (or some simple function of) those in the second: that is to say a position should be discoverable in which the factor in the second study will have a pattern which is the same as the first, but stepped up or down.

The parallel proportional profile principle has great intuitive appeal, exactly because it determines the orientation of component axes without recourse to assumptions about the shape of the loading patterns. In particular, there is no attempt to maximize simplicity or any other structural property *within* a given occasion. Instead, the proportional profiles criterion is based on a simple and plausible conception of how loadings of "true factors" would vary *across* occasions.

A further strength of the parallel proportional profile principle is that it is based on patterns in the data that could not have easily arisen by chance, or as a by-product of the way that the study was conducted. When it can be empirically confirmed that such proportional shifts are present and reliable, one has established an important fact about the data, that is, that a particular orientation of the factors is plausible (see Harshman & Lundy, 1984c, pp. 147–152, 163–168, for a more detailed presentation of the argument). The Parafac model is said to have the *intrinsic axis property*, which implies that, under 'reasonable' conditions, the orientation of the axes of components or axes of the model are determined by the specification of the model. For the Parafac model this means that the data structure must be trilinear, and the components must show distinct variation in its modes.

There are disadvantages of the Parafac model compared to models employing target rotations: no theories based on specific patterns in the components can be tested, nor can simple easily interpretable components be searched for. One way of attempting to get the best of both worlds is to use constrained Parafac models; see, for instance, Krijnen (1993) and Bro (1998c) for several ways in which this can be done. Furthermore, in practice, data sets very often only support a limited number of components with parallel profiles but these might be sufficient to model the systematic information in the data. Other three-mode models can also describe additional patterns, which do not conform to the parallel proportional profile principle. This is one of the important differences between the three-mode models discussed here.

One of the consequences of the parallel proportional principle is that the components have constant correlations across the remaining modes. It is therefore not possible to fit data in which the correlations between the components change over time. Generally, the correlations between subtests of an intelligent test are different for very young children compared to older children, which is an infringement of the parallel proportional profile principle. Such data cannot be modeled in a Parafac context. Should such differences in correlations exist and the researcher attempts to fit a Parafac model anyway, it might lead to degenerate solutions (for details and solutions see Section 5.6.5, p. 89).

Description. The Parafac model can be written as

$$\mathbf{X}_k = \mathbf{AD}_k\mathbf{B} + \mathbf{E}_k, \tag{4.18}$$

where $\mathbf{D}_k$ is an $S \times S$ diagonal matrix of weights for occasion k, $\mathbf{A}$ is the $I \times S$ scores matrix for individuals, and $\mathbf{B}$ is the $J \times S$ loading matrix for variables. By writing the Parafac model in this fashion, the principle of proportional parallel profiles in a single mode is emphasized. The relationships between component $\mathbf{a}_s$ and $\mathbf{b}_s$ is expressed through d_{ssk} for each k, so that for each occasion k the first-mode and second-mode components are the same (or parallel) apart from the proportionality constant d_{ssk}. Because the model is symmetric in its three modes, as is immediately evident from Eqs. (4.19) and (4.20), parallel proportional profiles exist for all three modes. Because there is a unique linking between the columns of each of the modes in the sense that the first columns of each mode are exclusively linked to each other, it is valid to assert that there only exists one set of components that have coefficients for each mode. Given proper scaling, one could interpret the model as having components with one set of principal coordinates (or loadings) for the variables and two sets of normalized coordinates (or scores) for the other modes. This aspect makes the models attractive from an interpretational point of view, as only one component interpretation is necessary, rather than a separate one for each mode, as is the case for the Tucker3 model.

The Parafac model can be written as a sum of rank-one arrays as (Fig. 4.9)

$$\mathcal{X} = \sum_s^S g_{sss}(\mathbf{a}_s \otimes \mathbf{b}_s \otimes \mathbf{c}_s) + \mathcal{E}, \tag{4.19}$$

so that we can see that to reconstruct the data array we only need S terms, compared to $P \times Q \times R$ for the Tucker3 model. For a Tucker model in which $P = Q = R = S$, this means going from S^3 to S terms. Thus, the Parafac model is indeed much simpler than the Tucker3 model. Of course, such simplification comes at a price. The model is far more restricted than the Tucker3 model, so that it might be more difficult to fit a data array with the Parafac model than with a Tucker3 model. In fact, any fallible data array can be exactly fit with a Tucker3 model given enough components, while there are few data arrays that can be so fitted with the Parafac model. However, in practical applications, it is not so much exact fitting but good approximations to the systematic part of the data that counts.

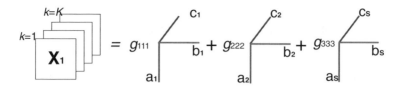

Figure 4.9   The Parafac model.

Using sum notation we get

$$x_{ijk} = \sum_{s=1}^{S} g_{sss}(a_{is}b_{js}c_{ks}) + e_{ijk}, \tag{4.20}$$

where $\mathbf{A}$ is an $I{\times}S$ matrix with component scores for the individuals, $\mathbf{B}$ is a $J{\times}S$ matrix with components for the variables, and $\mathbf{C}$ is a $J{\times}S$ matrix with components for the occasions. Each of these matrices have unit length columns so that the g_{sss} indicate the weights of the combinations. This form is called the basic form of the Parafac model. Multiplying the columns of any of the matrices with the g_{sss} allows these columns to be interpreted more or less like loadings rather than as scores (for details see Section 9.4.4, p. 221).

Properties.   The nature of the model is such that the components do not need to be constrained to orthonormality in order to find a solution, and the components are correlated in most applications. A further characteristic of the model is that, like the Tucker3 model, it is symmetric in all modes and that, unlike the Tucker3 model, the numbers of components in all modes are necessarily the same. The simple superdiagonal core array consisting of only the g_{sss} (see Section 4.8) facilitates the interpretation of the model compared to the Tucker models, but on the other hand the correlated components in the modes may complicate the interpretation.

Types of variations.   In Harshman and Lundy (1984c, p. 130) the parallel proportional profile principle is explained in terms of models or components underlying the data. In the *system variation model*, the components "reside in the system under study and through the system affect the particular objects; the [component] influences exhibited by particular objects would thus vary in a synchronous manner across a third mode. [...] In the second model, the *object variation model*, separate instances of the [component] can be found in each of the objects, and these within-object [components] would not be expected to show synchronous variation across levels of a third mode."[1]

Typically, object variation is not commensurate with the Parafac model. Harshman and Lundy (1984c) suggest that if the object variation is present in one mode only,

---

[1]Harshman and Lundy (1984c) use the word *factor* instead of *component*.

the fully crossed data can be converted to covariance matrices such that the object-variation mode "disappears", in the same way that individuals are no longer visible in a correlation matrix but only the variables. The resulting covariance matrices can be investigated for parallel proportional profiles in the other modes with a standard Parafac model.

Uniqueness. The Parafac model possesses some of the properties of the singular value decomposition, but not all. In particular, each component of a mode is uniquely related to only one component in each of the other modes. A very special characteristic of the Parafac model is its component uniqueness: any transformation of the components will affect the fit of the model to the data (called the "intrinsic axis property" by Harshman). More technically, the model is identified, generally even overidentified due to the restrictions, so that it is possible to test whether the model fits or not. This feature, together with the relative simplicity of the model, makes it extremely attractive, certainly from a theoretical point of view. Moreover, in those cases where explicit models of the Parafac form exist, such as analytical chemistry, signal processing, and EEG research, the uniqueness is a powerful property to estimate the parameters of the model; see also Section 5.6.4 (p. 88) for algorithmic aspects.

Degeneracy. One of the consequences of a model with restrictions is that the model does not fit all data. This is in itself not a problem, but in the case of Parafac algorithms the built-in restrictions may produce degenerate, nonconverging solutions because of the incompatibility. A relatively elementary treatment of degeneracy can be found in Sections 5.6.5 (p. 89) and 8.6.2 (p. 185). However, a full technical treatment is beyond the scope of this book, and the reader is referred to Harshman and Lundy (1984d, pp. 271-281), Kruskal, Harshman, and Lundy (1989b), Stegeman (2006), and, for degeneracy indicators, to Krijnen and Kroonenberg (2000).

Explicit models. In several disciplines there are explicit physical models that have the form of the Parafac model, so that the Parafac algorithm can be used directly for estimation of the parameters. Examples are analytical chemistry (e.g., see the thesis by Bro, 1998c), topographic component models where the model is used to analyze event-related potentials in the brain (Möcks, 1988a, 1988b), and signal processing (Sidiropoulos et al., 2000b, 2000c).

Direct and indirect fitting. In two-mode analysis one may solve the parameters of the principal component model either via least-squares fitting of the raw data (*direct fitting*) or via least-squares fitting of the correlation matrix (*indirect fitting*), and the results are the same. The terms derive from the fact that in the first option the model is directly fitted to the raw data, while in the second option the model is fitted to derived data (e.g., correlations). In multiway analysis this equivalence is no longer true, as is demonstrated in detail in Harshman and Lundy (1984c, pp. 133–139).

They also discuss the additional flexibility that fitting the Parafac model to sets of covariance matrices provides, compared to fitting the Parafac model to the raw data. In particular, it permits multiway models to handle data with object variation which is not possible in fitting the Parafac model to raw data. This indirect fitting comes at the cost of requiring orthogonal components for the mode over which the cross products are calculated, but in some circumstances this seems a small price to pay. Moreover, as we see in the next section, there are more sophisticated versions of the Parafac model, which relax the orthogonality requirement. An additional possibility for indirect fitting to covariance matrices is that certain common factor models can be fitted as well. Details can be found in Harshman and Lundy (1984c, pp. 125–147).

4.6.2   Parafac2 model

Almost simultaneously with the introduction of the Parafac model, Harshman (1972b) introduced the Parafac2 model for sets of covariance or similarity matrices . The model is less restrictive than the standard Parafac model because the proportionality constraints do not apply to one of the modes. Typically, for each occasion a congruence or correlation matrix is calculated, so that the nonproportional mode, in this case subjects, "disappears". The requirement for this case is that the congruence coefficients (or correlations) are constant for all occasions. Whereas Harshman (1972b) exclusively discussed the Parafac2 model for derived data, Ten Berge, Kiers, and Bro (Kiers, 1993a; Ten Berge & Kiers, 1996b; Kiers, Ten Berge, & Bro, 1999) revived the model, developed algorithms for it, and recast it into a form suitable for fully crossed three-way data sets rather than only for multiset data (Section 3.7.2, p. 39). Their model has the form

$$\mathbf{X}_k = \mathbf{A}_k \mathbf{D}_k \mathbf{B} + \mathbf{E}_k, \qquad (4.21)$$

with the restriction that the covariance matrix (often cross product or correlation matrix) of subjects scores $\mathbf{A}'_k \mathbf{A}_k$ is $\mathbf{R}$ for all k. Characteristic in this approach to the model is that the model is a direct extension of the classical PCA or factor models with scores (for subjects) $\mathbf{A}_k$ and loadings (for variables) $\mathbf{B}$. The third mode consists of conditions under which the relations between scores and loadings may vary, as expressed by varying weights and/or covariances (correlations) of the components.

The model was taken up by Timmerman (2001), in connection with longitudinal multiset data in which the roles of subjects and occasions are reversed compared to the way data are generally presented in this book. For this design, the occasions form the first mode in Eq. (4.21), and they can be different in number and time spacing for different subjects, who make up the third mode (see Fig. 3.8(b), p. 38). The Parafac2 model in this connection is called SCA-PF2, where SCA refers to simultaneous component analysis. In the present chapter, we consider Parafac2 primarily as a model for fully crossed three-way data, but simultaneous component analysis is applicable in a wider context, as is evident in Timmerman (2001).

Under fairly mild conditions and the presence of a sufficient number of levels in the third mode (four may already be enough), the parameters of the model are uniquely determined (Kiers et al., 1999). Bro, Andersson, and Kiers (1999b) and Dueñas Cueva, Rossi, and Poppi (2001) present applications in analytical chemistry, and Timmerman (2001) in psychology and education. The modeling can be done using the MATLAB N-way toolbox developed by Andersson and Bro (2000)[2] or with the MATLAB programs developed by Timmerman, or its Fortran version (called Simul-Comp) in 3WayPack[3].

Timmerman (2001) defined an additional two models which can be seen as constraint variants of the Parafac2 model. (1) SCA-IND: The components are required to be orthogonal; and (2) SCA-ECP: the covariance matrices are equal for all occasions. The Tucker1 model (Section 4.5.1) can be seen as the most general SCA model, and in the SCA context, it is referred to as SCA-P (see also Kiers & Ten Berge, 1994a).

Orthogonal Parafac2 model.   The orthogonal Parafac2 model, called SCA-IND, uses the same model equation as the Parafac2 model, that is, Eq. (4.21), but there is a stronger restriction on the covariance (or cross product) matrix $\mathbf{A}'_k \mathbf{A}_k$ of the component scores, as it now has to be diagonal $\mathbf{\Delta}$ for all k, implying that the components are orthogonal. Merging $\mathbf{\Delta}$ with the $\mathbf{D}_k$, the scores are shown to be orthogonal with different weights (or explained variability) at each occasion. An important property of this model is that its parameters are identified under relatively mild conditions.

Equal covariances Parafac2 model.   A final simplification of the Parafac2 model (SCA-ECP) is that all the $\mathbf{D}_k$ are equal as well, which implies that for all occasions the subject components have equal covariance (cross product or correlation) matrices. The disadvantage of this model is that, due to the extreme simplification, the uniqueness of the solution is lost and transformational freedom is reintroduced. This implies that the components are only determined up to a common nonsingular transformation.

4.7 PARATUCK2 MODEL

Harshman and Lundy (1996) presented ParaTuck2, a model in between the Parafac model and the Tucker2 model:

$$\mathbf{X}_k = \mathbf{A}\mathbf{D}_k^L \mathbf{\Gamma} \mathbf{D}_k^R \mathbf{B}' + \mathbf{E}_k, \qquad (4.22)$$

where there is a single core slice $\mathbf{\Gamma}$ representing the relationships between the components of the individuals and the variables for all occasions at the same time, but the weights of the components, both for the individuals and for the variables, can be different for each occasion. The technical interest in the model is that under not

---

[2] *http://www.models.life.ku.dk/source/nwaytoolbox/index.asp.* Accessed May 2007.
[3] *http://three-mode.leidenuniv.nl/.* Accessed May 2007.

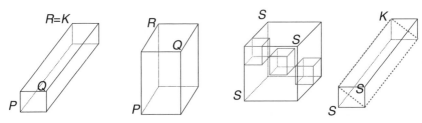

Tucker2 model:
Extended core array
$(R = K)$

Tucker3 model:
Full core array

Parafac model:
a. Superdiagonal full core array:
    $(S = P = Q = R)$
b. Slice-diagonal extended core array
    Dashed rectangle = component matrix **C**

Figure 4.10 Core arrays of several three-mode models.

too restrictive assumptions this model has uniquely determined components, unlike the Tucker2 model itself, while it is more flexible than the sometimes too restricted Parafac model. Kiers and Smilde (1998f) and Bro (1998c, p. 37–44) discuss the merits of this model in relation to the Parafac model (see also Bro, Harshman, & Sidiropoulos, 2007), and present several applications in analytical chemistry.

4.8 CORE ARRAYS

The three basic component models, Tucker2, Tucker3, and Parafac, each have their own form of the core array (Fig. 4.10).

4.8.1 Types of core arrays

Tucker2 core array. The Tucker2 model has an *extended core array*, $\mathcal{H}$, of size $P \times Q \times K$ and it is called extended because one mode of the core array has the same size, K, as the original data array. Of the three core arrays, the extended core array has the largest number of elements, that is, $P \times Q \times K$.

Tucker3 core array. The Tucker3 model has a *full core array* $\mathcal{G}$ of size $P \times Q \times R$. It can be calculated, along with the third component matrix **C**, from the extended core array, $\mathcal{H}$, by a singular value decomposition on the extended core array of the Tucker2 model. However, such a procedure would not give the optimal solution for the Tucker3 model. Its number of elements is $P \times Q \times R$, which is smaller than that of the extended core array, unless $R = K$.

Superdiagonal core array. The core array for the Parafac model is a *superdiagonal core array*, that is, a $S \times S \times S$ cube with the g_{sss} from Eq. (4.20) on the diagonal.

Parafac core array. A further type of core array which will be referred to as the *Parafac core array*, is the core array constructed from the data and the Parafac components according to the standard formula for calculating core arrays in the Tucker3 model (see Eq. (5.6)). However, in this instance, generalized inverses (indicated with a superscript + in Eq. (4.23)) have to be used because the Parafac component matrices are not orthogonal. The matrix formulation for the Parafac core array is

$$G = A'^{+}X(C^{+} \otimes B^{+}). \tag{4.23}$$

The Parafac core array is used to assess to what extent the Parafac model is an appropriately fitting model for the data as is explained in detail in Lundy, Harshman, and Kruskal (1989). An example of the use of the Parafac core array can, for instance, be found in Section 13.5.4 (p. 329).

Slice-diagonal extended core array. As indicated in Section 4.6.1, the Parafac model can also be written as a collection of two-way frontal slices (see Fig. 4.10),

$$X_k = AD_kB, \tag{4.24}$$

where D_k is an $S \times S$ diagonal matrix in which the weights g_{sss} are absorbed. Together the D_k form a *slice-diagonal extended core array*. The elements of the diagonal matrices scale the common components in A and B with a constant specific to each condition k. If we consider the first mode A as scores of subjects, and the second mode B as loadings for variables, then the diagonal entries in the D_k contain the amounts by which these components are scaled up and down, thereby illustrating the parallel proportional profiles principle. The diagonal elements of each D_k, d_{ssk}, ($s = 1, \ldots, S$) are equivalent to the rows of the component matrix C, thus $d_{ssk} = c_{sk}$ for all k and s.

4.8.2 Orthogonality and core arrays

Three-way orthogonality of the Tucker3 core array. In the basic form of the Tucker3 model, that is A, B, and C are orthonormal, the core array is three-way orthogonal. The three-way orthogonality was already implicit in the formulas presented in Tucker (1966), but was for the first time explicitly mentioned by Weesie and Van Houwelingen (1983). The orthogonality follows from the fact that columns of each component matrix are the eigenvectors of the data array reduced along the other two modes; for details see Tucker (1966, p. 289; formula 24) or Kroonenberg and De Leeuw (1980, Section 4). Thus, the columns of A are such eigenvectors and the corresponding eigenvalues ξ_p can be obtained by squaring the elements in each $Q \times R$ core slice G_p. These eigenvalues are the sums of squares accounted for by each column $\mathbf{a}_p$ of $\mathbf{A}$. In order to show the orthogonality of the core array for Mode A, first each horizontal $Q \times R$ core slice $\mathbf{G}_p$ is vectorized, that is $vec(\mathbf{G}_p) =$ the column vector $\breve{\mathbf{g}}_p = (\breve{g}_{11}^p, \breve{g}_{21}^p, \ldots, \breve{g}_{QR}^p)'$. The resulting $QR \times P$ matrix $\breve{\mathbf{G}}$ with columns $\breve{\mathbf{g}}_p$ is then orthogonal, that is, $\breve{\mathbf{G}}'\breve{\mathbf{G}} = \Xi$, where Ξ is the diagonal matrix with the eigenvalues ξ_p or explained variabilities of the components of the first mode.

Because the core array is three-way orthogonal, for each mode the product of the number of components of two modes, say, $P \times Q$, must be larger than or equal to the number of components of the remaining mode in this case R (see Wansbeek & Verhees, 1989, p. 545, for a formal proof). If not, the core array cannot be fully orthogonal and will be singular. Thus, due to the three-way orthogonality the minimum-product rule (see also Section 4.5.3) applies, i.e., $P \times Q \geq R$, $Q \times R \geq P$, and $P \times Q \geq R$ (see Tucker, 1966, p. 289). The situation is similar for core arrays for the multiway Tucker model. For instance, in the four-way case, $P \times Q \times R \geq T$. In other words, the $1 \times 1 \times 2 \times 4$ Tucker4 model does not exist, which is equivalent to the statement that it is the same as a $1 \times 1 \times 2 \times 2$ Tucker4 model.

Two-way orthogonality Tucker2 core array.   In the basic form of the Tucker2 model, that is $\mathbf{A}$ and $\mathbf{B}$ model are orthonormal, the extended core array $\mathcal{H}$ is two-way orthogonal. In particular, if we vectorize each lateral $P \times K$ slice, we get a $P \times K$ by Q matrix $\hat{\mathbf{H}}$ and $\hat{\mathbf{H}}'\hat{\mathbf{H}} = \Phi$, where Φ is a $Q \times Q$ diagonal matrix with the ϕ_{qq} the explained variabilities of the variables (second mode) on the diagonal in decreasing order. When we vectorize the horizontal slices of $\mathcal{H}$, we get a $Q \times K$ by P matrix $\breve{\mathbf{H}}$ and $\breve{\mathbf{H}}'\breve{\mathbf{H}} = \Psi$, where $P \times P$ Ψ is a diagonal matrix with the ψ_{pp} the explained variabilities of the individuals (first mode) on the diagonal in decreasing order. As a consequence of the two-way orthogonality also for the Tucker2 model the minimum-product rule applies, that is $P \times K \geq Q$ and $Q \times K \geq P$.

Superdiagonal core array.   A superdiagonal core array is obviously always orthogonal, but in general the g_{sss}^2 are not the variabilities explained by the associated components. The reason is that, especially in the standard Parafac model, which has a superdiagonal core array, none of the component matrices are orthogonal unless at least one of them is constrained to be so. For orthogonal Parafac models, the partitioning of the variability is valid (Kettenring, 1983a, 1983b).

4.9   RELATIONSHIPS BETWEEN COMPONENT MODELS

Central to several investigations is the choice between three-mode models in order to optimize the balance between fit and ease of interpretation. The best-fitting model may be extremely difficult to interpret due to the number of parameters, while the most easily interpretable model may be so restricted that it does not fit the data very well. To assist in the choice of an appropriate model, it is useful to construct model trees or model hierarchies. Several authors have discussed such hierarchies

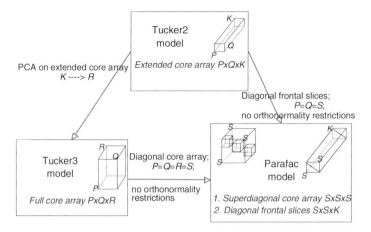

Figure 4.11   Three-mode model tree: Relationships between component models for three-way data.

for three-mode models, among others, Kroonenberg (1983c, p. 49ff.), Harshman and Lundy (1984d), and Kiers (1988, 1991a). We will first present the major relationships between the most important three-mode models, followed by an outline of the Kiers (1991) hierarchy. A full discussion of choosing between and within classes of models is presented in Chapter 8.

4.9.1   Primary three-mode model tree

In Fig. 4.11 the major three-mode models presented in the previous sections have been depicted in relation with each other, especially in connection with the shape of their core arrays (see Section 4.8). By putting more restrictions on the core array, one can go from the Tucker2 model with the largest number of parameters to the Parafac model with the smallest number of parameters.

4.9.2   The Kiers hierarchy for three-mode models

Kiers (1991a) presented two hierarchies of three-mode model, of which we will only take up that for fully crossed three-way data. Kiers concentrated on one of the modes, in particular, the third one, and built his hierarchy starting from occasions. The reason for this was that he started from the assumption that one usually collects a data matrix of individuals by variables on several occasions, concentrating, however, on the relationships between the variables. If, on the other hand, the main interest is on the development of the scores of the individuals over time, it might be more appropriate to start with the matrix having the individuals×occasions as rows and the variables as columns. This is the perspective we will present here. However, in Fig. 4.12 all

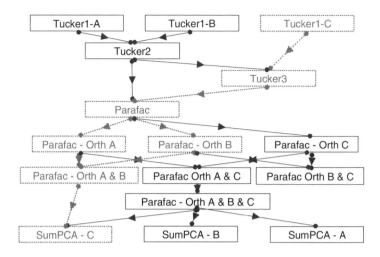

Figure 4.12   Hierarchical relationships between component models for three-way data. The models in the dotted-line grey boxes are those considered by Kiers (1991).

possible arrangements are included. In practice, it is probably advantageous to limit oneself as Kiers did, in order not to get involved in too many model comparisons.

At the top of the hierarchy are the Tucker1 models, referred to by Kiers as PCASup. The principal component analysis of these models proceeds on the tall combination-mode matrix called *supermatrix* by Kiers, hence the name of the technique. Because the Tucker1 model has many more parameters than any of the other Tucker and Parafac models, it is the starting point in the hierarchy as the least restricted model. Due to its asymmetry, there are three Tucker1 models, one for each mode. The next, more parsimonious, model is the Tucker2 model of which there are three variants (only one of which is shown), again one for each mode. This is followed by the Tucker3 model. One step down the line comes the Parafac model. The latter can be further restricted by placing orthogonality constraints in any of the three modes, and the very last models are simple PCAs on each of the three average or summed matrices (see Section 4.3.1). Kiers (1991a) explains in mathematical detail how these models can be put into a single framework with increasing restrictions on the components, except for the Tucker2 model, which is not mentioned in this hierarchy. In Section 8.7 (p. 186) we will present an empirical example of selecting from a hierarchy.

4.10   MULTIWAY COMPONENT MODELING UNDER CONSTRAINTS

In at least two situations, it is useful not only to fit a completely unrestricted model, but also to examine the fit of some constraint version of the model (see also Section 5.6.7, p. 94). In the first situation, the design of the data is such that it is natural to place constraints on the components, or at least determine whether such constraints are appropriate. For example, in time-series data it seems appropriate to require or test smoothness constraints or functional constraints on the time mode. Alternatively, when one or more of the modes are wavelengths or subtests of a test to measure a single construct, it seems sensible to require the components to be nonnegative, as negative wavelengths do not occur, and subtests of intelligence tests are all assumed to be positively correlated. For certain types of chemical data, for instance, those collected in chromatography, the outcome measures should be unimodal on theoretical grounds. The second, heuristic, situation occurs when one suspects that a solution obtained via a multiway model can be more parsimoniously described by imposing restrictions. The most drastic version of this is using another model that imposes inherently more restrictions on the components, but alternatively more subtle measures may be employed by requiring that certain components are constant, or by testing whether certain restricted components will deteriorate the fit.

In this section an overview is given of possibilities for imposing restrictions on components without going into much detail about fitting and indicating when and where these restrictions are most appropriate. A comprehensive treatment of constraints on components in multiway component models is given in Bro (1998c, Chapter 6), who presents algorithms as well as examples but primarily from chemometrics. Additional work is reported by Timmerman (2001) with an emphasis on smoothing and functional constraints for longitudinal data. A treatment within the Parafac context is given by Kroonenberg and Heiser (1998), on which this section is partly based.

Another set of restrictions that can be defined in multiway models and have been examined extensively in three-mode models is the imposition of structure on core arrays. As this topic is closely related to rotations and transformations of the core array, this subject is taken up in Chapter 10.

4.10.1   Orthogonality

This type of constraint requires that the components within a particular mode are pairwise orthogonal, that is, they form an orthogonal base in the (reduced) space of the components (see also Section 5.6.7, p. 95). In principle, restricting one of the modes to orthogonality is sufficient to obtain a partitioning of the total variability by components (e.g., see Kettenring, 1983a). The orthogonality restriction is not necessarily a logical constraint for the Parafac model, because the parallel proportional profile principle refers to the triplets $(\mathbf{a}_s, \mathbf{b}_s, \mathbf{c}_s)$ and does not explicitly refer to projections into lower-dimensional spaces. A more extensive discussion of this issue can be found in Franc (1992, pp. 151–155), who discusses the differences between a model that consists of the sum of rank-one arrays (such as Parafac), and a model which projects the modes of a data matrix from a high-dimensional space into a lower-dimensional one, as is the case for the Tucker models. For estimating the parameters of the Tucker models it is necessary to impose orthogonality in order to find a solution. Once a solution is found, any orientation of the axes within the subspace found is equally acceptable, and the orientation has no influence on the fit of the solution (subspace uniqueness). Harshman and Lundy (1984c) suggest using orthogonality constraints in order to avoid so-called degenerate (nonconverging) solutions for the Parafac model (see Section 5.6.5, p. 89).

4.10.2   Nonnegativity

Nonnegativity is a very natural constraint to impose on components in several cases, especially when quantities expressed in the components cannot be negative (see also Section 5.6.7, p. 95). In spectroscopy, components cannot have negative elements because negative values for absorption, emission, or concentration are nonsensical. Similarly, in the latent class model, tackled with three-way methods by Carroll et al. (1992), the estimated probabilities should be nonnegative. As a final example, it is known that subtests of intelligence tests generally correlate positively with each other, because they all measure intelligence in some way. Krijnen and Ten Berge (1992) make a case for imposing nonnegativity constraints on the components of the Parafac model in this context. Nonnegativity constraints are particularly useful to avoid so-called contrast components on which, for instance, the performance tests have negative coefficients and the verbal tests positive ones (see also Krijnen, 1993, Chapter 4).

4.10.3   Linear constraints including design variables

Design on the subject mode.   Suppose that subjects fall into specific groups, and that not their individual values but their group means are interesting. Carroll, Pruzansky, and Kruskal (1980) showed that such constraints can be handled by first averaging the original data according to the design, followed by a multiway analysis on the condensed data (see also Section 5.6.7, p. 94). However, they warned that in general on interpretational grounds this does not seem a good procedure for the subject mode in some models such as the INDSCAL model. DeSarbo, Carroll, Lehmann, and O'Shaughnessy (1982) employ such linear constraints in their paper on three-way conjoint analysis and also take up the issue of the appropriateness of restrictions on the subject mode in such situations.

Facet or factorial designs for variables.   Tests and questionnaires are sometimes constructed according to a facet or factorial design and, as above, these variables may be combined and then subjected to a standard Parafac analysis.

A priori clusters on the variables.   Another type of constraint is employed when variables belong to certain explicitly defined *a priori* clusters, which should be reflected in a simple structure on the components. In the three-way case one might like to fit such a constraint via the application of the Parafac model. The details of such an approach have been worked out by Krijnen (1993, Chapter 5); see also the work

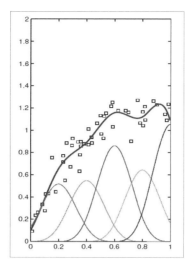

Figure 4.13 Example of smooth curve fitting with B-splines. Courtesy of Paul Eilers.

by Rocci and Vichi (2003a) and Vichi et al. (2007) on simultaneously clustering and component analysis.

A posteriori clusters on variables.   Rather than knowing the clusters beforehand, a Parafac solution might be desired in which a number of optimal nonoverlapping clusters are searched for. In practice, several different numbers of clusters may be tried to find an optimal solution (Krijnen, 1993, Chapter 6).

4.10.4 Smoothness

Especially when there is a natural order for the levels of one of the modes, as in longitudinal data, it is often a sensible requirement that the coordinates for that mode vary smoothly from one time point to the next. Timmerman (2001, Chapter 4) gives a detailed exposition of how smoothness constraints may be incorporated into Parafac and Tucker models. The smoothing is done using B(asis)-splines for general smoothing functions, and I(ntegrated)-splines for monotonically increasing or decreasing smoothing functions. For general treatments and overviews of splines, see Hastie and Tobshirani (1990) and Ramsay and Silverman (1997). In practice, the splines most used are pieces of third-degree polynomial functions, which are combined linearly to create a smooth well-fitting function through the available data points. This is achieved by leading the curves not exactly through all observations, but by finding a compromise between regularity of the curves and fit to the data points (for an example, see Fig. 4.13).

4.10.5   Functional constraints

As explained in Timmerman (2001, Chapter 5), an alternative approach to fitting the Tucker and Parafac models under constraints is to postulate that some functional relationship exists between the levels of that mode based on the theory behind the data or, in rare cases, based on earlier analyses. Browne (e.g, Browne & Toit, 1991) specified such an approach for (two-mode) learning data. The advantage of using a specific function is that the interpretation of the components is known beforehand. It is only the actual values of the parameter estimates that require additional interpretation. Moreover, the number of parameters to be estimated is significantly reduced by this method. Obviously, the success of using functions depends on the quality of their fit to the data.

4.10.6   Unimodality

Bro (1998c, pp. 151–152) discusses the idea of requiring components to have single peaks, in other words to be unimodal (see also Bro and Sidiropoulos (1998b) . The assumption is that the components have underlying fairly smooth functions with only a single peak so that the determination of the components should respect this unimodality. His discussion concentrates on chemical applications, but it might be possible to use such an approach in ecology, where abundances are generally assumed to be a unimodal functions of such variables as water and nutrients, while in psychology data which are generated by "single-peaked preference functions" (see Coombs, 1964) might possibly benefit from such an approach.

4.10.7   Constant components

Suppose one is uncertain whether the data at hand really exhibit individual differences in all components. For instance, one may notice differences in the first component of the subjects, but they are not very large. One could consider a three-mode analysis with the first component fixed at a constant value and compare the fit of the solutions with and without this restriction. In case of a negligible difference one may conclude that the differences were not worth bothering about. In the extreme case of no individual differences whatsoever, one will most likely average over subjects, as would be relevant in the study of semantic differential scales where the basic theory suggests that such individual differences are negligible (e.g, see Wiggins & Fishbein, 1969).

Inclusion of external information.   Most studies are part of a research tradition, so that almost always earlier results exist using the same type of data and/or the same variables. In such cases one might have a configuration available for one of the modes, and one might be interested in whether the older information is comparable to the information contained in the new three-way data set. By fixing the configuration of the mode, one can estimate the other parameters in the model within the context of the already available external information. By comparing the restricted with the unrestricted model, one may assess the differences between the study at hand and a previous study. For examples using the Tucker3 model, see Section 14.6.4, p. 364. Other examples are Van der Kloot and Kroonenberg (1985a) who studied implicit theories of personality, Kroonenberg and Kashima (1997) who compared patterns of showing respect exhibited by Australian and Japanese children, and Veldscholte, Kroonenberg, and Antonides (1998) who used external analysis to merge initially incompatible data sets.

Another situation might be that external (two-way) information on continuous variables is available. For instance, personality information is available for the subjects, and it is desired to explain the structure of the analysis in terms of these external variables. Within the analysis-of-variance context such procedures have sometimes been called factorial regression. A discussion of this approach for the two-way case, as well as references to examples, can be found in Van Eeuwijk and Kroonenberg (1995). Hardly any work has been done on this for higher-way data.

Fixing components for modeling purposes.   It is possible to decide for each component in a Parafac model whether one, two, or three ways should have constant values for this component. By doing this, one can, for instance, perform a univariate three-way analysis of variance with the Parafac model. Work is underway to develop a viable way of carrying out three-way analysis of variance with multiplicative terms for the interactions via a Parafac model as suggested by De Leeuw in Kroonenberg (1983c, p. 141). Further work on using the Parafac model with constant terms for analysis-of-variance purposes has been carried out by Bro (1998c, GEMANOVA; pp. 192–196, 248–253). His work is also discussed in Smilde et al. (2004, p. 340ff.). Published applications of their approach are Bro and Heimdal (1996b) and Bro and Jakobsen (2002a) dealing with enzymatic browning of vegetables and meat discoloration. Together with Andersson, Bro has also developed MATLAB programs collected in the *N*-way Toolbox to perform such analyses[4]. An important extension to the standard MATLAB functions to build these models is contained in the Bader and Kolda's Tensor Toolbox[5]. A prototype program (Triadic) also exists within 3WayPack[6].

4.10.8   Order constraints

Order constraints for autocorrelated data.   When one of the modes in the three-way array is a time mode, or two of the modes contain spatial coordinates, it may be desirable to use this design information during the analysis. One way to do this would be by requiring that the levels of a component obey certain order restrictions. One might also try to take this further and require equal intervals, but this would more or less take us back to the previous section.

---

[4]*http://www.models.life.ku.dk/source/nwaytoolbox/index.asp*. Accessed May 2007.
[5]*http://csmr.ca.sandia.gov/~tgkolda/TensorToolbox*. Accessed May 2007.
[6]*http://three-mode.leidenuniv.nl*. Accessed May 2007.

4.10.9   Constraints to create discrete models

Discrete and hybrid clustering.   Carroll and co-workers have developed procedures to perform clustering on multiset data, in particular, two-mode three-way similarity data by using constraints of the components. This approach demands that certain components can only have a limited number of integer values. As in the previous section, this is an example of using constraints to fit specific models. Details can be found in Carroll and Chaturvedi (1995).

4.10.10   Optimal scaling

Optimal scaling concerns constraints on the data rather than on the model. Detailed discussions of optimal scaling can be found in Gifi (1990), and the earlier references mentioned there (see also Section 5.6.8, p. 96). The fundamental idea is that any variable can be considered to be categorical and the measurement level of a variable is specified by the transformations that may be applied to the values of that variable without changing the meaning of the variable. Thus, for nominal variables the only requirement for the transformation function is that subjects in the same category remain together in a category. Any monotone transformation of the values of an ordinal variable is permissible as it will maintain the order of the values and thus their basic meaning. Sands and Young (1980) presented an optimal scaling version of the Parafac model, under the name ALSCOMP3, as well as an algorithm to compute its solution, but no further applications of this approach are known (see, however, some comments by Harshman & Lundy, 1984c, pp. 188–191).

4.10.11   Constraints for single elements

Missing data.   Missing data may lead to constrained estimation problems within three-mode component models. This can be handled either by a kind of EM-algorithm estimating the model and the missing data in turn, or by differentially weighting elements of the three-way data array, for instance, by weighting every valid data point with 1 and each missing data point by 0 (see Chapter 7).

Equality constraints.   In certain applications it might be interesting to query or test whether certain elements in a component are equal to one another, when it is not known beforehand what the size of these equal elements should be. Such constraints clearly operate at the level of the individual parameters in the model and are thus model constraints. No exposition of this problem is at present available in the literature.

4.11   CONCLUSIONS

This chapter provided an overview of component models for three-way data for fully crossed designs, in particular, the three most common models, the Tucker2 model, the Tucker3 model, and the Parafac model, and their variants were presented. It was shown how core arrays can be defined for all of them and how these core arrays are structured in the various models. In addition, an informal overview was given of their relationships, leaving the technical details to papers already in the literature. Finally, an overview was presented of restrictions on components.

Nearly all models mentioned in this chapter are treated in more technical detail in the book by Smilde et al. (2004) and in Bro's thesis (Bro, 1998c). These books require a solid background in linear algebra and present virtually only chemical applications. This does not really affect the exposition of the methods, but readers must be prepared to follow the chemical slant.

Arabie et al. (1987) is entirely devoted to three-way scaling and related techniques for two-mode three-way scaling data; a Japanese translation is also available Arabie, Carroll, and DeSarbo (1990). In several other books there are chapters on two-mode three-way methods, such as F. W. Young and Hamer (1987, Chapter 6; reprinted 1994) and Borg and Groenen (2005, Chapter 22). Two collections of articles on multiway data and methods are Coppi and Bolasco (1989) and Law, Snyder Jr., Hattie, and McDonald (1984).

CHAPTER 5

ALGORITHMS FOR MULTIWAY MODELS

5.1 INTRODUCTION

In this chapter the basic procedures and ideas for estimating the parameters of the multiway models used in this book are discussed. An overview of types of data can be found in Chapter 3 and a discussion of the models is contained in Chapter 4.

It is not the intention to go into the mathematical details of algorithms, such as convergence proofs or acceleration mechanisms. On the contrary, an attempt will be made to explain what an algorithm is, which types of algorithms are commonly encountered in multiway analysis, and what are the major characteristics of these algorithms. In this chapter "multi" means three and higher, so that the three-way case can serve as a leading case.

Why, one might ask, is it necessary to discuss algorithms at all, if the emphasis in the book is on applying multiway techniques to data? The reason for this is that even though an insight into the details of the calculations might not be necessary, one has to understand the basics of the methods in order to arrive at an interpretation and appreciation of the estimates of the parameters of the models. The nature of multiway

Applied Multiway Data Analysis. By Pieter M. Kroonenberg
Copyright © 2007 John Wiley & Sons, Inc.

analysis is such that with difficult data or complicated models, one has to ascertain whether the obtained solution is one that can reliably be interpreted or whether one has landed in some kind of trap or undesirable situation. For example, in regression analysis one has to understand that serious outliers can influence a solution, because the method to solve the parameters is based on *squared* deviations from the means, so that a value far away from the main body of the data has a large influence on the solution. For virtually all multiway algorithms this is true as well. Thus, the manner in which parameter values are computed, determines how one should view and evaluate an obtained solution.

In principle, we will here supply only that type of information that is necessary for understanding the results of an analysis, what the pitfalls are that one might come across due to imperfect workings of an algorithm, and how to recognize this. This implies that although we will attempt to keep the technical level as low as possible, some technical details are necessary in order to understand the problems one might run into.

5.2 CHAPTER PREVIEW

This chapter deals with the technical aspects of solving the parameters of multi-way models. We will primarily pay attention to the Tucker2 model, Tucker3 model (Tucker, 1966), and Parafac model (Harshman, 1970; Carroll & Chang, 1970). In addition, we will describe a technique developed in France under the name STATIS, which stands for "structuration des tableaux à trois indices de la statistique" or "structuring of three-way statistical tables" (L'Hermier des Plantes, 1976; Lavit, 1988) . This technique has a considerable following in France and is therefore presented in this chapter. The major reason why it has not been discussed before is that it is a method to analyze data, but it is not based on a model whose model fit can be explicitly evaluated against the data (see the discussion about models and methods in Section 2.3, p. 18).

As indicated earlier, our basic aim is to provide the information necessary to evaluate solutions, and to find out what the model does and does not tell us about the data. To begin with, we will introduce definitions of central concepts. This is followed by a discussion of general characteristics of algorithms for multiway component models. We then will deal with the major multiway models and methods in turn and discuss some of the features that one needs to know in order to appreciate their solutions. Often more detailed information is contained in other chapters that describe multiway models and methods, such as Chapter 4 on component models.

5.3 TERMINOLOGY AND GENERAL ISSUES

5.3.1 Algorithms

The following definition of an algorithm[1] is taken from Cormen, Leierson, and Rivest (1989, p. 1).

> Informally, an *algorithm* is any well-defined computational procedure that takes some value, or set of values, as *input* and produces some values, or set of values, as *output*. An algorithm is thus a sequence of computational steps that transform the input into the output.
>
> We can also view an algorithm as a tool for solving a well-specified *computational problem*. The statement of the problem specifies in general terms the desired input/output relationship.

In our situation the *input* consists of a three-way data set and the *output* is the set of estimated parameters of a model or method. The *computational problem* that we typically deal with is the minimization of a loss function f, which specifies the difference between the observed data (x_{ijk}), or a function of them, and the structural image or the estimated data computed using the model of interest $(\hat{x}_{ijk})$. What an algorithm has to do in this case is to find those estimates for the parameters of the model which minimize the difference between the original data and those estimated by the model. It is, of course, crucial how this difference is defined. Most commonly, the sum of the squared differences is used and the required parameters of the model are estimated with a *least-squares loss function*,

$$\mathcal{F}(\text{model parameters}) = \sum_{i=1}^{I}\sum_{j=1}^{J}\sum_{k=1}^{K}(x_{ijk} - \hat{x}_{ijk})^2. \tag{5.1}$$

It is important not to confound computational problems with the tools to solve them, that is, algorithms. Often one computational problem can be solved by several, fundamentally different, algorithms. Sometimes these algorithms can lead to different optimal solutions, while other algorithms, even though they work differently, will give exactly the same solutions. This opens up possibilities to tailor algorithms to different types of data, depending on the known or expected structure in the data, the different sizes of the levels of the modes, and so on.

5.3.2 Iteration

Algorithms can be extremely simple. For instance, the algorithm for estimating the mean is (1) sum all the available values (input) and (2) divide the result by the number

---

[1]The word *algorithm* comes from the name of a famous Persian, Abu Ja'far Mohammed ibn Mûsâ al'Khowârizmî (born around 800); e.g., see *http://www.disc-conference.org/disc2000/mirror/khorezmi/* for detailed history and references. Accessed May 2007.

of values. The result is an estimate for the mean (output). Characteristic in this case is that there is an explicit procedure (a formula) for the estimation. Unfortunately, in multiway analysis this is generally not the case. Mostly, there are no explicit formulas that can be used to obtain the desired estimates for the parameters. One of the main reasons for this is that the minimization of the loss function is very complex and often data dependent. In such cases, one must resort to *iterative* algorithms. Such algorithms start with a rational or random guess for the parameters, followed by the repeated execution of a procedure in order to find a better estimate for the parameters in a prescribed way. It is applied until no improvement in the fit between model and data can be found. Each application of the basic procedure is called an *iteration*. In the example below, we will illustrate the idea of iterative algorithms by having a blindfolded mountaineer trying to find the top of a mountain, which is clearly a maximization problem: he has to find the point with the greatest elevation above sea level. However, we might just as well have used the example of trying to find the lowest point of a lake, which is a minimization problem. Most of the algorithms in multiway analysis solve minimization problems, rather than maximization problems. *Optimization* refers to either maximization or minimization depending on the details of the algorithm.

5.3.3 Alternating least-squares iteration

Multiway component models generally consist of groups of parameters, in particular, a component matrix for each mode (and a core array). It is not possible or feasible to try and optimize the loss function in one single step. Instead, the parameters are estimated one group at a time. Thus, the coefficients of the first-mode components, **A**, are estimated assuming that the other component matrices (and the core array) are known, that is, they are held fixed. Once we have obtained estimates for **A**, we use those and the values of the third-mode components **C** (and the core array) to estimate the coefficients of the second mode components, **B**, and so on. By rotating the component matrices, we improve them one at a time and if all is well we improve the loss function at each substep. Procedures using least-squares loss functions that alternate the strike like this are called *alternating least-squares* algorithms.

Alternating least-squares algorithms in multiway data analysis work very similarly to *expectation–maximization (*EM*) algorithms* (Dempster, Laird, & Rubin, 1977). Expectation–maximization algorithms also alternate between estimating sets of parameters. The expectation (E) step computes expected values using a loss function (mostly a maximum likelihood one) by fixing the values of unobserved variables as if they were observed, and a maximization (M) step, which computes the estimates of the parameters by maximizing the expected values found in the E step. The parameters found in the M step are then used to begin another E step, and the process is repeated. The motivation of using EM-algorithms is that no explicitly algebraic solution exists for the loss function and that latent (thus unobserved) variables are involved. Because alternating least-squares algorithms also use an iterative algorithm with estimating sets of parameters in turn, they are often referred to as EM-algorithms as well, especially when the estimation of missing data is involved (see Section 7.3.2, p. 150).

5.4 AN EXAMPLE OF AN ITERATIVE ALGORITHM

5.4.1 Finding a mountain top blindfolded

As an example, suppose you were required to find the top of a mountain blindfolded. How should you approach this? The starting point is to ask someone else to set you at the base of the mountain. The procedure to be repeated over and over again is that you take a small step in each direction, or maybe in order to save time in only the eight compass directions. The decision criterion is that you will go in that direction that ascends the most. In this particular process, you may want to vary your testing step size so that if you think you are far from the top you proceed with big steps, but nearer to the top you start to fine-tune your stepping. You also need a stopping criterion, which will determine whether to take another step or not. The standard way to do this is by comparing your present position with the previous one and evaluating whether the difference in height is sufficiently large to warrant stepping further. Note that you cannot compare your present position with the highest point itself because you do not know how high it is (you are blindfolded, after all). The procedure is said to *converge* if you end up on some top or other. If it is a local top, but not the global one, the convergence is to a *local maximum*, while if you end up on the real top the convergence is to a *global maximum*. The convergence is *monotone*, when your height increases at each step, and a procedure is *nonconvergent* when you do not reach the top after an (infinite) number of steps, and, finally, the procedure is *divergent* when at each step you get further away from your goal, for instance, ending up in another mountain range.

5.4.2 Problems of iterative algorithms

This example immediately shows some of the problems of iterative algorithms.

1. *Stopping criterion.* You have to decide on a stopping criterion, that is, you have to decide when you consider yourself to be at the highest point.

2. *Local maxima.* You may end up on top of a locally highest point, but not the absolute highest point.

3. *Multiple solutions.* You may end up on a mountain with a completely flat top, so that you stop stepping as soon as you get to the plateau; however, if you start from somewhere else, you may end up on a different part of the plateau because in that case as well you stop as soon as you climb over the rim.

4. *Starting solution.* You have to choose a starting point that will lead you to the highest point.

5. *Slow convergence.* You may be on a very, very gentle and very long slope, so that with each step you get higher but only very marginally so. This contains the danger that you decide to stop because you are not improving enough, while you have still a lot of climbing to do. For instance, you might be on a local plateau but a bit further down the line there is again a steep slope upward.

6. *Maximum number of iterations.* You may never stop climbing. Say that you have decided you are willing to travel 10,000 steps or one whole day. However, due to the gentle slope and the fact that you are a long distance away from the top, you simply may not get there within the agreed number of steps or time. Think of this as running out of steam after the indicated number of steps, your energy is gone, your time is up.

7. *Step size.* You may have to choose a (varying) step size; big ones in the beginning and smaller ones toward the end.

5.4.3   Troubleshooting iterative algorithms

The importance of acquiring some general knowledge about algorithms, such as those used in multiway analysis, is that it is necessary to recognize when there are problems in practice, and to have some idea what might be done about it in practical circumstances. To this end, let us look at the problems mentioned earlier.

1. *Stopping criterion.* Generally, an algorithm optimizes a particular criterion, so there is no real choice there, apart from a possible choice between algorithms. However, in all iterative algorithms it has to be decided which difference between successive iterations is considered to be small enough. Based on this size, a decision is made whether there is sufficient improvement to continue with the iterations. All programs have a default value for this but in uncertain situations it might be worthwhile to choose a smaller difference. This should not be exaggerated because there is a limit to the numerical accuracy of complex algorithms.

2. *Starting solutions and local maxima and multiple solutions.* Many algorithms are only guaranteed to lead you to *a* top, but whether you get to the right one depends on the starting point. Often, algorithms are designed in such a way that they calculate their own starting solutions based on rational (mathematical) arguments. However, in order to maximize the chance of reaching the highest top, you may have to choose several starting points and compare the height of the respective end points. If you end up on the same top from wherever you start, there is a good chance you have reached your goal (i.e., the highest

point). If different starting points lead you to different end points, it is difficult to put great faith in any of them. In such a case, you might consider changing something more fundamental, such as the number of components, or even change the model and reexamine the data themselves to understand what is the problem.

3. *Slow convergence.* Some algorithms only proceed with very small steps with little improvement per step, so convergence might be a time-consuming process. Some programs offer acceleration devices, which might or might not help. One way to speed up is to extrapolate from the existing direction and take a long step in the direction you have been following for a while, a procedure called "overrelaxation". In alternating algorithms, doing this for one type of step might lead you a bit off course for one of the other types of steps, but if correct, you have proceeded more quickly than you would have otherwise. It could affect the monotone convergence in complex algorithms but you might get there faster in the end, even though there is no guarantee that you do. Especially in the standard Parafac algorithm, taking such long steps has proved to be effective in several cases. Some acceleration procedures are clever enough to return to the original algorithm when the convergence is too much affected.

4. *Maximum number of iterations.* Programs have defaults for the number of iterations and it is almost always possible to increase this number. Again, in the Parafac algorithm degenerate solutions may occur (see Section 5.6.5) that are nonconvergent in the loss function and divergent in the parameter (or component) matrices. In order to establish whether such a situation is present, it is often very useful to increase the number of iterations and to decrease the stopping criterion to see whether the situation gets worse or whether it is possible to converge after all.

5. *Step size.* Very few of the algorithms in multiway analysis require the researcher to choose a step size. So this is generally not an issue.

Surveying all these problems one might wonder whether it is at all worthwhile to start on the journey. Luckily, many algorithms behave very nicely and are guaranteed to converge to a local optimum, and by starting from different points it can be assessed whether these local optima are present. Generally, the more structure there is in the data, the bigger the chance that a global optimum can be reached. Thus, there is a premium for having good data. When a model is very restrictive, the chance is that it is more difficult to find a global optimum. The standard algorithm for the Parafac model is especially notorious for getting bogged down in case of minimal inclines and for encountering local optima, and much research has gone into improving it.

5.5 GENERAL BEHAVIOR OF MULTIWAY ALGORITHMS

Algorithms are presented here as nested boxes of which the innermost are black, and only the outer one(s) are transparent. What happens within the black boxes will not concern us here. Not uncommonly, alternative, improved algorithms have been proposed for the original basic ones, which do not fundamentally influence the interpretation of the outcomes. In this book such technical improvements take place within the innermost black box and do not concern us.

The standard two-mode principal component model can be solved directly with the singular value decomposition (see Section 4.4.1, p. 47, or Appendix B.2, p. 492) or, alternatively, via an iterative minimization using a loss function. However, most multiway models are so complex that, for real data with error, no direct estimation procedures exist, and nearly all models are estimated via iteration. However, for some multiway models explicit algebraic solutions exist for error-free data, and such solutions are typically used as rational starting solutions for iterative algorithms. Of course, in practical data analyses, error-free data which exactly conform to some model, do not exist.

The basic algorithms for the Parafac and Tucker models use an alternating least-squares approach in which the number(s) of components have to be specified beforehand and all components of a component matrix have to be determined at the same time. This is different from the two-mode case, where the components can be derived successively and estimating additional components has no effect on the previous ones. If, as in the case of two-mode PCA, the estimation of earlier components is not influenced by the presence or absence of later components, they are called *nested* solutions. Leurgans and Ross (1992, pp. 298–299), among others, have provided a simple proof that the components of the Parafac model cannot be found one by one (or recursively), except when explicit orthogonality constraints are imposed on the components of two modes. In that case, the solutions are nested as well (Harshman, pers. comm., 2007). In general, Tucker models are not nested either. Thus, the components of multiway models with fewer components than were estimated are not the same as the components at hand.

An important difference between the Parafac model and the Tucker models is that the former model is essentially a restrictive one. Not all data sets can be meaningfully fitted with a Parafac model, because they lack the parallel profile property (see Section 4.6.1, p. 57). On the other hand, the Tucker models are *decomposition models*, which means that in principle they can decompose any data set into components and a core array. This difference has important repercussions for algorithms. The restrictive nature of the Parafac model may make the estimation of its parameters impossible because the data do not conform to the model. This will show up in unusual behavior of the algorithms. Or even if the data do conform to the model, it may be difficult to find the correct solution. On the other hand, the lack of restrictions for the Tucker models means that algorithms will always find a solution, however, the fit of the model to the data may be poor. Moreover, because of the restrictive nature, it is possible that a Parafac model with too few or too many components will not lead to a solution, while algorithms for the Tucker models will always find a solution. This also translates itself into the sensitivity to local optima. Parafac algorithms are far more troubled by this, and it is always advisable to run more than one analysis from different starting points. On the other hand, the TUCKALS algorithms are much more likely to end up in the global optimum. The only real exception occurs when models with very large numbers of components are specified, in which case several alternative solutions may exist because the algorithm tries to fit random variability (Timmerman & Kiers, 2000). In the end, it comes down to the fact that a Parafac analysis is more confirmatory (or explanatory, as Harshman, 1970, has it), while a Tucker analysis is always exploratory.

Missing data can have a very large effect on the speed and ease of convergence of multiway algorithms, even to the point that an otherwise nicely behaved algorithm cannot cope with the amount and distribution of the missing data. The central problem is that information is lacking to steer the algorithm to the right solution. This problem can be compounded when the starting values inserted for the missing values are far away from "true" values. In such situations diverging solutions have been observed; see Sections 7.7.3 (p. 161) and 19.6.2 (p. 480) for examples.

5.6   THE PARALLEL FACTOR MODEL – PARAFAC

In order to describe the more common algorithms for our multiway models we will recapitulate the model descriptions for the three-mode case before presenting the algorithms themselves.

5.6.1   Model description

Harshman's Parafac model for a three-way array $\mathcal{X}$ with elements x_{ijk} has the form

$$x_{ijk} = \sum_{s=1}^{S} g_{sss}(a_{is}b_{js}c_{ks}) + e_{ijk}, \tag{5.2}$$

where the a_{is}, b_{js}, and c_{ks} are the elements of the unit-length component matrices **A**, **B**, and **C**, respectively, the g_{sss} are the elements of a superdiagonal core cube $\mathcal{G}$ (i.e., $g_{pqr} = 0$ if $p \neq q \neq r$), and e_{ijk} are the errors of approximation (for details see Section 4.6.1, p. 57). For the purpose of estimation, the g_{sss} may be absorbed into one of the component matrices, and after a solution has been found their values follow from setting the lengths of each component in all modes equal to 1. In order to estimate the component matrices (i.e., the parameters of the model) there are two basic approaches. If it is known that the model fits perfectly then there exists an explicit, algebraic solution; if not, one has to resort to iteration.

5.6.2 Algebraic solutions

Sands and Young (1980), using a suggestion by Yoshio Takane, presented a decomposition that will provide a solution to the Parafac model in the case of error-free data totally conforming to the model. Leurgans, Ross, and Abel (1993) provide an overview of similar proposals. In the case of approximate solutions, the Sands and Young procedure and its variants can be used as an initialization for an alternating least-squares algorithm, in the same manner as Tucker's (1966) Method I is used as initialization of the alternating least-squares algorithms for the Tucker models (see below).

5.6.3 Iterative algorithms

Basic Parafac alternating least-squares algorithm. The basic idea of the algorithm to estimate the parameters of the Parafac model for three-way data is to break up the model into three sets of parameters. In particular, each component matrix forms a set. If we pretend to know the values of two of the matrices, we can estimate the third via linear regression techniques. This algorithm is given in Table 5.1.

The basic algorithm for the Parafac model applied to three-way data outlined in Table 5.1 shows that within each step α of the iteration procedure three regression problems have to be solved. The regression equations arise because at each substep two of the component matrices are held fixed and only the third one has to be estimated (details can, e.g., be found in Smilde et al., 2004, pp. 114–115). As reported by most authors, convergence can be rather slow and alternative algorithms have been investigated. In his Parafac program, Harshman uses a type of overrelaxation method to speed up his algorithm (Harshman, 1970, p. 33). Appellof and Davidson (1981) investigated the performance of the so-called Aitken extrapolation method for the same purpose. Both are based on the idea that when the convergence is slow one takes somewhat bigger steps into the best direction in the hope to accelerate the convergence, and this is seen to work reasonably well.

An interesting alternative within the same framework is not to iterate on the component matrices but on the components themselves, that is, first the components a_1, b_1, c_1 are estimated, then a_2, b_2, c_2 and so on, until all S components have had a turn, and then the iteration turns to the first one again, and so on until convergence. Harshman mentioned to Kiers (1997b, p. 251) that he had developed a crisscross regression algorithm for the Parafac model generalizing Gabriel and Zamir's (1979) *dyadic* algorithm (see Section 7.3.1, p. 148). Carroll, De Soete, and Pruzansky (1989) did this as well. More recently, Heiser and Kroonenberg (1997) and Kroonenberg and Heiser (1998) presented a weighted variant of this algorithm for the Parafac model, calling this a *triadic* algorithm. Given the popularity of the Parafac model, many authors have proposed alternative algorithms to solve the parameter estimation and an evaluation of various algorithms can be found in Faber, Bro, and Hopke (2003) and Hopke, Paatero, Jia, Ross, and Harshman (1998).

Table 5.1 Basic Parafac algorithm

Set iteration counter
$\alpha = 0$.

Starting solution
Initialize $\mathbf{A}_0$, $\mathbf{B}_0$, $\mathbf{C}_0$, for instance, by using the Sands and Young (1980) or similar decomposition, or by using random start matrices.

Update iteration counter
$\alpha = \alpha + 1$.

A-substep
Fix $\mathbf{B}_{\alpha-1}$ and $\mathbf{C}_{\alpha-1}$, and solve the regression equation for $\mathbf{A}$ to obtain a new $\mathbf{A}_\alpha$.

B-substep
Fix $\mathbf{A}_\alpha$ and $\mathbf{C}_{\alpha-1}$, and solve the regression equation for $\mathbf{B}$ to obtain a new $\mathbf{B}_\alpha$.

C-substep
Fix $\mathbf{A}_\alpha$ and $\mathbf{B}_\alpha$, and solve the regression equation for $\mathbf{C}$ to obtain a new $\mathbf{C}_\alpha$.

Estimate missing data (Optional)
Use current estimates $\mathbf{A}_\alpha$, $\mathbf{B}_\alpha$, and $\mathbf{C}_\alpha$ and the data $\mathcal{X}$ to estimate the missing values according to the model equation (5.2).

Check convergence
If the differences between successive iterations with respect to the loss function
  $||\mathcal{F}^{\alpha-1} - \mathcal{F}^{\alpha}||^2$ are not small enough,
or the Euclidean norms of successive estimates of $\mathbf{A}$, $\mathbf{B}$, and $\mathbf{C}$ are not small enough,
then continue with a next iteration,
else end iterations.

A more technical description can be found in Smilde et al. (2004, pp. 114–118).

Iterative algorithms tend to be sensitive to local minima, so that one generally needs several runs from different starting positions to find the "best" minimum. Harshman and Lundy (1984c, 1984d) and Harshman and DeSarbo (1984b) discuss at length the possibilities and problems of analyzing three-way data with the Parafac model (see also the next sections).

Algorithm for multimode covariance matrices. A interesting variant of the basic Parafac algorithm for data sets with one large mode was proposed by Kiers and Krijnen (1991b). In this algorithm, the amount of multiplication involved in the step with the largest number of levels of I, J, and K can be circumvented by cleverly rearranging the computations, so that manipulation with the original data array $\mathcal{X}$ of, say, subject by variables by occasions is not necessary. In particular, they showed that one may convert the $I \times J \times K$ raw data into a $JK \times JK$ variable–occasion multimode covariance matrix. Given that the subject mode is much larger than the other two, considerable savings can be obtained because this large mode does not figure directly in the computations during the iteration procedure. The component matrix **A** can be recovered after convergence. An interesting feature of this estimation method is that if only a multimode covariance matrix is available, a Parafac analysis can still be carried out resulting in the same **B** and **C** component matrices as in the case that the full data were available. However, the first-mode coordinates **A** will not be available.

Conclusions. Several improvements and variants of the basic algorithm have been proposed, an overview of which can be found in Faber et al. (2003). However, they came to the conclusion that no recently proposed algorithms surpass the best alternating least-squares algorithm in its versatility and universal usefulness. Its primary drawback is its slow convergence. In a similar vein, Tomasi and Bro (2006) also discussed and evaluated a number of algorithms for estimating the Parafac model, but they also came to the conclusion that, apart from a number of difficult cases and when the quality of the solution is very important, "the standard ALS represents a good trade-off between computational expense and quality of the solution." For precise instructions when to use which algorithm for which data sets, one should consult their original paper. A later proposal was put forward by De Lathauwer (2006), who based his fundamentally new algorithm on simultaneous matrix diagonalization. Comparative evaluations other than in De Lathauwer's paper are not available yet.

5.6.4   Uniqueness of the Parafac model

Harshman (1970),Carroll and Chang (1970), and Kruskal (1989a) amongst others have shown that the Parafac algorithm, under fairly general conditions, has a unique solution (see also Section 4.6.1, p. 61). This means that the estimation procedures provide values for the parameters that cannot be changed without affecting the fit; in technical terms: the Parafac model is identified. It is especially its three-way characteristic that puts the restrictions on the parameters so that there is only one solution given the number of components. The uniqueness property makes the model extremely useful in many applications; see also Leurgans and Ross (1992) and Sidiropoulos and Bro (2000a) for detailed mathematical treatments. Whereas two-mode PCA and exploratory factor analysis suffer from rotational indeterminacy, that is, the solution may be rotated or nonsingularly transformed without affecting the quality of the model fit, such indeterminacy disappears in the three-way case due to its uniqueness. Uniqueness also comes at a price. Due to the multiway restrictions, some data may not fit the model, which can lead to an inadequate or degenerate solution; see Zijlstra and Kiers (2002), who discuss the relationship between uniqueness and degeneracy.

Krijnen (1993, Chapter 2) provided a further discussion of the concept of uniqueness. The major thrust of his argument is that one should distinguish weak and strong uniqueness. The solution is *weakly unique* if the obtained solution is surrounded by a number of nonunique solutions that fit almost as well as the unique one. This weakens Harshman's claim that the unique solution should be given preference over all other solutions. In particular, Krijnen argues that in such a case it might be advantageous to take a nearly as good solution if it is easier to interpret, for example, because via rotations the components can have a simple structure. A solution is *strongly unique* if there are no nonunique models that have almost as good a fit.

Harshman (1972a) shows that a necessary condition for the uniqueness of a solution is that no component matrix has proportional columns (see also Krijnen, 1993, pp. 28–29). Krijnen suggested to check for weak uniqueness by comparing the fit of a regular solution of the Parafac model with a *parallel solution* that the solution of a Parafac model with two proportional columns in any of the component matrices. If the difference is small, the solution is considered weakly unique (for an example, see Section 13.5.3, p. 323). A similar approach may be taken by comparing the fit of the regular unique Parafac solution with that of any other Parafac model, say, with orthogonality constraints on one of the component matrices (for an example, see also Section 13.5.3, p. 323). A recent contribution to the analysis of uniqueness was provided by De Lathauwer (2006), who derived a new and relatively weak deterministic sufficient condition for uniqueness. His proof is constructive and he derived an easy-to-check dimensionality condition that guarantees generic uniqueness. His paper came too late to our attention to be able to use his results in the practical applications presented in this book.

5.6.5 Degeneracy

As indicated previously, the Parafac model is a restrictive one, so that the data may not conform to the model. A mismatch between data and model will result in problematic behavior of any algorithm for estimating the parameters. Because the cause lies with the mismatch, in principle any algorithm will have difficulty coming up with a valid solution. The most common indication that something is wrong is the *degeneracy* of the solution; for the first detailed discussion of degeneracy see Harshman and Lundy (1984c). An important contribution is Kruskal et al. (1989b). Mitchell and Burdick (1994) discuss slowly converging solutions that get bogged down in what has become known as a "swamp" but are not necessarily degenerate. Paatero (2000) provides further insight into the nature of degeneracy, by explicitly creating data that lead to degenerate solutions (see also Stegeman, 2006, for a recent contribution). In addition, Zijlstra and Kiers (2002) postulate that only models with unique solutions can have degenerate solutions. A far reaching conference paper taking stock of the issue is Harshman (2004)[2] and an annotated bibliography is also available from him[3]. In Harshman (2005a, p. 48)[4] three types of degeneracies are mentioned. The first

---

[2] *http://csmr.ca.sandia.gov/~tgkolda/tdw2004/Harshman - Talk.pdf.* Accessed May 2007.
[3] *http://csmr.ca.sandia.gov/~tgkolda/tdw2004/Annotated bibliography on degenerate solutions.pdf.* Accessed May 2007.
[4] *http://publish.uwo.ca/~harshman/pftut05.pdf.* Accessed May 2007.

is *temporary degeneracy*, which is not a real degeneracy but it only looks like that because the algorithm is stuck in a swamp. This is comparable with the plateau the blindfolded mountain climber came across (Section 5.4.1). The second type, *bounded degeneracy*, is bounded but permanent; for examples see Sections 8.9.5 (p. 204) and 13.3.3 (p. 314). In other words, the solution is degenerate but stable in the sense that from different starting positions the same degenerate solution is reached. The third kind of degeneracy, *divergent degeneracy*, is the real and most troublesome one. The solution diverges because no optimal solution exists, and the fit of every solution can be improved by increasing the number of iterations that in fact make it more degenerate. The cause of the degeneracy is that the data do not conform to the model. In particular, the data have *Tucker variation*, that is, a component, say, $\mathbf{a}_s$, that should be exclusively linked with $\mathbf{b}_s$ also has variation in common with $\mathbf{b}_{s'}$. The Parafac model does not cater for such relationships but a Tucker model does. If the Tucker variation is too strong, it influences the algorithms, so that the third kind of degeneracy may occur (see also Kruskal et al., 1989b). It is not a "fault" of the algorithm but of the incompatibility of model and data; degeneracy will occur for all algorithms. In principle, degenerate solutions should not be interpreted, but action should be taken to resolve the discrepancy either by putting restrictions on the Parafac model via orthogonality of one of the modes, via nonnegativity of one or more modes, or by fitting a Tucker model.

Krijnen and Kroonenberg (2000) discuss a number of measures to assess whether an algorithm is tending toward a degenerate solution(see also Krijnen, 1993, p. 13ff.). Their approach was inspired by and an improvement on Harshman and Lundy (1984d, p. 271ff.). In order to assess degeneracy, we need the cosines between two components in a single mode. As an example, the cosine between the s and s' component of the first mode is $\cos(\alpha_{s,s'}) = (\mathbf{a}'_s \mathbf{a}_{s'})$. If we define f_s as the $I \times J \times K$ (column) vector of $\mathbf{a}_s \otimes \mathbf{b}_s \otimes \mathbf{c}_s$, then the cosine $\theta_{s,s'}$ between f_s and $f_{s'}$ is equal to

$$\cos(\theta_{s,s'}) = \cos(\alpha_{s,s'}) \times \cos(\beta_{s,s'}) \times \cos(\gamma_{s,s'}). \tag{5.3}$$

If the triple cosine product $\cos(\theta_{s,s'})$ is approaching -1, there is almost certainly a degeneracy. It signifies that the two components f_s and $f_{s'}$ have become proportional (see Fig. 5.1) and this is explicitly "forbidden" in the Parafac model. This conclusion can be further supported by creating an $S \times S$ matrix Θ of the $\cos(\theta_{s,s'})$ and inspecting its smallest eigenvalue. If it gets, say, below 0.50, a degeneracy might be present. In addition, one should assess the condition number (i.e., the largest eigenvalue divided by the smallest one) of the triple-cosine matrix Θ. If the condition number is getting large, say, larger than 5, a degeneracy is likely. Both the smallest eigenvalue and the condition number are indicators of whether the triple-cosine matrix is of full rank as it should be for a Parafac solution. Table 5.2 provides an example of the use of these indicators, and the results strongly suggest that the solution is divergent degenerate.

The best way to confirm such degeneracy is to run the same analysis again but with an increased number of iterations and possibly with a more stringent criterion.

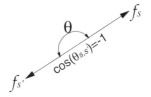

Figure 5.1 Two-component degeneracy of components f_s and $f_{s'}$.

If the indicators get worse, a divergent degeneracy is getting more and more likely. In such cases it is absolutely necessary to use several starting positions for the algorithm, because ending up with a degeneracy does not necessarily mean that there is no properly convergent solution; see especially Paatero (2000) for an enlightening discussion of this point.

The situation is more uncertain with a bounded degeneracy, because the indicators point toward a degeneracy but all different starting values end up at the same point. Increasing the number of iterations and decreasing the convergence criteria do not change the solution. Possibly the best way to proceed is putting restrictions on the columns of one or more of the component matrices and comparing the resulting solutions (for an example, see Section 13.3.3, p. 314).

With temporary degeneracies the situation is even more unclear, because it is more difficult to discern what is happening. One strategy might be to increase the number of iterations, decrease the convergence criterion, increase the number of different starts, and hope for the best. It might be wise to fit a Tucker model as well in order to see whether Tucker variation is likely.

Table 5.2 Example of a degenerate solution

| Components | Component cosines | | | Triple cosine product | Degeneracy check |
(,)	Mode A	Mode B	Mode C		
(2, 1)	−0.9810	0.9817	0.9999	−0.9631	——
(3, 1)	0.0486	−0.1377	0.2433	−0.0016	+
(3, 2)	−0.0339	−0.1256	0.2396	0.0010	+

+ = no degeneracy; -,–,—,—— = possible to serious degeneracy.

The other degeneracy indicators for this example are as follows:

1. Standardized lengths of the components: 1.65, 1.62, 0.64 (the first two > 1.00).

2. Triple-cosine products; smallest value is −0.96 (approaching −1).

3. Smallest eigenvalue of triple-cosine matrix: 0.52 (not yet < 0.50, but close to it).

4. Condition number of triple-cosine matrix: 26.6 (> 5).

In summary, if degeneracy is suspected, one should make sure by rerunning the analysis at least once with many more iterations, and check whether the indicators are getting more extreme. In addition, several starting solutions should be tried; the less clear the results are, the more starting solutions should be used. To eliminate the degeneracy, one could rerun the analysis with fewer components, reexamine the preprocessing procedure employed (see Chapter 6), or put constraints on the components to prevent degeneracy occurring. In particular, one of the modes should be kept orthogonal or nonnegative (see Harshman & DeSarbo, 1984b, for a detailed discussion of using such constraints in the case of degeneracy). If all else fails, it is always possible to choose a Tucker model.

5.6.6   Example: Parafac analyses with multiple minima

The first two parts of Table 5.3 illustrate the effect of using overrelaxation of acceleration for the Parafac algorithm. Four of the five analyses have the same value of the loss function, but in both algorithms one of the analyses has ended up with a lower local minimum (lack of fit). The accelerated version of the algorithm reaches the same convergence values in a one-third of the time, while it does not seem to add much to the time per iteration. Unfortunately, we have no guarantee that the lowest local minimum is also the global minimum. If one wants more security, the obvious thing to do is to run more standard and/or accelerated analyses. The results of 15 runs are shown in the last part of the table. There is now more evidence that there could be a global minimum since we have reached the lowest value twice, but the evidence is not yet very strong. Moreover, the second block of results shows that one may easily run into trouble due to nonconvergence or slow convergence. Which of the two is difficult to tell without increasing the number of iterations above 1000. The third block convincingly shows that there also exists a local minimum. In practice, one should examine all the various solutions to determine what exactly is happening here. The results were not affected by the acceleration because starting with the same values, the standard algorithm without acceleration produced the same results.

Figure 5.2 shows the effect of the acceleration on the successive differences between the values of the loss function, graphed for each substep separately. For example, for Mode A each point on the curve represents the difference between the loss function at step $\alpha - 1$ and its value α after Mode A has been updated, or

$$\mathcal{F}^{\alpha-1}_{\text{A-step}} - \mathcal{F}^{\alpha}_{\text{A-step}};$$

for details on the algorithm itself see Table 5.1. In the beginning, the convergence is all but monotone, but after a while this effect is no longer noticeable. What happens is that the acceleration for mode A throws the convergence for mode B off course, and the same happens for mode C after a B-substep. Thus, the large steps make the algorithm deviate a bit from its correct path, but the gains outweigh the losses. Note that in the beginning the loss function may have negative gains in one mode due to the previous substep. As mentioned before, (seeming) convergence of the loss function

Table 5.3   Two-component solution for a Parafac model. Sorted by lack of fit and execution times

Analysis number	Number of iterations	Proportional lack of fit	Execution time (s)
Basic analyses: 5 runs			
3	179	0.1192064075	4.34
2	172	0.1262702405	4.18
1	183	0.1262702405	4.51
5	186	0.1262702405	4.56
4	187	0.1262702405	4.62
Accelerated analyses: 5 runs			
3	51	0.1192064075	1.32
1	53	0.1262702405	1.32
5	54	0.1262702405	1.37
2	54	0.1262702405	1.43
4	54	0.1262702405	1.43
Accelerated analyses: 15 runs			
3	51	0.1192064075	1.26
12	53	0.1192064075	1.26
15	1000	0.1204105565	34.82
9	1000	0.1204219810	27.80
6	1000	0.1204221202	39.66
11	1000	0.1204277559	25.00
8	1000	0.1204301318	42.84
1	53	0.1262702405	1.32
7	53	0.1262702405	2.86
14	53	0.1262702405	1.76
2	54	0.1262702405	1.37
4	54	0.1262702405	1.54
5	54	0.1262702405	1.37
10	54	0.1262702405	1.38
13	54	0.1262702405	1.76

is not synonymous with having obtained a correct solution, because for degenerate solutions, too, the loss function may converge.

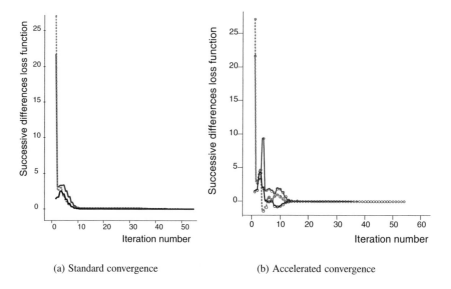

(a) Standard convergence   (b) Accelerated convergence

Figure 5.2   Convergence of the Parafac algorithm. Mode A = middle; Mode B = top; Mode C = bottom. Vertical axis: Successive differences of the loss function. Curves connect the successive differences of the loss function between the substeps of each mode. Horizontal axis: number of iteration step.

5.6.7   Constraints

Another aspect of the Parafac model is the possibility of imposing constraints on the parameters. Much work in this area has been done by Carroll and co-authors (Carroll et al., 1980, 1989; Carroll & Pruzansky, 1984b). An important more recent source for information is Bro (1998c, Chapter 6). In his book he describes a wide variety of constraints that may be imposed, especially on Parafac models. These include fixed parameters, nonnegativity, inequality, linear constraints, symmetry, monotonicity, unimodality, smoothness, orthogonality, and functional constraints, but only a few of these are discussed here (see also Section 4.10, p. 68).

Linear constraints.   Carroll et al. (1980) and Carroll and Pruzansky (1984b) deal with finding solutions such that specific parameters are linearly related to prespecified variables or to a design matrix (see also Section 4.10.3, p. 70). An efficient solution is found by using the basic Parafac algorithm on the data matrix that is reduced by the design matrix. This results in the same solution as that obtained from applying the design matrices to the components. The procedure was christened CANDELINC (canonical decomposition with linear constraints). In Fig. 5.3 we see the principle illustrated for a two-mode matrix. One may either analyze $\mathbf{X}$, in which case a PCA yields the component matrix $\mathbf{A}$, and then apply a two-group design matrix on $\mathbf{A}$, so as

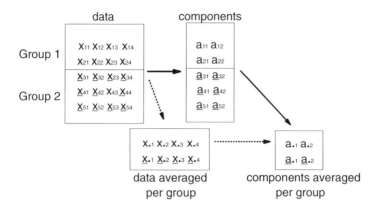

Figure 5.3 CANDELINC example for two-mode data.

to average the components over the two groups. Alternatively, one applies the design matrix directly in order to average the data matrix over the two groups (the reduced problem) in which case a PCA yields a component matrix that is the same averaged component matrix. A theorem due to Kruskal presented in Carroll et al. (1980) shows that a similar procedure could be incorporated in the alternating least-squares solutions of the Tucker3 and Tucker2 models as well.

Orthogonality constraints. In the general Parafac model, no restrictions are placed on the components, in contrast with the Tucker models where, for identification purposes, temporary orthogonality constraints are necessary to construct proper algorithms. However, it is possible to modify the basic Parafac algorithm to include such orthogonality restrictions (see also Section 4.10.1, p. 69). These restrictions lead to a loss of fit, but they allow for the partitioning of the fit per component and prevent certain kinds of degenerate solutions (for details see Section 5.6.5 and Harshman & Lundy, 1984c; Harshman, Lundy, & Kruskal, 1985; Lundy et al., 1989).

Nonnegativity constraints. A further contribution with respect to constraints on a Parafac configuration affects the core of the method, that is, its uniqueness property. Krijnen and Ten Berge (1991) raised the problem of using Parafac in the case of exclusively positively correlated variables, such as occur between subtests of an intelligence test. In the two-way case, a standard satisfactory solution is an oblique simple-structure orientation of the axes. A basic (principal component) solution generally provides one component with all positive loadings and a second *contrast component* with one group of negative and one group of positive loadings. Krijnen and Ten Berge (1991) indicated the undesirability of such a solution and showed that Parafac, too, can produce such contrast components. They argued that even though unique orientations of axes might be attractive in principle, for the positively correlated variables the resulting contrast components are not what is desired. To remedy the situation, they developed variants of the basic Parafac algorithm by putting nonnegativity constraints on the solution employing special least-squares regression algorithms from Hanson and Lawson (1974) (see also Section 4.10.2, p. 70). Durell, Lee, Ross, and Gross (1990) refer to programs for three-way and four-way Parafac models (C.-H. Lee, 1988), which also include nonnegativity constraints, and also Carroll et al. (1989) discuss the inclusion of nonnegativity constraints. Paatero (1999) developed a sophisticated program, the *multilinear engine*, based on a conjugate gradient algorithm, which includes nonnegative constraints.

Important applications of nonnegativity constraints can be found in analytical chemistry where spectra, concentrations, elution, and retention times are necessarily nonnegative (see Bro, 1998c; Smilde et al., 2004). Another important area in which extensive research has been carried out with multilinear models with nonnegative constraints is receptor modeling. These studies, mostly using the multilinear engine program by Paatero (1999), are concerned with identifying sources of airborne pollutants and providing quantitative estimates of the contributions of each of such sources of airborne particulate matter (Paatero & Hopke, 2002; Hopke, 2003); for an example on the identification of patterns in four-way aerosol data see Section 19.6. This source-identification work bears resemblance to signal processing where also nonnegative signals from different sources have to be disentangled (e.g., see Sidiropoulos et al., 2000b).

5.6.8   Optimal scaling

Sands and Young (1980) developed an *optimal scaling* algorithm for the Parafac model called ALSCOMP3 (see also Section 4.10.10, p. 74). Optimal scaling is an approach in data analysis in which variables with different measurement levels are transformed into numerical values in such a way that their measurement level is respected (e.g., ordinal values are transformed monotonically) and they optimize a specific criterion. Optimal scaling is combined with a particular (linear) model, such as a regression model, and simultaneously the regression parameters as estimated as well as the optimal transformations for the variables, such that the explained variability is as high as possible. The inclusion of an optimal scaling phase in the estimation of the Parafac model thus allows for a Parafac analysis of data with different measurement levels (e.g., see Gifi, 1990, for details on optimal scaling). Harshman and Lundy (1984c, p. 188ff.) give their opinion on the relative merits of the Sands–Young ALSCOMP3 procedure and Parafac. Unfortunately, at the time of writing, no other applications have appeared other than that in the original paper by Sands and Young (1980).

5.7 THE TUCKER MODELS

In this section we will discuss algorithms for two Tucker models: the Tucker3 model in which components are defined for all three modes, and the Tucker2 model with components for two of the three modes. With the latter model we will assume that the third mode is always the uncondensed one, but the other two models can be obtained by permuting the indices (see Fig. 4.12, p. 68). To start with, we will provide the model formulation to facilitate reading the algorithms later on.

5.7.1 Model descriptions

The Tucker models. The Tucker3 model is the decomposition of the three-way data array $\mathcal{X} = (x_{ijk})$, where we will assume that the first mode consists of subjects, the second of variables, and the third of conditions. This decomposition has the form

$$x_{ijk} = \sum_{p=1}^{P} \sum_{q=1}^{Q} \sum_{r=1}^{R} a_{ip} b_{jq} c_{kr} g_{pqr} + e_{ijk}, \tag{5.4}$$

where the coefficients a_{ip}, b_{jq}, and c_{kr} are the elements of the component matrices $\mathbf{A}$, $\mathbf{B}$, and $\mathbf{C}$, respectively, the g_{pqr} are the elements of the three-way core array $\mathcal{G}$, and the e_{ijk} are the errors of approximation collected in the three-way array $\mathcal{E}$. $\mathbf{A}$ is the $(I \times P)$ matrix with the coefficients of the subjects on the subject components. $\mathbf{B}$ is the $(J \times Q)$ matrix with coefficients of the variables, and $\mathbf{C}$ is the $(K \times R)$ coefficient matrix of the conditions.

The Tucker2 model only contains components for two of the three modes and can be written in sum notation as follows:

$$x_{ijk} = \sum_{p=1}^{P} \sum_{q=1}^{Q} a_{ip} b_{jq} h_{pqk} + e_{ijk}, \tag{5.5}$$

where the same definitions for the first and second mode components are as above, and $\mathcal{H} = (h_{pqk})$ is the extended core array.

5.7.2 Algebraic solution

Tucker (1966) was the first to propose what he called "three-mode factor analysis", but what is now called three-mode principal component analysis. He proposed three methods for solving the estimation of the parameters of his model. The first of which, Method I, is discussed here. The method he proposed is an algebraic solution in the sense that if an exact solution exists, this method will produce it. An approximate solution is obtained when not all components are required that are necessary to produce an exact solution.

His method to estimate the parameters starts by first matricizing (stringing out) the three-way data array in three different ways, of which two are shown in Fig. 2.1 (p. 16). On each of the strung-out matrices a two-mode principal component analysis is performed. The *tall combination-mode matrix* is used to estimate the components of the first mode, **A**, the *wide combination-mode matrix* supplies the components of the second mode, **B**, and the third arrangement provides the components of the third mode **C**. The core array $\mathcal{G}$ is then derived from these three component matrices as follows:

$$g_{pqr} = \sum_{i=1}^{I} \sum_{j=1}^{J} \sum_{k=1}^{K} a_{ip} b_{jq} c_{kr} x_{ijk}. \qquad (5.6)$$

Tucker's algorithm works reasonably well in many cases, but it is not a least-squares one. At present, it is primarily used as a starting solution for the alternating least-squares procedure discussed next. Another use is to determine the maximum size of the core array. Decomposing each of the matricized versions of the three-way array by regular PCA gives an indication of the maximum number of components for each of the modes and thus of the maximum size of the core array. Recently Kiers and Der Kinderen (2003), proposed to use the Tucker solution for dimensionality selection (see Section 8.5.1, p. 179).

5.7.3 Alternating least algorithms

Basic TUCKALS algorithm.   In this section we will discuss the basic algorithms for the Tucker models. We will primarily present the algorithms for the Tucker3 model. Algorithms for the Tucker2 model are essentially similar, in particular, for the basic solution. Any algorithm developed for the Tucker3 model can be used for the Tucker2 model by using the appropriate identity matrix for one of its component matrices, and holding this fixed during iteration. It is only in the output of programs implementing the algorithms that special provisions have to be made for the Tucker2 model.

Kroonenberg and De Leeuw (1980) (see also Kroonenberg, 1983c; Ten Berge et al., 1987) described an alternating least-squares loss function as well as an algorithm to estimate the parameters that minimize this function — the TUCKALS algorithm. Each substep of the algorithm consists of a standard eigenvalue–eigenvector decomposition, which can be solved with well-known techniques, and which is considered here as part of the black box. After convergence of the loss function and the norms of the component matrices, the core array is computed as in Eq. (5.6).

If we compute all the components, thus $P = I$, $Q = J$, and $R = K$, (or less if the product of two of them is larger than the remaining one; see Section 8.4, p. 177 for the maximum-product rule), one could decompose a data matrix exactly into its components. However, in practical applications one is just interested in the two, three, or four components that explain most of the variability. This generally precludes finding an exact decomposition of $\mathcal{X}$ into **A**, **B**, **C**, and $\mathcal{G}$. One therefore has to settle for an approximation, that is, one has to find $\mathbf{A}$, $\mathbf{B}$, $\mathbf{C}$, and $\mathcal{G}$ such that the difference between the model and the data is minimal according to the least-squares loss function. This means we are looking for a best approximate decomposition of the array $\mathcal{X}$ into $\mathbf{A}$, $\mathbf{B}$, $\mathbf{C}$, and $\mathcal{G}$ according to the Tucker3 model.

The Tucker models are overidentified because each of the component matrices is only determined up to a nonsingular transformation. In order to find a solution, one has to place restrictions on the component matrices during the iterative process. It is convenient to carry out the minimization under the restriction that $\mathbf{A}$, $\mathbf{B}$, and $\mathbf{C}$ are column-wise orthonormal, because this makes for an efficient and elegant algorithm. After estimates have been found, nonsingular transformations can be applied to $\mathbf{A}$, $\mathbf{B}$, and $\mathbf{C}$ without loss of generality, provided the core array $\mathcal{G}$ is multiplied with the corresponding inverse transformation matrices.

The basic algorithm presented here uses the Kroonenberg and De Leeuw (1980) approach to solving the estimation of the parameters. The first step is to express the core array in terms of the component matrices as in Eq. (5.6), and substituting this expression into the model equation (i.e., Eq. (5.4)). Then the loss function only depends on $\mathbf{A}$, $\mathbf{B}$, and $\mathbf{C}$, so that it is sufficient to first estimate $\mathbf{A}$, $\mathbf{B}$, and $\mathbf{C}$ and solve for $\mathcal{G}$ afterwards. The estimation proceeds as outlined in Table 5.4.

Kroonenberg and De Leeuw (1980) discuss the convergence properties of the basic algorithm. As in virtually all problems of this kind, only convergence to a local optimum is assured. When we skip the C-step in the TUCKALS algorithm and replace $\mathbf{C}$ by the $K \times K$ identity matrix, the algorithm can be used for the Tucker2 model as well. The Tucker2 model is not symmetric in its three modes but the Tucker3 model is, and therefore one has to make a decision which mode will remain uncondensed. To initialize the algorithm, $\mathbf{A}_0$, $\mathbf{B}_0$, and $\mathbf{C}_0$ are chosen in such a way that they will fit the model exactly if such an exact solution is available. This initial solution is Tucker's (1966) Method I solution. In the late 1980s, Kroonenberg introduced a missing data step into the algorithm of Table 5.4 using the expectation–maximization (EM) approach (Dempster et al., 1977), analogously to the procedure included in the Parafac algorithm (Harshman & Lundy, 1984c).

Algorithm for multimode covariance matrices. A major variant of the standard TUCKALS algorithm was proposed by Kiers, Kroonenberg, and Ten Berge (1992b). The principal innovation is that the multiplications involved in the step with the largest of I, J, and K are circumvented by cleverly rearranging the computations, so that during iterations manipulation with the original data matrix $\mathcal{X}$ of subjects $\times$ variables $\times$ conditions is not necessary. In particular, Kiers et al. (1992b) showed that one may convert the $I \times J \times K$ raw data into a variables by conditions multimode covariance matrix of order $JK \times JK$. If the subject mode is the largest mode, and much larger than the other two modes, considerable savings can be obtained because this large mode does not figure in the iterations. After convergence, the component matrix $\mathbf{A}$ can be recovered. An interesting feature of this estimation method is that if only a multimode covariance matrix is available, a three-mode component analysis can still

Table 5.4 TUCKALS algorithm

Set iteration counter: $\alpha = 0$.

Starting solution
Initialize $\mathbf{A}_0$, $\mathbf{B}_0$, $\mathbf{C}_0$ by using the Tucker (1966) Method I.

Update iteration counter: $\alpha = \alpha + 1$.

A-substep
Fix $\mathbf{B}_{\alpha-1}$ and $\mathbf{C}_{\alpha-1}$, and solve an eigenvalue–eigenvector problem for $\mathbf{A}$ to obtain a new $\mathbf{A}_\alpha$.

B-substep
Fix $\mathbf{A}_\alpha$ and $\mathbf{C}_{\alpha-1}$, and solve an eigenvalue–eigenvector problem for $\mathbf{B}$ to obtain a new $\mathbf{B}_\alpha$.

C-substep
Fix $\mathbf{A}_\alpha$ and $\mathbf{B}_\alpha$, and solve an eigenvalue–eigenvector problem for $\mathbf{C}$ to obtain a new $\mathbf{C}_\alpha$.
(This step is skipped for the Tucker2 model and $\mathbf{C}$ is replaced by $\mathbf{I}$, the $K \times K$
identity matrix.)

Estimate missing data (Optional)
Use current estimates $\mathbf{A}_\alpha$, $\mathbf{B}_\alpha$, and $\mathbf{C}_\alpha$ and the data $\mathcal{X}$ to estimate the missing values
according to the model equation (5.4).

Check convergence
If the differences between successive iterations with respect to the loss function
  $\|\mathcal{F}^{\alpha-1} - \mathcal{F}^\alpha\|^2$ are not small enough,
or the Euclidean norms of successive estimates of $\mathbf{A}$, $\mathbf{B}$, and $\mathbf{C}$ are not small enough,
then continue with a next iteration,
else end iterations.

Compute core array
Use Eq. (5.6) to calculate the core array $\mathcal{G}$ from the data and the estimated $\mathbf{A}$, $\mathbf{B}$, and $\mathbf{C}$.

---

be carried out, as is the case for two-mode data. Moreover, comparisons with structural equation models can be made for the same data (see Kroonenberg & Oort, 2003a). When there are missing data, one has to deal with them during the computation of the covariances, because the algorithm does not operate on the original data.

Murakami (1983) developed a variant of the TUCKALS algorithm for the Tucker2 model using multimode covariance matrices to which he referred as *quasi three-mode factor analysis*. Apart from the different Tucker model fitted, the major difference between this algorithm and that by Kiers et al. (1992b) is that Murakami used eigendecompositions, while Kiers et al. used Gram–Schmidt orthonormalizations. Apart from Murakami, it seems that few authors have explicitly investigated alternative algorithms for estimating the parameters in the Tucker2 model (Murakami, 1979, 1981).

Regression-based algorithms.   In 1983 an alternative for the TUCKALS3 algorithm, called GEPCAM (generalized PCA with missing data), was proposed by Weesie and Van Houwelingen (1983); see also De Ligny, Spanjer, Van Houwelingen, and Weesie (1984) and Spanjer (1984). They showed that it is possible to estimate the parameters using regression techniques rather than eigenvalue–eigenvector methods. This could be realized by making iteration steps not only for the component matrices $\mathbf{A}$, $\mathbf{B}$, and $\mathbf{C}$, but also for each slice $\mathbf{G}_r$ ($r = 1, \ldots, R$) of the core array $\mathcal{G}$ (see Table 5.5). At the same time, they showed how missing data procedures could be incorporated in the algorithm. An as yet unexplored extension of this algorithm was also indicated by Weesie and Van Houwelingen. Rather than using basic least-squares regression, they mentioned the possibility of using robust regression, such as ridge regression, to prevent the undue influence of outlying observations (see Sections 12.8–12.10 for robustness in multiway analysis).

On the basis of some experiments, Weesie and Van Houwelingen (1983) concluded that this scheme works best by recalculating the core array each time one of the component matrices was reestimated. Missing data are handled by the regression procedures rather than via a separate step. One solution for that is introducing a three-way weight array with the same dimensions as the original data array containing zeroes at the locations of the missing data and ones elsewhere (see Chapter 7 for a detailed discussion of missing data procedures in multiway analysis).

5.7.4   Constraints and rotational freedom

To solve the estimation of the Tucker models, Kroonenberg and De Leeuw (1980) put orthogonality constraints on the three component matrices in the model. However, this is only a temporary constraint for the purpose of estimation, because after a solution is obtained the component matrices may be rotated both orthogonally or obliquely without affecting the fit of the model. More specific constraints, such as smoothness of components, and latent curve fitting for the components were introduced by Timmerman (2001, Chapters 3, 4, and 5) within the context of fitting longitudinal data (see also Timmerman & Kiers, 2002). Extensive discussions of placing constraints on the components and/or core array are contained in Section 4.10 and Chapter 10.

5.7.5   Compression: Combining the Tucker and Parafac algorithms

When confronted with very large data sets, it can be very time consuming to fit the Parafac model, especially because it is necessary to make several runs with different starting values to ensure convergence to a global minimum. Probably Appellof and Davidson (1981, p. 2054) were the first to realize that a significant gain in computational efficiency could be made by first applying component analyses to the three

Table 5.5   GEPCAM algorithm

Set iteration counter: $\alpha = 0$.

Starting solution
Initialize $\mathbf{A}_0$, $\mathbf{B}_0$, $\mathbf{C}_0$ by using the Tucker (1966) Method I, and compute the core array according to Eq. (5.6)

Update iteration counter: $\alpha = \alpha + 1$.

A-substep
Fix $\mathbf{B}_{\alpha-1}$ and $\mathbf{C}_{\alpha-1}$, and solve a problem for $\mathbf{A}$ to obtain a new $\mathbf{A}_\alpha$.

G-substep
Fix all component matrices $\mathbf{A}_\alpha$, $\mathbf{B}_\alpha$, and $\mathbf{C}_\alpha$, then the core slices remain as unknowns. Solve the regression equation for each core slice $G_r, r = 1, \ldots, R$, in turn to get a new estimate for the entire core array.

B-substep
Fix $\mathbf{A}_\alpha$ and $\mathbf{C}_{\alpha-1}$, and solve a regression problem for $\mathbf{B}$ to obtain a new $\mathbf{B}_\alpha$.

G-substep
Fix all component matrices $\mathbf{A}_\alpha$, $\mathbf{B}_\alpha$, and $\mathbf{C}_\alpha$; leaving the core slices as unknowns. Solve the regression equation for each core slice $G_r, r = 1, \ldots, R$, in turn to get a new estimate for the entire core array.

C-substep
Fix $\mathbf{A}_\alpha$ and $\mathbf{B}_\alpha$, and solve a regression problem for $\mathbf{C}$ to obtain a new $\mathbf{C}_\alpha$.
(This step is skipped for the Tucker2 model and $\mathbf{C}$ is replaced by $\mathbf{I}$, the $K \times K$ identity matrix.)

G-substep
Fix all component matrices $\mathbf{A}_\alpha$, $\mathbf{B}_\alpha$, and $\mathbf{C}_\alpha$, then the core slice remain as unknowns. Solve the regression equation for each core slice $G_r, r = 1, \ldots, R$, in turn to get a new estimate for the entire core array.

Check convergence
If the differences between successive iterations with respect to the loss function
$\quad ||\mathcal{F}^{\alpha-1} - \mathcal{F}^\alpha||^2$ are not small enough,
or the Euclidean norms of successive estimates of $\mathbf{A}$, $\mathbf{B}$, $\mathbf{C}$, and $\mathcal{G}$ are not small enough,
then continue with a next iteration,
else end iterations.

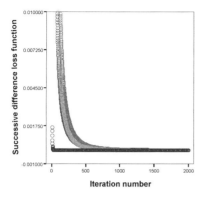

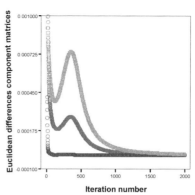

(a) Convergence loss function per substep. Mode A step = behind Mode B; Mode B step = middle curve; Mode C step = bottom curve; Missing step = top curve.

(b) Convergence Euclidean norms of component matrices. Mode A = middle; Mode B = bottom; Mode C = top.

Figure 5.4   Convergence of the TUCKALS algorithm.

modes "to reduce redundancy in the raw data matrix" (see also Section 8.4.1 on the maximum-product rule). In passing, they actually independently proposed Tucker's three-mode principal component analysis (Tucker, 1966) without knowing or realizing it. The proposal of first using the Tucker3 model before applying the Parafac model was later called *compression*. More sophisticated versions of this compression approach were proposed by Bro and Andersson (1998a) and Kiers (1998c). As reviewed by Tomasi and Bro (2006, p. 1711), compression based on the Tucker3 model "introduces additional complexity in the algorithm, this part eventually provides a large reduction in the number of operations. Furthermore, it needs only be done once, whereas several alternative PARAFAC models are usually fitted to find the most feasible one. Once the PARAFAC model has been calculated on the core extracted by the Tucker3 algorithm, the solution can be expanded to its original dimensions providing very good starting values for the PARAFAC-ALS standard algorithm, which is used only to refine it." Alsberg and Kvalheim (1993, 1994) proposed compressing Tucker3 algorithms using B-splines (for a brief discussion of B-splines, see Section 4.10.4, p. 71).

5.7.6   Example: TUCKALS algorithm with missing data

Figure 5.4 shows the convergence of the TUCKALS loss function and of the Euclidean norms of the three component matrices for a $95 \times 2 \times 12$ subjects by variables by concepts data set. The very first iterations are omitted because they show comparatively

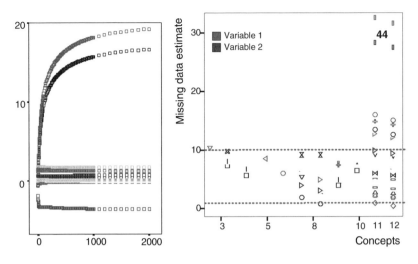

(a) Diverging estimates for 10 of the missing data. Horizontal axis: Iteration number; Vertical axis: Value of estimate.

(b) Estimates for missing data. Subject 44 has extreme estimates on concepts 11 and 12.

Figure 5.5   Problematic behavior of the TUCKALS algorithm with respect to the estimation of the missing data. The values in the left-hand panel are deviations from their means, while in right-hand panel the means have added.

very large differences. Even though the loss function decreases in a monotone fashion with decreasing differences between iterations, the norms decrease continuously but not in a monotone fashion. Around the 100th iteration the differences increase, indicating a real change in the components.

Figure 5.4 shows that something untoward had happened with respect to the estimation of the 81 missing data points (3.6%) in the data set. An indication for this was that after 2000 iterations the missing data step (see Table 5.4) still changed the loss function at each step, because the estimates still had not stabilized. As Fig. 5.5 shows, the estimates for some of the missing data were getting out of hand considering that the range of the variables was 1–10, and some estimates were approaching 30. The estimated missing data for subject 44 are a case in point.

The conclusion from this example is that careful inspection of all aspects of an obtained solution is necessary before one may conclude that it can be interpreted. This is especially true in the case of missing data. More detailed discussions of the effect of missing data on solutions, convergence, and so on can be found in Chapter 7.

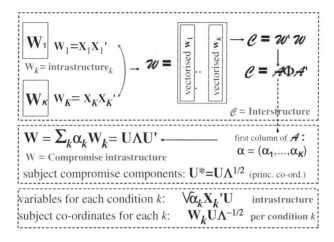

Figure 5.6   The STATIS method.

5.8   STATIS

This present section is somewhat different in that it presents a brief exposé of STATIS, a method to analyze three-way data, which does not have an underlying model. The limited presentation here cannot do full justice to the method and for more information in English, one should consult especially Lavit et al. (1994). English-language examples can, for instance, be found in Lavit and Pernin (1987); Lavit et al. (1994), and Meyners, Kunert, and Qannari (2000).

5.8.1   Method description

STATIS (Structuration des tableaux à trois indices de la statistique – structuring of three-way statistical tables) is a flexible method for handling three-way data (see L'Hermier des Plantes, 1976; Lavit, 1988; Lavit et al., 1994). The method is especially geared toward the situation where the same individuals have measures on a set of almost the same variables under a number of conditions, thus $\mathcal{X} = (\mathbf{X}_1, \ldots, \mathbf{X}_K)$. STATIS consists of a three-step method to analyze three-way data. We only give a simplified description here (see Fig. 5.6); full details are contained in Lavit (1988).

First, a global comparison between the conditions is made in order to assess the *interstructure* by constructing for each condition an $I \times I$ similarity matrix between subjects, $\mathbf{W}_k = \mathbf{X}_k \mathbf{X}'_k$. These similarity matrices for the conditions can then be compared by first vectorizing them into $I \times I$ by 1 vectors $\mathbf{w}_k$. These are then handled as one would normal variables: calculate their covariance matrix $\mathcal{C}$, perform an eigendecomposition on the covariance matrix $\mathcal{A}\Phi\mathcal{A}'$, and plot the principal coordinates $\mathcal{A}\Phi^{1/2}$ from which their similarities may be assessed.

The second step is to construct a *compromise solution* for the K data matrices $\mathbf{X}_k$. A straightforward option would be to average them, but this does not take into account that the data of some conditions resemble each other more than others, as was indicated by the interstructure. Therefore, the compromise solution $\mathcal{W}$ is determined by weighting each $\mathbf{X}_k$ by the value of its $\mathbf{W}_k$ on the first component of the interstructure solution; thus, $\mathbf{W} = \sum_{k=1}^{K} \alpha_k \mathbf{X}_k$. This compromise solution $\mathbf{W}$ is then subjected to an eigenanalysis to evaluate the similarity between the subjects or the *compromise structure*.

The third step is to evaluate to what extent the compromise structure is reflected in the individual data matrices $\mathbf{X}_k$. This can be done by projecting these data matrices into the space of the compromise solution, so that the *intrastructure* of the kth condition with respect to the common one can be assessed.

Because STATIS consists of several steps that cannot be fused into a single loss function, it is difficult to evaluate the overall solution. However, several fit measures, such as explained variances, can be derived both for the separate analyses and for the fit of the individual matrices to the common compromise structure.

Recent research into the method can be found in Meyners et al. (2000), Vivien and Sabatier (2004), and Vallejo Arboleda, Vicente Villardon, and Galindo Villardon (2007), while an application to the taxonomy of roses can be found in Raymond, Fiasson, and Jay (2000) and one on sensory perception in Meyners et al. (2000), to name but two examples. Routines in R have been developed within the ade4 package at the Laboratoire de Biométrie of the Université Lyon II by Chessel and Thioulouse[5]; see Thioulouse, Chessel, Dolédec, and Olivier (1996) for an English description.

5.9   CONCLUSIONS

In this chapter the basic characteristics of the algorithms for multiway models were outlined. Important aspects for discussion were the iterative nature of these algorithms, and the pitfalls one may come across during analysis. For algorithm development the information in this chapter is far from sufficient, but for most practitioners the information should be sufficient to evaluate whether any given solution is technically acceptable and can be interpreted.

[5] *http://pbil.univ-lyon1.fr/ade4html/statis.html*. Accessed May 2007.

DATA HANDLING, MODEL SELECTION, AND INTERPRETATION

In Part II practical issues in setting up, carrying out, and evaluating multiway analyses will be presented. In particular, we will pay attention to such issues as preprocessing (centering and normalization of the data), handling missing data, model and dimensionality selection, postprocessing (scaling components and the core array), transforming and rotating component matrices and core arrays, plotting the results of an analysis, internal validation of solutions via residuals, and robustness and stability issues.

CHAPTER 6

PREPROCESSING

6.1 INTRODUCTION

Suppose a researcher has finished collecting her subjects by variables by occasions data and is ready to analyze them in order to unravel the relationships among the three modes. She is interested in whether certain groups of subjects have specific patterns of scores on some but not all of the variables, given a particular set of occasions. For instance, her question might be: "Is there a certain type of subjects who tend to show particular kinds of coping strategies (say, avoiding tackling the problem, getting sad and reacting very emotionally) when confronted with a variety of circumstances (such as being bullied at work, being unable to do their assigned tasks) in contrast with another type of subjects who use similar strategies but in different circumstances?" Given the fact that the researcher is concerned with analyzing such three-mode relationships, the question arises which part of the data contains the information for answering such questions. Generally, the questions cannot be answered, unless the raw data are preprocessed in an appropriate way. With *preprocessing* we mean the application of techniques to a data set before a multiway model is fitted, especially

Applied Multiway Data Analysis. By Pieter M. Kroonenberg
Copyright © 2007 John Wiley & Sons, Inc.

centering, normalization, and standardization. This chapter is devoted to setting out what can and needs to be done to the data prior to a three-mode analysis proper, before the research questions can be answered. This chapter leans heavily on four sources: Kruskal (1981), Kroonenberg (1983c, Chapter 6), Harshman and Lundy (1984d, pp. 225–253), and Bro and Smilde (2003a).

6.1.1 Two-way data

Principal component analysis as a two-step procedure.   In the analysis of two-way data, say, subjects by variables measured on a single occasion, principal component analysis is routinely applied when looking for relationships between and among variables and subjects. Researchers generally do not (need to) give much thought as to which part of their data contains the required information for their research questions, as principal component analysis automatically takes care of this by first preprocessing the raw data. One may conceive of principal component analysis as a two-step procedure: first the raw data are standardized, and secondly the standardized scores are modeled by a set of products of subjects coefficients and variable coefficients. In addition, standardization is a two-step process in itself: it consists of both a centering step and a normalization step. Thus, combining these two, we get that principal component analysis takes the form

Data = Centering constants + normalization constants×multiplicative model + error.

In the most common case of profile data, that is subjects having scores on variables, this can be expressed as

$$x_{ij} = \bar{x}_j + s_j \times \sum_{s=1}^{S} a_{is}b_{js} + e_{ij}. \tag{6.1}$$

Thus, first the scores x_{ij} are centered across the subject mode for each variable j ($\tilde{x}_{ij} = x_{ij} - \bar{x}_{.j}$), so that they become deviation scores with mean 0, and subsequently the deviation scores are normalized such that their sums of squares become 1: $\hat{x}_{ij} = \tilde{x}_{ij}/s_j$, where $s_j = \sqrt{\frac{1}{I-1}\sum_i \sum_j \tilde{x}_{ij}^2}$ (see also Fig. 6.1).

Preprocessing and profile data.   The implication is that differences in means and scales in the variables are not of interest, obscure the view on the relationships among the variables, and obstruct in finding out how the subjects differ in their scores on these (groups of) variables. In other words, the required information can only correctly be modeled with the normalized deviation scores and not with raw scores. One reason why this is so is that the means contain no information on the individual differences, but the deviation scores do. In addition, variables are often measured on an interval scale so that the means contain no intrinsic information, and for a meaningful interpretation they should not be included in the analysis. Furthermore, when

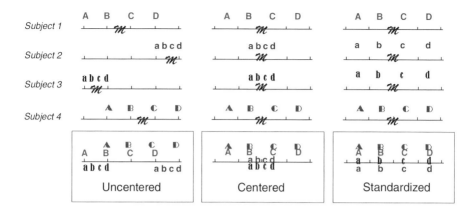

Figure 6.1   Effects centering and normalization.

the measurement units of the variables are different, or alternatively the differences in variabilities are idiosyncrasies of the variables rather than carriers of information about the differences between subjects, the scale of the variables should be eliminated before the multiplicative analysis.

Preprocessing and factorial data.   When the data have a factorial design with a single observation per cell, and the entries in the data matrix are the values of a single dependent variable, the standard procedure is to apply a two-way analysis of variance. Thus, the raw data are modeled in an additive fashion with main and interaction effects. Additive modeling is equivalent with centering the raw data. Multiplicative modeling is then used to analyze the interaction(s). An example of this approach occurs in plant breeding experiments, where the dependent variable is the yield from plots originating from different environments and planted with different genotypes. In such data, the main emphasis is on the genotype and environment means as well as the genotype by environment interaction, so that centering is part of the analysis. Normalization is not an issue as there is only one variable.

Preprocessing and interval scales.   A more technical consideration, especially with respect to centering, is that many multivariate analysis methods such as principal component analysis assume that the data have ratio rather than interval scales. Therefore, many social and behavioral science data cannot be properly analyzed in their raw form with such methods. Thus, activities like removing means and scale factors from the data before fitting the main analysis method are necessary to make the data amenable for analysis. Operations such as centering, which involve creating deviation scores, are often referred to as *additive adjustments*, and operations such as normalization, which involve multiplying each data point with a term not depending on all indices, is often called *multiplicative adjustment*.

6.1.2 Multiway data

Preprocessing is inherently more complex with multiway data, because there are many more means and scale factors to consider. For example, for three-way data there are three sets of one-way means, three sets of two-way means, and one overall mean, not all of them necessarily meaningful in any one data set. Moreover, one can define an overall normalization factor, three kinds of slice normalizations, and three kinds of fiber normalizations, again not all of them meaningful at any one time. There are two types of arguments in deciding for which modes means should be removed and which variables should be normalized: substantive or *content-based arguments* concentrating on which of the modes to consider, and *model-based arguments*, dealing with the technical appropriateness of preprocessing procedures. The ensuing discussion and examples will be primarily in terms of three-way data and three-mode models, but most arguments directly carry over to the multiway case.

6.2 CHAPTER PREVIEW

Based primarily on model-based arguments (see Section 6.4), the general consensus is that in most applications using three-mode designs, centering should take place per fiber across a mode (removing row, column, or tube means; see Fig. 6.2(a)), and normalization should take place per slice or within modes (equalizing the (mean) squares of lateral, horizontal, or frontal data slices; see Fig. 6.2(b)) . Harshman and Lundy (1984d) have developed extensive theoretical arguments why certain types of centering and normalizations are "appropriate" and some others are not (see also Bro & Smilde, 2003a). Which and how many of these are necessary in any particular case is very much dependent on content-based arguments, but model-based arguments can also play a role (Harshman & Lundy, 1984d). This chapter summarizes the arguments that lead to the above conclusions. This discussion will also lead us to briefly consider *triadditive* models, which are models with additive main effect terms and multiplicative interaction terms (Denis & Gower, 1994), which are called *extended three-mode models* by Harshman and Lundy (1984d). Such models are a combination of both analysis of variance and principal component analysis.

6.3 GENERAL CONSIDERATIONS

6.3.1 Types of preprocessing

We will primarily discuss two basic kinds of preprocessing: *centering* or *additive adjustments*, that is, "subtracting a constant term from every element so that resulting data values have a mean 0" (see Kruskal, 1981, p. 15), and *normalization* or *multiplicative adjustments*, that is, "dividing every element by a constant term, so as to achieve this result: the "scale" of the resulting data values has some fixed value

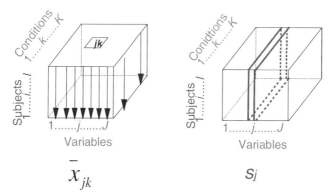

$$\overline{x}_{jk} \qquad\qquad S_j$$

J x K matrix of means removed J x 1 vector of scale factors removed

(a) Fiber centering   (b) Slice normalization

Figure 6.2   Common types of preprocessing for three-way data: Fiber centering and slice normalization.

(often chosen to be 1). "Scale" generally refers to some measure of variability, most often the standard deviation" (see Kruskal, 1981, p. 17). The process of centering and normalization, such that the resulting data values have a mean 0 and a standard deviation 1, will be called *standardization*. Not considered in this chapter are some other types of preprocessing, such as procedures to adjust for nonlinearity via data transformations, such as logarithms, procedures to create smoother components via spline transformations, and procedures to quantify categorical variables via optimal scaling.

Although we primarily look at centering and normalization, there are alternative operations with similar aims. One may, for instance, subtract the midpoints of scales, rather than the means. In addition, one may decide to equalize the range of the variables rather than their variance. An alternative to least-squares procedures such as removing means and normalization factors would be to use robust statistical equivalents (e.g., see Huber, 1996). However, in multiway analysis robustness research is relatively young; see Chapter 12 for the present state of affairs.

6.3.2 Selecting a type of preprocessing

Types of arguments.   There are two types of arguments why one should use a particular type of preprocessing for any one mode. The first *content-based* types of arguments are based on measurement characteristics of the variables, research questions, and research designs, which require or indicate a particular pretreatment of the data before a multiway analysis proper. The second *model-based* types of argu-

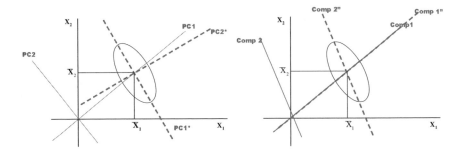

(a) Effect of centering on the first and second principal axes — PC1 and PC2 (solid lines): the new orientations (dashed lines) are entirely different from the original ones.

(b) Effect of centering on components with fixed directions — Comp1 and Comp2 (solid lines): new components (dashed lines) have the same orientation but are centered — Comp 1* and Comp2*.

Figure 6.3   Effects of removing the means on principal and on fixed components.

ments are based on technical considerations and characteristics, which also influence the kind of preprocessing that is appropriate. In particular, some models rule out preprocessing procedures considered to be inconsistent with the model's definition. Measurement characteristics of the data generally also play a fundamental role in some model-based arguments.

Reasons for preprocessing   First, it is necessary to take a closer look at the question of why one should need to consider preprocessing at all. There is a straightforward content-based answer to this: preprocessing is needed to provide a proper understanding of the relationships between elements of the three modes. Generally, the multiway component model does not immediately provide insight into the relationships when applied directly to the raw data, but it does for an appropriately transformed or scaled form. In particular, means may obscure the interactions between the elements of the modes. Thus, it is expected that the model will give an incorrect or imprecise description of the relationships in the data when applied to the raw data.

Reasons for centering.   The answer to the question of why one might get an improper description when certain means are not removed from the raw data follows from the definition of component analysis. In the space spanned by, say, the variables, the components derived by the technique represent directions along which successively and orthogonally the largest variations can be found. If the centroid, defined by the means of the variables, is located at a considerable distance from the origin of the variable space, the direction of the first component will almost certainly run from the origin through this centroid (see Fig. 6.3(a)). If, however, the main purpose of an analysis is to investigate the variations of the variables from the centroid, the means

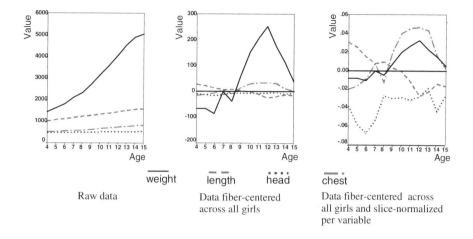

weight   length   head   chest

| Raw data | Data fiber-centered across all girls | Data fiber-centered across all girls and slice-normalized per variable |

Figure 6.4   Effect of standardization on the time series of a single girl.

of the variables should be removed before the component analysis, and they should be modeled separately.

Another interesting effect of removing means when the second mode is variables and the third mode consists of occasions or time points is that the base line of each of the variables is removed and the analysis is conducted on the deviations from this base line. Figure 6.4 displays the growth data (weight, height, head and chest circumference) of a single French girl between her fourth and fifteenth year taken from a study by Sempé (1987). Extensive analyses of the full data set can be found in the volume edited by Janssen, Marcotorchino, and Proth (1987, pp. 3–112), which also lists the data. In Chapters 11 and 15 these data are again used as an illustration. The left-hand part of Fig. 6.4 gives the raw data, and the middle panel gives the data for this girl after removing the means of all variables at all occasions taken over all girls. Whereas the mean levels of the variables, which are incommensurate, dominate the raw data and thus make analysis nearly impossible, the middle panel shows the growth of the girl with respect to the average growth curves of the group, or, said differently, with respect to the average girl's growth curves. It shows that after being a slight girl, around her ninth year she increased considerably in weight compared to the other girls, and that around her fifteenth year she was back to roughly the average weight of the group. In other words, she had a weight spurt much earlier than the other girls but they caught up with her later. Due to the centering the graph allows a serious comparison of the patterns in the variables.

Reasons for normalization.   The main (content-based) reason for normalization is that variables may have arbitrary or incomparable measurement scales, and that their scales therefore should not be allowed to influence the analysis.

Kruskal (1981, p. 18) used a model-based argument for normalization in connection with his discussion of the Parafac model. Paraphrasing his argument, we write the Parafac model as

$$x_{ijk} = \sum_{s=1}^{R} a_{is} b_{js} c_{ks} + \epsilon_{ijk}, \tag{6.2}$$

where the ϵ_{ijk} are the residuals. Because least-squares fitting is used to solve the estimation of the model, it is implicitly assumed that the standard deviations of the error terms are all equal or, as expressed by Bro and Smilde (2003a, p. 24), "[t]he ordinary least-squares fitting of a model is statistically optimal in a maximum likelihood sense if the errors are homoscedastic, independent and Gaussian". If the standard deviations were known, the x_{ijk} could be scaled to make the error terms as nearly equal as possible. As in practice they are not known, one has to fall back on seeking to make the total standard deviations of the elements of one of the modes equal (or use a similar type of normalization). A third reason is that all parts of the data should exercise an equal influence on the outcomes and this can be realized by normalization. Finally, it may be desirable to emphasize parts of the data at the cost of other parts. In that case, specific unequal weights are attached to different variables.

The combination of removing means and normalization of the variables can be useful to improve the interpretability of the parameters in a three-mode analysis and may facilitate comparisons within and across data sets. This aspect is taken up in Section 9.4 (p. 218).

As an example, let us look again at the data of the French girl. Due to the different variabilities of the variables as is evident in the middle panel of Fig. 6.4, the comparison between the variables is not yet optimal. In particular, it is difficult to say whether her growth pattern on all variables follows one pattern or several different ones. This can be resolved by a normalization of each variable as shown in the right-hand panel of the figure, where due to normalization the variabilities of the variables are now comparable, and patterns can be investigated in more detail. The collection of these curves for all girls and all variables was the subject of several three-mode analyses reported in Kroonenberg (1987a, 1987b) and Lavit and Pernin (1987); see also Chapters 11 and 15.

Content-based versus model-based.   Content-based arguments generally focus much more on the choice of modes that have to be preprocessed, and whether one or more particular modes should be centered or normalized. A thorough evaluation of the nature of the data and their variability will lead to a choice of what is to be done to which particular mode. However, sometimes content-based and model-based arguments suggest different types of preprocessing. In this book the point of view is taken that the valid content-based arguments should always prevail. Luckily, the substantive and model-based arguments generally do not contradict each other.

6.4 MODEL-BASED ARGUMENTS FOR PREPROCESSING CHOICES

In this section we will review in more detail the basic model-based arguments for choosing fiber centering and slice normalization as the preferred preprocessing options. The central argument for subtracting fiber means is that if a three-mode model holds for the raw data, subtracting fiber means will keep the model intact. Its only effect will be that the components of the mode across which has been centered will be centered as well. Similarly, using slice normalization will leave the shape of the model unimpaired. Note that these arguments are model-based, rather than content-based, and do not refer to what should be done with any mode in particular.

6.4.1 Offsets

Harshman and Lundy (1984d) introduced the idea of an *offset*, that is, the constant across a mode that functions as the zero point of an interval scale. Such an offset may be idiosyncratic to a person or it may be characteristic for the variable or scale. In the treatment by Harshman, it is assumed that the multiplicative model applies to the deviation from the offset, rather than to the deviation from the mean of a scale. Generally, the size of the offset is unknown, and in order to be able to analyze interval variables as if they are ratio-scale variables, a way must be found to handle the offsets in such a way that the multiplicative model can be correctly estimated. One of the suggestions by Harshman and Lundy (1984d) was to define a single model consisting of an additive part with offsets and a second part containing the multiplicative terms (for a further discussion of such a design see Section 6.7).

Two-way models with offsets. For example, for two-way data the complete model has the form

$$x_{ij} = h + h_i + h_j + \sum_s a_{is} b_{js} + e_{ij}, \qquad (6.3)$$

where the h terms are the offsets that depend on none or one of the indices and are constant across the other index. The fourth term is the multiplicative term (PCA model) and the e_{ij} are the errors of approximation. Writing an offset here does not imply that such offsets are necessarily meaningful for all data types. When variables (2nd mode) have different measurement units, the offset h_i should not even be considered because it implies adding over these measurement units. Even though this model looks like an analysis-of-variance model with an overall mean, two main effects and an interaction term, this is not the case, because the h terms represent zero points and not means.

The advantage of looking at preprocessing in terms of offsets is that preprocessing is no longer necessarily a two-step procedure of first preprocessing the data and then analyzing the preprocessed data with a component model, but the whole analysis can be discussed within a single framework.

Three-way models with offsets   In the three-way case, this *triadditive model* or *extended three-mode model*, as it is called by Harshman and Lundy (1984d), takes the form, using the Parafac model as an example,

$$
\begin{aligned}
x_{ijk} &= h + h_i + h_j + h_k + h_{ij} + h_{jk} + h_{ik} \\
&+ \sum_s a_{is} b_{js} c_{ks} + e_{ijk},
\end{aligned}
\tag{6.4}
$$

with six h terms. Again, not all h terms can be meaningfully defined for all data designs. As indicated by Harshman and Lundy (and reiterated by Bro & Smilde, 2003a) it is difficult and impractical to try and estimate the offset terms, and instead they propose to eliminate them via "appropriate" centerings. A particularly thorough study of this problem was carried out by Kiers (2006), who came to the conclusion on the basis of mathematical and simulation studies that "it can be concluded that the traditional approach for estimating model parameters can hardly be improved upon, and that offset terms can sufficiently well be estimated by the proposed successive approach, which is a simple extension of the traditional approach"(p. 231). The proposal comes down to first solving the three-mode model and estimating the offset term afterwards.

Without going into details here, it can be shown (see Section 6.4.3) that only *fiber centering* will eliminate all pertinent offsets without introducing additional terms into the model (see Kruskal, 1981, p. 18; Harshman and Lundy, 1984d, pp. 232–240; Bro and Smilde, 2003a, pp. 21–22). Therefore, Kruskal (1981), and in more detail Harshman and Lundy (1984d), argued that only fiber centerings are appropriate for three-mode models because only for such centerings do the components after centering bear a simple relation to the components before transformation. Put differently, such centerings do not "destroy the agreement with the model". The core of the argument is illustrated in Fig. 6.3(b) where we see that the centering depicted does not change the directions of the uncentered solution but only put them in deviation of their means. What Fig. 6.3 also illustrates is that even though the directions may remain the same, the relative explained variability of the uncentered and the centered solutions bear no relation to each other. Furthermore, the figure shows that the principal components for centered and uncentered solutions can be widely different, and should only be compared after rotating them to maximum agreement. Even then it might happen that due to the centering one of the original components is not included in the maximum variance components of the centered solution.

Because the theoretical arguments about handling offsets indicate that the proper way to treat them is by centering the data across modes (per fiber), it is not always necessary to consider such offsets further as long as one follows the model-based recommendation of fiber centering. Thus, content-based discussions about centering can be primarily couched in terms of means, which simplifies the discussion considerably.

Table 6.1   Centering proposals

Type of centering	Operation	Example	Formula
None			$\tilde{x}_{ijk} = x_{ijk}$
Overall	Across all data		$\tilde{x}_{ijk} = x_{ijk} - x_{...}$
Slice centering	Within a slice	Lateral	$\tilde{x}_{ijk} = x_{ijk} - x_{\cdot j \cdot}$
Fiber centering	**Across a mode**	**First**	$\tilde{x}_{ijk} = x_{ijk} - x_{\cdot jk}$
Double centering	**Across two modes**	**First &**	$\tilde{x}_{ijk} = x_{ijk} - x_{\cdot jk} - x_{i \cdot k} + x_{..k}$
		Second	
Triple centering	**Across three modes**		$\tilde{x}_{ijk} = x_{ijk} - x_{\cdot jk} - x_{i \cdot k} - x_{ij \cdot}$ $+ x_{i..} + x_{\cdot j \cdot} + x_{..k} + x_{...}$

Centerings in **bold** are considered appropriate centerings on model-based arguments.

6.4.2   Fiber centering

In this section we will present the various ways in which centering may be conceived. In particular, we will show that with fiber centering components are centered and possible offsets are removed.

In Table 6.1 an overview is given of centering possibilities for three-way data arrays. Cattell (1966, pp. 115–119) has coined terms for preprocessing of two-way matrices, but these will not be used here (see Kroonenberg, 1983c, p. 142, for a listing). Besides the four references on which this chapter is primarily based, Tucker (1966, p. 294), Bartussek (1973, pp. 180–181), Lohmöller (1979, pp. 156–158), and Rowe (1979, p. 78) also discussed the preprocessing of input data for three-mode models, and most of the schemes for centering have been proposed by at least one of them.

Figure 6.5 presents an overview of the consequences of various types of centering on the means in a data set. The general effect is that if centering takes place at a certain level, means at the lower connected levels will also be 0. Especially noteworthy is the case of double centering (e.g., across rows (per column centering) combined with across columns (per row centering)) because the only nonzero means remaining are $\bar{x}_{\cdot ij}$, that is, those contained in the two-dimensional average frontal slice of the data array. Note, however, that the one-dimensional marginal means of this slice are both 0 again. In other words, the average frontal slice of the data array is itself double-centered. For the triple centering indicated in Table 6.1, the grand mean and all one-dimensional and two-dimensional marginal means are 0; in other words, from a three-way ANOVA perspective all that is left is the three-way interaction. Harshman and Lundy (1984d, p. 235) stated that "[i]n theory, then, triple application of fiber-centering leaves the data in ideal condition, with all troublesome constituents [offsets] and one-way and two-way components removed". However, it should be realized that the situation is only ideal if all offsets and means can be meaningfully defined, which is only the case for three-way factorial designs.

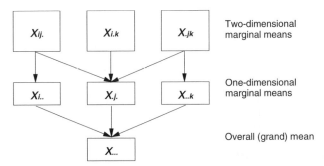

General rule: If means have been removed at a particular
level they will also have been removed at a lower level

Figure 6.5   Effects of centering.

Uncentered modes will have large first components, due to the presence of the
means. This should be taken into account when assessing the relative contributions of
the components of such an uncentered component. Such uncentered first components
are often highly correlated with the mean vectors, and also with the fitted sums of
squares of the elements of the mode, which may or may not be acceptable.

The effect of centering across a mode is that the components of this mode become
centered as well. Moreover, Harshman and Lundy (1984d) show that this removes
unwanted offsets as well without introducing further complications. For a full decomposition, the correctness of the statement follows directly from

$$
\begin{aligned}
\tilde{x}_{ijk} &= x_{ijk} - \bar{x}_{.jk} = \sum_{p,q,r} a_{ip}b_{jq}c_{kr}g_{pqr} - \frac{1}{I}\sum_{I}\left[\sum_{p,q,r} a_{ip}b_{jq}c_{kr}g_{pqr}\right] \\
&= \sum_{p,q,r} b_{jq}c_{kr}g_{pqr}\left[a_{ip} - \frac{1}{I}\sum_{I} a_{ip}\right] = \sum_{p,q,r} b_{jq}c_{kr}g_{pqr}\left[a_{ip} - \bar{a}_{.p}\right] \\
&= \sum_{p,q,r} \tilde{a}_{ip}b_{jq}c_{kr}g_{pqr}.
\end{aligned}
\tag{6.5}
$$

If only a limited number of components are used to model the data, we have to include
an error term, and the centering also causes the errors to be centered along the first
dimension.

Initially, it may come as a surprise that if the variables are centered for all conditions, it is the components of the subject mode that become centered. The algebraic
derivation shows that it is more of a terminological confusion, as both for the raw
data and in the model the centering is over the index i. In other words, one should
keep track of the index across which a centering is done. Part of the confusion is that,
especially in the social and behavioral sciences, components are often explained in

terms of variables and the subjects play no explicit part in the interpretation. In other disciplines, the emphasis is often more on the scores of individuals on the components, which are after all linear combinations of original variables. In standard principal component analysis the scores of subjects on the variables are deviation scores after centering, and so are the scores on the components.

Another problem that has been raised with respect to centering is the relationship between components derived from various centering procedures. An extensive literature pertaining to two-mode matrices exists, but will not be discussed here. The presence of three modes, and thus three component matrices, makes the matter considerably more complex (see Kruskal, 1981). When the assumption is correct that a Parafac model is underlying the data, the centered components are essentially the same as the original ones, due to the uniqueness of the model. However, when a principal component analysis is performed with less than the maximum number of components, there is no guarantee that the principal directions before and after centering are the same. This depends on the amounts of variability explained by the original principal components and those of the centered data (see also Fig. 6.2). Whether an investigation into the comparability of components in such a case will be useful for practical applications is rather doubtful. Some selected references with respect to this issue for the two-mode case are McDonald (1967, especially pp. 8–10), Gollob (1968b, 1968c), Tucker (1968), and Noy-Meir, Walker, and Williams (1975).

The consequence of removing any mean is that the amount of variation explained by a subsequent analysis will be smaller, and sometimes dramatically smaller than the uncentered analysis, because the sum of squares caused by nonzero means is often the largest one present in the data. Fiber centering for variables can, for instance, be interpreted geometrically as moving the centroid of the data to the origin (see Fig. 6.2), and so the sum of squares caused by the variables means (which is often not meaningful, due to different measurement scales or due to interval data) is removed. Therefore, the loss in (artificial) explained variation need not be regretted in such a case. A further aspect of centering is the interaction between centering and normalization, which will be taken up in Section 6.4.8.

6.4.3 Several fiber centerings

As mentioned in the previous section (see Table 6.1), one may combine different fiber centerings; applying Eq. (6.5) to each of the three modes shows that centering across one mode does not affect the centering in any other mode. Once a mode is centered, it cannot be uncentered by any other centering, which can be checked in Fig. 6.5. This also means that order in which the centering takes place is irrelevant.

6.4.4 Other types of centerings

Neutral points. The model-based argument for fiber centering — that is, that it eliminates unknown offsets and only centers the associated component — is not an argument that is relevant to all. One could also argue that the proper three-mode model to analyze is the one that is defined in deviation from the offset itself. Given that many bipolar scales consist of antonyms, one may postulate that in that case the offset is equal to the neutral midpoint of the scale. To remove the offset, all one has to do is to subtract the neutral point rather than the mean, and analyze the deviations from the neutral points by three-mode methods. The implicit assumption is that deviations of the observed means from these neutral points have substantive interest and should be modeled within the three-mode context.

Shifted models.   In analogy with the proposal by Seyedsadr and Cornelius (1992) (see also Gauch Jr., 1992, pp. 96–102), one could define a shifted multiplicative model for nonadditive three-way factorial designs[1]. This would come down to a model with a single overall offset h:

$$x_{ijk} = h + \sum_{p,q,r} a_{ip}b_{jq}c_{kr}g_{pqr}. \tag{6.6}$$

For the two-mode case, the proposers have gone into great detail to examine this model. The offset h is estimated so that the residual sum of squares of the loss function is as small as possible, and in general the offset is not equal to the overall average. In the two-mode case, it was proved that the shifted model "is more efficient than AMMI[2] and PCA in terms of capturing the SS in the fewest df" (Gauch Jr., 1992, p. 229) (see also Denis & Gower, 1994). The reason for this is that the principal axes of a PCA necessarily go through the centroid, while the origin of the axes for the shifted model can be located anywhere within the cloud points (Fig. 6.6). Thus, it is sometimes possible to get a smaller residual sum of squares using the shifted models than by using the overall mean. As pointed out by Gauch, it is not certain that the gain in explained variability — which he reports to be small in most two-mode cases (p. 229) — is really useful as it is mostly noise. Moreover, it does away with such useful concepts as variable means, which are generally easily understood and which will return in some disguised form in the components of the shifted model anyway. He continues to argue that biplots of the shifted model are often more difficult to interpret due to the presence of such means. Based on such considerations, it does not seem worthwhile to go deeply into an investigation of the merits of a shifted approach for three-way factorial data, unless first some explicit need for such models arises in real-life applications, or it can be shown to have distinct interpretational advantages.

Robust estimation.   Least-squares estimation is known to be sensitive to outliers, so it can easily be imagined that one or a few outlying observations could greatly influence the means, so that when they are subsequently removed, it might be difficult to unravel the underlying multiplicative patterns due to bias introduced by the biased

---

[1] The term shifted is used in a different sense by Harshman, Hong, and Lundy (2003)

[2] Additive Main effects and Multiplicative Interaction model - AMMI (Gauch Jr., 1992).

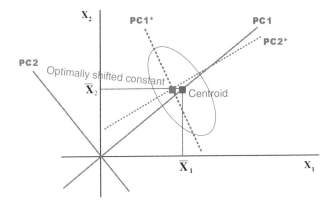

Figure 6.6   Principle behind the shifted multiplicative model.

means (see also Gnanadesikan & Kettenring, 1972, p. 107). The solution in such cases is to deal with the outliers in an appropriate way by using robust measures of centrality. Thus, one could replace the centering procedure via means by a procedure based on medians or other robust measures for location. The question of robustness is the more important, because sometimes (three-way) interactions arise not so much because of systematic patterns but due to outlying observations. Using other measures of location (and/or spread) also has implications for the interpretation of the outcomes of three-mode analyses. This area is, however, still in its infancy; for a discussion of robust procedures in three-mode analysis in general, see Section 12.8 (p. 294).

6.4.5 Normalization

The major aim of normalization is to differentially weight certain parts (mostly variables and/or subjects) of the data to change the way they influence the outcome of an analysis. This may be done because the measurement scales of the variables are not comparable, or with the aim to equalize their variances so as to equalize their influence on the solution. Not uncommonly, persons engaged in some kind of judgmental task have widely different scale usages, resulting in large differences in variances, so that subjects with larger variance dominate the outcome of an analysis. Depending on the context, one may or may not want to equalize such differences in scale. For instance, it is not necessarily sensible to equalize the variances of persons who do and persons who do not have an outspoken opinion. The latter will tick scale points around the midpoints of scales and produce largely random noise; the former use scales effectively and have mainly systematic variation. Alternatively, if the differences in variability are the result of response styles, that is, some persons tend to use the end points of the scale, whereas others never do, but both answer according to the same model or the same considerations, normalizing is probably a good thing to do.

Sometimes variables that are theoretically important may not have smaller variances due to the measurement scales used, and increasing their variability through a scaling factor may make them exert more influence on the analysis (e.g., see Bro & Smilde, 2003a).

Slice normalization   Earlier it was mentioned that the slice normalization is to be preferred on model-based arguments. To demonstrate this argument, suppose we have a normalization factor, s_j, based on the second-mode or lateral slices. Then we see that

$$\tilde{x}_{ijk} = s_j(x_{ijk} + e_{ijk}) = s_j \sum_{p,q,r} a_{ip} b_{jq} c_{kr} g_{pqr} + s_j e_{ijk} \tag{6.7}$$

$$= \sum_{p,q,r} a_{ip} [s_j b_{jq}] c_{kr} g_{pqr} + s_j e_{ijk} = \sum_{p,q,r} a_{ip} \tilde{b}_{jq} c_{kr} g_{pqr} + \tilde{e}_{ijk},$$

so that after normalization the model is the same as before, except for the scaling of one set of components, here the second mode. Such a simple effect can only be reached with slice normalization, because, for instance, fiber normalization involves a scale factor s_{jk}, which cannot be absorbed into the already existing model.

Normalization and units of measurement.   Another model-based argument for slice normalization is connected with the link between the original scale and that of the normalized one. Suppose that we use fiber normalization, s_{jk}, and that the data have already been fiber-centered, even though the latter is not necessary for the argument. When the standard deviations are different for each occasion, each standard score represents a different amount of units on the original scale, that is, in condition 1, s_{j1} may be equal to 3 units, while in condition 3, s_{j3} could be equal to 7 units. In general, such an inconsistent relationship between the original and the preprocessed scales seems undesirable. The same argument is used in structural equation modeling, where analyzing sets of covariance matrices is preferred over analyzing sets of correlation matrices because the latter arise as the result of fiber normalizations.

In certain circumstances, certain authors (e.g., Cooper & DeLacy, 1994) recommend *fiber normalization*, that is, per variable–occasion combination. This leads to equal sums of squares of each variable at each occasion, so that the influence of each variable on the analysis is the same at each occasion. This type of normalization is equivalent to using correlation matrices for each occasion and differences in variability across occasions no longer play a role in the analysis. However, if the multiway model was a valid one before the normalization, after the fiber normalization the model is no longer valid as the normalization cannot be absorbed in the components in the same way as this was possible for slice normalization.

Equal average diagonal normalization.   Harshman and Lundy (1984c, pp. 202–203) discussed the normalization for the case that the input data consist of a set of covariance matrices, $S_1, S_2, \ldots, S_K$. When the measurement levels are different between the variables, or when the differences in variances of the variables should not play a role, two options seem available. Either the covariance matrices are converted to correlation matrices, which is equivalent to applying fiber normalization with all its problems; or a normalization should be applied such that the average of the covariance matrices is a correlation matrix. The latter option is in line with slice normalization. An average correlation matrix is realized if each covariance entry $s_{jj',k}$ is normalized as follows:

$$\tilde{s}_{jj',k} = \frac{s_{jj',k}}{\sqrt{\bar{s}_{jj}}\sqrt{\bar{s}_{j'j'}}}, \tag{6.8}$$

where $\bar{s}_{jj} = (1/K)\sum_k s_{jj,k}$ is the average over the occasions of the variances of the jth variable.

6.4.6 Several slice normalizations

A fundamental difference between centering and normalization is that combinations of different centerings do not influence each other, whereas normalizations do (see Fig. 6.7). Normalization of one mode will generally destroy that of another. Iterative procedures have been devised to arrive at stable normalizations for two modes, but the final solution depends on the mode one starts with (see Cattell, 1966, p. 118, and earlier references cited by him).

Harshman and Lundy (1984d, pp. 247–248) and Harshman and DeSarbo (1984b, p. 606) introduce a procedure for simultaneous normalization both within subjects and within scales. Their approach needs an iterative algorithm, and interestingly, the procedure bears resemblance to the problem of iterative proportional fitting for contingency tables (e.g., Bishop, Fienberg, & Holland, 1975). This resemblance was exploited by Ten Berge (1989), when he showed that the process is guaranteed to converge. The advantage of the double normalization is that one deals simultaneously with different types of measurement scales and with undesired variability between subjects. However, there has been very little experience with this procedure (for one

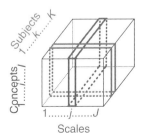

Figure 6.7   Two slice normalizations are always perpendicular to each other

of the very few applications see Harshman & DeSarbo, 1984b) and it is somewhat worrying that one eliminates a large amount of (possibly) informative variability from the data. Therefore, at present, it is difficult to recommend the procedure for routine applications.

6.4.7 Other types of normalizations

Standardization, correlations, and covariances. An argument put forward by, for example, Lohmöller (1979) in favor of fiber normalization, in particular, per occasion normalization, is that it creates correlations

$$
\begin{aligned}
r_{jk,j'k'} &= \frac{1}{I}\sum_i z_{ijk}z_{ij'k'}, \\
r_{jj'} &= \frac{1}{KI}\sum_k\sum_i z_{ijk}z_{ij'k}, \\
r_{kk'} &= \frac{1}{IJ}\sum_j\sum_i z_{ijk}z_{ijk'},
\end{aligned}
\tag{6.9}
$$

with z_{ijk} the standardized quantities based on fiber normalization. It has been argued that this normalization is advantageous because it allows for "the usual interpretation of the loadings" (Lohmöller, 1979, p. 158). The statement that $r_{jj'}$, and $r_{kk'}$, are correlations is, however, not completely correct, as they are only averages of correlations, e.g.,

$$
\begin{aligned}
r_{jj'} &= \frac{1}{K}\sum_k\left[\frac{1}{I}\sum_i\left\{\frac{(x_{ijk}-\bar{x}_{.jk})}{s_{jk}}\times\frac{(x_{ij'k}-\bar{x}_{.j'k})}{s_{j'k}}\right\}\right] \\
r_{jj'} &= \frac{1}{K}\sum_k r_{jk,j'k},
\end{aligned}
\tag{6.10}
$$

and averages of correlations are not necessarily correlations themselves. This is only the case when the s_{jk} are equal for each j across all k. With centering these problems do not occur because sums of covariances are again covariances and can be interpreted as "within sums of squares".

No normalization. In those cases where one is really interested in including the variability of variables with comparable measurement scales (such as rating scales with the same range), one could simply forego any type of normalization. A great advantage of not normalizing in this case is that it is possible to assess via the three-mode model whether certain variables have larger variances than others and whether such larger variances are fit well by the model (see Section 12.6.1, p. 289). When not normalizing variables, one should inspect their contributions to the solution very carefully to ensure that not one of them dominate the solution.

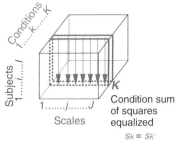

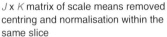

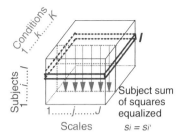

J x K matrix of scale means removed centring and normalisation within the same slice

J x K matrix of scale means removed; normalisation perpendicular to centring

Figure 6.8   Compatible centering and normalization (left) and incompatible centering and normalization (right).

Equalizing ranges.   Alternatively, when comparable scales have different ranges, say, mixture of five-point scales and seven-point rating scales, one might consider equalizing those ranges, rather than equalizing variances. It seems advisable that, in that case, the differences in variability should be discussed and interpreted.

6.4.8   Combining fiber centering and slice normalization

Combining centering and normalization may lead to complications because normalization over one mode may destroy the centering over another. Following the common procedure from two-mode analysis, we will always assume that centering precedes normalization. This makes sense because normalization does not uncenter a variable, but centering a normalized variable does affects its normalization. As portrayed in Fig. 6.8, slice normalization within the fiber-centered mode is a proper procedure leaving the centering intact. Dividing every value in a matrix by a constant does not influence the centering within a matrix. However, when the slice normalization is done "perpendicular" to the centering, the centering is seriously affected, and this is clearly not desirable. Should one insist on a "perpendicular" operation, an iterative procedure should be employed that cycles between centering and normalization.

Complications increase when one wants to have, for instance, a centering over one mode, a normalization of another, and a normalization of a third mode as Harshman and DeSarbo (1984b) do in an example. In that case all three operations have to become part of the iterative procedure. Harshman and DeSarbo report convergence for the procedure, but it is uncertain what the relationships are between the results of such a procedure and the raw data, the more so because in certain circumstances the order in which the various normalizations are performed also seems to have an effect on the solution.

6.4.9 Two-step and one-step estimation

An interesting aspect of using fiber centering is that, when all data are present, the estimates of the parameters are the same whether the estimation is done in two steps, that is, first preprocessing and then estimating the multiplicative part, or in one single step, that is, simultaneously estimating the additive and the multiplicative parts (Kruskal, 1981). Kiers (2006) performed a series of simulations in connection with this issue with the result that the theoretical superior procedure did not necessarily perform better in practice. As pointed out by Bro and Smilde (2003a), this equivalence breaks down when there are missing data, in which case only the simultaneous (one-step) procedure is adequate.

6.5 CONTENT-BASED ARGUMENTS FOR PREPROCESSING CHOICES

In this section we will discuss a number of data contexts that a researcher may encounter and examine possible preprocessing options and procedures that may be followed in these contexts. In particular, the question of arbitrary and/or uninterpretable means and variances will be considered.

6.5.1 Arbitrary means and variances

Many variables in the social and behavioral sciences have interval properties, and hence no natural zero point. In addition, the absolute size of the variances of these values is often arbitrary, in the sense that it is dependent on the number of possible values chosen rather than on any "true" range of the variable. As variation in the data due to arbitrary means is uninterpretable, it should not have an effect in a component analysis, and thus such means should be removed. In certain cases with comparable variables (e.g., sets of similar rating scales), the differences in arbitrary means across conditions of the same scale might be of interest, and they could be retained in the multiway analysis. In such cases the midpoints of the variables generally define some neutral point, which might be used for centering. However, most published applications that are not explicitly based on ratio-scale data eliminate some type of means, and analyze the variability in the means separately. This approach is also in accordance with the general recommendations formulated by Harshman and Lundy (1984d) (see Section 6.4) based on the removal of unknown offsets.

The situation with variances is similar. If variances are arbitrary, and variables have different ranges, then in order to avoid artifacts these variances should certainly be equalized before a multiway analysis is performed. Slice normalization is effective in eliminating the overall scale factor of each variable. In the case that all variables have the same measurement scales, one could consider normalizing the overall variance over the entire data set, but this has no influence on the outcome on a multiway analysis because all the data values are divided by the same constant (Kruskal, 1981, p. 17). Possibly, there might be some use for this when comparing data sets.

6.5.2 Incomparable means and variances

Consider the situation in which the scores of a number of subjects are available on a diverse collection of variables, each of which has its own measurement characteristics. The means of such variables are in themselves meaningful, but within the context of a multiway analysis they are incomparable, as are their variances. Therefore, it does not make sense to consider components that are influenced by these means and variances. In other words, these means and variances should be modeled separately, and not via a principal component analysis. These model-based arguments are typically automatically applied to variable modes.

However, as the further modes of multiway data are often conditions or years, the substantive question arises of whether one should remove the incomparable means per year (or condition), or over all years together. The argument for the normalization procedure in Section 6.5.1 was based on the idea that within one year the means and standard deviations across variables are arbitrary. However, in the present case, differences in means and scale factors over years are comparable for each variable, and one may decide to model the differences across years by the principal component analysis or model them separately outside the model, depending on the research questions one has in mind. Note that according to the model-based considerations, slice centering and fiber normalization are not preferred algebraically, and the content arguments should therefore be very strong if one wants to use these approaches (e.g., see Section 6.4.7). A specific example of such an argument can be found in the two-way factorial designs in plant breeding research. In particular, DeLacy and Cooper (1990) have argued that in the investigation of genotype–environment interaction one should perform fiber normalization, in their example "environment normalization", because in that way the ordination "will reflect the phenotypic correlation among environments, and, therefore, opportunities for exploiting indirect selection among environments" (p. 563). In Cooper, DeLacy, and Basford (1996), this argument is extended to the three-mode context of multiattribute genotype by environment investigations, which use multivariate two-factor designs.

Thus, one may have incomparable means across variables and comparable means within occasions in one data set, and the "best" way to treat them depends on one's view of the subject matter. It may, of course, happen that one has to perform more than one kind of scaling due to lack of insight in the data set itself, but after each type of centering or normalization the data analyzed contain a different type of variability, so that it is not always meaningful, let alone easy, to compare the results from different preprocessing options.

In other situations, all means or variances are both interpretable and comparable; for instance, all variables are monopolar scales of the same type and range. In that case, it is a substantive question whether means and/or variances should be modeled separately or not. The same applies more generally to data from multiway factorial designs.

6.6   PREPROCESSING AND SPECIFIC MULTIWAY DATA DESIGNS

In this section we will investigate some of the substantive considerations that may go into selecting an appropriate preprocessing procedure for a particular data set when the means and/or normalization factors are interpretable. It is not possible to do so in a general sense, because different research questions are asked of different data, and because different measuring instruments and research designs are used. However, it is possible to make specific recommendations in specific research areas, as has been demonstrated by Noy-Meir (1973), and Noy-Meir et al. (1975) for ecological ordination data, and by Cooper and DeLacy (1994) and McLaren (1996) for agricultural data. Here we will look at some general categories of multiway data for which general preprocessing statements can be made.

6.6.1   Three-way profile data

The most common way to handle three-way profile data with subjects in the rows, variables in the columns and conditions in the tubes is to center across subjects, that is, per variable–condition combination (jk) and to normalize per variable slice j so that the completely processed data $z_{ijk} = (x_{ijk} - \overline{x}_{.jk})/\sqrt{\sum_{ik}(1/IK)(x_{ijk} - \overline{x}_{.jk})^2}$ (see Fig. 6.2). We will call this *profile preprocessing*. With three-way data one could, of course, decide to fiber-normalize variables per occasion, but in practice this is very rarely done. In general, this is not recommended because it does have an unfavorable effect on the model, in that it does not agree with the general form of the three-mode models (see Section 6.4.5).

Fiber centering is recommended because then the scores on each variable at each occasion are deviation scores with respect to the average subject's profile, so that preprocessed scores z_{ijk} are in deviation of the variable means. The fiber centering has the additional advantage that if there is a real meaningful, but unknown, zero point of the scale, it is automatically removed from the scores; see Section 6.4.1. A further effect of this centering is that the component coefficients of the subjects are in deviation of their means. The normalizations over all values of each of the variables mean that the normalized deviation scores carry the same meaning for all occasions. Thus, a standard deviation has the same size in the original scale for each occasion. A similar argument is used in multigroup structural equation modeling in favor of using covariance rather than correlation matrices. The effect of the recommended preprocessing is that the structure matrix with variable–component correlations can be computed. The recommended preprocessing is the only way to achieve this interpretation.

6.6.2 Three-way rating scales: Centering

In many behavioral studies the variables have the form of graded response scales, that is, subjects have to indicate or rate which of a number of ordered alternatives apply to them. Such rating scales are often *bipolar*, that is, the anchors at the end points of the scales are two adjectives that are assumed to be antonyms, like GOOD and BAD or BEAUTIFUL and UGLY. Another common form of rating scales are monopolar scales where only one adjective determines the scale and the anchors at the end points indicate superlative degree of the adjective in the positive and negative sense, such as NOT SAD AT ALL and EXTREMELY SAD. Typical three-way designs using rating scales have the form of concepts by scales by subjects (or judges).

The choice of centering with bipolar scales is primarily determined by the nature of the scales. Which particular adjective is placed at which end of the scale is entirely arbitrary. For example, for a seven-point BEAUTIFUL–UGLY scale, there is no particular reason to code the response "very beautiful" as 7 and "very ugly" as 1, or the other way around. As the direction of a bipolar scale is arbitrary, the means after averaging across scales depend on the numeric coding of the end points of the scales, and thus, centering across scales is not a well-defined operation. Therefore, centering for bipolar scales can only be based on the multivariate two-way ANOVA decomposition per scale (see Section 6.7), and fiber centering per scale–subject combination across the concept mode is generally recommended. Harshman and DeSarbo (1984b, p. 606) presented an analysis of semantic differential scales but used double centering by including centering across scales. As argued here, this will in general not be advisable.

Centering monopolar scales, in some contexts called Likert scales, does not create the same problem as centering bipolar scales, because they are unidirectional, that is, they run from "not at all" to "extremely so'. Notwithstanding, one should still think carefully about the means and/or offsets when deciding whether centering across monopolar scales is a meaningful and appropriate operation.

6.6.3 Multiway rating scales: Normalization

For three-way rating scales consisting of concepts by variables by subjects data , there are three *options* for normalization: (1) *column/fiber normalization* such that the sums of squares of the scales are equal for each subject k, that is, dividing by $s_{jk} = \sqrt{\sum_i x_{ijk}^2}$; (2) *lateral slice normalization* such that the sums of squares of scales are equal across all measurements of the scale j, that is, dividing by $s_j = \sqrt{\sum_{ik} x_{ijk}^2}$; and (3) *frontal slice normalization* such that the sums of squares of the subjects are equal, that is, dividing each subject's data by the square root of its total sum of squares, $s_k = \sqrt{\sum_{ij} x_{ijk}^2}$.

After normalization, the data can have the following *properties*. (a) Differences in scale usage of subjects are equalized, in the sense that subjects who use the extremes of the scales and subjects who only score in the middle of the scales have the same sum of squares. Generally, the first component of the subjects then correlates highly with the model's goodness of fit to the subjects' data. However, fit comparisons can always be carried out in three-mode models, irrespective of the normalization, so that there is no particular reason to normalize to achieve this end. (b) Given that the proper fiber centering has been carried out, the elements of the component matrix of the scales, **B**, can be interpreted as correlations. (c) In cases where all scores on a scale are equal, their variance is 0, and normalized scores are undefined.

In the first option (fiber/column normalization), the columns of each data matrix $\mathbf{X}_k$ are normalized, that is, $\hat{x}_{ijk} = x_{ijk}/s_{jk}$. When this option is carried out, all three properties of normalization mentioned above do or may occur. This normalization is especially sensitive to property (c), because subjects frequently use the same scale value for each concept. Moreover, the procedure is sensitive to outliers, which occur when all concepts are rated (more or less) the same and a single concept is rated very differently. In addition, for each subject standard deviation units will have different sizes with respect to the original scale. These properties make the first option an unattractive choice, even apart from the general model-based considerations.

The second option, slice normalization within scales (lateral slice normalization), that is, $\hat{x}_{ijk} = x_{ijk}/s_j$, only has the desirable property (b). In addition, it has the property that differences in sizes of variances between scales do not affect the results of the analysis. However, if a small variance in a scale only reflects the unreliability of judgments, it may be harmful to enlarge it through this normalization, as this procedure increases the random component in the variation. The advantage over the first option is that differences in variability in scale usage by the subjects and their influence on the ratings remain in the analysis. Property (c) cannot occur because variables without variance will be eliminated from the outset.

In the third option (frontal slice normalization), the sums of squares of the complete concept by scale matrix produced by each subject, that is, $\hat{x}_{ijk} = x_{ijk}/s_k$, are equalized. This is, for instance, standard practice in individual differences scaling (Carroll & Chang, 1970), where subjects are also treated as judges. This option only has property (a) and thus eliminates to a certain extent response bias in extremity of scale usage. An argument against this option is that the variability of the scores of the subjects is related to their judgments of the relationships between concepts and scales, and therefore should be analyzed together with the rest of the data. Moreover, this normalization does not have property (b).

6.6.4   Centering and individual differences

Several parts of three-way profile data of subjects by variables by conditions pertain to differences between individuals (Sections 3.6.1 and Chapter 13). Often the idea behind a multiway analysis is to explicitly concentrate on these individual differences, and the question arises of how to bring these to the fore in a particular analysis. This will be illustrated with the Coping data, which are treated in detail in Section 14.4

Table 6.2   Coping data: Univariate three-way analysis of variance table

Source	SS	df	MS	F
Main effects				
Children (*Mode A*)	486.8	118	4.1	14.4
Emotions–strategies (*Mode B*)	1796.2	8	224.5	784.1
Situations (*Mode C*)	124.5	7	17.8	62.1
First order interactions				
Situations × Emot-Strat	537.7	56	9.6	33.5
Situations × Children	558.7	826	0.7	2.4
Emot-Strat × Children	895.8	944	0.9	3.3
Three-way interaction	1892.2	6608	0.3	
(Situations × Emot-Strat) × Children				
Total	6291.9			

(p. 354). In brief, the children in the study were confronted with a number of problematic and challenging situations at school, and they were asked to indicate how they dealt with them in terms of emotions and coping strategies (*Emot-Strat*). As all the scales had similar ranges and were monopolar, one can in principle treat the data as a three-way factorial design and construct a complete univariate three-way analysis-of-variance decomposition (see Table 6.2).

To eliminate all but the parts of the data explicitly related to individual differences, we have to eliminate the Situations and emotions–strategies main effects, and the Situation × emotions–strategies two-way interaction, which can be achieved by column (i.e., fiber) centering. Thus, a three-mode analysis will only contain the terms related to individual differences.

At the same time, we now have a frame of reference for the results for the children's mode, as the zero value on all scales for emotions and strategies is the behavioral profile of the average child. The centering procedure also centers the child components, so that in a paired-components plot (Section 11.4.1, p. 260) the origin represents the profiles of the average child and all individual differences can directly be related to those profiles.

This type of arrangement is typical of multiway profile data or three-way rating scale data in their profile arrangement; if the scales had not been comparable, lateral slice normalization might have been necessary. In the latter case, the three-way analysis-of-variance decomposition is not possible, and a multivariate two-way ANOVA table (see Section 6.8) would have been advisable.

When the Coping data and comparable data sets are viewed as three-way rating scales the children are seen as judges. Mostly the data will then be arranged as situations/concepts by scales by subjects (see Chapter 14). Viewed in this light, for the Coping data it is the relationship between the situations and the emotions and strategies that is of prime interest, as well as the individual differences in this relationship. For the factorial arrangement, double centering (across situations and across emotions and strategies) seems the most appropriate procedure. The double centering eliminates all but the Children main effect, the Situation×Emotions and Strategies two-way interaction, and the three-way interaction. Typically, this will give a strong first component for the children on which most children will have a positive score. These scores then primarily reflect the strength of the contributions of the Situations×Emotions two-way interaction to each child's score. Furthermore, the second and later components primarily reflect the individual differences between the children in the situation-emotions and strategies relationships. A similar situation as described here for the Coping data can be found within an agricultural context in De la Vega, Hall, and Kroonenberg (2002), where genotypes by environments by attribute data were double centered and the consequences of this centering were laid out. Another example is D. A. Gilbert, Sutherland, and Kroonenberg (2000), which deals with the relational communication skills of nurses.

Individual differences dissimilarity and similarity data.   For dissimilarity,similarity or proximity data (Section 3.8, p. 40), such as stimuli by stimuli by judges, the most common procedure is double centering. Since the subjects or judges in the third mode are assumed to be independent and we want to describe individual differences in the way the stimuli (variables) are treated by the subjects, the data should be centered per subject, or matrix-conditional (e.g., see F. W. Young, 1981). In the multidimensional scaling literature the data values, which are assumed to have distance-like properties, are often first squared before double centering, so that the double-centered values, $\tilde{z}_{jj'k}$, can be interpreted as scalar products between j and j' for each k (e.g., see Torgerson, 1958, p. 258). Alternatively, one could consider the observed values as being already squared, as Tucker (1972) did to demonstrate three-mode scaling. One of the effects of squaring is that the larger numbers carry even more weight than they already do in least-squares fitting procedures.

In similarity data, it is at times desirable to normalize matrix-conditionally (frontal slice normalization), for instance, in order to eliminate response styles. This normalization can be done without influencing the results from the common double centering, as it takes place within the same slice. It is the default approach in individual differences scaling (Carroll & Chang, 1970).

6.7   CENTERING AND ANALYSIS-OF-VARIANCE MODELS: TWO-WAY DATA

In this section we will discuss the intimate relationships between centering and analysis of variance models, in particular, we will show which models underlie which types of centerings.

6.7.1   Centering and biadditive modeling

To facilitate the discussion let us assume that we are dealing with scores of individuals on a series of tests, each having the same rating scale. The means of these tests are comparable, as are those of the individuals averaged across tests. Assuming that it makes sense to talk about this average performance of an individual over all tests, the question arises as to how the average performance should be modeled. Similarly, given that we have determined the averages of all tests, the question arises as to how they should be included in an analysis. One way to do so is to perform a standard two-mode principal component analysis or singular value decomposition (see Section 4.4.1, p. 47) on these averages.

An alternative way to treat these means (and the means of the individuals over tests) is to model them according to a model originally called the FANOVA (Factor ANalysis Of VAriance) model (see Gollob, 1968a, 1968b, 1968c), but here referred to as the *biadditive model* Denis and Gower (1994). This model combines the grand mean, row, and column effects with a singular value decomposition for the two-way interaction or residuals. Note that the assumption of the same scale for all variables is essential in this formulation. The derived components are "interaction-components" in that they describe the interactions of the deviations from the means of the individuals and the tests, respectively:

$$x_{ij} = \mu + \alpha_i + \beta_j + \epsilon_{ij} \text{ with } \epsilon_{ij} = \sum_{s=1}^{S} a_{is} b_{js} g_{ss}, \tag{6.11}$$

where μ, α_i, β_j, and ϵ_{ij} are the usual grand mean, row effect, column effect, and residuals from analysis-of-variance models with standard zero-sum assumptions for the effects (e.g., see Kruskal, 1981, pp. 6–7). One hopes, of course, that very few components (singular vectors) are necessary to describe the interactions. The biadditive model is thus a combination of an additive model (grand mean, row effect, column effect), and a multiplicative model (component decomposition of the remainder term). Note that model (6.11) has only one observation (or mean) per cell: there is no replication and also no within-cell variability. Within agricultural research this model is widely applied, especially in connection with genotype–environment interaction studies. It is mainly known under the name of AMMI (additive main effects and multiplicative interaction model) (Gauch Jr., 1992).

Essential to the application of the biadditive model is that it must be sensible to compute means across both row and column entities. When the columns contain variables with different measurement characteristics or units, the model does not make sense and averaging across items is inappropriate. In a way, we are then in a multivariate one-way design without replication, assuming we see the subjects or objects as levels of a factor. In the full biadditive model, we have a univariate two-way factorial design with one dependent variable without replication.

6.7.2 Modeling raw data or interactions

The main differences between the two ways of modeling, biadditive and singular value decomposition, are the treatment of the means and the interpretational differences connected with the components. Tucker (1968), for instance, contends that "the mean responses to various stimuli over a population of individuals are better conceived as describers of the population than as simple, fundamental describers of the stimuli" (p. 345), and continues that, therefore, such means should be included in a principal component analysis, that is, the analysis should be performed on the original measures. In this way the means are "equal to the measures that would be observed for a person at the centroid of the factor score distribution" (p. 350). The components then determine the original measures.

In contrast, the biadditive model first sets the means apart, and only then looks at components in the residuals. It gives a special *a priori* status to those means. It is a moot point whether this is just "a useful heuristic to use main effects as a point of reference from which to describe individual differences in patterns of subject responses" (Gollob, 1968c, p. 355), or whether in the biadditive model "the mean measure is considered as a basic characteristic of the responses of the individuals" (Tucker, 1968, p. 350). In the end, the subject matter will determine which of the two is the more correct interpretation of the mean responses, and the research questions will determine if it is more useful to model the means *a priori* (Gollob) or *a posteriori* (Tucker). When the means are expected to be the resultant of an agglomeration of influences that have to be disentangled, Tucker's view seems to be pertinent. However, when the means represent a "primary psychological construct", have intrinsic meaning in another way, or reflect a fundamental aspect of the data which needs to be explained, Gollob's view and model seem to be more appropriate. On the whole, the multiway community in practice has more often followed Gollob's point of view than that of Tucker.

6.7.3 Other views: Factorial designs

Whereas Gollob and Tucker discuss the biadditive model within the context of the problems of removing or maintaining means before performing a principal component analysis, the same model has been considered from a different angle by Mandel (1971), and in fact even earlier by Fisher and Mackenzie (1923) and N. Gilbert (1963). Mandel was looking for a way to model the interactions in a two-way analysis-of-variance design without replications and attempted to fit multiplicative interactions of row and column factors, ending up with the same model as Gollob. Gauch Jr. (1992) and Denis and Gower (1994) give important new insights since the original discussion in the late sixties. Incidentally, recent developments on robustifying the biadditive model can be found in Croux, Filzmoser, Pison, and Rousseeuw (2003).

Within this context there is no problem as to whether or not it is appropriate to remove means, as the primary focus is on modeling interaction terms after the additive main effects have already been investigated. Another and more fundamental difference between the two presentations of the model is the kind of data involved. Whereas Gollob considers observations of subjects on certain variables, and therefore looks at the relationships between variables, Mandel and his agricultural colleagues are dealing explicitly with one response variable and two predictor variables. Because of this fundamental difference not all considerations, tests, and so on from the analysis-of-variance side are relevant to the Gollob–Tucker discussion, and vice versa.

6.7.4   Biadditive models as completely multiplicative models

In connection with later expositions, we note that the biadditive model (Eq. (6.11)) can be written as a complete multiplicative model by using constant components, here taken to have a constant value of 1. Thus, 1_{i1}, for example, is the ith element of the first constant multiplicative component $\mathbf{1}$. For clarity we will drop the length-one restrictions for the components, so that terms like g_{ss} are not written.

$$x_{ij} = \sum_{s=1}^{S} a_{is}b_{js} = 1_{i1}1_{j1}\mu_{11} + \alpha_{i2}1_{j2} + 1_{i3}\beta_{j3} + \sum_{s=4}^{S} a_{is}b_{js}. \qquad (6.12)$$

Note that even though the models are equivalent in form and number of parameters, one might not really want to estimate the parameters via such a multiplicative model. Three-way analogues of this formulation are discussed in Section 6.8.3.

6.7.5   Degrees of freedom

Connected with modeling factorial designs is the question of the degrees of freedom to be associated with biadditive and triadditive models, and testing the statistical significance of the models and terms in the model. An overview of the state of affairs has been summarized by Van Eeuwijk and Kroonenberg (1995). In Section 8.4 (p. 177) degrees of freedom for multiway models are discussed.

6.8   CENTERING AND ANALYSIS-OF-VARIANCE MODELS: THREE-WAY DATA

6.8.1   Early additive plus multiplicative models

Lohmöller (1979) was probably one of the first to discuss additive and multiplicative models for three-way data. He suggested the following generalization of the biadditive model:

$$\hat{x}_{ijk} = \mu + \beta_j + \gamma_k + \beta\gamma_{jk} + s_{jk}\times e_{ijk} \qquad (6.13)$$

with the normal analysis-of-variance notation for the grand mean (μ), the variable effect (β_j), the condition effect (γ_k), and the combined variable–condition interaction effect ($\beta\gamma_{jk}$). The remaining fiber-centered, slice normalized e_{ijk} was decomposed with the three-mode principal component model, but the Parafac model can be used as well in this context. Lohmöller called this model the *standard reduction equation*. We will use the term *triadditive model* to refer to three-way additive-plus-multiplicative models, thereby generalizing the term "biadditive model".

A large number of triadditive models may be proposed, the form of which is determined by both model-based and content-based considerations. One possibility is the three-way main effects analysis-of-variance model for the additive part, and the three-mode principal component model for the multiplicative part. The triple-centering model could also be used in this way, especially for univariate three-way factorial designs. The latter perspective was taken up by Gower (1977). He described three-mode triadditive models that fit the overall mean and main effects additively and two-way interactions multiplicatively, thereby staying in the Mandel tradition.

$$\hat{x}_{ijk} \quad = \quad \mu + \alpha_i + \beta_j + \gamma_k + e_{ijk} \tag{6.14}$$
$$\text{with} \quad e_{ijk} = \breve{a}_i \breve{b}_j + \ddot{a}_i \ddot{c}_k + \breve{b}_j \breve{c}_k,$$

where $\breve{a}_i$, $\ddot{a}_i$, and so on refer to components, that are derived from separate singular value decompositions on the two-mode marginal matrices averaged over one subscript. It is assumed that all effects and multiplicative components sum to 0, and that there is no three-way interaction. The additive portion is fitted by the standard least-squares estimators, and the multiplicative part is based on the residuals. Gower (1977) also discussed setting components equal to each other, such as, for instance, $\breve{a}_i = \ddot{a}_i$. He also suggested adding a single term $a_i b_j c_k$ to model the three-way interaction.

6.8.2   Primary additive plus multiplicative models

Taking into consideration the preferred method of fiber centering and the distinction between the Gollob and Mandel approaches, the two main models for decomposing the variability in a three-way array $\mathcal{X}$ are the univariate three-way factorial design with a single observation per cell (the Mandel model) and the multivariate two-way factorial design (the Gollob model).

Triadditive models: Univariate three-way factorial design.   The univariate analysis of variance decomposition for the univariate two-way factorial design is (using the Parafac model)

$$x_{ijk} = \mu + \alpha_i + \beta_j + \gamma_k + \alpha\beta_{ij} + \alpha\gamma_{ik} + \beta\gamma_{jk} + \alpha\beta\gamma_{ijk} \qquad (6.15)$$

$$\text{with } \alpha\beta\gamma_{ijk} = \sum_{s=1}^{S} a_{is} b_{js} c_{ks} g_{sss},$$

where μ, α_i, β_j, and γ_k are the usual grand mean, row effect, column effect, and tube effect and the two-way and three-way interactions are similarly defined with each term having the standard zero-sum assumptions for the effects. Three-way monopolar rating scale data are typically handled this way, as are genotype by year by location yield data in agriculture. In terms of estimating this decomposition, we may express this in terms of means as

$$
\begin{aligned}
x_{ijk} = & \ \bar{x}_{...} + (\bar{x}_{..k} - \bar{x}_{...}) + (\bar{x}_{i..} - \bar{x}_{...}) + (\bar{x}_{.j.} - \bar{x}_{...}) \qquad (6.16) \\
& + (\bar{x}_{ij.} - \bar{x}_{i..} - \bar{x}_{.j.} + \bar{x}_{...}) + (\bar{x}_{i\cdot k} - \bar{x}_{i..} - \bar{x}_{..k} + \bar{x}_{...}) \\
& \mid (\bar{x}_{j\cdot k} - \bar{x}_{.j.} - \bar{x}_{..k} + \bar{x}_{...}) + \sum_{s=1}^{S} a_{is} b_{js} c_{ks} g_{sss}.
\end{aligned}
$$

Variations of this model occur when some of the additive terms are absorbed into the three-mode model or are (partly) decomposed themselves via singular value decompositions.

Triadditive models: Multivariate two-way factorial design.   The second way is to represent the data as a multivariate two-way factorial design again with a single observation per cell (the Gollob model).

$$x_{ik}^j = \mu^j + \alpha_i^j + \gamma_k^j + \alpha\gamma_{ik}^j \text{ with } \alpha\gamma_{ik}^j = \sum_{s=1}^{S} a_{is} b_{js} c_{ks} g_{sss},$$

where the superscript belongs to the mode of the criterion variables. Three-way profile data and three-way rating scales with bipolar scales would typically use these formulations.

To estimate the parameters of the multivariate two-way design, we may express it in terms of means as

$$x_{ijk} = x_{ik}^j - \bar{x}_{.k}^j = (\bar{x}_{i.}^j - \bar{x}_{..}^j) + (x_{ik}^j - \bar{x}_{i.}^j - \bar{x}_{.k}^j + \bar{x}_{..}^j). \qquad (6.17)$$

The second term on the right-hand side of Eq. (6.17), $(x_{ik}^j - \bar{x}_{i.}^j - \bar{x}_{.k}^j + \bar{x}_{..}^j)$, indicates the subject by concept interaction for each scale and describes the relative differences between the individuals in the concept usage for each scale. If it is sufficiently small, the equation is almost equal to Eq. (6.11) (p. 135), and the results will closely resemble the results of a two-mode analysis of the ratings averaged over all individuals. Therefore, it is the size of this term that is crucial for the meaningful application of three-mode analyses.

6.8.3   Mixed multiplicative models

One of the difficulties of the multiway factorial models discussed earlier, such as Eq. (6.14), is that separate decompositions seem to be necessary for the two-way and the three-way interaction terms. In principle, in such a setup all components for a mode are different. For example, the model with one component for each interaction becomes (see Gower, 1977)

$$\hat{x}_{ijk} \quad = \quad \mu + a_i + b_j + c_k + \breve{a}_i \breve{b}_j + \ddot{a}_i \ddot{c}_k + \breve{b}_j \breve{c}_k + \dot{a}_i \dot{b}_j \dot{c}_k. \qquad (6.18)$$

The essence of these models is that at the same time they contain multiplicative decompositions for the two-way interactions and the three-way interaction. Such models have the advantage of parsimony because it is not the complete interactions that are fitted but only their systematic part. The not-fitted (or residual) parts of all interactions together form the residual sum of squares, and they absorb most of the degrees of freedom associated with the complete interactions.

A problem is that there is no relationship between any of the components within a mode, nor is there a specific loss function. If one wants to bring more structure into Gower's model, it is not immediately obvious how to handle this situation within standard analysis of variance frameworks. De Leeuw (1982, personal communication), however, suggested using an extension of the Tucker3 model for such a purpose.

Mixed multiplicative Tucker3 model.   Let us assume that all component matrices **A**, **B**, and **C** of a three-mode principal component model have a constant first column consisting entirely all ones, $\mathbf{1}^3$. We may then write the triadditive model as a complete Tucker3 model, the simplest of which only needs two components for each mode:

$$
\begin{aligned}
x_{ijk} \quad = \quad & 1_{i1}1_{j1}1_{k1}g_{111} & (6.19) \\
& + a_{i2}1_{j1}1_{k1}g_{211} + b_{j2}1_{i1}1_{k1}g_{121} + 1_{i1}1_{j1}c_{k2}g_{112} \\
& + 1_{i1}b_{j2}c_{k2}g_{122} + a_{i2}1_{j1}c_{k2}g_{212} + a_{i2}b_{j2}1_{k1}g_{221} \\
& + a_{i2}b_{j2}c_{k2}g_{222}.
\end{aligned}
$$

If this model fits, a considerable decrease in interpretational complexity has been achieved because of the small number of multiplicative terms. By introducing additional terms, different components for different interactions can be accommodated.

---

[3]Technically, the columns should have a sum of squares of 1 to stay in line with the basic form of the model, but this would complicate the formulas unnecessarily.

Mixed multiplicative Parafac model.   An interesting alternative proposed by Bro (Bro, 1998c; Bro & Jakobsen, 2002a), under the name GEMANOVA, and also by Heiser and Kroonenberg (1997) is to use the Parafac model for the same purpose. An example of the Parafac formulation is

$$x_{ijk} = 1_{i1}1_{j1}1_{k1}g_{111} \qquad (6.20)$$
$$+ a_{i2}1_{j2}1_{k2}g_{222} + b_{j3}1_{i3}1_{k3}g_{333} + 1_{i4}1_{j4}c_{k4}g_{444}$$
$$+ 1_{i5}b_{j5}c_{k5}g_{555} + a_{i6}1_{j6}c_{k6}g_{666} + a_{i7}b_{j7}1_{k7}g_{777}$$
$$+ a_{i8}b_{j8}c_{k8}g_{888}.$$

For the Parafac model it is necessary to define different components for each term because the model does not allow reuse of the same component as the Tucker model does. Reusing components would immediately lead to proportional components within modes and this violates the basic assumptions for estimating the model and will lead to degeneracies (see Section 8.6.2, p. 185).

In matrix notation these models become slightly more transparent if we use the convention that a **1** indicates a constant component of ones. Then, the mixed multiplicative Parafac model using rank-one arrays becomes

$$\underline{\mathbf{X}} = g_{111}(1_I \otimes 1_J \otimes 1_K) \qquad (6.21)$$
$$+ g_{222}(\mathbf{a}_1 \otimes 1_J \otimes 1_K) + g_{333}(1_I \otimes \mathbf{b}_2 \otimes 1_K) + g_{444}(1_I \otimes 1_J \otimes \mathbf{c}_3)$$
$$+ g_{555}(\mathbf{a}_5 \otimes \mathbf{b}_5 \otimes 1_K) + g_{666}(\mathbf{a}_6 \otimes \mathbf{I}_J \otimes \mathbf{c}_6) + g_{777}(1_I \otimes \mathbf{b}_7 \otimes 1_7)$$
$$+ g_{888}(\mathbf{a}_8 \otimes \mathbf{b}_8 \otimes 1_8)$$

Estimation.   Because Eq. (6.21) is a complete loss function for these mixed multiplicative three-mode models, the estimation of the parameters can be solved via an adaptation of the standard alternating least-squares algorithms. The major difficulty in such models is the imposition of the proper identification restrictions. However, this is an advanced and not completely solved issue, which will not be pursued here.

6.9   RECOMMENDATIONS

The recommendations presented here should be seen as a general guide to what can be done with data sets. Especially in situations in which means can be modeled, much more content-specific information is needed. Harshman and Lundy (1984d, pp. 257–259) present a set of comparable recommendations.

When means cannot be interpreted or when they are incomparable within a mode, they should be eliminated, that is, set equal to 0 via subtraction. Furthermore, when means are interpretable and comparable within a mode, and when it is evident that the differences have been caused by influences not connected with the multiway data itself, they had best be modeled and explained separately outside the three-mode model.

For multiway profile data of subjects (objects) by variables by conditions (situations and/or time points), one will generally use *profile preprocessing*, that is column (i.e., fiber) centering (across subjects) and lateral slice normalization per variable. Only in rare cases, lateral-slice centering is employed. The choice for the latter needs to be based on specific arguments for why one does not want to have fiber centering with its nice modeling and interpretational properties. After fiber centering, the data will contain all individual differences, that is, the subject by variable interaction, the subject by condition interaction, and the subject by (variable by condition) three-way interaction.

When three-way ratings scales are not handled as profile data but as rating scale data, mostly arranged as concepts (situations, conditions) by scales by subjects, serious consideration should be given to double centering, in particular, fiber centering across concepts (i.e., per variable) and fiber centering across scales (i.e., per row), provided that the scales are not bipolar. The result will be that the data will primarily contain the concept-by-scale interaction and that individual differences are only present in the three-way interaction.

In the case of a multiway factorial model, the choice of centering is primarily determined by the interactions that need to be modeled simultaneously, and this will often be only the multi(three)-way interaction. In the latter case, the two-way interactions are often modeled separately with two-way decompositions. So far, experience with simultaneously modeling the two-way and three-way interactions with only a few components has been limited.

Finally, as different centerings lead to different solutions, it is preferable to determining *a priori* which centering is appropriate. Although solutions based on different centering procedures can be inspected in order to decide *a posteriori* which of them is better, one can easily lose sight of the difference between *a priori* hypotheses and *a posteriori* conclusions. In the end, it will be considerations of the purpose of the research, the data design, and the subject matter which should decide the appropriate preprocessing, but the choice is never an easy or automatic one.

With respect to normalization, the standard recommendation in the case of incomparable or uninterpretable scale factors is to normalize the slices, but in agricultural experiments, for instance, a case has been made for environment (or fiber) normalization in both the two-way and the three-way cases. In this respect it is easier to compare solutions with different normalization, because the procedure does not so much change the data, as weight different parts of the data in different ways. The effect of this is easier to evaluate across models.

CHAPTER 7

MISSING DATA IN MULTIWAY ANALYSIS

"[a]nalyzing data that you do not have is so obviously impossible that it offers endless scope for expert advice on how to do it." (Ranald R. MacDonald, University of Stirling, UK.)[1]

7.1 INTRODUCTION

7.1.1 Occurrence of missing data

It is rare in any empirical study not to have at least a few data points for which the values are unknown. There can be a variety of causes for such missing data, such as accidents of some sort, flaws in the data collection, the impossibility to obtain measurements because of a too low signal, or by design. There may be no pattern to the missing data, so that they are scattered through the entire data array, or they may show specific patterns because certain data could not be collected, were no longer

---

[1] *http://www.psychology.stir.ac.uk/staff/rmacdonald/Missing.htm.* Accessed May 2007.

Applied Multiway Data Analysis. By Pieter M. Kroonenberg
Copyright © 2007 John Wiley & Sons, Inc.

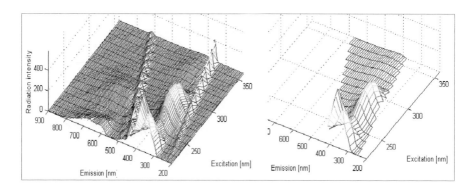

Figure 7.1   Creation of missing data in fluorescence emission-excitation spectroscopy . Relative intensities of combinations of excitation and emission wavelengths (left-hand panel) have been set to missing (right-hand panel) because they are unrelated to the phenomenon under study.

collected, or part of the data collection started later than that of the main body of the data. For some data, certain combinations of levels simply cannot occur, such as pregnant males; see, for instance, Schafer and Graham (2002, for a full discussion). The origin of the missing data can influence the way they are handled in an analysis, and the way the results of the analyses are interpreted. In particular, one may want an analysis in spite of the presence of the missing data, or one may set up an analysis with the specific purpose to estimate the missing data themselves, or a combination of both. It should be noted that methods differ in their appropriateness for both purposes.

Missing data may also be produced on purpose, in order to perform analyses of data sets in which outlying data values have been identified or are suspected. In such situations the outliers can be designated as missing, and analyses with and without these outliers can be performed to assess their influence on the solution. A comparable situation can arise when one suspects contaminations from irrelevant sources to be present in one particular year, and this needs to be investigated. For instance, one may suspect that a three-way interaction is caused by a single or by only a few data points. By treating the outliers as missing, it is possible to establish whether they are responsible for the three-way structure or interaction. If so, the complexity of the analysis may be reduced so that there may no longer be a need for a multiway analysis of the data. An alternative is to use robust methods of estimation; see Sections 12.8–12.10. A final example can be found in certain experiments in analytical chemistry, when one knows that a certain unwanted variation exists, and produces, for instance, a signal in wavelengths unrelated to the processes under investigation. One may choose to eliminate that part of the signal, as Fig. 7.1 illustrates (for a more detailed explanation, see Section 7.4.2).

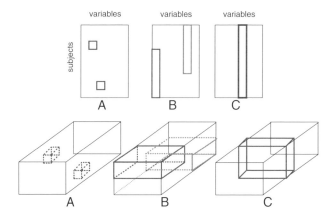

Figure 7.2   Patterns of missing data in two-way and three-way data. Top: *Two-way data*, A: single observations missing, B: partial columns missing, C: complete column missing. Bottom: *Three-way data*, A: single observations missing, B: part of observations have no observation in the beginning, some have all their observations, some have observations at the end — cohort design, C: complete slice missing.

7.1.2   Missing data structures

Presenting guidelines in the presence of missing data is difficult because the appro-priateness of an analysis is dependent on the mechanism underlying the occurrence of the missing data. The common distinction is between *missing completely at random* — MCAR, *missing at random* — MAR, and *missing not at random* — MNAR. The ramifications of these types of missing data are discussed at length in Schafer and Graham (2002). In what follows we will assume that the MAR assumption is reason-able for the examples discussed here, that is, the probability of being missing may depend on the observed data but not on the missing data themselves. In longitudinal data this means that dropout may depend on any or all occasions *prior* to dropout, but not the data of the occasion itself. Planned missing values also fall under the MAR mechanism. For instance, in long-term field trials such planned dropout is a common practice by plant breeders, as old cultivars are faded out in a trial and new cultivars or varieties are introduced. In the social and behavioral sciences, complex cohort studies are sometimes conducted in which different cohorts start at different times and thus create missing data by design.

7.1.3   Missing data in three-way designs

One of the interesting aspects of analyzing multiway data in the presence of missing values is that due to the presence of more ways, the data are much richer than in the two-way case. The additional information makes the analyses easier and often more

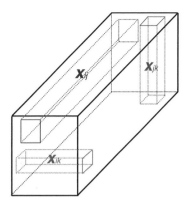

Figure 7.3   Three types of missing fibers: rows ($\mathbf{x}_{ik}$), columns ($\mathbf{x}_{jk}$), and tubes ($\mathbf{x}_{ij}$)

reliable, even when a considerable amount of data are missing (e.g., see Fig. 7.2, bottom row B). However, there must be sufficient information at the right places in the data set to allow for estimation of the missing data. For instance, completely missing slices are effectively no part of the data set because no information about the slice is available. The frontal slice in Fig. 7.2C (bottom row) may serve as an example. It may happen, for instance, that there are no data on any of the variables available in a particular year. Unless one has an analysis with specific constraints on the components, say, smoothness constraints, it does not make sense to estimate the data in the missing slice, and the analysis has to proceed without it. Completely missing fibers pose a problem in that the final estimates of the missing data may depend on starting values in the algorithm. Missing fibers occur regularly: given a subject by variable by occasion array, a respondent i may have forgotten to reply to a question j at each time point because it was situated on the last page, so that a complete tube $\mathbf{x}_{ij}$ is missing; if a respondent i was not present at a particular time point k, a complete row $\mathbf{x}_{ik}$ is missing; or if at a particular time point k the interviewer forgot to ask all respondents question j, the column $\mathbf{x}_{jk}$ is missing (see Fig. 7.3).

7.1.4   Missing data and stochastics

Certain multiway data consist of a set of random variables that are measured under several conditions on a number of subjects. For such data, it is sufficient to compute variable means per condition as well as a multivariable–multioccasion covariance matrix. There are other data sets, in which the stochastic framework is lacking and none of the modes consists of random variables, because the subjects, objects, and so on do not represent a sample from a population. For instance, in an agricultural variety trial the "subjects" are specially selected varieties or all viable varieties of a crop. Similarly, the multiway data set may consist of the three-way interaction of a large experiment with three factors, each with many levels. In such cases, no clear stochastic framework is present as an estimate of the variability cannot be made from the single interaction value per cell.

Without a stochastic framework, imputing missing data with advanced techniques such as multiple imputation, bootstrap analyses, and maximum likelihood estimation is virtually impossible, because each of these techniques assumes that there is one stochastic mode. Therefore, not all techniques for handling missing data may are appropriate in all situations.

7.1.5 Basic approaches

There is now a general consensus that there are three major approaches for handling missing data, which produce good results in many cases: (1) *full-information methods*, which use all data and their inherent structure to estimate the missing data via expectation–maximization (EM) methods; (2) *multiple imputation methods*, which generate a number of alternative data sets with varying values for the missing data, each of which is analyzed according to the intended analysis procedure; and (3) *likelihood-based methods*, which assume that one of the modes consists of random variables and that the observations are random draws from (mostly normal) distributions.

The drawback of the expectation–maximization methods is that its missing value estimates reduce the variability in the data because they exactly conform to the model. Moreover, the estimates are model dependent and are generally also dependent on their starting values. To overcome the starting-value dependence, one can use several different starting values for the missing data before the iterative model-dependent estimation starts, and compare the stability of the solutions. With multiple imputation, one can overcome the error reduction and some of the model dependence. In multiway analysis, however, not much work has been done in this area.

All procedures discussed below share the basic assumption that the missing data actually belonged to the data set and would have conformed to the model should the data have been available. This assumption is not trivial, as it presupposes that the reason that these data points were not observed is not related to the processes that generated the data themselves, that is they are missing at random. For instance, it is assumed that the missing data do not come from a totally different population. Furthermore, the methods aim at using all available information in the data set in order to provide a reasonable estimate of these missing data.

7.2 CHAPTER PREVIEW

Several papers have treated the handling of missing data within the context of (two-mode) principal component analysis, but there is only a limited number of papers addressing this issue within the context of multiway analysis. In this chapter we will describe such procedures for the two-mode case (in particular, EM-estimation and multiple imputation) and then apply extensions of these proposals to multiway data. The emphasis in the exposition will be on analyses with the Tucker3 model.

A special problem, possibly first pointed out by Grung and Manne (1998), is the combination of incomplete data and preprocessing (see Chapter 6 for an overview of preprocessing). As nearly all multiway data sets, especially in an exploratory context, require some form of preprocessing before the analysis proper, this combination nearly always arises and introduces algorithmic complications. Extensive discussions of applications with missing data within chemistry using the Parafac model can be found in Smilde et al. (2004).

Because at the time of writing only limited experience was available about especially multiple imputation in multiway analysis, in places this chapter has more the character of a reconnaissance than an in-depth treatment of available proposals.

7.3   HANDLING MISSING DATA IN TWO-MODE PCA

In this section we first look at some procedures in use for handling missing data in two-mode component analysis, such as crisscross regression, expectation–maximization procedures, maximum-likelihood component analysis, multiple imputation, and optimal scaling. We will also pay attention to preprocessing in relation to missing data in principal component analysis.

7.3.1   Crisscross regression

The basic discrepancy function $\mathcal{F}$ for PCA is

$$\mathcal{F} = \sum_{ij} \left(x_{ij} - \sum_{s=1}^{S} a_{is}b_{js} \right)^2. \tag{7.1}$$

Gabriel and Zamir (1979) discussed estimating the parameter vectors $\mathbf{a}_s$ and $\mathbf{b}_s$ via crisscross regression, as an alternative for using the standard eigenvalue–eigenvector approach. In the latter case, the component matrices $\mathbf{A}$ and $\mathbf{B}$ are estimated in a single step by using the singular value decomposition. In the regression approach, the estimation is solved by taking the components of the row and column space in pairs, first $(\mathbf{a}_1, \mathbf{b}_1)$, and subtracting the resulting components from the data, that is, by deflation. Thus, the steps consist of solving for the s' pair $(\mathbf{a}_{s'}, \mathbf{b}_{s'})$ after the previous $s' - 1$ pairs have already been dealt with:

$$\mathcal{F}^1 = \sum_{ij} \left(\left[x_{ij} - \sum_{s<s'}^{S} a_{is}b_{js} \right] - a_{is'}b_{js'} \right)^2. \tag{7.2}$$

Equation (7.2) is solved by taking a starting value for, say, $\mathbf{a}_{s'}$, finding $\mathbf{b}_{s'}$ via regression, using the new $\mathbf{b}_{s'}$, and solving for $\mathbf{a}_{s'}$ until convergence. Thus, rank-one regression steps are iterated, hence the name crisscross regression. The deflation is possible because the successive pairs $(\mathbf{a}_{s'}, \mathbf{b}_{s'})$ are orthogonal. In contrast, the singular value decomposition handles all S components of the two modes together.

An indicator or weight matrix $\mathbf{W}$ can be defined with $w_{ij} = 0$ if x_{ij} is missing, and $w_{ij} = 1$ when it is not. The discrepancy function $\tilde{\mathfrak{F}}$ including the weight matrix becomes

$$\tilde{\mathfrak{F}} = \sum_{ij} w_{ij} \left(x_{ij} - \sum_{s=1}^{S} a_{is} b_{js} \right)^2. \tag{7.3}$$

The standard way to solve Eq. (7.1) via a singular value decomposition is not possible in this case due to the missing data. In addition, it is also impossible to solve the equation by simple deflation, because due to the missing data the successive components are no longer necessarily orthogonal.

One solution is not to do successive deflation but to subtract from the data all S components except for s':

$$\tilde{\mathfrak{F}} = \sum_{ij} w_{ij} \left(\left[x_{ij} - \sum_{s \neq s'}^{S} a_{is} b_{js} \right] - a_{is'} b_{js'} \right)^2, \tag{7.4}$$

and then estimate the $a_{is'}$ and $b_{js'}$. For the next step, subtract again from the data all components, but this time excepting those for $s' + 1$, and continue. This procedure is done for each s in turn and then repeated until convergence. Gabriel and Zamir call this process *successive dyadic fitting*.

An alternative solution is to perform weighted regression over the whole component matrix $\mathbf{A}$ with a prespecified number of columns S given $\mathbf{B}$, then do the same for $\mathbf{B}$ given the newly computed $\mathbf{A}$, and so on. In this case, the per-matrix regressions are carried out on a per-row basis. For instance, the regression for the first mode takes the form

$$\left[\left(\sum_{i}^{I} w_{ij} a_{is} a_{is'} \right) \right] \mathbf{b}_j = \left[\left(\sum_{i}^{I} w_{ij} a_{is} x_{ij} \right) \right]. \tag{7.5}$$

Convergence in both algorithms is assured but, not necessarily to the global minimum.

In addition, Gabriel and Zamir (1979) show that the iterative procedure may not converge to the closest fit. "In particular, when some weights were 0 [as is the case with a missing-data weight matrix] these iterations occasionally converged well away from the least squares fit." (p. 490). They show that convergence of the loss function can be reached with the estimates for the missing values increasing without bound, because these increasing estimates do not affect the fit due to their zero weights. They tried to solve this phenomenon by a suitable choice of starting values. The possible cause might be the use of too many components in combination with starting values far away from the global minimum of the function, so that random error present in higher-order components influences the recovery of the lower-dimensional structure.

The same problem is the likely cause for a similar phenomenon in multiway analysis, as demonstrated in Section 7.7.3.

Van Houwelingen (1983) independently developed a similar algorithm combining successive and simultaneous alternating least-squares, which was first described and applied in De Ligny, Nieuwdorp, Brederode, Hammers, and Van Houwelingen (1980). He also mentioned the problem of local minima, the need for several starting solutions, and the possibility of diverging estimates. In addition, under the assumption of normal distributions with equal variance for the variables, Van Houwelingen developed confidence intervals for the estimated data irrespective of whether or not they are missing. De Ligny et al. (1980) placed a strong emphasis on estimating the missing data and their accuracies, rather than on estimating the parameters.

7.3.2   Expectation–maximization: Single imputation

Dempster et al. (1977) proposed a general approach for estimating parameters of models when the data are incomplete. As summarized by Kiers (1997b, p. 252), this approach "consists of setting the missing data elements initially at arbitrary values, and in the loss function, assigning zero weights to the residuals that belong to these elements. Such loss functions can be minimized by a 'missing data imputation' approach, which is a special instance of the EM-algorithm (Dempster et al., 1977): By alternately fitting the model to the full data set (including estimates for the missing values), and replacing the missing elements by the current model estimates for these elements, the weighted loss function is decreased monotonically (and assumed to be minimized at least locally)." Thus, in this case, the iteration is performed by solving the complete Eq. (7.1) at each step, but with the estimated values for the missing data readjusted at each step. The changes in the estimates of the missing data, of course, also affect the components themselves, so that they will also change between iteration steps (see also Grung & Manne, 1998).

In his paper Kiers (1997b, p. 254) also proposed a majorization approach to solve weighted least-squares fitting using ordinary least-squares algorithms called iterative ordinary least squares (OLS). He showed that in the special case where the weight matrix is binary (as is the case for missing data where 0 is missing and 1 is present), the "iterative OLS algorithm reduces to a missing data imputation procedure", that is, the EM approach.

7.3.3   Comparison of EM and crisscross

Kiers then compared the iterative OLS algorithm for fitting PCA with the crisscross algorithm of Gabriel and Zamir (1979). On the basis of several simulation experiments, Kiers (1997b, p. 255) observed that (1) "in 29 cases the two algorithms yielded virtually equal function values, and in only one case the iterative OLS algorithm led to a function value that was more than 1% lower than that resulting from the crisscross algorithm"; (2) the two "methods hardly differ in terms to sensitivity to local minima, and it can be expected that this difference will further diminish when more than one additional randomly started run is used"; (3) the methods differed considerably in computational efficiency, as the iterative OLS (or alternating least-squares) algorithm required three to four times more floating point operations than the crisscross algorithm.

7.3.4 Maximum likelihood PCA

Maximum likelihood PCA (MLPCA) (Wentzell, Andrews, Hamilton, Faber, & Kowalski, 1997) is a generalized form of principal component analysis, which allows estimates of measurement uncertainty to be incorporated in the decomposition. The starting point is that each observation x_{ij} for subject i on a variable j is the result of fallible measurement, and that the observation is the sum of a nonstochastic 'true' score x_{ij}^{true} and a stochastic measurement error ϵ_{ij}. For each observed variable $\mathbf{x}_j$ we have an I vector of true scores x_j^{true} and an I vector of measurement errors ϵ_j. Ψ_j is the covariance matrix of the measurement errors for the jth variable. Thus, in the case of uncorrelated errors for the subjects Ψ_j is diagonal. However, this assumption is typically not true when the 'subjects' are time points, that is, for time-series and similar time-and space-dependent observations. Note that for each variable there may be a different error covariance matrix. This proposal takes its lead from standard regression analysis with fixed predictors or effects, in which case the randomness is not in the model part but in the error part. MLPCA assumes that the error (co)variances are known, but this is not generally the case so that they have to be estimated from the observed scores. In essence, MLPCA thus becomes a two-step procedure in which first the error covariances are estimated and then the parameters of the model. Most authors would prefer some kind of simultaneous procedure, but a disadvantage is that in that case more complicated algorithms are needed. Earlier similar proposals were made by G. Young (1940) and Whittle (1952) (cited in Tipping & Bishop, 1999).

In maximum likelihood PCA a weighted least-squares loss function is used rather than an unweighted one. Thus, instead of $\sum_j (\mathbf{x} - \hat{\mathbf{x}}_j)^2$ the weighted loss function $\left(\sum_j (\mathbf{x} - \hat{\mathbf{x}}_j)' \Psi_j^{-1} (\mathbf{x} - \hat{\mathbf{x}}_j) \right)$ is employed. Missing data can be accommodated by assigning very large variances to the missing measurements prior to the analysis proper. The effect of this is that they are seriously downweighted so that they do not seriously contribute to the analysis (Andrews & Wentzell, 1997).

Tipping and Bishop (1999) proposed a maximum likelihood procedure for PCA (called probabilistic PCA) by assuming stochastic latent variables, rather than assuming nonstochastic true scores. Moreover, they assumed that the errors are uncorrelated latent variables with unknown, but identical (or isotropic) variance, in contrast with the assumptions in Wentzell et al. (1997). The parameter estimation is solved via an EM-algorithm. Given that an EM-algorithm is already used, the treatment of missing values can be incorporated in a natural way as a single imputation step (e.g., see Skočaj, Bischof, & Leonardis, 2002) .

7.3.5 Optimal scaling and missing data in PCA

So far, we have only looked at missing data procedures for PCA assuming the data were continuous. However, principal component analysis for data with mixed measurement levels (i.e., nominal, ordinal, and/or numerical) has undergone sophisticated development, and through its availability in major software packages can be routinely applied; see Meulman, Van der Kooij, and Heiser (2004) for an introductory account of the optimal scaling approach (se also Section 4.10.10) toward PCA including applications. Within this context, subjects with missing observations can be dealt with by defining an extra category for a variable to which all subjects with missing values on that variable are allocated, or by defining for each variable separately a category for each subject who has a missing value for that variable. A description of the details and comparisons for these approaches can be found in Meulman et al. (2004).

7.3.6 Multiple imputation and PCA

When employing single imputation for missing data using a particular model, the variability in the completed data set with respect to the model is reduced in two different ways. In the first place, an *observed* score consists of a part fitted by the model plus a certain amount of error, while for an *imputed* missing data point the error is lacking, because it was "created" by the model itself. The effect of this is that the model seems to fit the data better than would have been the case if all data had been present. Second, as their true values are unknown, it is also unknown how these missing data points would have influenced the parameter estimates had they been present. Thus, the accuracy of the estimates of the parameters is overstated.

Multiple imputation attempts to remedy this situation by creating several data sets with different imputed values for the missing observations. Because of that parameter estimates are (slightly) different in the analysis of each imputed data set. In this way, the dependence of the parameters on the missing values can be assessed. When each of these data sets is analyzed by means of a principal component analysis, it is possible to assess the influence of the imputed missing values on the parameters of the component model, and on their variability. When there are very few missing data, say, 5% or less, multiple imputation might not be necessary, but with larger percentages of missing values, multiple imputation can ensure better estimates and an assessment of their variability.

There are two types of variability estimates that are central to multiple imputation. One type is *within-imputation variance*, which is based on the average of the squared standard errors taken over the imputed data sets. The other type is *between-imputation variance*, which is the sample variance of the parameter estimates. A specific problem in principal component analysis is the possible absence of a stochastic mechanism underlying the analysis. If in two-mode analysis the subjects are seen as a random sample from a population, and an estimate can be obtained for the standard errors of the parameters similar to probabilistic or maximum likelihood PCA, then multiple imputation seems a good option. In that case the final standard errors of the parameters are properly based on both types of imputation variances. However, if no stochastic framework is present for the multiway data, the situation is more difficult, because then the within-imputation standard errors are not available. However, there is no problem with the between-imputation variance as that is based on the estimated parameters themselves.

So far, very few studies have considered multiple imputation in the context of two-mode principal component analysis. Recent exceptions are Ho, Silva, and Hogg (2001), who compared multiple imputation and maximum likelihood PCA, and D'Aubigny (2004), who combined multiple imputation and (two-mode) maximum likelihood PCA, and used subsequently three-mode analysis to assess the results.

7.3.7 Centering, normalization, and missing data

Similar to analysis of variance, the estimations of means, normalization factors, and parameters of the principal component model in the case of incomplete data can no longer be done independently. This is true for two-way and multiway analyses. Without missing data, the means and normalization factors are removed first from the data before the analysis proper. However, when missing data are present this can no longer be done, because the continuous reestimation of the missing data during the iteration, will uncenter the data; see Grung and Manne (1998), Bro and Smilde (2003a), and Smilde et al. (2004, pp. 231, 251). If only centering is necessary, one may perform a recentering step after the missing data have been imputed from the then current model. This will not give information on the actual values of the means, but the model estimation can proceed in an efficient fashion. However, when normalization is also involved and/or estimates of the means are desired, each main iteration step should include a step for estimating the means and the normalization factors.

7.3.8 Bootstrap, missing values, and PCA

Rather than using a maximum likelihood approach to assessing the variability of the parameters in a principal component analysis with missing data, Adams, Walczak, Vervaet, Risha, and Massart (2002) propose to use bootstrap procedures in combination with an EM-algorithm for PCA. They investigated two procedures for this. The first procedure consists of generating bootstrapped data, including missing data, then imputing each of the bootstrapped data matrices, and subsequently analyzing each imputed and bootstrapped data set with EM PCA. This procedure is a form of multiple imputation as each of the bootstrap samples generates different estimates for the missing data. The major difference with the standard multiple imputation is that in the latter case all values except the missing data are the same, while in the former all values in the data matrix are different irrespective whether they are missing or not. The second procedure is to first impute the missing values and then generate bootstrap versions of the completed data. The authors supply two examples and did not find many differences, but further validation of their suggestions is required. An interesting addendum to their work would be to combine the bootstrapped data matrices, either before or after imputation into a single three-mode analysis following D'Aubigny's (2004) suggestion.

7.3.9 Conclusions from two-mode PCA

Handling incomplete data within the context of two-mode principal component analysis has seen a varied development and several procedures have proved feasible in handling the estimation. Of the procedures discussed, however, only a limited amount have been extended to three-way data.

7.4 HANDLING MISSING DATA IN MULTIWAY ANALYSIS

7.4.1 Introduction

Going from incomplete two-way data to incomplete multiway data is stepping into relatively uncharted theoretical territory. Several proposals for handling missing data in the multiway context have been put forward, discussed, and widely applied, but not many systematic comparative analyses have been carried out. Several analyses using the Parafac model are available in chemistry, for instance in the fields of chromatography (for an overview, see Smilde et al., 2004). It should be noted that missing value routines based on the EM-principle have for a long time been incorporated in most programs for three-mode component analysis mostly without accompanying information evaluating the procedures. Notwithstanding, many applications exists in which these missing data procedures have been routinely employed.

7.4.2 Algorithms

Probably the earliest known generalization of Gabriel and Zamir's (1979) crisscross algorithm to three-way data comes from Weesie and Van Houwelingen (1983) (see also De Ligny et al., 1984; Spanjer, 1984, pp. 115–121). They developed an algorithm and program for the Tucker3 model called GEPCAM — Generalized principal component analysis with missing data (see Section 5.7.3, p. 101). It uses a weighted regression based algorithm rather than the algorithm based on eigendecompositions proposed by Kroonenberg and De Leeuw (1980) (see Section 5.7.3, p. 98).

Harshman mentioned to Kiers (1997b, p. 251) that he had developed a crisscross regression algorithm for the Parafac model generalizing Gabriel and Zamir's (1979) approach. Carroll et al. (1989) presented a similar algorithm. Heiser and Kroonenberg (1997) and Kroonenberg and Heiser (1998) have developed a weighted crisscross algorithm for the Parafac model referring to it as a *triadic algorithm*. Tomasi and Bro

(2005) compare a Levenberg–Marquardt algorithm with the basic Parafac algorithm with imputation. It turned out that in most cases their capabilities in recovering the correct solution during simulation experiments was comparable, but that in difficult problems the Levenberg-Marquardt appeared to be slightly better and certainly faster.

Bro (1997, p. 157) mentions EM and crisscross regression, demonstrating the use of missing data to eliminate nonlinear Rayleigh scatter, which corrupts Parafac analyses of spectra resolutions. Bro (1998c, p. 235) adds: "For weak solutions fluorometric data can theoretically be described by a PARAFAC (CP) model [...] with the exception that for each sample the measured excitation–emission matrix [...] has a part that is systematically missing in the context of the trilinear model [...]. Emission is not defined below the excitation wavelength and due to Rayleigh scatter emission slightly above the excitation wavelength does not conform to the trilinear PARAFAC model. As the PARAFAC model only handles regular three-way data it is necessary to set the elements corresponding to the 'nontrilinear' area to missing, so that the fitted model is not skewed by these points" (see also Section 7.7 and Fig. 7.1). As an aside it should be remarked that recently other procedures have been developed to detect and handle irregular emissions and scatter. Generally, they are designed to deal with the scatter before estimating the signal with the Parafac model (Bahram, Bro, Stedmon, & Afkhami, 2007; Engelen, Frosh Möller, & Hubert, 2007a; Engelen & Hubert, 2007b).

Furthermore, Bro's (1998, p. 148) experience was that "[u]sing direct weighted least-squares regression instead of ordinary least-squares regression is computationally more costly per iteration, and will therefore slow down the algorithm. Using iterative data imputation on the other hand often requires more iterations due to the data imputation (typically 30–100% more iterations). It is difficult to say which method is preferable as this is dependent on implementation, and the size of the data, and size of the computer. Data imputation has the advantage of being easy to implement, also for problems which are otherwise difficult to estimate under a weighted loss function."

A special case not treated here is the three-mode common factor model (Bentler & Lee, 1978b, 1979; Oort, 1999), which is based on multimode covariance matrices. The reader is referred to the standard literature dealing with estimation of covariance matrices for incomplete data (e.g., see Bollen, 1989, p. 369ff.).

Paatero (1999) has developed a sophisticated program based on a conjugate gradient algorithm for estimating virtually any multilinear model, called the *multilinear engine* which includes missing data handling as well as downweighting procedures of both individual observations and errors.

7.5   MULTIPLE IMPUTATION IN MULTIWAY ANALYSIS: DATA MATTERS

7.5.1   Nature of the data

The first consideration is whether one should treat the variables in the data set as random variables or not. This has consequences for decisions on the acceptability of using multiple imputation, what the status is of the scores of the subjects, and so on. It may occur that the data are a random sample, but one does not want to treat as such or vice versa.

7.5.2   Data arrangements for multiple imputation

Given that present procedures for multiple imputation are designed for two-mode data, it is necessary to matricize three-way data if we want to apply such procedures. There are two options (see Section 2.1, p. 15): (1) construct a wide combination-mode matrix of I subjects by $J \times K$ variables $\times$ measurement times, or (2) construct a tall combination-mode matrix of $I \times K$ subjects $\times$ measurement times by J variables. The first option seems more closely related to the purpose of the multiple imputation, because each variable at each time point is treated separately. In the second option, dependence between "subjects" is introduced due to the multiple measurements on each subject, while the assumption for multiple imputation is that they are independent. To investigate the effect of the two different arrangements for imputation, we may use both methods to compare the results but no systematic analyses exist that explore this.

In Sections 7.7 and 7.8 two examples of handling three-way data with missing elements are presented. It is, of course, difficult to generalize from these two finger exercises, and a serious study should be undertaken to shed more light on the best way to proceed. It is, for instance, likely that if the number of observations is small, the tall combination-mode matrix might be able to give more appropriate results. On the other hand, in case there are large differences in means between the levels of the third mode, the tall combination-mode matrix might introduce more variability than is necessary. In such cases, it could be advantageous to first center the data and then do the multiple imputation. A further consideration is whether for any variable–measurement combination values are available. It is not uncommon in three-mode analysis that at some time point no observations are available on a particular variable, but that they are available at other time points. In such a case, it might even be necessary to start with a tall combination-mode matrix, in order to avoid getting into mathematical problems during the estimation. When both options can be used and the results are comparable, we provisionally suggest using multiple imputations for the wide combination-mode matrix, which conceptually seems most appropriate.

7.5.3   Imputation and out-of-range values

When inspecting the imputed data sets, it may turn out that imputed values are negative and outside the range of the possible values of the variables, for instance, when they take on only positive values. The reason for the negative imputed values is that the imputation is carried out with normal distributions, whereas observed distributions are often rightly skewed with 0 as a lower bound. However, one should not round the negative imputed values to 0 as this will bias the parameters of the model. In particular, the mean will be increased, and the variance will be reduced. It should be borne in mind that, as suggested by Von Hippel (2007), the purpose of most imputation is parameter estimation rather than missing data estimation. If one wants to obtain estimates for the missing data, they should be based on the *means* of the imputed values.

7.6   MISSING DATA IN MULTIWAY ANALYSIS: PRACTICE

In this section a series of considerations will be presented, which can help in deciding what to do when confronted with incomplete multiway data. It is, however, not meant to be prescriptive, but aims at providing a framework for deciding how the missing data can be handled.

Aim of the analysis.   When confronted with missing data, one might desire (1) to estimate the missing data in order to complete the data or to acquire information as to what these values might have been, had they been measured, or (2) to carry out an analysis in spite of the fact that there are data missing. With respect to the second option one may carry out a single imputation while simultaneously estimating missing values and the model, so that the missing data values and the model concur with each other. This will lead to a reduction in error variance because the missing-data estimates fit the model perfectly. The alternative is to perform the estimation of the missing data externally via multiple imputation and then estimate the model. This will incorporate more error variance in the estimation and will provide an estimate of the variability in the parameters due to the uncertainty about the real value of the missing data. It will also provide estimates of the missing data and the accuracy of their estimation.

Estimation of missing data and preprocessing.   When the primary aim is to estimate the missing data within the multiway model, there are two options. The first is to estimate the missing data directly within a multiway model *without* preprocessing. Alternatively, the data are preprocessed first, and then the missing data are estimated with a well-fitting multiway model. In the first option a purely multiplicative model is used for the data so that the means are involved in a multiplicative manner; in the second option a combined additive and multiplicative (or triadditive) model) is used so that the means enter in an additive way. When the variables have different measurement units, some kind of preprocessing seems inevitable. Which of the approaches mentioned is better in which circumstances has not been investigated yet. The crucial factor is to establish in which way it is possible to obtain more accurate estimates, especially in the presence of large amounts of error, as is typical for data in many disciplines.

Model selection.   The selection of an appropriate model is in general independent of the nature of the missing data. However, an interesting, as yet unresearched, question is whether models that can be estimated with different algorithms (e.g., a $2\times2\times2$ Tucker3 model is in many cases equivalent to a Parafac model with two components) lead to different estimates for the missing data.

Single imputation.   Single imputation is carried out via an iterative expectation–maximization procedure in which the missing data step is incorporated in the model estimation procedure. (Clever) initial estimates for the missing data are chosen, the model is estimated, the missing value is estimated from the model, and using the newly completed data set the model is estimated again, and so on. After convergence, one has estimates for both the missing values and the parameters of the model, but no information on the standard errors. Most algorithms for estimating parameters of multiway models include such an estimation procedure. For the Tucker models, the evidence so far is that the weighted least-squares regression approach of Weesie and Van Houwelingen (1983) performs a bit better and faster than the eigenvalue–eigenvector approach of Kroonenberg and De Leeuw (1980) (see Section 7.7.3, p. 161).

Multiple imputation.

In multiple imputation the missing data estimation is separated from the model estimation. Thus, several different data sets are created and the multiway analyses are carried out on the completed data sets, providing the estimates of the variability of both the missing data and the parameters. A fundamental assumption is that the estimated missing data in the completed data sets are reasonable from the point of view of the multiway model to be fitted. How this may be guaranteed is difficult to say. Most imputation methods estimate means and covariance matrices, sample the covariance matrices, and use these estimates with normally distributed disturbances to find values for the missing elements. To what extent such a procedure will be reasonable in multiway analysis among other things depends on the stochastic framework assumed to underlie the data (see Section 7.1.4, p. 146).

Bootstrap procedures.

An alternative to getting estimates for the variability is to perform a single imputation, which optimizes the missing data estimates with respect to the model, possibly with different starting values for the missing data to ensure optimal convergence. Then the completed data set is subjected to a bootstrap analysis (see Section 8.8.1, p. 188) in order to find the uncertainty for the parameters. If the missing data estimates are very unstable, or the solution is unstable due to the presence of the missing data, this will show up as a result of the bootstrap analysis. For details how to proceed in this case, see Section 9.8.2 (p. 233).

Maximum likelihood methods.

So far maximum likelihood methods for analyzing multiway data have only been developed for the Parafac model (Mayekawa, 1987; Vega-Montoto & Wentzell, 2003, 2005b; Vega-Montoto, Gu, & Wentzell, 2005a). Bro, Sidiropoulous, and Smilde (2002b) proposed a maximum likelihood type of procedure but according to Vega-Montoto and Wentzell (2003) its statistical properties are not very clear. Very few applications of the maximum-likelihood procedures have appeared so far, but further testing will have to bring its general practical usefulness for missing data analysis to light.

7.7   EXAMPLE: SPANJER'S CHROMATOGRAPHY DATA

7.7.1   Stage 1: Objectives

Spanjer's chromatography data were published in Spanjer (1984) (see also De Ligny et al., 1984). The authors studied chromatographic retention characteristics of several chemical compounds (Fig. 7.4). The basic purpose of this type study is to determine which chemical compounds are present in a chemical mixture. To do this, the mixture is dissolved in a liquid, the *eluent* or *mobile phase*. This solution is entered into a vertical glass tube or *column* packed with solid material, the *stationary phase*. Different compounds or *solutes* travel at different speeds through the column, depending both on the phases and the affinity of the solute to these phases. The time it takes to travel through the column is called the *retention time*, and this differs with phases and solutes. It takes some time for a solute to emerge from the column but there generally is a single-peaked function indicating the amounts of a solute exiting the column over time. The instrument used to measure the relative concentration over time is called a chromatograph. Missing data can occur because some compounds take too long to travel through the column, so that their peaks do not emerge until after the experiment is concluded.

   Spanjer (1984) used the data set to estimate the missing data themselves and we will also present such an analysis here. In particular, we will explore the effect of starting solutions, different algorithms, varying numbers of components, and effects of centering. In this case, we will only employ a single imputation using the Tucker3 model, leaving the multiple imputation and model comparisons for the second data set. Even though the original authors were primarily interested in estimating the

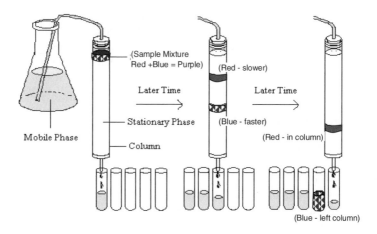

Figure 7.4   Setup for a column chromatography experiment with a two-compound mixture.   One compound (blue - patterned) leaves the column first, the other compound (red - unpatterned) has not left the column yet.   Source: Previous website of Dr.   G. O. Gray, Southwest Baptist University, Bolivar, Missouri (present site: *http://www.sbuniv.edu/chemphys/faculty/GaryGray.htm*; Accessed May 2007); reproduced and slightly adapted with kind permission from Dr. Gray.

missing values, we will also briefly present the results of the three-mode analysis itself.

7.7.2   Stage 2: Data description and design

The data set contains data on phase equilibrium constants of transformed retention data for 39 solutes with 2 eluents (mobile phase) and 3 adsorbents (stationary phase), and of a possible 234 measurements, 21 (= 9%) of which were missing. For six of the solutes only 1 out of the 3 possible values were measured for one of the two eluents.   Retention volumes (V_N) were measured for 39 solutes (monosubstituted phenols, anilines, and pyridines), 2 eluents (35% v/v methylene chloride in n-hexane and pure methylene chloride), and 3 silica-based adsorbents (octadecyl-silica, N-cyanoethyl-N-methylamino-silica, and aminobutyl-silica).   The retention volumes were transformed in order to get retention rates with respect to the adsorbents: Y = $\log(V_N/W)$, where W is the weight of the adsorbent. Full technical details can be found in the original publication. It is important to note that the values in the data set are all comparable without further preprocessing, as they are all retention rates, thus they can be conceived as multiway factorial data. Thus, there is no explicit need for normalization. Even centering is not automatically necessary, and following Spanjer (1984) this was not done in the present analysis.

There is no stochastic mode in the data as the solutes, adsorbents, and eluents all have fixed levels. It is better to see them as, for instance, *true scores*, where the model is fixed but the data points themselves are subject to random error due to measurement inaccuracy, random fluctuations in the chemical processes themselves, and so on. In that sense, they are similar to the data that Wentzell et al. (1997) deal with (see Section 7.3.4). At the same time, these data could lend themselves to multiple imputation if the random element is modeled via the addition of random error to each missing data point. This requires that an estimate of the accuracy of the retention volumes can be made independently, so that an error distribution can be defined. However, we will not pursue this here.

7.7.3   Stage 3: Model and dimensionality selection

Effects on model selection and convergence.   The data have been analyzed in a number of ways, which made it possible to assess the influence of different starting values for the missing elements, different numbers of components, different algorithms, and different starting values for the parameter matrices. In particular, the starting values for the missing data were estimated using (1) the grand mean (G), (2) a main-effects three-way ANOVA model (M), (3) an ANOVA model with all two-way interaction terms included (I), and (4) the final estimates from the original analysis by Spanjer (1984) (S). Both the regression-based GEPCAM algorithm of Weesie and Van Houwelingen (1983) ($\mathcal{G}$) and the alternating least-squares TUCKALS algorithm of Kroonenberg and De Leeuw (1980) ($\mathcal{T}$), were used. Spanjer (1984) used a $4 \times 3 \times 2$ model consisting of 4 solute components, 3 adsorbent components, and 2 eluent components. Based on the deviance plot (Fig. 7.5), we added a $3 \times 3 \times 2$ model and the even simpler $3 \times 2 \times 2$ model (see Section 8.5.2, p. 181, for an explanation of deviance plots). The parameter matrices were either started with random uniform numbers (R) or with a rational start based on the Tucker (1966) method for estimating the three-mode PCA model (T).

The results from the analyses are summarized Tables 7.1. One striking result is that the $4 \times 3 \times 2$ model with unspecific starting values (here, the grand mean for all nonmissing values) can easily lead to nonconvergent solutions. What happens for this model using the EM-algorithm and a grand-mean start is shown in Fig. 7.6. Some of the estimates diverge rapidly as the number of iterations increase. By using more specific starting values for the missing data, this lack of convergence can be remedied. This dependence on starting values disappears when the number of components is reduced. Spanjer (1984, p. 106) suggested that "[t]he ability of a statistical model to fit observations is not its most useful property. Far more important is its ability to predict accurate values for missing data.", but he qualifies this statement by pointing out that his observation is particularly true for (physical) sciences where there are precise measurements, and he goes on to say in the social and behavioral sciences that fit to the data is probably more important.

Table 7.1 Chromatography data: Proportional fit and execution times

Proportional fit

WLS algorithm (GEPCAM - $\mathcal{G}$)

	Grand mean start - G		Main effects - M		Two-way interactions - I	
	Random $\mathcal{G}$RG	Tucker $\mathcal{G}$TG	Random $\mathcal{G}$RM	Tucker $\mathcal{G}$TM	Random $\mathcal{G}$RI	Tucker $\mathcal{G}$TI
322	0.998	0.998				
332	0.998	0.998	0.998	0.998	0.998	0.998
432	0.999	0.999 †	0.999	0.999	0.999	0.999
	0.999 †					

EM-*algorithm* (TUCKALS3 - $\mathcal{T}$)

	Grand mean start - G		Main effects - M		Two-way interactions - I	
	Random $\mathcal{T}$RG	Tucker $\mathcal{T}$TG	Random $\mathcal{T}$RM	Tucker $\mathcal{T}$TM	Random $\mathcal{T}$RI	Tucker $\mathcal{T}$TI
322	0.998	0.998				
332	0.998	0.998	0.998	0.998	0.998	0.998
432	0.999 †	0.999 †	0.999 †	0.999 †	0.999	0.999
	0.999 †					

Execution times

WLS algorithm (GEPCAM - $\mathcal{G}$)

	Grand mean start - G		Main effects - M		All two-way - I	
	Random $\mathcal{G}$RG	Tucker $\mathcal{G}$TG	Random $\mathcal{G}$RM	Tucker $\mathcal{G}$TM	Random $\mathcal{G}$RI	Tucker $\mathcal{G}$TI
322	3.90	4.39				
332	4.61	5.11	5.99	5.61	4.78	4.18
432	17.69		11.59	11.09	13.13	11.54
	671.30†	659.57†				

EM-*algorithm* (TUCKALS3)

	Grand mean start - G		Main effects - M		All two-way - I	
	Random $\mathcal{T}$RG	Tucker $\mathcal{T}$TG	Random $\mathcal{T}$RM	Tucker $\mathcal{T}$TM	Random $\mathcal{T}$RI	Tucker $\mathcal{T}$TI
322	5.00	5.38				
332	5.60	5.27	4.94	4.39	3.68	4.23
432	83.10†	95.02†	90.18	89.20*	5.87	5.82
	83.70†					

Two separate random starting solutions were used for the $4\times3\times2$ model (column $\mathcal{G}$RG). All convergent GEPCAM analyses did so in 51 iterations or less. All convergent TUCKALS3 analyses took between 134 and 190 iterations. Nonconvergent solutions (number of iterations > 2500) are indicated by a †. Model indication (PQR): P = $\mathcal{G}$epcam/$\mathcal{T}$uckals algorithms; Q = **R**andom/**T**ucker Method I initialization; R = **G**rand Mean/**M**ain effects/**T**wo-way **I**nteractions initial estimates for missing data.

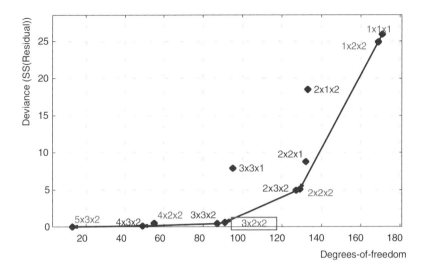

Figure 7.5   Deviance plot for EM-algorithm (including nonconvergent solutions).

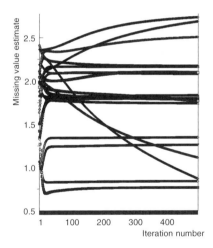

Figure 7.6   Changes in the estimates of the missing data. A $4 \times 3 \times 2$ Tucker3 model with grand-mean initial estimates for the missing data and a Tucker Method I start for the parameter matrices: $\mathcal{T}$TG.

On the basis of these observations, Spanjer (1984) opted for an extremely well-fitting model with nearly as many components as can be extracted from this data set. The maximum solution is a $6 \times 3 \times 2$ model that fits perfectly and therefore is not useful for estimating missing data, because with any serious starting solution, the data set will be fit perfectly including the starting values for the missing data. The reason for the perfect fit for this number of components is the minimum-product rule, which states that the product of the number of components of two modes must always be larger than or equal to the third one (see Section 4.8, p. 64). In the present case the second and third mode have all the components they can have, respectively, 3 and 2, so that this puts a limit of 6 on the first mode. From the point of view of fitting the data, a far simpler model does suffice, and one would generally settle for a $3 \times 2 \times 2$ model (see also the Deviance plot in Fig. 7.5; see Section 8.5.2, p. 181, for an explanation of such plots).

Estimates of the missing data. From Table 7.2 it is clear that a nonconverged solution cannot be trusted, as either the estimates run away or lead to unacceptable values. Note that the values of the estimates have no influence on the fit function because the missing data have zero residuals by construction .

Another question is whether all converged solutions produce the same estimates for the missing values, and the answer from Table 7.2 seems to be affirmative. It also shows that an extreme accuracy of approximation is not necessary, as the estimates for the $3 \times 3 \times 2$ and the $3 \times 2 \times 2$ are all close enough to the original Spanjer solution, certainly if the confidence intervals produced in Spanjer (1984) are to be believed. Moreover, the solutions with fewer components lead to the same solutions, irrespective of the algorithm, starting values for the missing data, and the starting solutions for the parameters, implying that solutions with less components are more stable and less influenced by random fluctuations. Finally, it is also clear that the ANOVA estimates themselves, used as starting values for the missing data, are not sufficiently close to the other estimates to be acceptable.

The technical conclusions from the analyses of Spanjer's chromatography data are that the weighted least-squares GEPCAM algorithm needs fewer iterations than the TUCKALS EM-algorithm, but the iterations are more costly . However, for this example, the total computing times are comparable. The number of iterations is fairly independent of the starting solutions for the missing data and the type of starting configurations for the parameter matrices, even though a more elaborate ANOVA model for the starting values reduces the number of the EM-algorithm somewhat more than for the weighted least-squares algorithm. The EM-algorithm seems more sensitive to starting values away from the correct solution, but it is an open question whether converged solutions will always be on the mark.

It should also be noted that in the case of profile data with different measurement levels for the variables, the closest one can get in estimating starting values for the missing data is a multivariate two-way ANOVA main effects model for each variable. Whether this will generally be sufficient is a subject for investigation. The conclusion must be that several different starting solutions for the missing values are imperative, especially if one wants to have as high an explained variability as Spanjer, De Ligny, Van Houwelingen, and Weesie (1985) were aiming for. On the other hand, both

Table 7.2 Chromatography data: Estimates for the missing data

No.	Location			Starting		Spanjer 432	$\mathcal{G}$TI 432	$\mathcal{G}$RM		$\mathcal{G}$RG2 432	$\mathcal{T}$RM 432
	i	j	k	Main	Int.			332	322		
1	1	1	1	1.50	1.31	0.96	0.98	0.83	0.85	0.75	0.77
2	2	1	1	1.40	0.88	0.81	0.82	0.84	0.82	0.83	0.84
3	13	1	1	1.95	1.89	1.78	1.80	1.77	1.76	1.79	1.78
4	15	1	1	1.89	1.89	1.74	1.76	1.73	1.74	1.74	1.74
5	16	1	1	2.05	2.00	1.79	1.81	1.77	1.79	1.80	1.79
6	17	1	1	2.19	2.35	2.08	2.10	2.02	2.05	2.10	2.07
7	18	1	1	2.43	2.42	2.17	2.20	2.19	2.23	2.17	2.17
8	28	1	1	1.50	1.99	1.85	1.85	1.81	1.81	1.79	1.80
9	33	1	1	1.61	2.28	2.13	2.14	2.10	2.08	2.09	2.09
10	34	1	1	1.64	2.19	2.12	2.13	2.13	2.09	2.15	2.16
11	35	1	1	1.09	1.31	1.29	1.30	1.32	1.32	1.28	1.27
12	36	1	1	1.08	1.29	1.32	1.32	1.37	1.35	1.36	1.35
13	38	1	1	1.35	1.86	1.84	1.84	1.87	1.85	1.85	1.84
14	11	3	1	1.74	2.24	2.58	2.59	2.61	2.61	7.16	2.98
15	12	3	1	2.01	2.78	3.31	3.29	3.21	3.26	*.**	−0.83
16	13	3	1	1.89	2.56	2.91	2.91	2.88	2.88	*.**	1.71
17	14	3	1	2.00	2.60	3.12	3.06	3.05	3.06	*.**	2.37
18	15	3	1	1.82	2.21	2.44	2.45	2.42	2.44	*.**	1.46
19	16	3	1	1.99	2.45	2.73	2.74	2.69	2.72	*.**	1.25
20	17	3	1	2.13	2.54	2.83	2.84	2.73	2.79	*.**	0.00
21	18	3	1	2.37	2.75	3.14	3.14	3.13	3.18	*.**	2.67

Starting = Starting values: main-effects; all two-way interaction. Spanjer 432 = Spanjer solution as starting values. Model indication (PQR): P = $\mathcal{G}$epcam/$\mathcal{T}$uckals algorithms; Q = **R**andom/**T**ucker Method I initialization; R = **G**rand Mean/**M**ain effects/Two-way **I**nteractions initial estimates for missing data. *.** = < −150.00; the last two columns refer to nonconverged solutions.

algorithms will serve in most contexts, even though the WLS seems to have a slight edge, assuming that the results from a single data analysis can be extrapolated .

Spanjer et al. (1985) present standard errors for their estimates, but we are too unsure about their status with other data and in situations with a much lower explained variability, so we decided not to present these here or investigate them further. One possibility to evaluate them would arise if the accuracy of the measurements themselves was known, so that there would be a gauge against which to measure the estimated standard errors for the missing data.

7.7.4 Stage 4: Results and interpretation

In contrast with Spanjer et al. (1985) we will also present a full three-mode analysis of his data using the $3\times2\times2$ model with 3 components for the solutes, 2 for the

Table 7.3   Chromatography data: Adsorbents and eluents components

Adsorbents (silica)	1	2	Eluents	1	2
Aminobutyl	1.06	0.31	35% v/v methylene chloride	1.21	0.11
Octadecyl	0.91	−0.24	100% v/v methylene chloride	0.70	−0.19
N-cyanoethyl -N-methylamino	0.93	−0.13			
Proportion explained variability	0.942	0.057		0.974	0.024

adsorbents, and 2 for the eluents, which in the previous section was shown to be satisfactory from several points of view.

From Table 7.3 we can deduce that the adsorbents primarily work in the same way for all solutes, and that the major difference in the working of the adsorbents is that between aminobutyl-silica versus the other two adsorbents. The effect on the retention rate is stronger for 35% v/v methylene chloride than for pure methylene chloride; the second component indicates a slight contrast between the two eluents. The large differences in explained variability between the two components are caused by the presence of the means in the first components, and the relative sizes of the means correspond by and large to the values on the first components.

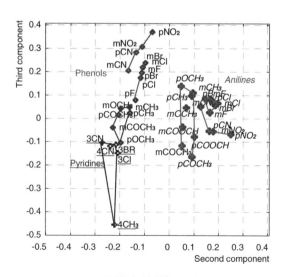

Figure 7.7   Solute component space for the 3×2×3-Tucker3 model for the Chromatography data. Shapes and connecting lines indicate groups of similar solutes. Phenols in normal typeface; Anilines in italics; Pyridines underlined

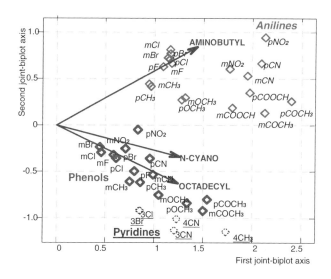

Figure 7.8   Joint biplot of the solutes and the adsorbents for the $3\times2\times3$-Tucker3 model for the Chromatography data with respect to the first eluent component. Phenols in normal typeface; Anilines in italics; Pyridines underlined.

To get a view of the differences in retention rates between the solutes, we inspect the solute space (Fig. 7.7) in a paired-components plot (see Section 11.4.1, p. 260), where we see that the major difference (on the first axis) is between the phenols and pyridines versus the anilines. Many additional systematic differences can be observed, but these will not be discussed here.

The crucial information necessary for the interpretation of the influence of the adsorbents and eluents is the relationship between the three modes. This is portrayed in the joint biplot for the display modes – solutes and adsorbents (Fig. 7.8) associated with the first component of the reference mode – eluents (see Section 11.5.3, p. 273, for an explanation of joint biplots). This component, as we have seen in Table 7.3, indicates the influence of the eluents on the retention times with a nearly twice as large relative retention time for the 35% compared to the 100% methylene chloride.

Again the division between the three groups of solutes can clearly be seen. From the position of the adsorbent vectors we may deduce that aminobutyl-silica provides higher retention rates for the anilines than for the phenols and pyridines. We may examine further details portrayed in this joint biplot, as well as the joint biplot associated with the second component of the reference mode. This leads, for instance, to the conclusion that for both the anilines and the phenols, the carbonyl substituents lead to high retention rates, and that for the anilines with a carbonyl substituent this is true for all adsorbents, as is indicated in the joint biplot by their long projections on the adsorbent vectors. However, such details will not be pursued further here.

The conclusion from the analyses of the chromatography data presented in this section is that we can get acceptable estimates for the missing data, that these can be acquired with both algorithms discussed here, but that for this data set the weighted least-squares algorithm seems somewhat more stable, that several different starting solutions for the missing data should be tested to ensure finding a good solution, that only converged solutions carry useful information, and that a detailed insight into the relationships between the three modes can be had at the same time.

7.8 EXAMPLE: NICHD CHILD CARE DATA

The data for this section were chosen because they are a sample from a larger population. Because of this, the incomplete data set will be analyzed using multiple imputation. In the chromatography example this was not a reasonable option as there was no clear stochastic element in the data.

7.8.1 Stage 1: Objectives

The NICHD Study of Early Child Care and Youth Development is a still continuing longitudinal study conducted by the Early Child Care Research Network (ECCRN), a consortium of researchers from 10 different universities in the United States of America and since 1989 financed by the National Institute for Child Health and Human Development (NICHD). Participants in the study were recruited from 10 sites around the USA throughout 1991. Potential participants were selected from among mothers giving birth during selected 24-hour sampling periods, who were screened to determine their eligibility for the study (NICHD Early Child Care Network, 1994); see the project website[2] for details on the study itself and a list of publications with analyses of these data.

7.8.2 Stage 2: Data description and design

For the present illustration, we have only included the 150 black participating families who had valid data on at least two of the four measurement periods. Furthermore, primarily background variables were selected, which were measured with the same or comparable instruments in each of the four periods and which were expected to be related to child care. Preliminary analyses led to the elimination of five variables that had insufficient correlations with the main variables or insufficient individual differences.

The final data set consisted of 150 black families, eleven variables, and measurements at 6, 15, 24, and 36 months. The eleven variables are listed in Table 7.4 with their means at the four times of measurement. The data are thus multiway longitudinal data (Chapter 15), but they will be treated as ordinary multiway profile data

[2] *http://secc.rti.org/*. Accessed May 2007.

Table 7.4 NICHD data: Variables and their means at the four times of measurement

Abbreviation	Description	06	15	24	36
HrWrkM-xx	Hours/week mother works—all jobs	17.1	21.3	21.0	20.9
Satisf-xx	Mother satisfied with own work schedule	3.6	3.8	3.6	3.5
Depres-xx	Maternal depression	11.9	11.2	13.1	11.9
Suppor-xx	Social support	5.0	4.8	4.6	4.7
PStres-xx	Parenting stress*	51.0	34.3	35.7	34.7
HealtM-xx	Health of mother	3.2	3.1	3.0	2.9
HealtB-xx	Health of baby	3.3	3.1	3.2	3.2
HrCare-xx	Hours/week in care	23.6	26.1	24.4	26.8
Financ-xx	Financial resources	9.3	9.3	9.2	9.4
Income-xx	Log total income	9.7	9.7	9.8	9.9
Need -xx	Log income to need ratio	0.3	0.2	0.4	0.4

xx (= 06, 15, 24, or 36) indicates observed xx months after birth. *Parenting stress at 6 months was measured with a different instrument explaining the differences in means.

(Chapter 13). Accuracy checks of the data necessitated corrections of some data. For instance, the data of one participant indicated that she worked 122, 134, 105, and 20 hours per week, respectively. The first three values were reduced to 80 hours per week to bring the scores in line with the working hours of other mothers. In addition, for some participants it was recorded that mothers worked zero hours per week and that the child went to child care zero hours per week, while all other values were missing. Such values were set to missing. In the end, the data set of 6600 values contained 567 or 8.6% missing data. The most common pattern was that all observations at a particular time point were not available. We have assumed that the missingness is unrelated to the variables in the study, but this might not be entirely valid because more black families and low income families dropped out of the study over time than white families.

As suggested in Section 7.5 the multiple imputation was carried out with the wide combination-mode matrix, thus on the subjects by variables×time points two-way matrix. Except for the logarithmically transformed ones, the variables could only take on nonnegative values. Missing data estimates in the imputed data sets showed several "impossible" negative values, but as explained in Section 7.5.3, this was not considered a problem for the analysis.

Given missing data are present, the best first step is to acquire a set of good starting values for the missing data, generally derived from an analysis-of-variance model. Because the measurement units of the present variables are very different, normalizing the variables is imperative (see Section 6.4.5, p. 124). Furthermore, we have to find starting values per variable slice, and an obvious choice is the mean of each variable–measurement combination when observations are present in the column. If not, the overall variable mean should be substituted as starting value

Table 7.5   NICHD data: Fit information on the solutions

Solution	SS(Fit)	Proportional fit per component		
		1	2	3
EM solution	0.415 (0.382)	0.229	0.123	0.063
Constant time component	0.414 (0.381)	0.228	0.122	0.063
Imputation 1	0.388	0.220	0.109	0.059
Imputation 2	0.383	0.216	0.110	0.058
Imputation 3	0.380	0.211	0.113	0.056
Imputation 4	0.382	0.212	0.111	0.059
Imputation 5	0.390	0.218	0.113	0.058

The SS(Fit) entries in parentheses in the first two rows are the proportional fit values for the solutions excluding the missing data points; the values without parentheses are those for the completed data. SS(Fit) values are based on the squared fitted data points themselves.

7.8.3   Stage 3: Model and dimensionality selection

Once the starting values are obtained, it seems best first to perform an EM-estimation of the missing data within the definition of the model (as was described for the chromatography example). In that case the missing data are estimated to concur as closely as possible with the model used to describe the data. However, in such an analysis, we lack information on the variability and uncertainty due to the missing data. The EM-analysis provides us with the estimates of the missing values and allows for a search for the appropriate number of components. For the NICHD data it was decided to settle for a $3\times3\times1$ solution for the subjects, variables, and time-points components. The single time-point component indicates that the structure of the subject–variable relationships stayed the same over time, varying only in the weights allocated to each time point. It was noted that the differences between the weights of the time points were very small, so that they could also be replaced by the same value at each time points. In other words, for interpretational purposes, one may just as well average the data over time points with only a minimal loss of fit and only interpret the relationships between persons and variables for the averaged data. However, for the purpose of illustration, we will continue to use the three-mode results.

7.8.4   Stage 4: Results and their interpretation

Fit measures.   The five imputed data sets were subjected to the same analyses as the original data set, and the summary of fit results of all analyses is listed in Table 7.5. Because there is only one component for the third mode, the proportional fit per component is the same for the person and variable modes (see Ten Berge et al., 1987). In general, the solutions from imputed data sets should be optimally rotated to one another in order to be sure the proportional fits refer to the same component. In

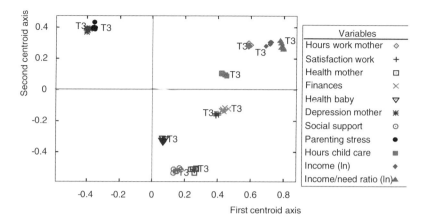

Figure 7.9   NICHD Data: Variables – Dimensions 1 and 2. TUCKALS3 EM solution (T3, placed directly to the right/left of its marker) and five multiple-imputation solutions aligned using a generalized Procrustes analysis.

the present case, it turned out not to be very important. A third point is that for the base EM solution two possible fit measures can be derived: one for the solution ignoring the missing data in both the fitted sum of squares and the total sum of squares, and one for the completed data. As can be observed in Table 7.5, there is a clear difference between the two. Considering that the fit in the imputed data sets is also based on a completed data set including the missing data estimates, it would seem best to use the completed fit for comparison. However, the table shows that for this example the fit of the available data is closer to those of the imputed data sets. Clearly, some further investigations are necessary to settle which of the two measures should be used in general.

Comparison of solutions.   The classic way to represent the results of a multiple imputation procedure is to provide the values of the estimates of the parameters and their standard errors based on the within-imputation and between-imputation variance. Apart from exceptional cases, in three-mode analysis only the between-imputation variance is available. However, in principal component analysis, we are not primarily interested in the specific parameter estimates for each and every coefficient, but more in the spatial configurations of the variables. Therefore, we want to compare the closeness of the solutions from the imputed data sets. This was assessed via the closeness of the configurations to their centroid derived via a generalized Procrustes analysis (Gower & Dijksterhuis, 2004). The proportional fitted sums of squares of the imputed configurations to the centroid ranged from 0.9976 to 0.9990; that of the EM solution was 0.9968, indicating very close agreement as can be seen in the coordinate space of the centroid solution (see Fig. 7.9).

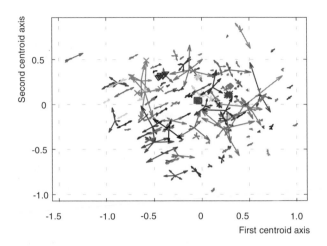

Figure 7.10   NICHD data: Subjects. TUCKALS3 EM solution and five multiple-imputation solutions aligned using a generalized Procrustes analysis. Each subject is depicted as a star emanating from its centroid coordinates. For some subjects the variability is so small that they appear as a single dot.

Notwithstanding the number of missing data (586 — roughly 9%), the configurations of all imputed data sets are very close together and close to the Tucker3 or EM solution (marked by T3). As higher-order dimensions are inherently more unstable than the lower dimensions, the differences in the third dimension between the positions of the variables for the imputed solutions (not shown) are slightly larger, but still close together. The fit of the solutions to the data is not very high (Table 7.5), but this does not have a great effect on the configurations themselves. All in all, we can have confidence in the EM solution for the variables.

The subjects show, as expected, more variability (Fig. 7.10). Each subject is represented by a star starting from the position of the centroid coordinates. The coordinates of the EM solution are similar to the those of the imputed solutions. In addition, all solutions have largely the same shape for their time modes (see Fig. 7.11). The EM solution lies above the imputed ones due to the slightly larger explained variance for the completed data in the EM solution.

7.9 FURTHER APPLICATIONS

7.9.1   Test equating in longitudinal designs

Intelligence tests have different equivalent subtests for different age groups, but not *all* subtests are different. The need for different subtests arises from the fact that

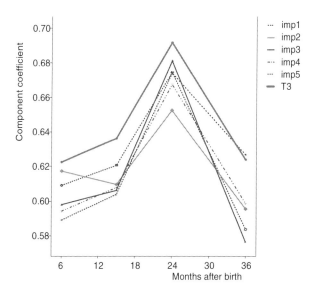

Figure 7.11 NICHD data: Time dimension. TUCKALS3 EM solution (T3) and five multiple-imputation solutions.

certain items have ceiling or floor effects for specific age groups. The parallel tests are designed to measure the same construct, but at a another level of competence.

Exploratory test equating in a longitudinal setting could rest on subject by tests by years data, where the test mode will contain all the parallel tests as separate variables. Such designs result in large blocks of missing data. However, some tests are present in virtually all years, supplying the coherence of the data. These designs require that the same persons are followed over the years, and that only a small number of them drop out or come into the program at a later stage. In this context Harshman and Lundy (1994b, pp. 65–67) call this a *linked-mode analysis*. The multilinear engine developed by Paatero (1999) can be used to estimate this design. As can be seen in Fig. 7.12, it is possible to carry out an analysis even when there are two groups of subjects who have no test in common. However, a successful analysis depends on a well-defined structure in the variables.

7.9.2 Variety trials in agriculture

Variety trials in agriculture testing old versus new variants of specific plants (crops) contain so-called check varieties which are kept in all years as a gauge for the performance of the other varieties and for continuity. Some test varieties may be deleted after a number of years due to inadequate performance, while new test varieties enter the trials at a later date. This gives large blocks of systematically missing data. Due to

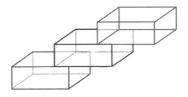

Figure 7.12   Systematic missing data: Not all subjects take all tests, but at least some subjects take tests from two of the blocks, in order to provide links between the three blocks.

the three-way character of the data and the presence of the check varieties, an analysis of the entire data set is still possible, even if, in addition, more or less random missing data are present as well (Fig. 7.13).

7.10   COMPUTER PROGRAMS FOR MULTIPLE IMPUTATION

Multiple imputations can be carried out with Schafer's (1999) program NORM[3] and with the program MICE[4], which is based on different principles (see Van Buuren, Brand, Groothuis-Oudshoorn, & Rubin, 2006). Schafer and Graham (2002, p. 167) indicate that variables should be normally distributed but it seems that this is not a very stringent one for two-way data.

---

[3] *http://www.stat.psu.edu/~jls/misoftwa.html.* Accessed May 2007.
[4] *http://web.inter.nl.net/users/S.van.Buuren/mi/hmtl/mice.htm.* Accessed May 2007.

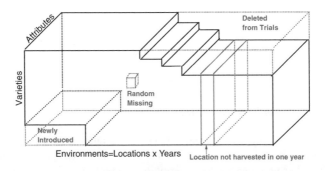

Figure 7.13   Types of missing data in plant breeding experiments: Random missing data, discontinued varieties; newly introduced varieties; crops not harvested.

CHAPTER 8

MODEL AND DIMENSIONALITY SELECTION

8.1 INTRODUCTION

Since Tucker introduced three-mode analysis in the 1960s, many more models have been developed and used in practice (see Chapter 4). Clearly, this complicates model selection. Whereas initially only the, already difficult, choice of components for each mode had to be made (*within-model selection* or dimensionality selection), now also the model itself has to be chosen (*between-model selection*). In many cases, the type of model is dictated by the nature of the data; in other cases, one particular class of models is required, but generally which model within a class is the most appropriate is not *a priori* clear. Exceptions are cases where previous experience has shown which model is most suitable for the data, or where an explicit (physical) model is available as in several chemical applications. To find a proper model and its dimensionality, ideally considerations of parsimony, stability, interpretability, and expected validity in new data sets have to be taken into account. Various tools for evaluating models both for the within-model selection and the between-model selection will be presented and illustrated in this chapter.

Applied Multiway Data Analysis. By Pieter M. Kroonenberg
Copyright © 2007 John Wiley & Sons, Inc.

8.2 CHAPTER PREVIEW

In this chapter we will concentrate on component models and their selection. For this class of models, methods are available that pay attention to the relative fit of models with different numbers of components, and to the fit of different but comparable models. Most of these methods come down to calculating the fit of each model as well as the fit of separate components within a model. In the selection process, systematic, heuristic, and interpretational criteria are considered to decide which model with which number of components is most appropriate for the situation at hand. Obviously, no absolute judgments about this can and will be made. Criteria such as the DifFit procedure (Timmerman & Kiers, 2000), which translates into a multiway version of Cattell's scree criterion, the st-criterion (Ceulemans & Kiers, 2006), and related plots such as the multiway deviance plot (e.g., see Kroonenberg & Oort, 2003a) will be employed.

Bootstrap resampling and the jackknife are especially useful for dimensionality selection and to assess the predictive validity when the search is already narrowed down to a limited number of candidate models. Pioneering work was done by Harshman and DeSarbo (1984b) in introducing split-half procedures for the Parafac model, and this work was taken up and extended by Kiers and Van Mechelen (2001) (see Section 8.8.1). Both bootstrap resampling and jackknife procedures can also be used for assessing the stability of the parameters of the models, a topic that will be taken up in Section 9.8.2 (p. 233). In this chapter we will concentrate on component models, but in virtually all multiway techniques resampling and jackknife techniques can be introduced; however, this has not been done for all available techniques.

Before going into model selection proper, we will have to discuss briefly the degrees of freedom in multiway component models. However, because multiway models do not introduce new notions compared to the three-mode models, the exposition will be primarily in three-mode terms.

8.3 SAMPLE SIZE AND STOCHASTICS

Choosing the sample size for multiway analysis can be a tricky problem. If the data set is approached as a population or if one sees the multiway analysis as a purely descriptive technique for the sample at hand, then sample size is not necessarily relevant, be it that in small data sets the description of the majority of the data can be more easily distorted due to outlying data points, a specific level of a mode, or both. Interpreting the solution as a descriptive analysis of the sample seems a valid procedure in the case where there is no obvious stochastic mode. If one of the modes, usually subjects or objects, can (also) be interpreted stochastically, it is in the interest of the researcher to have as large a sample as possible, in order to have as stable estimates as possible for the variable and occasion parameters. Given that distributional properties are generally unknown, the bootstrap seems the best way to examine this stability (see Section 8.8.1 and Section 9.8.2, p. 233). Split-half procedures discussed in Section 8.6.1 also assume that there is one mode that can be considered a stochastic one, as do the imputation procedures for generating multiple data sets in the case of missing data (see Section 7.5, p. 156). In multiway profile data as discussed in Chapters 13 and 19, there generally exists one mode, which can (also) be considered a stochastic one, so that these procedures can be applied to those types of data.

8.4   DEGREES OF FREEDOM

Gower (1977) proposed to use Gollob's (1968a) idea of simply counting the number of free parameters and subtracting them from the number of independent data points to derive the degrees of freedom for multiway component models — an approach that will be followed in this book. For a further discussion of the difficulty of the concept of degrees of freedom in multiway models, see Smilde et al. (2004, pp. 96–97). There has also been extensive discussion of the degrees of freedom for two-way models; for an overview see Van Eeuwijk and Kroonenberg (1995).

8.4.1   Degrees of freedom of multiway models

Weesie and Van Houwelingen (1983, p. 41) were the first to use the degrees of freedom in fitting three-mode models. They discussed the degrees of freedom for the Tucker3 model, but the principle can readily be extended to the Tucker2 and Parafac models.

$$
\begin{aligned}
df \quad = \quad & \text{no. of data points} \; - \; \text{no. of means removed} \\
& - \text{no. of sums of squares equalized} \; - \; \text{no. of missing data} \\
& - \text{no. of free parameters} \, ,
\end{aligned}
$$

with the number of free parameters, f_p,

$$
\begin{aligned}
f_p \quad = \quad & \text{no. of independent parameters in the component matrices} \\
& + \text{no. of elements in the core array} \\
& - \text{rotational indeterminacy for component matrices (Tucker)} \; or \\
& - \text{no. of unit length restrictions (Parafac).}
\end{aligned}
$$

The number of data points looks like a straightforward criterion: simply count the number of observations. For three-way profile and rating data this is obviously $I \times J \times K$. However, when calculating degrees of freedom, we should take the *maximum-product rule* into account. This rules states that "in cases where the size of one of the modes is larger than the product of the other two, special adjustments must be made [to the calculation of the number of free parameters], because in such cases (see Kiers & Harshman, 1997d), the biggest mode can be reduced considerably without loss of information essential for the components of the two other modes and the core. Specifically, when $I > JK$, the data can be reduced to a $JK \times J \times K$ data set. Therefore, in such cases, in the computation of the number of free parameters f_p, I should be replaced by JK" (Ceulemans & Kiers, 2006, p. 139). A consequence of this, for instance, is that to perfectly fit a data array $I \times J \times K$ with $I > JK$, we only need JK, J, K components for the three modes to fit the data perfectly. Thus, in calculating degrees of freedom we have to take this into account.

For various multiway models we get the following number of free parameters:

$$
\begin{aligned}
\text{Parafac}: f_p &= I \times S + J \times S + K \times S + S - 3S, \\
\text{Tucker2}: f_p &= I \times P + J \times Q + P \times Q \times K - P^2 - Q^2, \\
\text{Tucker3}: f_p &= I \times P + J \times Q + K \times R + P \times Q \times R - P^2 - Q^2 - R^2, \\
\text{Tucker4}: f_p &= I \times P + J \times Q + K \times R + L \times T, \\
&\quad + P \times Q \times R \times T - P^2 - Q^2 - R^2 - T^2.
\end{aligned}
\tag{8.1}
$$

8.4.2   Degrees of freedom of a three-way interaction

For testing multiplicative models for three-way interactions, little theory has been developed so far. Boik (1990) presented a likelihood ratio test for the first multiplicative term, including a table of critical values for comparatively small three-way arrays. In the cases not covered by Boik, one may follow the common practice for two-way tables, that is, attributing degrees of freedom in the same way as in Eq. (8.1). However, it should be kept in mind that in three-way interactions the maximum number of independent data points is already reduced to $(I - 1) \times (J - 1) \times (K - 1)$ due to earlier estimation of the main effects and two-way interactions. Van Eeuwijk and Kroonenberg (1995) provide an overview; Dias and Krzanowski's (2003) review also provides further discussion of this issue.

8.4.3   Degrees of freedom and rank of an array

In the two-way case there is a direct relationship between the size of a core matrix and its rank. However, the multiway case is considerably more complex because the same size core array can have different ranks, for instance, a $2 \times 2 \times 2$ core array can have either rank 2 or rank 3, which casts some doubt on what the real degrees of freedom are in that case. The whole question of the rank of multiway arrays is very complicated, but steady progress made has been in determining the ranks of three-way arrays (see especially Ten Berge, 2004). The topic of the rank of multiway arrays will not figure prominently in this book, and only in Chapter 10 will we pay some, but not much, attention to this issue. In several chemical applications, such as second-order calibration and curve resolution, the rank of the multiway data array is of crucial importance, for instance, to determine the number of compounds present in a mixture; a discussion of this issue can be found in Smilde et al. (2004, Sections 2.5 and 2.6). A pragmatic approach will be taken here, and only the above formulas will be used, warts and all.

8.5 SELECTING THE DIMENSIONALITY OF A TUCKER MODEL

8.5.1 Timmerman–Kiers DifFit procedure

An important aspect of dimensionality selection for Tucker models is that not all combinations of dimensions are feasible. In particular, for the Tucker3 model the minimum-product rule (see Section 4.5.3, p. 56) applies and $PQ \geq R, QR \geq P$, and $RP \geq Q$. Similarly, in a Tucker4 model, for instance, $PQR \geq T$. The reason for this is that in each step of the algorithm, the matrix for which the eigenvalues are calculated has the rank of the product of the numbers of components of the other modes. If that rank is lower than that of the current mode, the algorithm will fail. Another way of looking at this is to note that the core array is three-way orthogonal, and this can only be the case if the products of the component numbers are larger or equal to the number of components in the remaining mode (see Section 4.8.2, p. 65, for further details, and Wansbeek and Verhees (1989, p. 545) for a formal proof).

Given a Tucker3 model, Timmerman and Kiers (2000) suggested a dimensionality-selection procedure analogous to Cattell's scree plot for two-mode component analysis. Within the class of the Tucker3 models with the same total number of components $S = P + Q + R$, they suggested to select the model that has the highest proportion fitted sum of squares, V_S, or equivalently the smallest residual sum of squares or deviance. To compare classes with different S, they compute $dif_S = V_S - V_{S-1}$. Only those dif_S are considered that are sequentially highest. Timmerman and Kiers defined a *salience value*: $b_S = dif_S/dif_{S*}$, where dif_{S*} has the next highest value after dif_S. They proposed to select the model for which b_S has the highest value, and they called this the *DifFit-criterion*. Finally, the authors defined a lower bound for dif_S below which models should not be taken into account. The dif_S should be greater than the average proportion explained variability taken over all feasible values of S ($S_{\min} = \min(I, JK) + \min(J, IK) + \min(K, IJ) - 3)^1$. This is equivalent to Cattell's eigenvalue-larger-than-1 criterion in two-mode PCA, where 1 is the average value of the variance explained.

To visualize the procedure proposed by Timmerman and Kiers, it is useful to construct a version of Cattell's scree plot, the *multiway scree plot*, in which the residual sums of squares for each set of dimensionalities is plotted versus the sum

---

[1]In Timmerman and Kiers (2000), inadvertently, the word "max" is printed instead of "min".

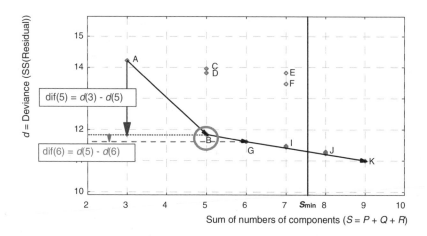

Figure 8.1 Multiway scree plot: Deviance versus sum of numbers of components.

of numbers of components S (Fig. 8.1)[2]. The DifFit procedure is essentially a way to define the convex hull in the multiway scree plot, and models that are taken into consideration are those on the convex hull. In Fig. 8.1 these are the models A, B, G, and K. Models I and J just miss the mark, but it is not unreasonable to consider such models as well given the presence of random error. The lower bound, S_{min}, has been drawn in the multiway scree plot as a vertical line and all models to the right of that line fail to make the minimum value (equivalent to the eigenvectors with eigenvalues smaller than 1 in the two-mode case). The multiway scree plot shows that from the models with a total of $S = 5$ components, the models C and D are much worse than B, and for $S = 7$ models E and F are much worse than I, while models with totals of $S = 8$ and $S = 9$ fail to make the cut. Note that increasing the total number of components from 3 to 5 gives a real reduction in the residual sum of squares, but beyond that the decrease is much slower, and thus the DifFit criterion points to model B as having the "correct" dimensionalities. Timmerman and Kiers (2000) present their procedure for the Tucker3 model, but it works entirely the same for other Tucker models.

Timmerman and Kiers (2000) proposed to calculate the residual sum of squares for each model by computing the Tucker3 model for each set of dimensionalities with the standard ALS-algorithm. In a later paper, Kiers and Der Kinderen (2003) showed that it is sufficient to approximate the models via a single computation using Tucker's original nonleast-squares method, and to calculate the residual sum of squares for the models under consideration using the core array. This constitutes an enormous saving in computer time.

---

[2]The choice was made to portray the residual sums of squares or deviance d rather than the fitted sums of squares, to mirror as much as possible the standard two-mode scree plot.

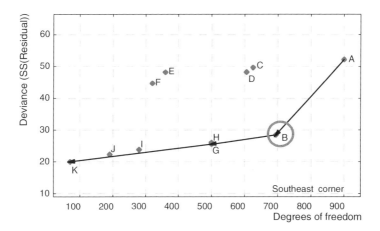

Figure 8.2   Deviance plot: Deviance versus degrees of freedom

Counting the number of components has the disadvantage that a component of a mode with many levels plays an equally important role as a component of a mode with only a few levels, so that models with an equal total number of components may have very different numbers of parameters. On the other hand, the size of S is uniquely defined and independent of centerings, normalizations, and numbers of missing data. Moreover, via simulation studies Timmerman and Kiers (2000) showed an excellent recovery rate of their procedure.

8.5.2   Deviance analysis

An alternative to the Timmerman–Kiers approach is the evaluation of the residual sum of squares or deviance together with the degrees of freedom (see Section 8.4). Again a plot can be constructed with the residual sum of squares d of each model plotted but now versus the degrees of freedom df (*deviance plot* – see Fig. 8.2). Also in this plot, a convex hull can be drawn to connect favored models; here A, B, G, and K (I just misses out). The general idea is that models clearly situated within the hull (such as C, D, E, F) are disfavored compared to the *hull models* who have similar or better fit and equal df (here E and F compared to I), equal fit with more df (here C and D compared to A), or a combination of both. Again, stochastic considerations may advise against automatically discarding models close to the convex hull (here H, I, and J).

   In deviance plots, model evaluation may be aided by drawing straight lines between the loci of models with constant (say, k) d/df. If the origin is included in such a plot, these lines can be seen to fan out from the origin with a slope of k. If $k = 1$, then each parameter added or degree of freedom lost leads to a gain of one unit of deviance. If two models on the convex hull are connected by a line having a steep slope (i.e., k is large) (here A and B), then, in exchange for a few degrees of freedom (adding a few extra parameters), one can get a large decrease in deviance or a considerably better fitting model. This happens typically at the right-hand side of the plot. On the other hand, if the slope is very flat (i.e., k is small) (here from B to G to K), one needs to sacrifice many degrees of freedom (or add many parameters) to get a small decrease in deviance or an increase in fit. The principle of parsimony, which can be expressed here as a preference for models with good fit and few parameters or many degrees of freedom, suggests that models in the southeast (bottom right) corner of the plot are the first to be taken into consideration for selection. Of course, substantive considerations play a role as well. Sometimes the statistically preferred models may be too simple for the research questions at hand, especially when individual differences between subjects are of prime importance.

Even though the deviance plot was designed to assist in choosing between Tucker models differing in dimensionalities, it is equally possible to include models from another class. In other words, it might be advantageous to add, the related Tucker2 models and appropriate Parafac models in a Tucker3 deviance plot as well. For Parafac models, the number of degrees of freedom is not always clear cut, but given a decision has been taken on that point they can be included in the deviance plot; for applications see Kroonenberg and Van der Voort (1987e) and Fig. 14.7, p. 363. Adding Parafac models to the multiway scree plot is even easier because no relatively arbitrary decisions on the degrees of freedom have to be taken. How to combine models of different numbers of component matrices in a multiway scree plot is less obvious.

8.5.3   Ceulemans–Kiers st-criterion

Ceulemans and Kiers (2006) proposed a numerical heuristic to select models in deviance plots, sometimes referred to as a numerical convex-hull-based selection method. The selection criterion is the st-criterion, which (in our notation is based on the deviance and degrees of freedom rather than sum of squares fit and number of free parameters) is defined as:

$$st_i = \frac{d_{i-1} - d_i}{df_{i-1} - df_i} \bigg/ \frac{d_i - d_{i+1}}{df_i - df_{i+1}}, \tag{8.2}$$

where d_i is the deviance and df_i the degrees of freedom of model i. The heuristic is formulated in our notation as follows:

1. Determine the df and d values of all component solutions from which one wishes to choose.

2. For each of the n observed df values, retain only the best-fitting solution.

3. Sort the n solutions by their df values and denote them by s_i $(i = 1, \ldots, n)$.

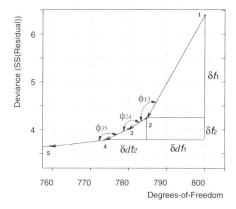

Figure 8.3 Ceulemans–Kiers heuristic for selecting a model in the deviance plot.

4. Exclude all solutions s_i for which a solution $s_j (j < i)$ exists such that $d_j < d_i$.

5. Consecutively consider all triplets of adjacent solutions. Exclude the middle solution if its point is located above or on the line connecting its neighbors in the deviance plot.

6. Repeat Step 5 until no solution can be excluded.

7. Determine the st values of the convex hull solutions obtained.

8. Select the solution with the smallest st value.

In summary, steps 1–6 serve to determine which models lie on the *convex hull*, and steps 7 and 8 serve to find smallest angle $\phi_{i-1,i+1}$ between the lines connecting a model i with its previous $i-1$ and subsequent $i+1$ models, as is shown in Fig. 8.3.

As Ceulemans and Kiers (2006) indicate that model selection according to their criterion should not be followed too rigorously and that substantive criteria can also play a role in selection, so that a model just inside the convex hull may sometimes be chosen because of its interpretational qualities. Their paper discusses the advantages of the *st*-criterion over the DifFit method (Timmerman & Kiers, 2000; Kiers & Der Kinderen, 2003) for different types of models, while also citing the virtual equivalence of the two methods in the case of a single type of Tucker model.

An additional consideration in selection on the basis of the convex hull is also illustrated in Fig. 8.3. The hull shows two places where the convex hull has a relatively sharp angle with $\phi_{13} < \phi_{35}$ and a near 180 deg angle for ϕ_{24}. In such a case, it might be advantageous in practice to select the less parsimonious Model 4, which provides more detailed information about the data. Model 3 is not to be preferred because its angle is larger than both Model 2 and Model 4.

8.6   SELECTING THE DIMENSIONALITY OF A PARAFAC MODEL

In principle, selecting the dimensionality of a Parafac model is much simpler than selecting the dimensionalities of a Tucker model because all modes have the same number of components. In psychology it is not unusual that at most two or three Parafac components can be supported by the data. However, when physical models exist for the data, which is fairly common in chemical applications, often larger models can be fit. For instance, Março, Levi, Scarminio, Poppi, and Trevisan (2005) reported a six-component Parafac model for chemicals in hibiscus flowers, and Murphy, Ruiz, Dunsmuir, and Waite (2006) analyzing the chemical content of ballast water exchange by ships ended up with a nine-component Parafac solution. When one has a limited number of components and a relatively bad fit, one has to establish whether a model with a few more components is merely difficult to fit, or whether there is no further systematic information that has parallel proportional profiles. In line with Cattell's philosophy about the reality of parallel proportional profiles, it seems reasonable to try and fit the largest possible Parafac model, whereas with Tucker models one is more inclined to seek an adequate model for describing the systematic variability, because in principle the data can be decomposed into all their components. Incidentally, in multiway analysis, one aims to fit the systematic variability rather than total variability, and this need not be a very high percentage. Moreover, as one of the purposes is to investigate how much systematic variability is present in the data, it is seldom possible to give a guideline as to what constitutes an adequate percentage variability explained.

Harshman (1984a) gives a whole catalogue of considerations on how to search for the maximal Parafac model, among others using a type of scree plot and split-half analyses. This catalogue covers both fit and component solution aspects of a model search. However, in this section, we will concentrate on fit assessment and fit stability rather than on the stability of the parameters of the model itself.

8.6.1   Split-half procedures

One solution to evaluate the dimensionality of a Parafac solution is to assess the stability of the solution by splitting the data set in half and to perform a separate analysis on both parts. If there is a true underlying solution, it should show up in both analyses. Important restrictions on this procedure are that first there must be a stochastic mode which can be split. Such a mode almost always exists in multiway profile data, in particular, the subject mode, but not necessarily in factorial and rating scale data. When there is no stochastic framework, splitting might not be possible; for instance, in some typical experiments in analytical chemistry split-halves might not make sense. The other caveat is that there must be sufficient subjects in the mode that is being split. In other words, both splits must be large enough to minimize the influence of the idiosyncrasies of specific individuals. How much is "large enough" is difficult to say in general, because much depends on the noise level of the data and the clarity of the underlying structure. Harshman and DeSarbo (1984b) discuss and demonstrate the split-half procedure in great detail with illustrative examples. They also suggest making two (or more) orthogonal splits. For split-half analyses, this entails randomly splitting the data into four more or less equal parts, say, A, B, C, and D, which are combined into four new data sets: (A+B and C+D), and (A+C and B+D). This is the procedure followed in the example in Section 13.5.3 (p. 325). Kiers and Van Mechelen (2001) suggest a similar approach for the Tucker3 model.

8.6.2   Degeneracy

As discussed in Section 5.6.5 (p. 89), the dimensionality selection is complicated by the fact that Parafac algorithms can easily end up in local minima, so that several analyses from different starting positions need to be carried out. Moreover, it is not always easy to discern whether the algorithm slowly converges to a real solution or is on its way to a degeneracy. Such issues clearly complicate dimensionality selection for the Parafac model. Section 5.6.5 (p. 89) also contains a discussion on degeneracy indicators. Such indicators can be of great help in determining an appropriate dimensionality.

8.6.3   Core consistency

An interesting approach to selecting the number of Parafac components was proposed by Bro (1998c, pp. 113–122). He proposed the principle of *core consistency* to assess how many Parafac components the data can sustain (see also Bro & Kiers, 2003b, who referred the principle as CORCONDIA or core consistency diagnostic). The principle is based on assessing how far away the core array derived from the Parafac components is from the ideal, that is, a superdiagonal core array $\mathcal{G}$ (see Section 4.8, p. 64).

A Parafac core array, that is a core array calculated from the Parafac components, is superdiagonal if for, say, a three-mode model only the $g_{111}, g_{222}, g_{333}, \ldots$ have sizeable values and all other core elements are near zero (see Section 9.7.1, p. 232). In Bro's approach, the superdiagonal core array is scaled in such a way that the diagonal elements are equal to 1 (i.e., $\mathcal{G} = \mathcal{I}$), the superidentity array. For a three-mode Parafac model, the discrepancy from the ideal then becomes

$$\text{core consistency} = 1 - \frac{\sum_{p=1}^{S} \sum_{q=1}^{S} \sum_{r=1}^{S} (g_{pqr} - i_{pqr})^2}{\sum_{p=1}^{S} \sum_{q=1}^{S} \sum_{r=1}^{S} (i_{pqr})^2}. \tag{8.3}$$

The denominator will be equal to S for standard Parafac models. The core consistency diagnostic will become seriously negative when components within a mode are highly correlated. If this is deemed undesirable, i_{pqr} can be replaced by g_{pqr} in the denominator; the measure is then called *normalized core consistency*. Whether this would make any difference with respect to decisions about the number of components is doubtful, as it is core consistencies close to 1 that are informative. The *degree of superdiagonality* may also be used for the same purpose. This measure is equal to the sum of squares of the superdiagonal elements divided by the total sum of squares of the core elements. All three measures will be equal to 1 in the superdiagonal case.

Bro (1998c) (pp. 113–122) also proposed to construct a *core consistency plot* that has the values of the core elements on the vertical and the core elements themselves on the horizontal axis with the superdiagonal elements plotted first. In such plots, one can either fit a line through the points, as Bro does, or connect the core elements, as is done in this book. In Bro and Kiers (2003b) an especially interesting (chemical) example is given of the use of this plot in the presence of a large amount of noise and other disturbances. In their application, a three-component Parafac model is known to exist for the pure data, and it is shown that by removing various impurities (i.e., removing the non-Parafac structure from the data) the core consistency dramatically improves. Examples of the core consistency plots are shown in Sections 8.9.5 and 13.5.3 (p. 328).

8.7   MODEL SELECTION FROM A HIERARCHY

In two papers Kiers has discussed the comparison between several different three-mode component models. In Kiers (1988), he primarily examined models that have a relation with the Tucker2 model, while in Kiers (1991a) he presented two hierarchies in which the Tucker3 model had its place; several French proposals were also included in the between-model selections. The latter paper contains one hierarchy for models for direct fitting and one for models for indirect fitting. The first hierarchy is the subject of this section. Ceulemans and Kiers (2006, Fig. 1) also present a more complete hierarchy for the Tucker models (see also Fig. 4.12).

Kiers (1988) explicitly shows how each of the models fits into the hierarchy, but here we will limit ourselves to presenting it and refer the readers to the original paper for the mathematical underpinning. The sequence of models is summarized in Fig. 8.4.

The starting point of the hierarchy is that one mode is singled out as the *reference mode*, consisting of "occasions" in Kiers's paper (see Section 11.5.3 (p. 273), for a discussion of the concept of a reference mode). First, the number of components is determined, which can adequately describe the variability in this mode using a Tucker1 analysis of that mode, that is, a PCA on the tall combination-mode matrix of subjects × variables by occasions (see Section 4.5.1, p. 51 for a discussion of Tucker1 models). The fit of this solution is the standard against which the other models are held, and the other models are evaluated regarding the extent to which they succeed in approaching this fit with far less parameters.

Most of the models in the hierarchy are discussed in more detail in Chapter 4; however, the sumPCA model requires a bit more comment. Initially, one might think that simply averaging and performing a singular value decomposition on the averaged matrix would suffice. Even though this procedure does find the appropriate parameters for the variables and the individuals, it does not provide us with the fit

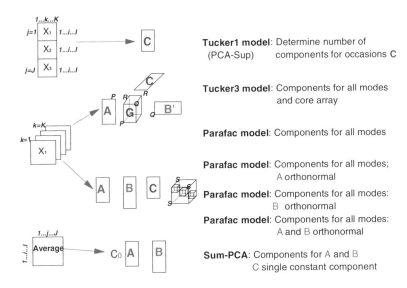

Figure 8.4   Kiers's (1991) hierarchy of three-mode component models for direct fitting.

of these results to the three-way data set. To evaluate this fit, one has to compute how the data of each occasion are fitted by the results for the averaged matrix. The parameters and the fit can easily be found via a Tucker3 analysis with one component for the occasions which has equal values for all its elements. Given orthonormality, this comes down to setting the values equal to $1/\sqrt{K}$, where K is the number of occasions. This procedure closely parallels the computation of the *intrastructure* in STATIS (see Section 5.8, p. 105).

Finally, note that the uniquely determined Parafac models are sandwiched in between models that allow for transformations of the components. The unique orientation of the Parafac models might be an additional consideration in choosing a model in the hierarchy.

8.8 MODEL STABILITY AND PREDICTIVE POWER

In classical statistical modeling, distributional assumptions form the basis for deriving standard errors for the estimated parameters. In multiway models, generally no distributional assumptions are or can be made, and therefore standard errors cannot be determined in the same way. Moreover, usually it is not even known which theoretical distributions would be appropriate. Since the development of high-speed computers, computer-intensive procedures have been developed to find nonparametric estimates of the parameters of arbitrary distributions and their standard errors. In this section we will discuss such methods in so far as they apply to fit measures, but they have further importance in assessing the stability and predictive power of the component matrices and core arrays (see Section 9.8.2, p. 233).

8.8.1 Bootstrap procedures

The fundamental principle of the bootstrap procedure, that is, the observed distribution is the best estimate of the population distribution, was established by Efron (1979). Given this principle, it is easy to see that sampling from the observed distribution is the best one can do, bar sampling from the population distribution itself. Therefore, in order to get an estimate for a population parameter, one repeatedly samples with replacement from the observed distribution with equal probability for all data points. For each such sample, the value of the estimator for the parameter is calculated and the mean value of the estimates is the best estimate for the population parameter and the standard deviation of the estimates is its standard error. Such an approach can be extremely time consuming, especially in multiway methods, where the estimation of the model parameters is already iterative. However, it is the only way to get standard errors and at the same time gives information on the position of the observed results with respect to the sampling distribution. In multiway analysis, applying bootstrap techniques is far from straightforward, but serious inroads into the problem have been made by Kiers (2004a) (see Section 9.8.2, p. 233).

In this section, we will concentrate on the mean of the fit measures and their standard errors, but since during the bootstrap procedure all model parameters are estimated, also the standard errors for the elements in the component matrices and the core array can be determined. However, deriving these in a proper manner is not straightforward and in Chapter 9 we will discuss Kiers's proposals in some detail.

8.8.2 Jackknife procedures and predictive power

Another concern in model building is the question of whether the model found will stand up in new samples. Apart from having the data of a similar study, the replication issue has to be solved internally in the data set itself. The basic idea behind *jackknife* is to develop the model on a part of the data and then estimate the values of those data points not involved in the estimation. Subsequently, the estimated value is compared with the originally observed value; when across the whole data set such differences are small, the parameter estimates are said to have good *predictive power*. Of course, such an approach does not do anything for the sampling inadequacies of the original sample, but at least it gives a certain amount of information on how well the model will stand up in other samples from the same population. Note that the estimates from jackknife procedures are assumed to have been acquired from random samples. Thus, it is the stochastic mode of which certain parts are left out of the estimation of the model.

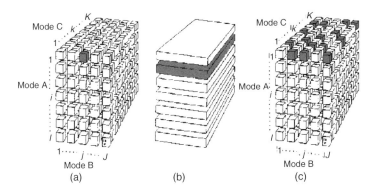

Figure 8.5   Three implementation methods of the jackknife: (a) leave out a single data point at a time (Louwerse et al., 1999); (b) leave out a complete slice at a time (Riu & Bro, 2003); (c) leave out a random half of a slice at a time (Kroonenberg, 2005c).

Jackknife in multiway analysis.   There are several ways to carry out jackknife procedures in multiway analysis. For instance, Riu and Bro (2003) deleted a complete level at a time (Fig. 8.5(b)). Their aim was to use the jackknife estimation of standard errors and to search for outliers, but not to assess predictive power of the model at hand. To do so would have been difficult in their case because no reliable and stable estimates can be calculated for the parameters in the component matrix associated with the removed level. A more subtle version of this proposal was put forward by Louwerse, Smilde, and Kiers (1999), based on Eastment and Krzanowski (1982). They also removed complete slices, but developed a sophisticated way to combine the results of leaving out each (group of) slices in turn so that the predictive power could be assessed. In that paper it was also proposed to remove each data point in turn, or several of them at the same time, by declaring them to be missing (Fig. 8.5(a)). They then used the basic missing data expectation–maximization (EM) algorithm to estimate the missing data points, and compared such estimates with the observed values (for details on EM-algorithms to estimate missing data, see Section 7.3.2, p. 150). In principle, this procedure works as intended; the only drawback is that leaving out single data points can be very time consuming for large data sets. The procedure illustrated in this chapter (see also Kroonenberg, 2005c) is a compromise in that it maintains the idea of leaving out an entire level of the stochastic mode, but not deleting all elements of the slice associated with this level together but only a random half of them (Fig. 8.5(c)). The other half of the data points in that level are deleted in the next step, so that all elements of the data array are eliminated once. In this way, there are always data points available in a slice to estimate the associated parameter values in the component matrix. The estimation procedure is the same as in Louwerse et al. (1999), that is, via an EM-approach, and for consistency their notation is used.

Predictive residual error sum of squares – PRESS.   The predictive power of a model is generally estimated by the predictive residual error sum of squares and is calculated by comparing the values of all original data points with their estimated values on the basis of models without the data point included. Thus, for the three-way case, x_{ijk} is compared with $\hat{x}_{ijk}^{PQR}$ and the predictive power, PRESS_{PQR}, is

$$\text{PRESS}_{PQR} = \sum_{i=1}^{I}\sum_{j=1}^{J}\sum_{k=1}^{K}\left(\hat{x}_{ijk}^{PQR} - x_{ijk}\right)^2. \tag{8.4}$$

In addition, Louwerse et al. (1999, p. 499) developed a statistic $W_{PQR,(P-1)QR}$ to gauge the relative difference in PRESS_{PQR} between two models differing in a single component, here the first mode. The W-statistic is defined as

$$W_{(P-1)QR,PQR} = \frac{(\text{PRESS}_{(P-1)QR} - \text{PRESS}_{PQR})/(df_{(P-1)QR} - df_{PQR})}{\text{PRESS}_{PQR}/df_{PQR}}, \tag{8.5}$$

where the df are the degrees of freedom (see Section 8.4). When checking models with different numbers of components in other modes than the first, one should adjust the formulas accordingly. If centering or normalization has been applied to a data array, the IJK should be suitably adjusted (see again Section 8.4). To avoid negative values, which may occur when the decrease in degrees of freedom is not accompanied by a sufficient decrease in PRESS, a lower bound of 0 could be introduced.

Model selection with PRESS.   Louwerse et al. (1999) also suggested a strategy to minimize the number of models that have to be inspected to find models with sufficiently low PRESS values. The essence of their strategy is that given a model with a specific number of components, the next set of models to be inspected are those with one more component in each of the modes in turn. The best of these models is then taken as a basis for the next iteration. This process stops if the PRESS increases or does not sufficiently decrease. The process is illustrated in Fig. 8.6.

8.9 EXAMPLE: CHOPIN PRELUDE DATA

8.9.1 Stage 1: Objectives

The primary aim of this section is to discuss how between-model and within-model comparison and selection can be carried out for the three most common three-mode models, the Tucker2, Tucker3, and Parafac models. The model selection and comparison will take into consideration the fit of the models as a whole, as well as that of different levels of the modes. First, analyses of the original data will be considered. Then, in Section 8.9.7, resampling procedures will be used to assess the stability of

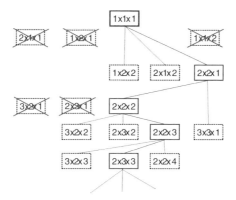

Figure 8.6 Model selection procedure using PRESS. The drawn lines indicate a minimum PRESS path. For all models that are not crossed out, PRESS is calculated. The crossed-out models are not permissible according to the minimum-product rule. (see Section 8.4).

the Tucker3 solutions, as well as the replicability via the predictive residual sums of squares measure, PRESS. Interpretability of results also plays an important role in model selection, but because it is so data dependent a discussion of that aspect will only be considered in the chapters with applications. The Chopin prelude study (see Murakami & Kroonenberg, 2001) will serve to illustrate the model selection procedures. This study also featured in the Introduction (Section 1.5, p. 7).

8.9.2 Stage 2: Data description and design

Music appreciation and the evaluation of the characteristics of musical pieces have frequently been researched by means of semantic differential scales. In the present study, Chopin's Preludes were judged on semantic differential scales by 38 Japanese university students (21 males and 17 females). For students to be eligible for the study, they had to be already familiar with classical music so that they could be expected to provide appropriate judgments about the music. The concept mode consisted of the 24 short piano Preludes making up Op. 28 by Frédéric Chopin, played by Samson François on Angel, AA-8047. They were copied on a cassette tape and edited for the experiment. Eighteen pairs of adjectives were selected from previous research on semantic differential studies of music. It was expected that the students would judge rather objectively on all but two scales on the basis of key (mainly, major or minor) and tempo of each prelude. In contrast, it was expected that they would express subjective evaluations on the scales UNATTRACTIVE–ATTRACTIVE and UNINTERESTING–INTERESTING. Before the final analysis, the codings of some scales were reversed to induce as many positive correlations among the scales as possible. Details of the experimental procedure can be found in Murakami and Kroonenberg (2001).

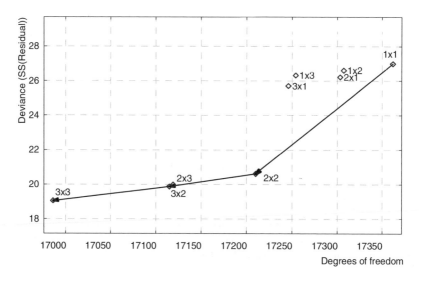

Figure 8.7 Chopin prelude data: Deviance plot for Tucker2 model. In the plot $P \times Q$ refers to a Tucker2 model with P components for the preludes and Q components for the scales.

8.9.3 Stage 3: Model selection

In this section we will inspect the results of a model search for the Chopin prelude data for each of the three major three-mode models. First, the deviance plot and the multiway scree plot will be investigated for the Tucker models in combination with the Timmerman and Kiers (2000) DifFit procedure, and the Ceulemans and Kiers (2006) *st*-procedure. For the Parafac model, such plots are less useful because the number of models that can be considered is very small in comparison, that is, only those with 1, 2, 3, or 4 components; orthogonal variants are also relevant in the present context. These procedures will be used to select a number of candidate models for further inspection. The selected Tucker models will be inspected for fit of the components in the model, and the fit of the elements of the core arrays. In addition, the fit for the levels of each mode will be inspected to see how well these levels are fitted by the final model.

Tucker2 analysis. Both the deviance plot (Fig. 8.7) and the multiway scree plot (Fig. 8.8) show very clearly that models with a single component for the preludes or the scales underfit the data. The model with two components each is a clear improvement over all of them. In the deviance plot, the angle for the 2×2 model (*st*-criterion) is the smallest, and in the multiway scree plot the DifFit-criterion points to the same model, so that this would be the preferred model according to both criteria. The convex hulls show that more complex models have less gain per degree of freedom or number of components. In both plots, the convex hull after the 2×2 model is more

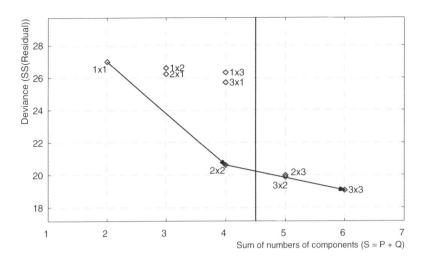

Figure 8.8   Chopin prelude data: Multiway scree plot for Tucker2 model.In the plot $P \times Q$ refers to a Tucker2 model with P components for the preludes and Q components for the scales.

or less linear. Moreover, with respect to the fit it does not matter if a component is added for the preludes or the scales. Thus, on the basis of the overall fit, it cannot be decided whether it is better to add a prelude or a scale component if one wants to go beyond the 2×2 Tucker2 model. Because it was desired to include more individual differences than could be accommodated by the 2×2 model, it was decided to further investigate the 2×3 model, the 3×2 model, and the 3×3 model. Note that all three are ruled out by the DifFit-criterion (i.e., lie to the right of the DifFit minimum).

Tucker3 analysis.   Both the deviance plot (Fig. 8.9) and the multiway scree plot (Fig. 8.10) are considerably more complex because components have to be computed for all three modes. Again it is clear that models with a single component for the preludes and scales underfit the data. A model with a single component for the students seems a reasonable proposition, but this implies that the only individual difference in the model is one of degree, expressed by the coefficients of the students on the first subject component. As we are interested in more subtle individual differences, the $2 \times 2 \times 1$ model and the $3 \times 3 \times 1$ model are not included in our selection. If individual differences are not important, these two models are the obvious choices. In both plots, the convex hulls after the $2 \times 2 \times 1$ model are more or less linear, and again with respect to the fit, it does not matter much if a component is added for the prelude or the scale mode; however adding a prelude component has a slight edge. However, on the basis of the overall fit, again it cannot be unequivocally decided whether it is better to add a prelude or a scale component.

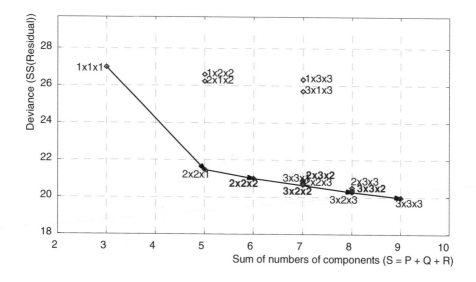

Figure 8.9   Chopin prelude data: Deviance plot for the Tucker3 model. In the plot $P \times Q \times R$ refers to a Tucker3 model with P components for the preludes, Q components for the scales, and R components for the subjects.

The DifFit criterion suggests that the $2 \times 2 \times 1$ model is the best model followed by the $2 \times 2 \times 2$ model and then by the $3 \times 2 \times 3$ model. The st-criterion also suggests that the $2 \times 2 \times 1$ model is the best model. On the basis of this information, the results from the plots, and the desire to investigate individual differences, it was decided to single out the $2 \times 2 \times 2$, $2 \times 3 \times 2$, $3 \times 2 \times 2$, and $3 \times 3 \times 2$ models for further consideration.

Parafac analysis.   As mentioned in Section 8.9, the search for Parafac models is in one sense much easier, because there are so few models to consider as all modes have the same number of components, but finding an appropriate converging model is more difficult because of the restrictiveness of the model itself. In Table 8.1 an overview is given of the fitting information of several Parafac models.

From Table 8.1 it follows that the number of models that can be considered is very limited, since for models with more than two components degeneracies occur (see Section 5.6.5, p. 89, for a full treatment of degeneracies in Parafac analyses). The degeneracies could be kept at bay by imposing orthogonality constraints on the scales, but still the four-component solutions take a large number of iterations to converge.

From a fitting point of view, one could single out the models with 2 and 3 components for further detailed inspection. However, for rating scale data with the centering as proposed here (i.e., centering per scale across concepts), the Parafac model has a distinct disadvantage when used as a multiway rating scale model, because the sub-

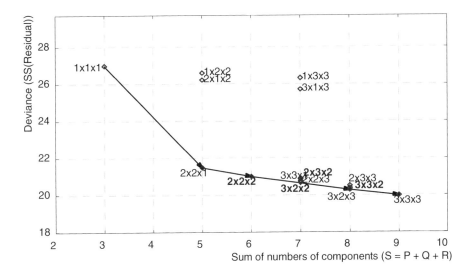

Figure 8.10   Chopin prelude data: Multiway scree plot for the Tucker3 model. In the plot $P \times Q \times R$ refers to a Tucker3 model with P components for the preludes, Q components for the scales, and R components for the subjects.

ject mode is not centered as it usually is for three-way profile data (see Section 6.6.1, p. 130 for preprocessing multiway profile data, and Section 6.6.4, p. 132 for pre-processing multiway rating scale data). Therefore, the first component will generally be very dominant in terms of explained variability and have a high correlation with the total sum of squares per subject. The lack of orthogonality constraints on the subject

Table 8.1   Chopin prelude data: Fit information for Parafac models

Number of components	Number of iterations	SS(Fit)	Comments
1	32	0.2897	Solutions identical to Tucker2 and Tucker3 models
2	37	0.4474	Solutions identical to $2 \times 2 \times 2$ Tucker3 model
3	1456	0.4695	Convergent solutions for 3 out of 5 runs
3	>7000	0.4708	Degenerate solutions in 2 out of 5 runs
3^{ortho}	284	0.4675	All three solutions identical
4	>7000	0.4913	All solutions degenerate
4^{ortho}	6506	0.4861	All solutions identical

Five runs with different starting values were made for each model. The Parafac model with 3 and 4 components were also analyzed with an orthogonal scale space: 3^{ortho} and 4^{ortho}.

mode will yield high congruence coefficients for the subject components (see below for the definition of this coefficient). In the present case, in all analyses the subject components had congruence coefficients of 0.90 or higher. In a sense this concurs with the results from the Tucker models, where the $2 \times 2 \times 1$ and $3 \times 3 \times 1$ models are strong contenders. The high congruence between the subject mode components can be seen as Parafac's way to mimic such solutions. Perfect equality, however, is not possible, given the definition of the model. Individual differences in these data cannot be modeled very well by Parafac, as without restrictions such as orthogonality it cannot go beyond the dominant structure between the scale and preludes. For comparisons with the other models, we will use only the two-component solution.

To make the Parafac model work for such data, one (also) has to center across subjects. This will, however, eliminate the consensus configurations of preludes and scales common to all students. At the same time this centering will bring individual differences much more to the fore. If it is used, the removed means should be analyzed separately in order to describe the consensus configuration.

Comparisons between components.   Given the fact that we have selected a number of solutions from a variety of models, a common basis for comparison must be found. The most common measure for comparing components from different solutions is Tucker's congruence coefficient (Tucker, 1951; Ten Berge, 1986). The congruence coefficient is comparable to the correlation coefficient, except that no centering is involved, that is, the congruence ϕ between two components x and y is:

$$\phi_{xy} = \frac{\sum x_i y_i}{(\sum x_i^2)^{1/2} (\sum y_i^2)^{1/2}}. \tag{8.6}$$

In an empirical study, Lorenzo-Seva and Ten Berge (2006) found that subjects judged components to have a fair similarity when ϕ was in the range 0.85–0.94, while two components with a ϕ higher than 0.95 were considered to be identical with respect to their interpretation.

The solutions for the models chosen will be compared by carrying out a singular value decomposition on the components of all analyses for each mode. An alternative would have been to use a generalized Procrustes analysis (e.g., see Borg & Groenen, 1997, Chapter 21; or Gower & Dijksterhuis, 2004). For the present purpose, the chosen approach proved to be adequate.

All components have been normalized, if they were not already, so that the matrix of congruence coefficients (Eq. (8.6)) is analyzed. Due to the centering applied to the data, both the scale components and prelude components are in deviation from their means, so that the analysis can be carried out with a standard principal component analysis. However, the subject mode is not centered so that most packaged PCA programs cannot be used due to their automatic centering. The principal axes of the congruence matrices have been rotated by standard varimax procedures. In the sequel we will refer to the rotated axes.

Table 8.2   Chopin prelude data: Rotated component loadings of a PCA on the covariance matrix of the components of Scales and Preludes, respectively

		Preludes					Scales		
Model	No.	1	2	3	Model	No.	1	2	3
T222	1	**1.000**	−0.010	−0.002	T222	1	**1.000**	−0.025	0.000
T322	1	**1.000**	−0.018	0.001	T322	1	**1.000**	−0.008	−0.002
T232	1	**0.999**	−0.042	0.004	T232	1	**1.000**	−0.018	0.000
T332	1	**1.000**	−0.013	−0.004	T332	1	**1.000**	−0.022	0.001
T333	1	**0.999**	−0.043	0.005	T333	1	**1.000**	−0.016	0.006
T22	1	**1.000**	−0.012	0.002	T22	1	**1.000**	−0.016	0.007
T23	1	**0.999**	−0.040	0.000	T23	1	**1.000**	−0.015	−0.006
T32	1	**1.000**	−0.012	0.003	T32	1	**1.000**	−0.015	−0.004
T33	1	**0.999**	−0.038	−0.001	T33	1	**1.000**	−0.014	−0.007
PF2	1	**0.862**	0.506	0.000	PF2	1	**0.969**	0.248	−0.005
PF2	2	−0.320	**0.947**	0.004	PF2	2	−0.649	**0.761**	−0.015
T222	2	0.010	**1.000**	0.004	T222	2	0.024	**0.999**	−0.019
T322	2	0.018	**0.999**	−0.026	T322	2	0.008	**0.999**	−0.034
T232	2	0.042	**0.999**	0.022	T232	2	0.018	**1.000**	−0.001
T332	2	0.013	**1.000**	−0.007	T332	2	0.022	**0.999**	−0.022
T333	2	0.043	**0.999**	−0.014	T333	2	0.015	**1.000**	−0.004
T22	2	0.012	**1.000**	−0.015	T22	2	0.016	**1.000**	−0.006
T23	2	0.040	**0.999**	−0.010	T23	2	0.015	**1.000**	−0.008
T32	2	0.012	**1.000**	−0.014	T32	2	0.016	**1.000**	−0.007
T33	2	0.038	**0.999**	−0.009	T33	2	0.014	**1.000**	−0.008
T322	3	0.000	0.025	**0.960**	T232	3	−0.001	−0.001	**0.980**
T333	3	−0.004	0.013	**0.970**	T333	3	0.007	0.004	**0.968**
T32	3	−0.002	0.015	**0.954**	T23	3	−0.005	0.009	**0.989**
T33	3	0.001	0.008	**0.956**	T33	3	0.008	0.007	**0.989**
T332	3	0.000	0.019	0.405	T332	3	0.001	−0.006	0.063

No. = Component number; $TPQR$ = Tucker3 model with P, Q, and R components for the modes, respectively; TPQ = Tucker2 model with P and Q components for the modes, respectively; PFS = Parafac model with S components.

Scales.   Table 8.2 shows that the first two axes of all solutions of the Tucker models are identical. The two Parafac components look somewhat different but theoretical considerations (see Ten Berge, 1991) show that in many cases a Tucker3 $2 \times 2 \times 2$ solution and a two-component Parafac solution span the same space. In the present case, this was empirically verified by regressing the Tucker components on both Parafac components. The multiple correlations of 1 confirmed for both scales and preludes that this was indeed the case.

With respect to the scale components, it is clear that model choice is not vitally important for interpretation, as all models lead to the same components. Note, however, that the third scale component of the $3 \times 3 \times 2$ solution is different from all other scale components.

Preludes. The components of the preludes from the various solutions are also identical. Again, the only real deviation is the third prelude component of the $3 \times 3 \times 2$ solution.

Subjects. The solutions for the subjects (Table 8.3) also show the first components to be identical. However, there are two additional types of subject components. Both are well separated in the $3 \times 3 \times 3$ solution; the second component of the $2 \times 3 \times 2$ solution is of one type and the second component of the $3 \times 2 \times 2$ solution is of the other type. In particular, the second subject component is associated with additional differences by the subjects in scale usage, and the third subject component is associated with additional differences in the evaluation of the preludes. Moreover, the additional differences in the scales and preludes are the results of judgments by different persons, since the two subject components are clearly different. The second subject components of the $2 \times 2 \times 2$ solution and the $3 \times 3 \times 2$ solution turn out to be a mixture of both types. The Parafac subject components show the point made in Section 8.9.3 that Parafac is not really suited to this kind of data, or at least not if they are not centered across subjects. The two Parafac components are virtually indistinguishable and in a Tucker model could be represented by a single component. Adding a third component leads to a third subject component, which is also virtually the same. Without additional constraints of different centerings, it does not seem possible to tease out one of the two other kinds of subject components using Parafac.

8.9.4   Stage 3: Tucker models — dimensionality selection

Partitioning variability: Fit of the three modes.   On the basis of the above comparisons of models, the most comprehensive analysis is found to be the $3 \times 3 \times 3$-Tucker3 solution, and all the models with fewer components are nested in this model, except for the $3 \times 3 \times 2$ one. In this section, we ask the question: To what extent do the models we have considered actually fit separate parts of the data? In particular, we look at the quality of fit or rather lack of it for each of the levels for all three modes.

This can be done because, after convergence of the least-squares algorithm to fit the model, the residual sum of squares can be partitioned for each level of each mode (Ten Berge et al., 1987); for example, for the levels i of mode A:

$$\text{SS(Total}_i \text{ of Mode A)} = \text{SS(Fit}_i \text{ of Mode A)} + \text{SS(Residual}_i \text{ of Mode A)},$$

and similarly for the other two modes. Thus, $\text{SS(Residual)}_i/\text{SS(Total)}_i$ is the proportional or relative lack of fit of the ith level of mode A. By comparing the relative

Table 8.3   Chopin prelude data: PCA loadings for covariances of the Subject
components

Model	No.	Subjects		
		1	2	3
T222	1	**1.000**	0.002	0.003
T322	1	**1.000**	0.001	0.004
T232	1	**1.000**	0.001	0.005
T332	1	**1.000**	0.009	0.004
T333	1	**1.000**	−0.003	0.005
PF2	1	**0.988**	0.112	0.102
PF2	2	**0.975**	−0.161	−0.143
T322	2	−0.001	**0.975**	0.074
T333	2	0.004	**0.987**	0.003
T332	2	−0.010	*0.821*	*0.517*
T222	2	−0.005	*0.722*	*0.647*
T232	2	−0.004	0.374	**0.910**
T333	3	−0.005	−0.013	**0.991**

No. = Component number; $TPQR$ = Tucker3 model with P, Q, and R components for the modes, respectively; PFS = Parafac model with S components.

residual sum of squares (referred to here as *relative residual*) within and across models, one can gauge to what extent the model adequately fits individual parts of the data. Alternatively, such an investigation can demonstrate that certain levels of a mode are dominating the solution more than is acceptable, given their role in the investigation.

Because not only the relative residual sum of squares but also the total sum of squares are known, one may distinguish between two cases when the relative residual is large. In the first case, a large relative residual is coupled with a small total sum of squares. Then there is little to worry about, because it means that probably the level was not very important in terms of variability and did not contribute much anyway. In the second case, when there is a large amount of variability but it is not fitted very well, it means that there is considerable variability that the model could not capture. There may be several reasons for this: the model does not have enough components, the data do not conform to the model, or there is simply a large amount of random noise. In any of these cases, further investigation might be called for. Van der Kloot and Kroonenberg (1982) have done such an analysis for data on implicit theories of personality.

Table 8.4 Chopin prelude data: Residual sums of squares for preludes

No.	SS(Total)	Tucker2 models			Tucker3 models				
		2×3	3×2	3×3	$2\times2\times1$	$2\times3\times2$	$3\times2\times2$	$3\times3\times2$	$3\times3\times3$
23.	1.96	0.32	0.34	0.30	0.39	0.35	0.36	0.38	0.34
15.	2.08	0.36	0.37	0.34	0.37	0.37	0.37	0.38	0.36
18.	1.86	0.35	0.36	0.35	0.38	0.38	0.37	0.37	0.38
7.	2.36	0.41	0.41	0.40	0.42	0.41	0.42	0.39	0.40
3.	1.86	0.42	0.43	0.41	0.46	0.44	0.45	0.45	0.44
22.	1.55	0.43	0.44	0.42	0.45	0.45	0.44	0.43	0.45
11.	1.54	0.42	0.45	0.43	0.47	0.46	0.47	0.46	0.46
16.	2.02	0.43	0.44	0.41	0.51	0.47	0.48	0.46	0.44
20.	1.72	0.49	0.48	0.44	0.49	0.48	0.48	0.50	0.46
2.	1.98	0.49	0.51	0.48	0.54	0.52	0.53	0.51	0.50
6.	1.66	0.48	0.53	0.48	0.55	0.52	0.54	0.52	0.50
5.	1.45	0.51	0.53	0.50	0.56	0.53	0.56	0.55	0.52
19.	1.33	0.51	0.48	0.49	0.55	0.53	0.54	0.55	0.51
24.	1.77	0.50	0.49	0.43	0.54	0.53	0.50	0.52	0.49
9.	1.68	0.57	0.58	0.56	0.62	0.59	0.59	0.52	0.57
14.	1.37	0.60	0.60	0.57	0.62	0.61	0.61	0.60	0.59
12.	1.29	0.63	0.64	0.63	0.67	0.66	0.67	0.65	0.66
4.	1.67	0.62	0.64	0.62	0.71	0.68	0.68	0.64	0.65
13.	1.26	0.62	0.58	0.56	0.70	0.68	0.57	0.64	0.59
21.	1.02	0.73	0.72	0.73	0.75	0.73	0.75	0.74	0.74
8.	1.25	0.83	0.82	0.83	0.83	0.83	0.83	0.77	0.83
1.	1.29	0.89	0.70	0.62	0.87	0.88	0.80	0.84	0.72
10.	1.09	0.83	0.79	0.82	0.88	0.88	0.81	0.79	0.84
17.	0.95	0.92	0.88	0.89	0.94	0.93	0.87	0.90	0.88

Tucker models: Relative residuals for preludes. For the Chopin prelude data, we first turn to the study of the relative residuals of the preludes. In Table 8.4 we have displayed the relative residuals for all models singled out previously in the comparisons. A general patterns which can be observed is that the relative residuals for the Tucker2 models are lower than those for comparable Tucker3 models due to the larger number of parameters.

More in details Table 8.4 shows that the models fit most of the preludes in a reasonable way, with the largest relative residual for No. 10 (around 88%) and No. 17 (around 90%) compared to 50% fit overall. The preludes with the smallest relative residuals are No. 15 (around 35%) and No. 23 (around 37%). However, note that the total sums of squares for the best fitting preludes are about twice as large as the those of the worst fitting preludes (and this is part of a trend), indicating a general aspect of least-squares fitting procedures, that is, larger variabilities are fitted better.

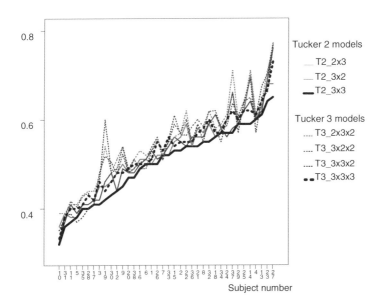

Figure 8.11   Chopin prelude data: Relative residual sums of squares for subjects; ordered by 3×3 Tucker2 model.

Tucker models: Relative residuals for subjects.   There are quite considerable differences between subjects with respect to the fit of the models to their data. The best-fitting subject has a relative fit on the order of 0.65, whereas for the worst-fitting subject the relative fit is more than twice as small (about 0.35). The general trend is that with a increasing overall relative fit, most subjects tend to fit better as well. The patterns and outliers, however, are difficult to spot from listing their fitted sums of squares in a table.

In Fig. 8.11 we have shown the relative residuals for all models versus the subject numbers. We do find that the best fitting 3×3 Tucker2 model has lower relative residuals for virtually all subjects, and a careful inspection shows that the patterns for the Tucker2 are somewhat more alike than for the Tucker3 models. This has undoubtedly been caused by the fact that components for subjects have been computed for the Tucker3 models and not for Tucker2 models. It can also be seen that this effect starts to wear off when three components are present for the subjects, as in the $3 \times 3 \times 3$ model.

Relative residuals for scales.   With respect to the fit of the scales, there are three scales with large relative residuals (see Table 8.5). In particular, LARGE (82%) and, as expected, the two evaluative scales ATTRACTIVE (83%) and INTERESTING (87%), in contrast with HEAVY (33%) and FAST (30%). The implication is that the well-fitting scales have been used to make differential judgments about the preludes and that the

Table 8.5   Chopin prelude data: Residual sum of squares for scales

Scale	SS(Total)	Tucker2 models			Tucker3 models				
		2×3	3×2	3×3	2×2×1	2×3×2	3×2×2	3×3×2	3×3×3
Fast	3.26	0.30	0.29	0.28	0.30	0.30	0.31	0.29	0.30
Heavy	3.32	0.32	0.30	0.30	0.36	0.34	0.34	0.33	0.33
Severe	2.02	0.39	0.37	0.36	0.42	0.41	0.38	0.40	0.38
Restless	2.33	0.39	0.36	0.37	0.44	0.42	0.40	0.42	0.39
Dark	2.09	0.48	0.46	0.45	0.52	0.50	0.48	0.45	0.46
Vehement	2.09	0.48	0.46	0.46	0.53	0.50	0.48	0.48	0.47
Noisy	1.53	0.49	0.48	0.47	0.51	0.50	0.49	0.49	0.50
Gloomy	1.80	0.49	0.46	0.46	0.53	0.50	0.48	0.46	0.47
Loud	1.71	0.49	0.47	0.48	0.53	0.52	0.49	0.51	0.48
Dramatic	2.82	0.54	0.52	0.52	0.56	0.56	0.54	0.55	0.55
Cloudy	1.78	0.57	0.55	0.53	0.60	0.58	0.59	0.58	0.55
Sad	1.92	0.58	0.56	0.55	0.60	0.59	0.56	0.53	0.56
Strong	1.71	0.58	0.57	0.57	0.62	0.60	0.59	0.56	0.58
Hard	1.64	0.61	0.59	0.59	0.66	0.63	0.62	0.62	0.61
Cold	1.21	0.67	0.65	0.65	0.72	0.70	0.66	0.66	0.66
Thick	1.23	0.73	0.75	0.70	0.79	0.75	0.76	0.76	0.72
Coarse	1.26	0.77	0.76	0.75	0.77	0.77	0.77	0.73	0.75
Large	1.35	0.79	0.81	0.79	0.83	0.81	0.82	0.78	0.82
Attractive	1.66	0.81	0.96	0.72	0.99	0.89	0.97	0.97	0.83
Interesting	1.34	0.84	0.96	0.78	0.97	0.92	0.97	0.97	0.87

ill-fitting scales were difficult to use consistently, so that little systematic variability is present. This is probably true especially for the evaluative scales.

Figure 8.12 shows the *sums-of-squares plot* (or *level-fit plot*) for the scales in the $2 \times 3 \times 2$ model, which has an overall fit of 45%; see also Section 12.6, p. 287. From the plot we can gauge to what extent the model succeeds in fitting the variability of a level. In the plot a dashed line is drawn from the point (0,0) through the point with the average SS(Fit) and the average SS(Residual). Points above the line fit worse compared to the overall model fit and the points below the line fit better than the overall model. The further away from the line the worse (or better) the fit. The negatively sloping solid line (and all lines that could be drawn parallel to it) represents points with an equal SS(Total), and the further away a line is from the origin, the larger the total sum of squares. Finally, points on lines fanning out from the origin have the same *fit/residual ratio*, where the fit/residual ratio = SS(Fit)/SS(Residual). Note that, SEVERE and RESTLESS have the same fit/residual ratio, but RESTLESS has a larger SS(Total).

Keeping this in mind, it is obvious that the scales HARD, FAST, and DRAMATIC have an almost equal SS(Total), but that the first two fit much better than the last one.

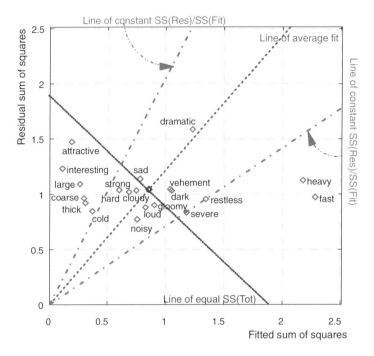

Figure 8.12   Chopin prelude data: Sums-of-squares plot for scales for $2 \times 3 \times 2$-Tucker3 model.

The lack of fit of the evaluative scales ATTRACTIVE and INTERESTING is noticeable. We also see that they have a low fit coupled with a normal SS(Total), indicating that the students' evaluations are rather different, but that only part of this is fitted by the model. Having more evaluative scales might be a solution if the intention is to find out what it is that students appreciate in music. The scale COLD is low in fit and residual, which means that all students judged the preludes more or less equally cold or warm. Given there is little variability, it means that the students scored this scale roughly at the mean. However, for this analysis the mean values were removed by centering, so that what is left might primarily be random variation.

Dimensionality selection Tucker models: Concluding remarks.   In this section, we have concentrated on the dimensionality of Tucker models rather than on that of Parafac models, primarily because for Tucker models we have to make a numerically based subjective decision as to which model provides an adequate fit to the data. For the Parafac model the question is more how many components exist that show a parallel proportional profile. This is a rather different decision with different considerations, as we have outlined in Section 8.9.3. In the next section, we will take a further look at model selection for Parafac, but primarily to illustrate core consistency

Table 8.6 Chopin prelude data: Degeneracy indicators for three Parafac models

	Triple cosine products			Smallest eigenvalue	Condition number	Standardized weights		
No.	$\theta_{(1,2)}$	$\theta_{(1,3)}$	$\theta_{(2,3)}$			1	2	3
2C	-0.085	——	——	0.99	1.01	0.29	0.20	——
3C	-0.574	0.020	-0.116	0.62	1.87	0.45	0.24	0.19
3D	-0.997	0.152	-0.176	0.25	5.74	42.46	42.81	0.25

No. = Number of components; C = convergent solution; D = Divergent solution.

and degeneracy. Often, Parafac analyses are carried out for different centerings to sort out which variability in the data has a Parafac structure. In a sense, one could say that finding an insightful Parafac analysis requires a much more subtle separation of the different sources of variability, while the Tucker models are used in a more direct fashion.

We have looked at the fit partitioning of the components, but not at that of the core array when the components are orthonormal (as is the case in most Tucker analyses). However, this question is closely linked with the interpretation of the core array and rotating it to maximum simplicity. All these points will be taken up in the chapter on core arrays (Chapter 10).

8.9.5 Stage 3: Parafac models — dimensionality selection

We will again analyze the Chopin prelude data, now using only the Parafac model. The main purpose is to illustrate the ideas of core consistency and of degeneracy. In particular, we will look at a Parafac model with two components and at two solutions with three components.

Degeneracy information. Table 8.6 shows that the two-component solution is well-behaved. There are two three-component solutions, one that shows heavy degeneracy and one that is well-behaved. Of five runs, two converged to the same solution and three did not converge within 5000 iterations. The good solution, using an acceleration mechanism, essentially converged in around 500 iterations to an accuracy of 10^{-6}, but it took another 1000 iterations to get the norms for the differences between successive component matrices within a similar tolerance.

The degenerate solution shows all the signs of indeed being divergent degenerate (for a discussion of degeneracy indicators see Section 5.6.5, p. 89). One of the triple cosines is nearly -1 ($\theta_{1,2} = -0.997$), the smallest eigenvalue of the triple cosine matrix is 0.25, and the condition number is 5.74, which is not overly large but certainly suspect. The standardized component weights, that is, the $g^2_{sss}/\text{SS(Total)}$, have grown beyond reasonable doubt. As proportions, these values should be below 1. Further iteration with a more stringent convergence criterion will show an even further increase in size, if the algorithm's computational accuracy will allow it.

Table 8.7   Chopin prelude data: Core consistency for three Parafac models

No. of components		Core consistency		Degree of superdiagonality
		Bro's	Normalized	
2	Convergent	1.000	1.000	1.000
3	Convergent	0.646	0.717	0.758
3	Degenerate	−5173.150	−0.010	−0.005

Core consistency and superdiagonality.   Even though a convergent three-component Parafac model exists, it is not self-evident that this provides an adequate model for the data. To check this, we can investigate the Parafac core array computed from the Parafac component matrices. Table 8.7 indicates that a proper two-component Parafac model can be found, as all indicators are 1.00. With three components, the measures do not show a perfect core consistency, but the indicators show that the model captures most information. The degenerate three-component solution clearly shows the absurdity of the results.

To gain further insight into the core array and its core consistency, we have constructed the core consistency plots for the three models (Fig 8.13); for an explanation of such plots, see Section 8.6.3. The adequacy of the two-component model is self-evident, the converged three-component solution looks reasonable, but there are some medium sized off-diagonal elements, and the nonconverged solution is terrible.

Table 8.8   Chopin prelude data: Kiers's hierarchy for three-mode component models

Model	Standardized SS(Fit)		
	$u = 1$	$u = 2$	$u = 3$
Tucker1 (Mode C) — PCA-sup	0.469	0.500	0.534
Tucker3	0.299	0.447	0.474
Parafac	0.299	0.447	-.-
Parafac — A orthonormal	0.299	0.447	0.469
— B orthonormal	0.299	0.446	0.468
— A and B orthonormal	0.299	0.444	0.463
Sum-PCA	0.281	0.430	0.445

Tucker3 models have $1 \times 1 \times 1$, $2 \times 2 \times 2$, $3 \times 3 \times 3$ components. -.- = Degenerate solution for Parafac with three components.

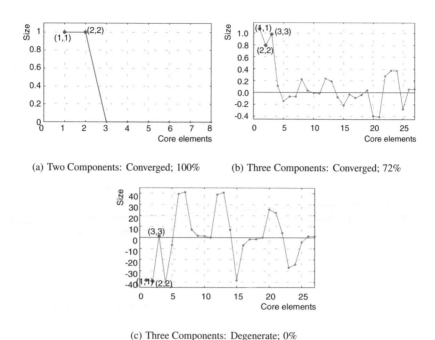

(a) Two Components: Converged; 100%   (b) Three Components: Converged; 72%

(c) Three Components: Degenerate; 0%

Figure 8.13   Chopin prelude data: Core consistency plots for three Parafac solutions.

8.9.6   Stage 3: Model selection — selecting from a hierarchy

Table 8.8 shows the results of Kiers's between-model selection procedure (Kiers, 1991a, p. 463); see also Section 8.7. None of the models fit anywhere near the Tucker1 fit for one component. For two components for all modes, the SUMPCA solution averaging over subjects fits quite well (Standardized SS(Fit) = 0.430). However, this solution only describes the structural characteristics of the relationships between the preludes and scales without modeling any individual differences. A Parafac model with both the scale and prelude spaces orthonormal provides a small increase in fit over the SUMPCA solution (Standardized SS(Fit) = 0.444), and none of the other models substantially improves on that. For three components, again the SUMPCA solution takes care of most of the fit (Standardized SS(Fit) = 0.445), the double-orthonormal Parafac solution adds nearly 2%, while the Tucker3 model adds another percent. On purely statistical grounds it seems that a double-orthonormal Parafac model with two components is the preferred solution.

Looking outside the hierarchy a $2 \times 2 \times 1$-Tucker3 solution has a fit of 0.435. This is an improvement of only 0.005 over the two-component SUMPCA solution, so that adding the one subject component to allow for individual differences in size does not contribute much to the fit to the data. In the same sense, not much is gained

Table 8.9   Chopin prelude data: Bootstrap and jackknife results

Model	S	No. param. model	SS(Fit)	Bootstrap SS(Residual)	Standard error	Standard score SS(Fit)	$1 - $ PRESS
$1 \times 1 \times 1$	3	80	0.290	0.296	0.016	-0.41	0.159
$2 \times 2 \times 1$	5	121	0.435	0.439	0.015	-0.28	0.346
$2 \times 1 \times 2$	5	139	0.310	0.319	0.015	-0.60	0.178
$1 \times 2 \times 2$	5	135	0.300	0.308	0.016	-0.50	0.161
$2 \times 2 \times 2$	6	160	0.447	0.451	0.015	-0.28	0.352
$2 \times 2 \times 3$	7	197	0.454	0.461	0.013	-0.55	0.357
$2 \times 3 \times 2$	7	179	0.451	0.460	0.014	-0.63	0.352
$3 \times 2 \times 2$	7	183	0.456	0.468	0.015	-0.81	0.359
$3 \times 3 \times 1$	7	160	0.450	0.458	0.016	-0.51	0.358
$2 \times 3 \times 3$	8	218	0.460	0.469	0.013	-0.61	0.357
$3 \times 2 \times 3$	8	222	0.466	0.475	0.012	-0.68	0.375
$2 \times 2 \times 3$	8	232	0.458	0.462	0.014	-0.33	0.362
$3 \times 3 \times 3$	9	246	0.474	0.488	0.015	-1.05	0.375
2×2	4	232	0.458	0.466	0.016	-0.58	0.361
2×3	6	323	0.474	0.482	0.014	-0.28	0.367

$S =$Total number of components; No. parm model $=$ number of parameters in the model; SS(Fit)$=$ Fitted sum of squares of original analysis; Bootstrap SS(Fit) $=$ Mean of SS(Fit) 100 bootstrap samples; Standard error $=$ Standard deviation of the 100 bootstrap SS(Fit); Standard score SS(Fit) $=$ Standard score of original SS(Fit) in the bootstrap distribution. The number of independent data points $= 20 \times 24 \times 38 - 20 \times 38$ (removed means) $- 38$ (removed SS of subjects) $= 17442$.

with a $3 \times 3 \times 1$-Tucker3 solution, since its fit of 0.450 is only a small gain over the 0.445 fit of SUMPCA with three components. These two models with an extra subject component do not fit into the hierarchy but are clearly more complex than the SUMPCA models.

8.9.7   Stage 5: Validation — model stability and predictive power

In Section 8.8 the bootstrap and the jackknife were introduced as methods to estimate the standard error of the SS(Fit) and the predictive residual error sum of squares (PRESS), respectively. In Table 8.9 both measures are presented for a subset of Tucker models. Also included are the bootstrap estimate of the fitted sums of squares and the difference between the original analysis and the bootstrap one[3]. In the table $1-$PRESS is shown so that all quantities refer to the fit, rather lack of fit, of the models.

[3]The consistently higher values for the bootstrap SS(Fit) compared to the original SS(Fit) are striking and unexplained as yet.

From the overall standard error of around 0.015, one may conclude that all models beyond the $2 \times 2 \times 1$ are more or less equivalent with respect to their fit to the data. In our case, we had two content-based arguments to include extra components. The first was that insight in individual differences was required and the second was that the evaluative scales INTERESTING and ATTRACTIVE had to play a role. With respect to the individual differences, a $2 \times 2 \times 1$ model only accommodates differences in the size of the combined scale–prelude configuration via the first subject component. To look at more relevant individual differences, a second subject component is necessary. In Table 8.5 we noticed that the smallest models fitting some of the variability of the two evaluative scales are the Tucker3 $2 \times 3 \times 2$ model and the Tucker2 2×3 model. The SS(Fit) of the former is just about two standard errors higher than the $2 \times 2 \times 1$ model, and thus is more or less acceptable. However, on statistical grounds it does not stand out from similar models with more components than the $2 \times 2 \times 1$ model. The Tucker2 2×3 model has no clear advantage as its fit is similar, but it uses somewhat more parameters.

With respect to the predictive residual error, the overall trend is that there is a shrinkage of around 0.10 in fit for all models so that we should expect some 10% less fit for any model when the experiment is replicated. There is a serious gain in replicability in going from the $1 \times 1 \times 1$ model to the $2 \times 2 \times 1$ model, but after that, the gain in predictability from more complex models only comes a single percentage point at a time.

The overall conclusion for the Chopin prelude study is that at least a $2 \times 2 \times 1$ model should be chosen, but no clear statistical advantages are gained from complicating the model further. However, we have seen (in Section 8.9.3) that more complex models in the present case introduce new components without seriously affecting the earlier ones. Thus, delving deeper into the data may lead to further insights, which, however, should be approached with more caution and are more in need of confirmation than the main structure.

8.10   CONCLUSIONS

Model selection is very much a judgmental process but the choice of a particular model should not only be based on numerical arguments and investigations, but also on content-based arguments. Proper regard should be paid to the information from all parts of the process of selection, so that a reasoned decision can be made. Luckily, there are a large number of procedures and indicators that can assist the researcher in making this decision.

CHAPTER 9

INTERPRETING COMPONENT MODELS

9.1 CHAPTER PREVIEW

Substantive interpretation of the parameters of component models is seldom straight-forward. First of all, there are several component matrices, and in the Tucker models a core array as well. The most interesting questions to be answered by means of multiway analysis are complex ones involving all modes, which also makes the answers complex. Thus, it is an illusion to expect simple univariate answers from a multiway analysis. The only exception occurs when a multiway analysis shows that a multiway analysis is not necessary because the data do not contain any multiway structures. In order to arrive at an insightful substantive interpretation of the outcomes of a multiway analysis, careful consideration has to be given to the statistical interpretation of the parameters in the model. In particular, everything should be done to enhance interpretation of the components in such a way that their values correspond to clearly defined statistically meaningful quantities, such as regression coefficients, correlations, scores on latent variables, and so on. A complication in outlining procedures to realize these statistical interpretations is that there are a number of fundamental

Applied Multiway Data Analysis. By Pieter M. Kroonenberg
Copyright © 2007 John Wiley & Sons, Inc.

differences between components of the Tucker models and the Parafac model, both in a technical and a conceptual sense, which precludes completely general statements.

Several interpretational devices are relevant for both Tucker and Parafac models, while others pertain to only one of them. The basic tools for interpretation fall broadly into four (related) categories:

1. *scaling of components* (sometimes also centerings) to give the coordinates special meaning, such as variable–component correlations;

2. *scaling of core arrays* to allow for interpretations in terms of explained sums of squares, regression coefficients, and idealized quantities;

3. *transformations*, such as rotations of components and/or core arrays to simplify the patterns to be interpreted by creating a limited number of large values and many small ones (preferably zeroes); and

4. *graphical displays*

In this chapter we will discuss the first two procedures and leave the other two for Chapters 10 and 11.

Before we go into these types of tools, we will touch upon a number of general aspects related to the data being modeled and which are directly relevant for interpretation. In particular, we have to explain the differences between model-based and content-based considerations and the difference between the multivariate point of view and the multidimensional point of view (Gifi, 1990, pp. 49–50).

The sections on interpreting components in this chapter lean heavily on several sources: Kroonenberg (1983c, Chapter 6), Harshman and Lundy (1984c, Appendix 5–1), and Murakami (1998a).

9.2 GENERAL PRINCIPLES

9.2.1 Model-based versus content-based considerations

In Chapter 6, we made the distinction between model-based and content-based arguments for preprocessing. A similar distinction can be made with respect to interpretational devices, such as component scaling and rotations. Model-based arguments are used to suggest procedures to make the estimated parameters of the model more interpretable, while content-based arguments pertain to choosing which mode should be handled in which manner. Thus, while in principle each mode may be rotated, typically not all modes will be rotated in the same way because of the entities represented by the modes. Most of this chapter will concentrate on model-based arguments, but content-based arguments will be discussed in passing.

9.2.2 Multivariate versus multidimensional point of views

Multidimensional point of view. In the multidimensional point of view we are primarily concerned with the data as a population; stochastic considerations, distributions of variables, and so on do not play a role. In essence, we analyze the high-dimensional spaces defined by the levels of the modes via appropriate low-dimensional subspaces, which are generally maximum-variance projections. Thus, we aim to represent the entities in a mode as well as possible in a low-dimensional space, preferably with as few as two or three dimensions. This view is explicitly expressed by Franc (1992), who shows that in the Tucker model the subspaces are the unknowns and not the coordinate axes in the subspaces, that is there is subspace invariance and not component invariance. Furthermore, no particular *a priori* status or priority is given to any mode with respect to the interpretation. A similar argument can be found in Bro, Smilde, and De Jong (2001).

Within this conceptualization of multiway analysis, it is not obvious or necessary to interpret the principal components as latent variables, even though one may search in the low-dimensional subspace for directions that could function as such. Only the first dimension has a clear interpretation as the direction of maximum variance, but the other components only have maximum variance after the first one, which is not a basis for interpretation. In general, this will mean that the interpretation of the subspaces, here referred to as *subspace interpretation*, will be based on the arrangement in space of the levels of the modes, rather than on specific directions.

Multivariate point of view. The multivariate point of view concerns "systems of random samples or random samples of such systems" (see Gifi, 1990, pp. 49–50). The multivariate point of view is based on the asymmetric roles of the modes. For instance, we may consider the situation that we have J random variables measured on K samples of I individuals each who may or may not be the same, or alternatively that there are $J \times K$ random variables with a sample of I subjects. The Parafac model fits very nicely in the multivariate point of view, but the Tucker models can also have valid interpretations within this conceptualization.

It should be noted that even though multivariate point of view is concerned with random samples, in our expositions the subjects are often analyzed in their own right and can be examined for individual differences without reference to the stochastic framework. Subjects in social and behavioral sciences (or their equivalents in other disciplines) have a special status in many three-way data sets because they are the *data generators*. Because of this the subject mode is often treated differently from the other modes.

9.2.3 Parafac components

Nature of Parafac components. From the multivariate point of view, in standard two-mode PCA each component s has one set of coefficients for the subjects, $\mathbf{a}_s$, and one set of coefficients for the variables $\mathbf{b}_s$. For each s, $\mathbf{a}_s$ is exclusively linked to $\mathbf{b}_s$, but not to any other set of coefficients $\mathbf{b}_{s'}$, $s \neq s'$. This unique link ensures that there is only one single interpretation for each component s, which belongs to both modes. The same situation is true for the Parafac model, so that a component exists "in each mode", and only one interpretation for each of the s components is possible. The values of the coefficients of all modes contribute to the construction of this interpretation. We will call this kind of interpretation of components, a *component interpretation* in contrast with a subspace interpretation. Because of the component interpretation of the Parafac model, the values of all three modes for the same component s can be plotted together in a per-component or line plot (see Section 11.5.1, p. 271).

Note that here we have used the word "component" slightly ambiguously, as both sets of coefficients in $\mathbf{a}_s$, $\mathbf{b}_s$, and $\mathbf{c}_s$ can be designated as components, but one might just as easily talk about a single component $\mathbf{f}_s$ ($= \mathbf{a}_s \otimes \mathbf{b}_s \otimes \mathbf{c}_s$), which has expressions in all modes. If one wants to stay closer to two-mode parlance, a component may be defined for the subject mode; the subjects have scores on such a component, and the entities in the other modes weight or have loadings for this component. Generally, we will use the word *component* in a loose sense, meaning either $\mathbf{f}_s$ or each of its constituent parts. When it is necessary to make a distinction we will refer to the $\mathbf{a}_s$, $\mathbf{b}_s$, and $\mathbf{c}_s$ as *coefficient vectors*.

Component uniqueness. A specific feature of an optimum solution of a Parafac model is that the components are uniquely identified in the sense that they cannot be rotated without deteriorating the fit of the Parafac model to the data, which is consistent with the view that the components are the unknowns not the subspaces they span. This *intrinsic axis property* gives a special status to the directions of the components in the subspace. Moreover, for the interpretation of these directions one should make use of all modes simultaneously. Because the Parafac model is one with constraints, it does not fit all three-way data. Therefore, it may not be possible to find (meaningful) unique components for all data (see Section 5.6.4, p. 88).

True Parafac models. In several cases in chemistry and signal processing, the data follow a physical or chemical model that already has the form of a Parafac model, so that the uniqueness property allows for unique estimation of the parameters of these models. In other words, the meaning of the components is not something to be determined after the analysis, but is known exactly beforehand. In that case, it is only the values of estimated parameters that need to be interpreted.

9.2.4 Tucker components

Nature of Tucker components. In the Tucker models components are separately defined for each mode. In particular, a set of components constitute a coordinate system for the (maximum-variance) subspace, which is a projection from the original high-dimensional data space. The components of each of the modes are determined separately and should therefore also be interpreted separately (i.e., independently of the other modes). Therefore, if a component interpretation in the Tucker model is used and it is necessary to distinguish the components from different modes, often terms are used as latent variables, idealized subjects, prototype conditions, or time trends. These terms should be taken at face value as they do not necessarily express "real" underlying quantities.

Role of the core array. The strengths of the links between the sets of components is expressed via the core array, which contains the information on the relative importance of the combinations of components from all modes. In particular, if, in the three-way case, the g_{pqr} of the model term $g_{pqr}(a_{ip}b_{jq}c_{kr})$ is large compared to the other terms in the core array, the pth, qth, and rth components of the first, second, and third mode, respectively, contribute much to reconstruction of the data on the basis of the model, or the *structural image* of the data. If it is around 0, the combination of components is unimportant.

The independent construction of the component spaces puts a heavy burden on finding an adequate interpretation for the core array. One can hardly expect to be able to easily interpret the links between components, if one does not know what these components mean. Therefore, any tools one can use to enhance the interpretation of components should be exploited. First among them are rotations of the components. However, one complication is that rotations directly influence the sizes of the core elements, as they make them more equal and thus smaller. This is the same effect as the redistribution of the variance of rotated components in two-mode PCA.

Nonuniqueness — rotational freedom. Because in the Tucker models it is not the axes that are unique but the entire low-dimensional space (subspace uniqueness), singling out directions for interpretation is a more arbitrary process, making a subspace interpretation more attractive. One may apply rotations to any of the three component spaces in order to search for new, more easily interpretable directions. The irony is that, while the interpretation of the Tucker models leans more heavily on the conceived meanings of the (various sets of) components because the relationships between them have to be examined via the core array, the meaning of the components themselves is more difficult to pin down because any rotated version of them can be used as well. In contrast, in the Parafac model the relationships between components of different modes are explicitly specified in the definition of the model (the intrinsic axis property).

9.2.5 Choosing between Parafac and Tucker models

Difference between Parafac and Tucker models. It has often been suggested that a Parafac model is nothing but a Tucker3 model with an equal number of components for the three modes and with a superdiagonal core array. However, as pointed out by Harshman and Lundy (1984c, pp. 170, 171) the concept of a component in the two models is different, and the difference can be linked with the multidimensional versus multivariate distinction; see also the previous section. A similar but more algebraic and precise statement was formulated by Franc (1992, pp. 154–155) in his (French) doctoral thesis, which unfortunately was never published[1]. Bro et al. (2001) make a similar point about the difference between the two models.

Given the properties of the components of the two types of multiway models, one may wonder why one bothers about the Tucker models at all. After all, their interpretation rests on the relationships between arbitrarily oriented components from different modes (to put it a bit bluntly), while the Parafac model has only one set of components, present in all modes, which as an added bonus are uniquely defined as well. The reasons for also considering and often preferring the Tucker model are various.

Availability of a solution.   In the first place, there is always a solution with the Tucker model. Said differently, high-dimensional data can always be represented in lower-dimensional spaces, so that (approximations of) the patterns present in the data can be inspected to any desired degree of accuracy. Therefore, one never ends up with empty hands, or forces patterns upon the data that barely exist or do not exist at all. In contrast, the uniqueness of the Parafac components is valuable to the extent that the data actually have parallel proportional profiles. If this is not the case, it is possible that the parameters of the Parafac model cannot be estimated. Moreover, the model is a (sometimes very) restricted one, and it may happen that only a limited number of components with low fit can be found that fulfill these restrictions, while more systematic patterns are present in the data but of another kind.

Finding a solution.   Another problematic aspect of the Parafac model is that it is often difficult to fit. Especially when a component in one of the modes is nearly constant, and/or the components have Tucker structure (components of one mode are linked with more than one component of another mode), Parafac algorithms can have great difficulty in reaching satisfactory solutions. Similarly, because of the restrictions built in the Parafac model, many different starts may be necessary to achieve a, hopefully globally optimum, convergent solution. Due to such problems, the existence of such solutions is not guaranteed.

Rotating a solution.   The lack of rotational freedom is mostly hailed as an attractive feature of the Parafac model, but it may lead to complex components, while slightly less well-fitting models without the uniqueness property but with rotational freedom can sometimes be more easily interpreted due to their simpler patterns of the component coefficients.

Number of components.   A more mundane point is that because in the Parafac model there is essentially only one set of components present in all modes, the dimensionality

[1] This thesis and its English summary are, however, available from *http://threemode.leidenuniv.nl/* (Three-mode books). Accessed May 2007.

of each mode is the same. Tucker models are more flexible and component spaces of different modes can have different dimensions, which is attractive if the numbers of levels in the data array are very different.

Partitioning the fit of a solution. An attractive feature of Tucker models, which is shared only by *orthogonal* Parafac models, is that a partitioning of the fitted and residual sums of squares is possible at both higher levels (components and core array) and lower levels (individual entities in each mode) of the model (for technical details, see Ten Berge et al., 1987).

Summary. If the data support a Parafac model with a reasonable number of components, which can be found without too much hassle, and which lend themselves to clear interpretations, it seems the model of choice. If, however, the data are more complex, the numbers of levels are very different, and considerable detail is required, and, moreover, a flexible approach toward modeling is desired, the Tucker model has more to offer. In the end, of course, it always pays off to be eclectic and investigate the behavior of both models for any one data set.

9.3 REPRESENTATIONS OF COMPONENT MODELS

Before discussing the various ways in which the solutions for the Tucker models and Parafac models can be enhanced and presented to facilitate interpretation, we will first describe the *basic form* of the most common multiway component models, which is also the form in which the parameters of these models are usually supplied by computer programs solving the estimation of the models. Such components have unit lengths, and in the case of the Tucker models, the components of a mode will also be mutually orthogonal, so that they are orthonormal. In the basic form of the Parafac model, the components have unit lengths, but no orthogonally is assumed or likely to be present. Models in their basic forms will provide the starting point for the rest of this chapter.

9.3.1 Common views of two-mode PCA

In this subsection we pay attention to terminology and two different ways to represent the two-mode principal component model: the multivariate and the multidimensional point of views.

Components and the multivariate point of view. Within the multivariate point of view there are two common forms for the coefficients in component analysis. The *covariance form*, places the primary emphasis on the variables and aims to represent the correlations between the variables and the components and is the standard form in the behavioral sciences. Moreover, it is closely related to the basic concepts of factor analysis. The *distance form*, is primarily concerned with representing and plotting the individuals in such a way that the Euclidean distances between subjects are preserved in the full decomposition and are represented as well as possible in the reduced space. The latter view is the predominant one in other disciplines and follows naturally from the definition of a component as a linear combination of the observed variables. Further discussion about these different views can, for instance, be found in Legendre and Legendre (1998, p. 391ff.), Ten Berge and Kiers (1997), and Kiers (2000a).

Covariance form.   Within the context of multivariate point of view, a clear distinction is made between component scores and component loadings. Scores refer to coefficients of unit-length components and are usually associated with subjects. Loadings are usually associated with the variables and refer to components that have lengths equal to their variances, so that the coefficients incorporate the size of the data. The *covariance form* for a full decomposition of a data matrix, $\mathbf{X}$, is

$$x_{ij} = \sum_s g_{ss}(a_{is}b_{js}) = \sum_s [a_{is}][g_{ss}b_{js}] = \sum_s a_{is}f_{js}, \qquad (9.1)$$

where $\mathbf{A}$ is orthonormal, $\mathbf{A}'\mathbf{A} = \mathbf{I}_s$, and $\mathbf{F}$ is orthogonal, $\mathbf{F}'\mathbf{F} = \mathbf{G}^2$, with the g_{ss}^2 the eigenvalues equal to the squares of the singular values g_{ss}. The usual terminology for the $\mathbf{a}_s$, the row coefficients with normalized coordinates, is *component scores* , and for the $\mathbf{f}_s$, the column coefficients with principal coordinates, *component loadings*. We will use the term "loadings" in a much looser sense and use the term "principal coordinates" for components scaled to have their lengths equal to their eigenvalues. When correlations between variables and components are meant, this will be mentioned explicitly. In two-mode analysis the f_{js} are variable–component correlations when the columns of $\mathbf{X}$ have been standardized (mean = 0; standard deviation = 1), because in that case, $\mathbf{R}_{(\text{data,components})} = \mathbf{X}'\mathbf{A} = \mathbf{F}$, or $r_{js} = \sum_s x_{ij}a_{is} = f_{js}^2$. Note that in the full space the sum of the squared loadings, $\sum_s^J f_{js}^2$, is equal to the variance of the jth variable, $\text{var}(\mathbf{x}_j)$. In the reduced S-dimensional space, $\sum_s^S f_{js}$ is the amount of variance of $\mathbf{x}_j$ explained by this reduced space. The correlation interpretation of the variable coefficients is no longer valid after nonorthonormal rotations of the loading matrix and we have to refer to the coefficients as *standardized regression coefficients*— standardized because the component scores and the variables have been standardized.

Distance form.   In many sciences the emphasis is not exclusively on the variables, and it is usual to plot the subjects as well. In order to do this in such a way that the distances between subjects in the full-dimensional space are appropriately approximated in the reduced space, the subject components have to be in principal coordinates, so that we have to work with and plot the $a_{is}^* = g_{ss}a_{is}$.

$$x_{ij} = \sum_s g_{ss}(a_{is}b_{js}) \sum_s [g_{ss}a_{is}][b_{js}] = \sum_s a_{is}^* b_{js}, \tag{9.2}$$

where $\mathbf{B}$ is orthonormal, $\mathbf{B'B} = \mathbf{I}_s$, and $\mathbf{A}^*$ is orthogonal, $\mathbf{A}^{*'}\mathbf{A}^* = \mathbf{G}^2$, with $\mathbf{G}^2$ a diagonal matrix with the eigenvalues or the squares of the singular values g_{ss}.

In the sequel, we will thus use the term *coefficients* as a generic one. The term *coordinates* will be used in a similar manner, especially with respect to geometric representations of components. At times, the term *weights* without further qualifications is used by other authors in this context, but this term is used in this book for other purposes. Note that our terminology is different in parts from, for example, Harshman and Lundy (1984c) and Smilde et al. (2004), who use loadings as a generic term.

Components and the multidimensional point of view. When we move to the multidimensional point of view and thus do not treat the rows and columns asymmetrically, the notion of interpreting coefficients as regression coefficients can be taken one step further. Even without specific scaling, Eq. (9.1) shows that if we know the values of the $\mathbf{a}_s$, the $\mathbf{b}_s$ are the regression coefficients for the multivariate regression of $\mathbf{x}_j$ on the $\mathbf{a}_s$ and vice versa if the $\mathbf{b}_s$ are known. This shows that ordinary PCA can be seen as two multivariate linear regression equations with latent predictors. Because of this conceptualization, the principal component model is also called a *bilinear* model; it is linear in both sets of its parameters. Thus, a bilinear model with two sets of *regression coefficients* is another general interpretation for the principal component model.

9.3.2 Basic forms for three-mode models

Even though the formulations of the basic three-mode models have been reviewed earlier in Chapter 4, they are briefly reiterated here, because we need the formulas in a later part of this chapter.

Tucker2 model. The Tucker2 model in sum notation has as its basic form

$$x_{ijk} = \sum_{p=1}^{P}\sum_{q=1}^{Q} h_{pqk}(a_{ip}b_{jq}) + e_{ijk}, \tag{9.3}$$

and its matrix notation, writing the model per frontal slice k, is

$$\mathbf{X}_k = \mathbf{AH}_k\mathbf{B'} + \mathbf{E}_k, \tag{9.4}$$

where the component matrices for the first and second modes, $\mathbf{A} = (a_{ip})$ and $\mathbf{B} = (b_{jq})$, are taken to be orthonormal, that is, $\mathbf{A'A} = \mathbf{I}_p$ and $\mathbf{B'B} = \mathbf{I}_q$. This can be done without loss of generality, because the orthonormality is not a fundamental restriction. Any nonsingular transformations $\mathbf{S}$ and $\mathbf{T}$ may be applied to the component matrices $\mathbf{A}$ and $\mathbf{B}$, giving $\mathbf{A}^* = \mathbf{AS}$ and $\mathbf{B}^* = \mathbf{BT}$, provided the core slices $\mathbf{H}_k$ are countertransformed, such that $\mathbf{H}_k{}^* = \mathbf{S}^{-1}\mathbf{H}_k(\mathbf{T}^{-1})'$. Note that in the basic form the size of the data is contained in the core slices $\mathbf{H}_k$.

Tucker3 model. Similarly, the Tucker3 model in sum notation has as its basic form

$$x_{ijk} = \sum_{p=1}^{P}\sum_{q=1}^{Q}\sum_{r=1}^{R} g_{pqr}(a_{ip}b_{jq}c_{kr}) + e_{ijk}, \tag{9.5}$$

which in matrix notation using Kronecker products reads

$$\mathbf{X} = \mathbf{AG}\,(\mathbf{C}' \otimes \mathbf{B}') + \mathbf{E}, \tag{9.6}$$

where $\mathbf{C} = (c_{kr})$ is the orthonormal component matrix for the third mode. Again the component matrices are orthonormal matrices, and the size of the data is only reflected in the core array, $\mathbf{G}$. The nonsingular transformations applied to the Tucker2 model can be applied in the same fashion.

Parafac model. Finally, the Parafac model in sum notation has as its form

$$x_{ijk} = \sum_{s=1}^{S} g_{sss}(a_{is}b_{js}c_{ks}) + e_{ijk}, \tag{9.7}$$

which in matrix, or rather vector, notation becomes

$$\mathcal{X} = \sum_{s=1}^{S} g_{sss}(\mathbf{a}_s \otimes \mathbf{b}_s \otimes \mathbf{c}_s) + \mathcal{E}. \tag{9.8}$$

Unlike most forms of the Parafac model, the scaling or size factor is explicitly indicated by a separate parameter, but during the computations the size coefficients, g_{sss}, are absorbed in the components. The explicit representation of the scaling factor goes hand-in-hand with unit-length components. Because of the restricted nature of the Parafac model, no rotational freedom exists and, in addition, the components of a mode are generally oblique rather than orthogonal.

9.4 SCALING OF COMPONENTS

The reason why components are often not interpreted in terms of their basic form is that multiplication with an appropriate scaling factor can give additional meaning to the coefficients, make the values comparable across modes, or allow for plotting in an appropriate way. There are three common procedures for adjusting the lengths of the components

1. *normalized coordinate scaling*: unscaled standard components, unit-length components ;

2. *unit mean-square scaling*: multiplying coefficients by $1/\sqrt{I}, 1/\sqrt{J}$, or $1/\sqrt{K}$, where I, J, and K are the number of levels of the first, second, and third mode, respectively;

3. *principal coordinate scaling*: multiplying the coefficients with the square root of the appropriate eigenvalue.

For demonstration purposes, we will refer primarily to the pth component (or column) of $\mathbf{A}$, that is, $\mathbf{a}_p = (a_{ip})$, and to $\tilde{\mathbf{a}}_p = (\tilde{a}_{ip})$ as its scaled version.

Compensatory scaling core array.   An interesting question concerns the necessity of compensating the scaling of components by the inverse scaling of the core array or other components. In order to approximate the data by a multiway model, one has to carry out such compensations so as not to affect the structural image of the data. Furthermore, if one wants to interpret the components and core array jointly in a coherent fashion, the compensatory scaling is necessary as well. However, if one wants to investigate either the components or the core array on their own, such consistency is not always necessary or desirable. A characteristic example is the plotting of the separate component spaces. If one wants to use Euclidean distances for interpretation of plots of component spaces, then the principal coordinate scaling is required (see Kiers, 2000a), and this can be done independently of what happens elsewhere in the model.

9.4.1 Normalized coordinates

Most computer programs produce normalized components for the subjects, that is, $\sum_i a_{ip}^2 = 1$. In two-mode analysis, these scores are generally only interpreted via the meanings given to the components on the basis of the variable loadings. However, in three-mode PCA, this is no longer possible and the components of the subject space have to be interpreted in their own right (see Section 9.2.5). In several designs this is possible, but in quite a few others, especially in the social and behavioral sciences, this can be problematic without additional (external) information. Furthermore, absolute comparisons between scores on different components, say, $\mathbf{a}_p$ and $\mathbf{a}_{p'}$, are not immediately possible due to the unit-length scalings. However, when normalized components are centered, they can be interpreted as standard scores, and their coefficients may be referred to as *standard coordinates*. Whether components are centered depends on the preprocessing of the data (see Section 6.4.2, p. 119).

Interpreting components in terms of theoretical constructs is necessary if the elements of the core array are to be interpreted as measures of the importance of the links between the components from various modes. To designate core elements as measures of importance, the size of the data must be exclusively contained in the core array, and this assumes normalization of the components of all modes. For the Tucker models, it will seldom be the case that all components can be explicitly linked to theoretical constructs without previously being rotated to some more meaningful position (see Section 10.3, p. 241).

9.4.2 Unit mean-square components

To get a mean square of 1 for components of a mode, each coefficient of a component has to be divided by the square root of the number of elements in a mode, so that for instance, $(1/I) \sum_i \tilde{a}_{ip}^2 = 1$. This makes the coefficients in the components independent of the number of levels in a mode, so that these coefficients become comparable across modes. This type of scaling of the components is especially useful for the Parafac model, because one often wants to plot the sth components of the three modes in a single per-component plot (see Section 11.5.1, p. 271); it is Harshman and Lundy's (1984a) scaling of choice for two of the three component matrices. The third matrix may then scaled by the component weights g_{sss}.

9.4.3 Principal coordinates

If we take ξ_p to be the unstandardized component weight of component $\mathbf{a}_p$ then the coefficients $\tilde{a}_{ip}$ are in principal coordinates if $\sum_i \tilde{a}_{ip}^2 = \xi_p$. The component weight is the eigenvalue described in Section 4.8.2 (p. 65). The result of this type of scaling is that the relative importance of the components within a mode is now reflected in the values of their coefficients, so that a comparative assessment within a component matrix is possible. Moreover, the scaled components have the proper metric for plotting (see Section 11.3.1, p. 259).

Content-based choice of principal coordinate scaling. A substantive or content-based question is how many modes should be simultaneously represented in principal coordinates. As principal coordinates are necessary to preserve Euclidean distances in low-dimensional component spaces (see Section 11.3.1, p. 259), one might argue that all modes should be scaled in this way, and that the core array should be countertransformed with the inverse of the eigenvalues to compensate for this. Such a proposal was actually put forward by Bartussek (1973). However, it seems difficult to envisage substantive arguments why all modes should conceptually be treated on an equal footing. In a two-mode PCA of profile data, the rows (often subjects) are treated differently from the columns (often variables). The situation that comes to mind in which the modes are all comparable is when we are dealing with factorial data, or three-way interactions from analysis of variance or three-way contingency tables. Thus, although several or all component matrices may be represented in principal coordinates for plotting purposes, for model description this seems less desirable. A similar situation exists in two-way correspondence analysis where often the two modes are represented in principal coordinates when plotted separately, but only one of the two is in principal coordinates when they are represented in a single plot.

9.4.4 Regression coefficients

In Section 9.3.1 we introduced the idea that in two-mode PCA the coefficients of one mode can be interpreted as regression coefficients given the coefficients of the other mode. This idea can be directly extended to multiway models. To facilitate the discussion, we will primarily take the multivariate point of view and assume that we are dealing with a three-mode Parafac model and with subjects by variables by occasions data. Let us first write the Parafac model as a structured two-mode PCA, hiding the scaling constants in the second-mode or variable component matrices (following the exposition by Harshman & Lundy, 1984c, p. 195ff.) .

$$x_{\ell j} \;=\; \sum_{s=1}^{S} u_{\ell s} b_{js} + e_{\ell j} \tag{9.9}$$

$$x_{[ik]j} \;=\; \sum_{s=1}^{S} [a_{is} c_{ks}] b_{js} + e_{[ik]j}.$$

Presented in this way, the Parafac model is a two-mode PCA with regression coefficients b_{js} and *structured component scores* $u_{\ell s}$ for each subject on each occasion. The scores $\mathbf{u}_s$ are structured as a product of coefficients for the subjects and coefficients for the occasions. If both the components $\mathbf{a}_s$ and $\mathbf{c}_s$ have unit lengths or unit mean-squared lengths, so have their products $\mathbf{u}_s$ so that the latter are properly scaled.

$$1/\ell \sum_{\ell} u_{\ell s}^2 \;=\; 1/IK \sum_{i} \sum_{k} (a_{is} c_{ks})^2 = 1/IK \sum_{i} \sum_{k} a_{is}^2 c_{ks}^2 \tag{9.10}$$

$$= \; 1/I \sum_{i} a_{is}^2 \left(1/K \sum_{k} c_{ks}^2 \right) = 1/I \sum_{i} a_{is}^2 \times 1 = 1 \times 1 = 1.$$

From a multidimensional point of view, the model is symmetric in all three modes, so that one may reformulate Eq. (9.9), in two similar ways, illustrating the *trilinearity* of the Parafac model.

Standardized regression coefficients.   Given a proper preprocessing of the variables, the interpretation of the variable components in terms of regression coefficients can be extended to standardized regression coefficients and even to variable–component correlations. In particular, if the original data have been column-centered across subjects, that is, $\tilde{x}_{ijk} = x_{ijk} - \overline{x}_{\cdot jk}$, then the components of the first mode $\mathbf{a}_s$ are also centered (see Eq. (6.5), p. 120), as well as the structured component scores $u_{\ell s}$:

$$1\ell \sum_{\ell} \tilde{u}_{\ell s} = 1/IK \sum_{k} \sum_{i} \tilde{a}_{is} c_{ks} = 1/K \sum_{k} c_{ks} \left(\frac{1}{I} \sum_{i} \tilde{a}_{is} \right) \quad (9.11)$$
$$= 1/K \sum_{k} c_{ks} \times 0 = 0,$$

where $\tilde{a}_{is}$ are the centered component scores of the first mode. The same is true for tube centerings across the occasions. As the structured component scores also have length 1 (see Eq. (9.10)), they are standardized, which leads us back to Section 9.3.1 where it was shown that in that case the regression coefficients are standardized. Thus, all common interpretations for standardized regression coefficients can be applied in this case, recognizing that the components are, of course, *latent predictors*.

Variable–component correlations. When, moreover, the u_s are orthogonal, the variable component matrix **B** contains variable–component correlations. The u_s are orthogonal if either **A** or **C** or both are orthogonal:

$$\mathbf{u}'_s \mathbf{u}_{s'} = (\mathbf{a}'_s \otimes \mathbf{c}'_s)(\mathbf{a}_{s'} \otimes \mathbf{c}_{s'}) = (\mathbf{a}'_s \mathbf{a}_{s'} \otimes \mathbf{c}'_s \mathbf{c}_{s'}) \quad (9.12)$$
$$= (\mathbf{0} \otimes \mathbf{c}'_s \mathbf{c}_{s'}) = 0 \text{ if } s \neq s'.$$

9.4.5   Scores and loadings in the Tucker2 model

A similar exposition to the above can be made for the Tucker2 model. However, we will here present some aspects of making the components more interpretable by reformulating the Tucker2 model itself. This presentation is based on a paper by Murakami (1998a), which contains an in-depth exposition and several other proposals for increasing the interpretability of the Tucker2 model. The scaling of component matrices he proposed will be referred to as the *Murakami form* in contrast with the basic form.

First-order and second-order components. Standard PCA for the column-centered and column-normalized **X** has the covariance form $\mathbf{X} \approx \mathbf{UB}'$ where **U** is the $I \times Q$ matrix with normalized component scores, and **B** is the $J \times Q$ the matrix with regression coefficients for predicting **X** from the components. If the component scores are orthogonal, the b_{jq} are variable–component correlations. With $\approx$ we indicate the least-squares fit when the number of components, S, is less than is necessary for a perfect fit.

When we have several $I \times J$ data matrices $\mathbf{X}_k, k = 1, \ldots, K$, the most straightforward extension is to model each data slice independently as $\mathbf{X}_k \approx \mathbf{U}_k \mathbf{B}_k'$. However, there is likely to be a large redundancy in the $K \times Q$ components of the $\mathbf{B}_k$. The first simplification is to replace the individual $\mathbf{B}_k$ by a common *first-order loadings* **B**.

From this point of view, the $\mathbf{U}_k$ are then *first-order component scores* per slice k. It is not unlikely that there will also be a large redundancy in these component scores, so that a further simplification can be achieved by defining the $I \times P$ *second-order component scores* $\mathbf{A}$ and modeling the first-order scores as $\mathbf{U}_k = \mathbf{A}\mathbf{H}_k$, where the $P \times Q$ matrices $\mathbf{H}_k$ are *second-order loadings*. In this way we end up with the Tucker2 model,

$$\mathbf{X}_k \approx \mathbf{A}\mathbf{H}_k\mathbf{B}', \tag{9.13}$$

where we have a complete interpretation of the parameters in terms of first-order and second-order component loadings and second-order component scores.

It should be noted that there is an important difference with the more common formulation used, for instance, by Kroonenberg (1983c) and Kiers (2000b) in that here $\mathbf{B}$ is orthogonal, rather than orthonormal, and that this is reflected in the scaling of the extended core array $\mathcal{H}$, that is $(1/K)\sum_{k=1}^{K}\mathbf{H}_k'\mathbf{H}_k = \mathbf{I}_Q$. The idea of casting three-mode models as higher-order component models was also discussed by Bloxom (1984) in the context of three-mode common factor analysis. As discussed in Murakami (1983, pp. 31–34) with this formulation of the Tucker2 model it becomes possible to distinguish between changes in component scores and changes in component loadings by checking the patterns in the $\mathbf{H}_k$ (see also Murakami & Kroonenberg, 2003).

Orthonormalizing the second-order component scores will scale the first-order component scores to have unit mean-squared lengths:

$$\mathbf{A}'\mathbf{A} = \mathbf{I}_P \text{ and } (1/K)\sum_{k=1}^{K}\mathbf{H}_k'\mathbf{H}_k = \mathbf{I}_Q \text{ so that } (1/K)\sum_{k=1}^{K}\mathbf{U}_k'\mathbf{U}_k = \mathbf{I}_P. \tag{9.14}$$

Furthermore, the component matrix $\mathbf{B}$ then becomes an average correlation matrix

$$\mathbf{B} = (1/K)\sum_{k=1}^{K}\mathbf{X}_k\mathbf{U}_k', \tag{9.15}$$

and finally $\mathbf{H}_k = \mathbf{A}'\mathbf{U}_k$, so that $\mathbf{H}_k$ is a kind of higher-order structure matrix, whose elements are the covariances between the first-order and second-order components. For a discussion of the Tucker3 model as a set of nested linear combinations of components and thus of higher-order components see Van de Geer (1974) and see also Kroonenberg (1983c, p. 18), and the discussion in Smilde et al. (2004, pp. 68–69).

The above formulation shows that the Tucker2 model can be redefined as a model with both first-order and second-order components and associated structure matrices. The only application known to us of the Murakami form of the Tucker2 variant (Murakami, 1998a) is contained in the original paper. He also shows that the structure interpretation requires that the data have been fiber-centered (fiber centering per column across subjects) and slice normalized (per or within a variable slice), which is equivalent to profile preprocessing, the standard preprocessing option for three-way profile data.

9.4.6 Scores and loadings in the Tucker3 model

The multivariate point of view can also be made to bear on the interpretation of the Tucker3 model. Of specific interest is the interpretation in relation to three-mode common factor analysis and structural equation modeling by Bloxom (1984), Bentler and Lee (1978b, 1979), Bentler et al. (1988), and later Oort (1999) and Kroonenberg and Oort (2003a).

We start again from two-mode component analysis, that is, $X \approx UL'$, where U are the row (subject) coefficients and L the column (variable) coefficients. Then there are three options to develop a system with scores and loadings for the Tucker3 model (see Section 9.3.1, p. 215): (1) the standard PCA scaling with orthonormal scores and orthogonal loadings — the covariance form; (2) the alternative PCA scaling with orthogonal scores and orthonormal loadings — the distance form (which will not be considered here); and (3) the Murakami form (Murakami, 1998a).

Covariance form. The standard PCA form gives $U = A$ and $L = (C \otimes B)G_a$, so that U is an orthonormal component score matrix and L is the orthogonal *structured loading matrix*. The orthogonality follows from:

$$L'L = G'_a(C' \otimes B')(C \otimes B)G_a = G'_a I_{QR} G_a = \sum G'_r G_r = \Xi_P. \quad (9.16)$$

The next question is whether further interpretations are possible by decomposing L into an orthonormal part $(C \otimes B)$ and an orthogonal part G_a. At present, it is not clear how one could get a further useful interpretation out of this split, and we will not pursue this further.

Murakami form. Let us first define $\tilde{U} = AG_a$ and $\tilde{L} = (C \otimes B)$, so that again $X \approx \tilde{U}\tilde{L}'$ (this is in fact the analogue of the distance form). Then, taking our cue from Section 9.4.5, we will define $U^* = \tilde{U}\Xi^{-1} = AG_a\Xi^{-1}$ and $L^* = \tilde{L}\Xi = (C \otimes B)G'_a\Xi$, where $G'_a G_a = \Xi^2$ with Ξ a diagonal matrix, so that G_a is orthogonal. The ξ_p^2 are sums of squares associated with the P first-mode components, and after centering they are the associated variances. When divided by the total sum of squares, they are the proportions explained sum of squares by the components.

The effect of these scalings is that U^* is an orthonormal matrix,

$$U^{*\prime}U^* = \Xi^{-1}G'_a A' A G_a \Xi^{-1} = \Xi^{-1}G'_a G_a \Xi^{-1} = \Xi^{-1}\Xi^2\Xi^{-1} = I_P, \quad (9.17)$$

and can thus be interpreted as a score matrix. L^* is orthogonal,

$$L^{*\prime}L^* = G'_a(C' \otimes B')(C \otimes B)G_a = G'_a I_{QR} G_a = \sum G'_r G_r = \Xi_P, \quad (9.18)$$

and

$$
\begin{aligned}
\mathbf{X}'\mathbf{U}^* &= (\mathbf{C} \otimes \mathbf{B})\mathbf{G}'_a\mathbf{A}'\mathbf{U}^* = (\mathbf{C} \otimes \mathbf{B})\mathbf{G}'_a\mathbf{A}'\mathbf{A}\mathbf{G}_a\mathbf{\Xi}^{-1} \qquad (9.19) \\
&= (\mathbf{C} \otimes \mathbf{B})\mathbf{G}'_a\mathbf{G}_a\mathbf{\Xi}^{-1} = (\mathbf{C} \otimes \mathbf{B})\mathbf{\Xi} = \mathbf{L}^*,
\end{aligned}
$$

so that $\mathbf{L}^*$ contains indeed the variable–component correlations. To what extent this form of the Tucker3 model is useful, is as yet untested.

9.5 INTERPRETING CORE ARRAYS

In this section we will discuss several ways in which the elements of the core arrays of the Tucker2 and Tucker3 models can be interpreted, and wherever feasible we will comment on the Parafac model as well. There seem to be at least seven not unrelated ways to interpret the core array.

1. strength of the link between components of different modes or the weight of the component combination;

2. regression weights;

3. percentages of explained variation;

4. three-way interaction measures;

5. scores of idealized or latent elements;

6. direction cosines (Tucker2 model);

7. latent covariances (see Section 15.4.2, p. 381).

In several of these interpretations it is crucial to know how the component matrices are scaled. We will start with the basic form of the three-mode models, that is, all Tucker component matrices are orthonormal, and all Parafac component matrices have unit (column) lengths. Later on, we will use other scalings for the component matrices as well.

9.5.1 Regression coefficients and explained sums of squares

Given the basic form of the Tucker models, the core array contains the linkage between the components from different modes. In particular, the core array indicates the weight of a combination of components, one from each reduced mode. For instance, the element g_{111} of the Tucker3 core array (see Table 9.1) indicates the strength of the link between the first components of the three modes $(a_{i1}b_{j1}c_{k1})$, and g_{221} indicates the strength of the link between the second components of the first and the second

Table 9.1   Notation for three-mode core arrays

		Third-mode components				
		1		2		3
		2nd-mode comp.		2nd-mode comp.		2nd-mode comp.
Components		1	2	1	2	1	2
First-mode	1	g_{111}	g_{121}	g_{112}	g_{122}	g_{113}	g_{123}
	2	g_{211}	g_{221}	g_{212}	g_{222}	g_{213}	g_{223}
	3	g_{311}	g_{321}	g_{312}	g_{322}	g_{313}	g_{323}

mode in combination with the first component of the third mode $(a_{i2}b_{j2}c_{k1})$. By this we mean that when the data are reconstructed on the basis of the model (i.e., the structural image is computed),

$$\hat{x}_{ijk} = \sum_{pqr}[g_{pqr}](a_{ip}b_{jq}c_{kr}), \tag{9.20}$$

a g_{pqr} is the weight of the term $(a_{ip}b_{jq}c_{kr})$. The same can be seen to apply in the Tucker2 model, where each term has the form $[h_{pqk}](a_{ip}b_{jq})$. In the Parafac model the terms are $[g_{sss}](a_{is}b_{js}c_{ks})$.

We can also view the elements of the core arrays as regression weights, where the predictors are of the form $f_{ijk,pqr} = (a_{ip}b_{jq}c_{kr})$. Thus, only three-way interaction terms or triple products of components are included in the regression, and we have the regression equation

$$x_{ijk} = \hat{x}_{ijk} + e_{ijk} = \sum_{pqr}[g_{pqr}](a_{ip}b_{jq}c_{kr}) + e_{ijk} = \sum_{pqr}g_{pqr}f_{ijk,pqr} + e_{ijk}. \tag{9.21}$$

If all components of a mode are orthonormal, so are their triple products, $f_{ijk,pqr}$ (see Eq. (9.12)). This means that we have a regression equation with orthonormal predictors so that the squared regression coefficients (i.e., the g_{pqr}^2) add up to form the explained sum of squares of the regression equation. Each squared coefficient g_{pqr}^2 indicates the explained variability[2] is used instead of its predictor. Therefore, we may write

[2]The term *variability sum of squares*, or *variance*, because centerings and normalizations determine the exact nature of the measure of variability.

$$\sum_{pqr} g^2_{pqr} \quad = \quad \text{SS(Fit)} \quad \text{(Tucker3 model)}; \tag{9.22}$$

$$\sum_{pqk} h^2_{pqk} \quad = \quad \text{SS(Fit)} \quad \text{(Tucker2 model)};$$

$$\sum_{s} g^2_{sss} \quad = \quad \text{SS(Fit)} \quad \text{(Parafac model with at least one orthogonal mode)}.$$

Thus, the fitted sum of squares can be completely partitioned by the elements of the core array. Dividing the g^2_{pqr} by the total sum of squares, we get the *proportion of the variability explained* by each of the combination of components, for which the term *relative fit* will be used, and the *relative residual sum of squares* is similarly defined.

This interpretation of the core array can be used to assess whether additional components in any one mode introduce sizeable new combinations of components, or whether the gain in explained variability is spread over a larger number of additional elements of the core array. In the latter case, one might doubt whether it is really useful to add the component; in the former, something interesting might emerge from a careful examination of the results.

Only when one of the modes is orthogonal does the Parafac model have orthogonal predictors in the equivalent equation to the regression of Eq. (9.21), and its partitioning of the fitted sum of squares. In all other cases the correlation between the predictors $\mathbf{f}_s$ prevents the partitioning.

An interesting side effect of this partitioning of the overall fitted sum of squares is that by adding the g^2_{pqr} over two of their indices we get the explained variability of the components themselves; that is, for the component $\mathbf{a}_p$ the explained variability SS(Fit)$_p$ is $\sum_{qr} g^2_{pqr}$. By dividing them by the total sum of squares, we again get the explained variability per component, or *standardized component weights*. Similar expressions can be derived for the Tucker2 core array. For the orthogonal Parafac model they are the $g^2_{sss}/$SS(Total).

Thus, the core array represents a partitioning of the overall fitted sum of squares into small units through which the (possibly) complex relationships between the components can be analyzed. In the singular value decomposition (see Section 4.4.1, p. 47) the squares of the singular values (i.e., the eigenvalues) partition the fitted variation into parts that are associated with each component. The core array does the same for the combination of the components of different modes, and in this respect multiway principal component analysis is the multiway analogue of the singular value decomposition.

Treating the core elements as regression coefficients for the triple product of orthonormal components is very different from the regression equations we presented earlier (see Eq. (9.9)). There, some of the components were treated as predictors and (combinations of the) other components with or without a core element as regression coefficients. The treatment in this section is entirely symmetric in the three modes and corresponds to the multidimensional perspective on data analysis, with the proviso that we explicitly investigate components rather than subspaces.

9.5.2 Scores of idealized elements

In the discussion of the core array in the previous section, the content of the modes did not play a role. But if do take the content into account, the core array can be seen as a miniature data box. Instead of real data of subjects on variables in several conditions, we now have a diminutive derived data box, where each element represents a combination of "latent entities". In particular, each core element g_{pqr} is the score of an "idealized subject" (or subject type) p on a latent variable q in a prototype condition r (Tucker, 1966). Thus, we consider the components to be unobserved constructs of which the observed levels of the modes are linear combinations: a real person is a linear combination of subject types, a real variable is a linear combination of latent variables, and a real condition is a linear combination of prototype conditions. Note that now the core array contains scores and the component coefficients in the model serve as weights in the linear combinations. One way to visualize this is by writing the model equation as a set of nested linear combinations (for an earlier discussion see Kroonenberg, 1983c, pp. 14–20) as follows:

$$x_{ijk} \approx \sum_p a_{ip} \left[\sum_q b_{jq} \left(\sum_r c_{kr} g_{pqr} \right) \right] \qquad (9.23)$$

$$x_{ijk} \approx \sum_r a_{ip} \left[\sum_q b_{jq} h_{pqk} \right]$$

$$x_{ijk} \approx \sum_r a_{ip} w_{pjk}.$$

The rightmost linear combination in the first line says that the scores of an idealized subject p in condition k are a linear combination of the scores of the R prototype conditions on latent variables. Furthermore, the scores on the real variables are linear combinations of the scores on the Q latent variables, and finally all real subjects are linear combinations of the subject types. Note that this form can also be viewed as a series of nested ordered loadings, in particular, third-order scores with first-order, second-order, and third-order loadings as in the Murakami form.

This interpretation of the core array fits most easily into the distance form of the model in which the subject scores are scaled to reflect the variance in the data and the other coefficients are orthonormal, as, for instance, is commonly done in mathematical statistics, where components are explicitly considered to have maximum variance. Bartussek (1973) proposed a scaling to concur with the orthonormal coefficients for scores and loadings for variables, but this idea was not taken up in the literature (see Kroonenberg, 1983c, Chapter 6, for a discussion of this proposal).

It depends very much on the applications as to whether the idealized subject or subject type interpretation is useful in practice. The reification or naming of components for all modes seems to be indispensable in such a setup, because core elements link the components of the different modes, and therefore these must have a substantive interpretation. It is probably always necessary to rotate the components in order to name them appropriately. When one rejects the idea of labeling components, the idealized-entity interpretation loses some of its charm. In examples with very few variables and conditions, the labeling of components is a rather risky business, and other interpretations should be preferred. In other applications, especially when the labeling of the components is firmly established, the approach in this section can be useful.

9.5.3 Linking components and elements of core arrays

The structure in a three-mode core array is often not simple, in the sense that there are many nonzero and often medium-sized elements in the core array. This occurs even more frequently after rotations of the components. The issue of simple structure and rotations of both components and core arrays will be treated in Chapter 10, but here we will go into some aspects of the relationships between components and core arrays which are relevant independently of rotations of the components or the core arrays or both.

Interpretational complexities arise in core arrays of both the Tucker3 and Tucker2 models, because the same components appear in several combinations with components of other modes. Parafac models are much simpler because each component of one mode is only linked to one component of the other modes.

Combining component coefficients and core elements. One of the complications in assessing the influence of a combination of components on creating a structural image of the data is the quadrilinear form, $f_{ijk,pqr} = g_{pqr}(a_{ip}b_{jq}c_{kr})$, of each term of the model. The meaning of such a term can be explained via the following example. We have observed that child i with an unshakable poise has a high score on extraversion item j when in a high-stress condition k. It is reasonable to assume that child i sides with other unruffled children on the child component p for such children, say with a coefficient $a_{ip} = +0.5$. Items which measure extraversion will have high coefficients on the extraversion component q, say in this case $b_{jq} = +0.7$. Finally, high-stress conditions form together the stress-condition component r, and our condition has a high coefficient on this condition component, say $c_{kr} = +0.6$. The product of the three component coefficients, $a_{ip}b_{jq}c_{kr} = (+0.5)\times(+0.7)\times(+0.6) = +0.21$, is positive and contributes a positive value to the structural image $\hat{x}_{ijk}$ of x_{ijk}, provided the weight of this (p, q, r)-combination is also positive (say, $g_{pqr} = 10$). If this weight or element of the core array is comparatively large compared to the other elements of the core array, the entire term $g_{pqr}(a_{ip}b_{jq}c_{kr})$ contributes heavily to the structural image of x_{ijk}.

To take this a bit further, first suppose we have been talking about the first child, the first variable and the first condition, that is about the data value x_{111}. In addition, assume that we were referring to the first components of all modes. Then $g_{111}a_{11}b_{11}c_{11}$ is the first term in the structural image. For a $2\times2\times2$-Tucker3 model, the structural image for x_{111} is built up in the following way.

$$
\begin{aligned}
\hat{x}_{111} &= \mathbf{g_{111}a_{11}b_{11}c_{11}} + g_{211}a_{12}b_{11}c_{11} + g_{121}a_{11}b_{12}c_{11} + g_{221}a_{12}b_{12}c_{11} \\
&\quad + g_{112}a_{11}b_{11}c_{12} + g_{212}a_{12}b_{11}c_{12} + g_{122}a_{11}b_{12}c_{12} + g_{222}a_{12}b_{12}c_{12} \\
&= 10\times0.5\times0.7\times0.6 + \text{the other terms} \\
&= 2.1 + \text{the other terms.}
\end{aligned}
\tag{9.24}
$$

Because all component coefficients and the corresponding core element are positive the first term has a positive contribution of 2.1 to the structural image of x_{111}. A further positive contribution to $\hat{x}_{111}$ could come from the fact that variable $j = 1$ has a large negative loading on introversion $q = 2$ (say, $b_{12} = -0.7$), condition $k = 1$ has a large negative loading on the play-condition component $r = 2$ $c_{12} = -0.6$. Then the term $a_{11}b_{12}c_{12} = (+0.5)\times(-0.7)\times(-0.6) = +0.21$ is positive, and depending on the value g_{122} the contribution to the structural image is positive or negative. Thus, the final effect of the components on the structural image of a data point depends not only on the sizes but also on the combinations of the signs of all four parts of the quadrilinear term $g_{pqr}(a_{ip}b_{jq}c_{kr})$. This generally requires quite a bit of juggling to get a properly worded statement. Part of the problem is that the components are continuous rather than discrete entities, unlike interactions between levels of factors in analysis of variance and categories in contingency tables. Moreover, one needs names or meanings for all the components in all modes, which are not always easy to devise.

Keeping track of the signs.   Suppose that g_{pqr} has a positive sign and that the product of the pth, qth, and rth components of the first, second, and third modes, respectively, is also positive; then the overall effect of the term, $g_{pqr}(a_{ip}b_{jq}c_{kr})$, is positive. A positive contribution of this term can be the result of four different sign combinations for the elements of the three modes: $(+,+,+)$; $(+,-,-,)$; $(-,+,-)$; $(-,-,+)$, in which a plus (minus) on the pth, qth, and rth place in (p, q, r) refers to positive (negative) coefficients on the pth component of the first mode, qth component of the second mode, and rth component of the third mode, respectively. A parallel formulation is that for a positive contribution one needs positive coefficients on p to go together with coefficients of the same sign of q and r: $(+,+,+)$ and $(+,-,-)$, and negative coefficients on p to go together with coefficients of opposite signs on q and r: $(-,+,-)$ and $(-,-,+)$. A negative g_{pqr} with any of the combinations $(+,-,+)$, $(+,+,-)$, $(-,+,+)$, and $(-,-,-)$ results in a positive contribution to the structural image. The mental juggling with combinations of positive and negative loadings of different components obviously requires care during interpretation. In some data sets, certain components have only positive coefficients, which simplifies the process, as the number of combinations is reduced by a factor 2. Sometimes certain core elements are so small that they need not be interpreted, which also simplifies the evaluation of the results. In Section 19.6.4 (p. 484) we have given a detailed analysis of a complex four-way core array as an example of how to deal with the problem of interpreting the combinations of negative and positive coefficients on different components.

Using conditional statements. A good strategy to simplify interpretation is to make *conditional* statements by only making statements about elements that have, for instance, positive coefficients on a component. The core array then represents only the interaction between the coefficients of the two other modes, "given" the positive coefficients on the third. The joint biplots and nested-mode biplots discussed in Sections 11.5.3 and 11.5.4, respectively, are examples of such an approach. In practice we have observed that it is most useful to use a "conditions" mode for "conditioning", and to keep this mode in normalized coordinates.

9.6 INTERPRETING EXTENDED CORE ARRAYS

The above discussion was primarily concerned with the core array of the Tucker3 model. The interpretation of the extended core array of the Tucker2, $\mathcal{H} = (h_{pqk})$, largely follows the lines set out above, be it that in many ways the interpretation is simpler because there are only two, rather than three, sets of components. The basic approach is to discuss the linkage between the sets of components in $\mathbf{A}$ and $\mathbf{B}$ per (frontal) slice of the third mode. In particular, if we write the Tucker2 model in its frontal slice form, that is, $\mathbf{X}_k \approx \mathbf{A}\mathbf{H}_k\mathbf{B}'$, we see that each slice of the extended core array indicates how strongly the components of $\mathbf{A}$ and $\mathbf{B}$ are linked to each other. If one combination h_{pqk} is zero or near zero, the combination of $\mathbf{a}_p$ and $\mathbf{b}_q$ is not important in creating the structural image of the data of k. By comparing the h_{pqk}, one can assess the relative importance of the combination of components for the levels of the third mode. The Murakami form in terms of higher-order scores and loadings is another basis for interpretation (see Section 9.4.5).

If $P = Q$ and, for a specific k, $\mathbf{H}_k$ is diagonal, we are in the same situation as in ordinary PCA, but for the fact that the $\mathbf{A}$ and $\mathbf{B}$ are common for all K. If all $\mathbf{H}_k$ are diagonal, the Tucker2 model is a Parafac model or one of its orthogonal variants depending on the restrictions we have imposed on $\mathbf{A}$ and $\mathbf{B}$ (see Section 4.6.1, p. 59, for a completer exposé).

When $P \neq Q$, the core slice is a selection array that provides information about which component of $\mathbf{A}$ goes together with $\mathbf{B}$, especially if there are many near-zero elements in a slice; see also Figure 14.1 (p. 352) and Murakami and Kroonenberg (2003) for a discussion of this point in connection with semantic differential scales.

9.6.1 Direction cosines

In those cases where the entities in the first two modes are equal and the input data are a block of symmetric slices, an additional interpretation of the extended core array is possible. Within the context of multidimensional scaling of individual differences (see Section 2.4.5, p. 23), the input similarity matrices satisfy these conditions. A special interpretation has been developed in terms of correlations and direction cosines of the axes of the space common to the first two modes; see, for instance, Tucker (1972, p. 7) and Carroll and Wish (1974, p. 91).

When the entities of the first two modes refer to the same space, it makes sense to consider the angle between the first and second components of the common space. This angle can be derived from the off-diagonal elements of the core slices, and they are the cosines or correlations between component p and component q, provided the h_{pqk} are scaled by dividing them by $h_{ppk}^{1/2}$ and $h_{qqk}^{1/2}$, and the components are normalized. The cosine indicates the angle under which the kth condition "sees" the axes or components of the common space.

This interpretation of symmetric extended core arrays originates with Tucker's three-mode scaling proposal (Tucker, 1972) and is contained in Harshman's Parafac2 model (Harshman, 1972b); see also Section 4.6.2, p. 62) and the discussion by Carroll and Wish (1974) and Dunn and Harshman (1982).

9.7 SPECIAL TOPICS

9.7.1 Parafac core arrays

Lundy et al. (1989) were the first to present a way of employing core arrays to evaluate problems emerging during some Parafac analyses. The essence of the problem is that some data do not fit the Parafac model well because they do not obey the basic tenet of parallel proportional profiles. The Parafac model specifies that each component should have an expression in all modes and the linkage is an exclusive one. On the other hand, the Tucker models are more promiscuous: each component can be linked with any component of the other modes. In real Tucker-type situations it is impossible to find an adequate solution for the Parafac model, and its algorithms will not converge and degenerate solutions will occur. One solution is to constrain one mode to be orthogonal, which will prevent such degeneracies; see Section 5.6.5, p. 89, for a detailed discussion of this problem. Even though the degeneracy has then been avoided, diagnostically one is none the wiser. However, by calculating a Parafac core array that is a core array based on the components from a Parafac analysis (see Section 4.8.1, p. 65), such diagnostics become available. The core array computed from Parafac components will generally have sizeable nonsuperdiagonal elements indicating which components have links with more than one component from another mode. An example of this phenomenon is presented in Table 13.5 (Section 13.5.4, p. 332).

9.8 VALIDATION

In validating multiway results, one may look internally toward the analysis itself and investigate the statistical stability of the solution via bootstrap analyses (see also Section 8.8.1, p. 188). In addition, one may investigate the residuals to see which parts of the data do not fit well and how much systematic information is left (see Chapter 12 on the treatment of residuals). Another approach to validation is to look at information external to the data set to see whether this can help to shed further light on the patterns found.

9.8.1 Internal validation: Assessing fit

As in the case of multiple regression, one should investigate the overall fit as a proportion of the explained variability in order to investigate whether the model has a reasonable or acceptable fit. For the Parafac model with its correlated components, assessing the fit of individual components is not very useful, but one can compare the relative sizes of the g_{sss}. However, in the Tucker models the overall fit can be partitioned in several ways: (1) per orthonormal component, (2) per level of each mode, (3) per element of the core array, and (4) per individual data point (see Chapter 12 for details). When just one of the component matrices is restricted to be orthonormal in the Parafac model, the partitioning per component is also available for this model.

9.8.2 Internal validation: Stability of parameter estimates

Because distribution theory for many multiway situations is nonexistent, the stability of parameter estimates must proceed via a bootstrap analysis. In Section 8.8.1 (p. 188) the basic ideas behind the bootstrap are introduced and attention is paid to the results from bootstrap analyses with respect to fit measures. In this section we will briefly summarize the major paper in this area (Kiers, 2004a) in connection with finding stability estimates for the parameters using a bootstrap procedure.

Crucial in applying any bootstrap in multiway analysis is the stochastics of the situation. If there is one mode, mostly subjects, which can be considered a sample from a population, the resampling can take place on that mode. Note that the result of the bootstrap sampling is that the data slices of subjects may appear once, several times or not at all in each of the newly created data sets. As pointed out by Kiers (pp. 35–36), if there is no stochastic mode in a multiway data set, the only course left for establishing parameter stability is to take bootstrap samples of the residuals, but this presupposes the validity of the multiway model chosen.

Even though the bootstrap is conceptually simple, in multiway analysis as in several other dimensional analyses, a number of practical issues make the application of the technique nontrivial. In particular, the component matrices from each of the bootstrap analyses should be transformed in such a way that their orientations are optimally the same before the bootstrap standard errors can be calculated. This problem was also encountered when comparing the results from multiple imputations in a missing-data analysis (see Section 7.8.4, p. 171), where generalized Procrustes analysis was chosen as the solution for the alignment problem; see for an example of a similar approach in categorical principal component analysis Linting, Meulman, Groenen, and Van der Kooij (2007).

Kiers (2004a) provides two other procedures to find the optimal orientation for the bootstrap solutions using transformations to a target solution: (1) a procedure based on optimizing the congruences of the components from the bootstrap solutions to the basic form of the original three-mode model, and (2) a procedure based on transformations or rotations toward the original component matrices . The outcome of extensive simulations to evaluate the quality of his proposals is that the proposed procedures yield intervals that are "fairly good approximations to the 95% confidence intervals in most cases" (p. 34).

9.8.3 External validation

One of the more valuable ways of validating outcomes of multiway analyses is the use of external variables, especially if additional background information on the subjects is available. Similarly, when there exists a design on one of the other modes, this can be used for validating the outcomes, for instance, if the variables have a facet structure or some kind of factorial design. Furthermore, when one of the modes is an ordered one, it can be established whether the original rank order also occurs in the components. In all cases, one attempts to use single variables or two-mode information to enhance the multiway analysis; this is especially true for the subject mode. Interpretational aids are especially useful for the Tucker models because the meaning of the components has to be ascertained for each mode separately.

Basically, there are two ways to handle external information. The first is to validate the results of a multiway model after the analysis; the second is to incorporate the external information into the model fitting itself. The latter procedure was, for instance, developed by Carroll et al. (1980) for the Parafac model under the name of CANDELINC. Timmerman (2001) (see also Timmerman & Kiers, 2002) discussed procedures for including external information (such as order information and functional information) in the Tucker3 model (see also Klapper, 1998).

A straightforward procedure when external two-mode information is available is to treat the standardized external variable as the dependent variable in a regression with the standardized (or orthonormal) subject components as predictors. In such a case, the regression coefficients can be used as direction cosines in the subject space and the line of largest variability for the external variable can be drawn in the plot of that space. The multiple correlation coefficient then indicates how well the external variable is represented in the subject space. An example is Fig. 11.3 (p. 261).

9.9 CONCLUSIONS

In this chapter we have discussed various ways of interpreting components and core arrays as well as ways to validate the results of an analysis. One *leitmotif* is the difference between the linear algebra in the multidimensional point of view, in which none of the modes has a special status, and the multivariate point of view in which different modes play different roles. The latter approach leads to interpretations that are direct descendants of those of standard two-mode PCA including the construction of scores and loadings within the three-mode context.

Interpretation is, however, failing in clarity when one solely relies on unrotated components and on plain numbers. In the next chapters we will take the interpretation one step further by considering rotations and transformations in general of both components and core arrays. In the chapter after that, various plotting procedures will be discussed, which make all the difference in interpreting the solutions of three-mode models.

CHAPTER 10

IMPROVING INTERPRETATION THROUGH ROTATIONS

10.1 INTRODUCTION

The topic of this chapter is the rotation of component matrices and core arrays to enhance the interpretability of the results of multiway analysis and in particular, three-mode analyses. This enhancement is realized by increasing the number of large coefficients and decreasing the number of small coefficients in the sets of parameters, the idea being that such rotated solutions are simpler to interpret. Thus, the quest for an increase in the interpretability of the solutions of component models is a search for simplicity of the components, the core array, or both. This quest is particularly relevant for the Tucker models because they have parameter matrices, which can be transformed without loss of fit of the model to the data. In contrast, the other major multiway model, the Parafac model, is defined in such a way that rotating its solution will lead to a loss of fit. Therefore, the chapter will primarily deal with Tucker models.

Rotations versus setting parameters to zero. A separate matter is whether one is willing to sacrifice a certain amount of fit in favor of more clear-cut solutions by effectively setting small coefficients to zero. Setting specific parameters of a model to zero introduces restrictions on the solution, the model itself changes, and such changes are essentially different from rotations that operate within the chosen model. It is possible to fit models with restrictions on the component configurations, but applications of such restrictions do not feature very prominently in this book.

When restrictions are introduced, a step toward confirmatory models is made, because with restrictions the models become testable and generally require special algorithms. There is a class of multiway models, that is, multimode covariance structure models, that are clearly confirmatory in nature; see Bloxom (1984), Bentler and Lee (1978b, 1979), Bentler et al. (1988), Oort (1999), and Kroonenberg and Oort (2003a). However, the confirmatory approach in multiway data analysis is a topic that falls outside the scope of this book.

Transformations and rotations.   In order to avoid confusion later on, a brief remark about the word "rotation" is in order. Most mathematicians associate the word with orthonormal transformations. However, especially in the social and behavioral sciences, expressions like "oblique rotations" crop up, referring to nonsingular transformations (mostly with unit-length columns) that transform orthogonal axes into nonorthogonal or oblique axes. Note that orthogonal rotations of other than orthonormal matrices also lead to nonorthogonal axes. Here, we will use the terms rotation and transformation interchangeably as generic terms and specify which kind of transformation is intended. However, we will generally refer to rotating components rather than transforming components, because the former expression seems less ambiguous.

Rotation for simplification.   Simplifying components has a different rationale than simplification of core arrays. Few large coefficients and many small ones in component matrices make it easier to equate a component with a latent variable with a narrowly defined meaning. A small number of large elements in a core array indicates that certain combinations of components contribute substantially to the structural image, while combinations of components associated with very small or zero core entries do not. In other words, if we succeed in rotating the core array to such a form that there are only a few large elements, we effectively reduce the size of the model itself, because many of the multiplicative combinations do not seriously contribute to the model.

Influence of simplifications.   Simpler structures in one or more of the sets of parameters allow for easier interpretation of that set. However, simplicity in one part of the model will generally create complexity in another part, so that the interpretation of those parts is impaired. In order to obtain coherent descriptions of the results, rotation in one part of a three-mode model should be accompanied by a counterrotation or inverse transformation of an appropriate other part of the parameters in the model. Mostly, this may mean that if a set of easily interpretable components is realized through rotations, the counterrotations of the core array equalize the sizes of the core entries in the same way that rotations in two-mode PCA redistribute the variance among the components. Equalizing the elements of the core array means that there are fewer large and small core elements and more medium-sized ones, which means that more component combinations have to be interpreted. On the other hand, a simple core array with only a few large elements generally yields complex and correlated components, which tend to be difficult to interpret. Correlated components destroy the variance partitioning in the core array, which may or may not be convenient. On the other hand, correlated variables are the bread and butter of the social and behavioral sciences. However, transformations that create an extremely simple core array may also introduce such strong correlations between components as to make it nearly impossible to designate them as separate components.

Model-based and content-based arguments for rotations. Two types of arguments may be distinguished when deciding which parts of a model should be rotated in which way. Model-based arguments prescribe which type of rotations are appropriate with respect to the model, and content-based arguments determine which modes have to be rotated with which type of rotation. Furthermore, much research has gone into the investigation of the number of zero entries a full core array can have given its size by making use of the transformational freedom. It turns out that mathematical theory shows that some core arrays can be extremely simple in this sense. In other words, some simple cores are not the result of a particular data set, but are the result of the mathematical structure of the problem. In such cases the simplicity itself cannot be used as an argument in support of a substantive claim (see Section 10.5).

Rotations and conceptualization of three-mode model. Within the multivariate point of view rotations are strongly linked to the (factor-analytic) idea of latent variables. However, not in all applications is a latent variable a useful concept. Moreover, within the multidimensional point of view one may be satisfied with finding the unique low-dimensional subspaces and may look at patterns among the entities (subjects, variables, etc.) in these low-dimensional spaces without attempting to find simple axes in the space. Instead, one could look for directions (possibly more than the number of dimensions of the space) in which the variables, subjects, and so on show interesting variation. The interpretation will then primarily take place in terms of the original quantities (levels of the modes), and rotations may only be useful to find a convenient way of looking around in the low-dimensional space, especially if it has more than two dimensions.

Rotations are two-way not multiway procedures. Even though we speak about rotations of core arrays and seem to imply that we handle the core array in its entirety, in fact, nearly all rotations discussed in multiway modeling pertain to rotations of matrices, that is, two-way matrices rather than three-way arrays. The latter are handled for each mode separately by first matricizing in one direction, then applying some (standard) two-way rotational procedure, and reshaping the rotated matrix back into an array again. This may be done for each mode in turn. Exceptions to this approach are techniques that explicitly attempt to maximize and minimize specific elements of a core array via specific optimization algorithms (i.e., not via explicit rotations).

Rotations do not change models. In all of what follows, it should be kept in mind that rotations do not change the information contained in the model but only rearrange it. Thus, all rotations are equally valid given a particular model. The rotations do not alter the information itself but only present it in a different way. Nevertheless, certain forms of the models may lend themselves more easily to theoretical development in a subject area than others, and therefore some forms may be preferred above others on grounds external to the data.

There are some model-based aspects as well. For instance, the statistical properties of the coefficients in the original solution may not be the same as the coefficients of the rotated solution. For instance, correlations between components may be introduced where they were not present before.

10.2   CHAPTER PREVIEW

In this chapter we will briefly review procedures for rotation common in standard PCA, in particular, the varimax and the Harris–Kaiser independent cluster rotation of component matrices. Then, we will discuss rotations for the core of both the Tucker2 and Tucker3 models. In particular, we will consider procedures to achieve simplicity in the core itself as well as procedures to get as close as possible to Parafac solutions of the same data. Finally, techniques that combine simplicity rotation for the core array and the components will be examined.

The material dealing with rotations is primarily described in a series of papers by Kiers (1992a, 1997a, 1998b, 1998d) in which rotations of components, rotations of full core arrays, and joint rotations of components and core arrays in the Tucker3 context are discussed. Earlier discussions of rotating Tucker2 extended core arrays toward diagonality can be found in Kroonenberg (1983c, Chapter 5) and Brouwer and Kroonenberg (1991), who also mention a few earlier references. Andersson and Henrion (1999) discuss a rotation procedure to diagonalize full core arrays.

The discussion of rotating core arrays is closely tied in with the problem of the maximum number of elements that can be zeroed in a core array without loss of fit (a maximally simple core array). This problem is discussed in Kiers's papers mentioned above and in Murakami, Ten Berge, and Kiers (1998b). There are also important links with the rank of a three-way array but these aspects will not be discussed here (see Section 8.4, p. 177, for a brief discussion and references).

10.2.1   Example: Coping at school

Most rotational procedures will be illustrated by means of a subset of the Coping data set discussed in Section 14.4 (p. 354). In brief, we have 14, specially selected, primary school children who for 6 potentially stressful situations mainly related to school matters have indicated to what extent they would feel emotions (like being angry, experiencing fear) and to which extent they would use a set of coping strategies to handle the stressful situation (Röder, 2000). The data will be treated as three-way rating scale data in a scaling arrangement (see Section 3.6.2, p. 35). Thus, we have placed the 14 children in the third, the 8 scales (4 emotions, 4 strategies) in the second and the 6 situations in the first mode. The data are centered across situations for each variable–child combination. If we normalize, it will be per variable slice (for more details on preprocessing choices see Chapter 6). We will emphasize the numerical aspects but introduce some substantive interpretation as well.

10.3 ROTATING COMPONENTS

In standard principal component analysis, it is customary to rotate the solution of the variables to some kind of "simple structure" or to "simplicity", most often using Kaiser's (1958) varimax procedure in the case of orthogonal rotations and oblimin (e.g., Harman, 1976) or promax (Hendrickson & White, 1964) in the case of oblique rotations. These and other rotational procedures have occasionally been applied in three-mode principal component analysis. In this book we will not review the technical aspects of these procedures as good overviews can be found in the standard literature (e.g., Browne, 2001), but only touch upon those aspects of rotating components that are different or specifically relevant for three-mode analysis.

Several authors have advocated particular rotations of component matrices from specific types of data. For instance, Lohmöller (1978a) recommends rotation of time-mode component matrices to a matrix of orthogonal polynomials as a target, a proposal also put forward by Van de Geer (1974). Subject components are often not rotated because the first component (especially in research using rating scales) can be used to describe the consensus of the subjects about the relationships between the other two modes. Sometimes subject-mode components are rotated so that the axes pass through centroids of clusters of individuals. On the strength of content-based arguments, Tucker (1972, pp. 10–12) advocated that the "first priority for these transformations should be given to establishing meaningful dimensions for the object space", which concurs more with a multidimensional point of view.

10.3.1 Rotating normalized versus principal coordinates

One of the confusing aspects in multiway analysis when discussing rotations to achieve simplicity is that the coefficients of the component matrices do not automatically have the characteristics of scores and variable–component correlations they have in the covariance form of two-mode component analysis (see Section 9.3.1, p. 215). The reason for this is that in the covariance form of two-mode component analysis, variables are automatically standardized and the subject scores are commonly presented such that they are standardized as well, while the variable coefficients are, at least before rotation, variable–component correlations. In three-mode analysis, there is nothing automatic about the preprocessing or the scaling of the components themselves (see Chapter 6). The basic form of the Tucker models is that all component matrices are orthonormal (i.e., in normalized coordinates), and the size of the data is contained in the core array. Whatever scaling or preprocessing is appropriate has to be explicitly decided by the researcher (see Chapter 9), and when it comes to rotations the type of rotation and the mode to which it has to be applied has to be a conscious decision based both on content and on model arguments.

For the Tucker3 model, three rotation matrices can be defined (two for the Tucker2 model) and after the components of a mode have been rotated the inverse transformation has to be applied to the core array. Thus, if we applied the orthonormal matrix $\mathbf{S}$ to $\mathbf{A}$, that is, $\mathbf{A}^* = \mathbf{AS}$, then the S by $T \times U$ core array $\mathbf{G}_a$ has to be counterrotated by $\mathbf{S}^{-1}$, that is, $\mathbf{G}_a^* = \mathbf{S}^{-1}\mathbf{G}_a$. This can be done independently for each component matrix. Starting from orthonormal component matrices, the procedure is the same for all modes.

Complications arise when one has decided to imitate the two-mode situation and scale one or more component matrices in principal coordinates, because then the difference between rotated normalized and principal coordinates starts to play a role. Especially in the social and behavioral sciences with its emphasis on the covariance form of PCA the subject mode has standard coordinates and the variable mode has principal coordinates. Because generally a simple structure for the variable mode is desired, a transformation matrix $\mathbf{T}$ is applied to the variable loadings $\mathbf{F}$ and the inverse transformation is applied to the scores $\mathbf{A}$, that is, if $\mathbf{X} \approx \mathbf{AF}'$, then after rotation $\mathbf{X} \approx \mathbf{A}^*\mathbf{F}^{*\prime}$ with $\mathbf{F}^* = \mathbf{FT}$ and $\mathbf{A}^* = \mathbf{AT}'^{-1}$.

Suppose that one wants to rotate the variable mode in a three-way array of subjects by variables by conditions, and one wants to imitate the two-mode situation. In that case we first have to matricize the Tucker3 model as $\mathbf{X}_b = \mathbf{BG}_b(\mathbf{A}' \otimes \mathbf{C}')$, followed by scaling the orthonormal components $\mathbf{B}$ to principal coordinates. This can be done using the fact that $\sum_p \sum_r g_{pqr}^2 = \phi_q^2$, so that the variable components in principal coordinates become $\tilde{\mathbf{B}} = \mathbf{B\Phi}$, where $\mathbf{\Phi}$ is the diagonal matrix with the ϕ_q on the diagonal. The structured and scaled scores $\tilde{\mathbf{G}}_b = \mathbf{\Phi}^{-1}\mathbf{G}_b(\mathbf{A}' \otimes \mathbf{C}')$ now have unit (row) lengths. The rotated component matrix then becomes

$$\tilde{\mathbf{B}}^* = \mathbf{B\Phi T}. \tag{10.1}$$

To get model consistency, the structured and scaled scores $\tilde{\mathbf{A}}_b^*$ are inversely transformed as

$$\tilde{\mathbf{A}}_b^* = \mathbf{T}^{-1}\mathbf{\Phi}^{-1}\mathbf{G}_b(\mathbf{A}' \otimes \mathbf{C}'). \tag{10.2}$$

Whether the $\tilde{\mathbf{B}}^*$ are variable–component correlations depends on the scaling of the data and the components. If the variables are fiber-centered per column and normalized per variable slice, then the structured and scaled scores are orthonormal and the scaled variable components are variable–component correlations. In that case their rotated versions are also correlations, provided we have applied an orthonormal transformation.

In the multidimensional point of view, it is assumed that the three-mode model has subspace uniqueness without explicit reference to scores and loadings or regression coefficients (see Chapter 6). Therefore, normalized components may be rotated, because all we are seeking is a new orientation of the orthonormal axes in the space. However, this is no longer true if we want to interpret Euclidean distances in the space; see Section 11.3.1, p. 259.

10.3.2   Types of rotations

The orthogonal rotation used most often is Kaiser's (1958) varimax procedure, which is part of the orthomax family (Jennrich, 1970); for an overview of orthogonal rotations within the context of three-mode analysis, see Kiers (1997a). With respect to oblique rotations, Kiers (1993b) concluded on the basis of several simulation studies that "Promax and HKIC have often been found to be the best techniques for oblique simple structure rotation" (p. 75), where HKIC is the abbreviation for the Harris and Kaiser (1964) independent cluster rotation. In other studies on oblique rotations, Kiers also suggested using the Harris–Kaiser procedure, partly because it yields the same pattern for component weights and for component patterns (see Kiers & Ten Berge, 1994b), partly because it fits naturally into the basic form of the Tucker models, which have three orthonormal component matrices. Given this situation, we will restrict ourselves in this book primarily to the varimax procedure for orthogonal rotation of components and the Harris–Kaiser independent cluster procedure for oblique rotations. Given that the varimax has been well described in virtually any textbook on component and/or factor analysis, we will not discuss its technical aspects. However, we have added a section on the Harris–Kaiser procedure because it is relatively unknown and some information about it is necessary in order to apply it correctly.

10.3.3   Harris–Kaiser independent cluster solution

In this section (largely based on Kiers & Ten Berge, 1994b), we will discuss the Harris and Kaiser (1964) proposal. The discussion is presented within the Tucker3 framework. The basic idea behind the Harris–Kaiser independent cluster rotation is to use an orthonormal transformation to realize an oblique one. The trick is to scale the orthogonal component matrix in principal coordinates to an orthonormal matrix in normalized coordinates, rotate the orthonormal matrix via varimax, and transform it back into principal coordinates after rotation. Thus, one needs a scaling matrix Φ_{pre} for the first step, an orthonormal rotation matrix Θ, and a scaling matrix Φ_{post} for the last step.

In the basic form of the Tucker models, the component matrices are already orthonormal so that step 1 is not necessary. However, assuming that we want to transform the variable (or second mode) components B, the scaling matrix Φ_{pre}, which contains the square roots of the explained variabilities ϕ_q^2, can be found by squaring the core entries and adding them over the first and third mode components: $\phi_q = \sum_p \sum_r g_{pqr}^2$. The orthonormal rotation matrix Θ follows from a varimax on $\mathbf{B}$, so that we get a rotated component matrix $\tilde{\mathbf{B}} = \mathbf{B}\Theta$. Then in order to ensure that $\tilde{\mathbf{B}}$ has principal coordinates, thus we need to multiply $\tilde{\mathbf{B}}$ by $\boldsymbol{\Phi}_{\text{post}}$. Thus,

$$\tilde{\mathbf{B}}^* = \tilde{\mathbf{B}}\boldsymbol{\Phi}_{\text{post}} = \mathbf{B}\Theta\boldsymbol{\Phi}_{\text{post}}. \tag{10.3}$$

$\boldsymbol{\Phi}_{\text{post}}$ needs to be chosen in such a way that the transformed $\tilde{\mathbf{B}}^*$ is in the proper principal coordinates and the structured and scaled coefficients

$$\tilde{\mathbf{L}}_b = \boldsymbol{\Phi}_{\text{post}}\Theta'\mathbf{G}_b(\mathbf{A}' \otimes \mathbf{C}') \tag{10.4}$$

have length one (see also Eq. (10.2)). This can be achieved by requiring that

$$\text{diag}\left(\sum_r \boldsymbol{\Phi}_{\text{post}}\Theta'\mathbf{G}_r\mathbf{G}'_r\Theta\boldsymbol{\Phi}_{\text{post}}\right) = \mathbf{I}_q, \tag{10.5}$$

or that

$$\text{diag}(\boldsymbol{\Phi}_{\text{post}}\Theta'\left(\sum_r \mathbf{G}_r\mathbf{G}'_r\right)\Theta\boldsymbol{\Phi}_{\text{post}}) = \text{diag}(\boldsymbol{\Phi}_{\text{post}}\Theta'\boldsymbol{\Phi}_{\text{pre}}^2\Theta\boldsymbol{\Phi}_{\text{post}}) = \mathbf{I}_q, \tag{10.6}$$

which is realized if $\boldsymbol{\Phi}_{\text{post}} = \text{diag}(\Theta'\boldsymbol{\Phi}_{\text{pre}}^2\Theta)^{1/2}$.

Applying only Kaiser's varimax procedure on the orthonormal or normalized matrices from the basic form of the Tucker models can thus be seen as an equivalent to using the Harris–Kaiser independent cluster procedure. However, because no rescaling is done afterwards, the coefficients cannot be interpreted as variable–component correlations or covariances even if the data have been adequately preprocessed.

10.4 ROTATING FULL CORE ARRAYS

The emphasis in the earlier literature was on rotating component matrices rather than the core array, because was such a procedure is unnecessary in standard principal component analysis. Therefore, until recently there was a lack of attention and procedures for rotating core arrays in multiway analysis. As indicated in Section 10.1, the function of rotating core arrays is different from rotating components. In addition, the interpretation of core arrays is different from that of components. Concentrating on the simplicity of the core array means attempting to reduce the complexity of the model as a whole by reducing the number of multiplicative terms, so that fewer interactions between components need to be interpreted.

Ten Berge, Kiers, and co-workers have instigated an investigation into various properties of core arrays in which topics have been studied such as the rank of three-way arrays, the uniqueness of core arrays, the maximum number of zero elements, and algorithms to transform core arrays to maximum simplicity; for an overview see Ten Berge (2004). Only some of their results will be discussed here (see Section 10.5).

10.4.1 Simplicity of two-mode core arrays

In the singular value decomposition of two-mode matrices, $\mathbf{X} = \mathbf{AGB'}$, $\mathbf{G}$ is diagonal and the S orthonormal eigenvectors in $\mathbf{A}$ and those in $\mathbf{B}$ are pair-wise linked, as can be seen when writing the SVD in sum notation:

$$x_{ij} = \sum_{s=1}^{S} g_{ss} a_{is} b_{js}, \tag{10.7}$$

or rank-one arrays

$$\mathbf{X} = \sum_{s=1}^{S} g_{ss} (\mathbf{a}_s \otimes \mathbf{b}_s). \tag{10.8}$$

The diagonal matrix $\mathbf{G}$ contains the singular vectors, which indicate strength of the link between the components of the first mode $\mathbf{A}$ and those of the second mode $\mathbf{B}$. For two-mode data $g_{ss'} = 0$ if $s \neq s'$, so that there is no link between the eigenvectors $\mathbf{a}_s$ and $\mathbf{b}_{s'}$ but only between corresponding pairs $\mathbf{a}_s$ and $\mathbf{b}_s$. In other words, the basic form in the two-mode case already gives a simple core array.

In contrast, rotations of the components in two-mode PCA destroy the simplicity of the core array, and the off-diagonal elements of $\mathbf{G}$ become nonzero. Given proper preprocessing, the elements of $\mathbf{G'G}$ can be interpreted as the correlations between the components. After rotation Eq. (10.7) becomes:

$$x_{ij} = \sum_{s=1}^{S} \sum_{s'=1}^{S} g_{ss'} a_{is} b_{js'}, \tag{10.9}$$

so that the $g_{ss'}$ are no longer 0, and thus, rotation of the components results in complexity of the core.

10.4.2 Simplicity of three-mode core arrays

The linkage structure between the components of three modes can be, and usually is, far more complex than the two-mode case. The parallel structure to the two-mode situation for the basic form is a superdiagonal $(S \times S \times S)$ core array, so that $\mathbf{a}_p$, $\mathbf{b}_q$, and $\mathbf{c}_r$ only have a link if $p = q = r = s$. This is the situation described by the Parafac model,

$$x_{ijk} = \sum_{s=1}^{S} g_{sss} a_{is} b_{js} c_{ks}. \tag{10.10}$$

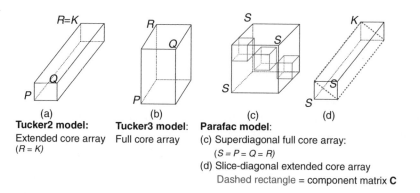

(a)
Tucker2 model:
Extended core array
(R = K)

(b)
Tucker3 model:
Full core array

(c)
Parafac model:
(c) Superdiagonal full core array:
    (S = P = Q = R)
(d) Slice-diagonal extended core array
    Dashed rectangle = component matrix **C**

Figure 10.1   Various core arrays and their relationships. From left to right: (a) Tucker2 extended core array, $\mathcal{H}$; (b) Tucker3 full core array, $\mathcal{G}$; (c) superdiagonality of full core array; and (d) slice diagonality of extended core array. The Parafac model corresponds with both a superdiagonal full core array and a slice-diagonal extended core array. In the latter case, the diagonal plane constitutes the third-mode component matrix **C** in the Parafac model.

A cubic core array with only nonzero superdiagonal elements has the most "simple" structure a core array can have. However, the standard Parafac model has nonorthogonal component matrices. To create a completer analogy with the two-mode case, an orthogonality constraint should be put on all component matrices. This is an extremely restricted Parafac model as well as a restricted Tucker3 model. From a modeling point of view, the interpretation can be relatively straightforward because each component is exclusively linked with only one component of the other modes (see Fig. 10.1(c)).

In a $p \times p \times u$ full core array, one can still have a relatively "simple structure" if $g_{pqr} \neq 0$ if $p = q, r = 1, \ldots, R$, and 0 elsewhere. In this case all frontal slices, $\mathbf{G}_r$, of the core array are diagonal. When $R = K$, we have a Tucker2 extended core array with diagonal frontal slices, $\mathbf{H}_k$ (see Fig. 10.1(d)).

A similar simple structure is sometimes found for $R = 2$: $g_{pqr} \neq 0$ for the core elements $(p, q), p = q, r = 1$; $g_{pqr} \neq 0$ for the elements $(p = P - (q - 1), q), r = 2$ and 0 elsewhere. In such a case, the first frontal slice is diagonal and the second is *anti-diagonal*, that is, the diagonal running from the bottom left-hand corner to the upper righthand corner is nonzero (Fig. 10.2)anti-diagonal.

Kiers (1998a) has shown that a $3 \times 3 \times 3$ core array, which has a structure as in Fig. 10.3, is unique, which means that different arrays with zeroes in the same places cannot be rotated to each other. The author indicates that "[p]ractical experience has suggested that models employing smaller core sizes are nonunique (except, of course, the models corresponding to CANDECOMP/PARAFAC)." (p. 567). Thus, in general, simplicity of core arrays has to be found by rotations. Kiers (1998b) remarked that using his oblique core rotation procedure (see Section 10.4.4), "we end up with at

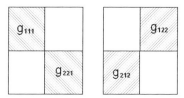

Figure 10.2   Simple $2 \times 2 \times 2$ core. Diagonal and antidiagonal frontal slices.

least as many as 18 zeros" (p. 322), that is, 9 nonzero entries. For higher-order core arrays virtually no uniqueness results are available. However, Kiers and colleagues have also shown that in many core arrays elements can be set to 0 without affecting the fit. This clearly shows that in general there is much to gain by attempting to rotate the core to simplicity, and that many elements can be eliminated or reduced in size by rotation.

10.4.3   Rotation toward a specific target

The earliest attempts at rotating full core arrays were directed toward achieving some form of diagonality in the core array. For the Tucker3 model, this was typically *superdiagonality*, that is, only the g_{sss} are nonzero (see Fig. 10.1(c)). A limitation in applying such a procedure is that the core array has to be cubic, that is, all dimensions should be equal. Such a search is in fact an attempt to approximate a Parafac solution after a Tucker3 analysis has been carried out.

A variant on this idea is to search for *slice diagonality*, which is most appropriate for the Tucker2 extended core array (see Fig. 10.1(d)). The earliest proposals date back to H. S. Cohen (1974) and MacCallum (1976). Kroonenberg (1983c, Chapter 5) proposed to use Parafac to find nonsingular transformation matrices for the first and second modes of the extended core array in a Tucker2 model as indicated with dashes in Fig. 10.1(d). When an extended core has slice diagonality, this is equivalent to superdiagonality of the full core array. In particular, when an extended core is slice-diagonal, its nonzero part has order $K \times S$. We may identify this diagonal slice with the component matrix C, so that we are back in a Parafac model with S components. Depending on the type of rotations, this is the standard Parafac model,

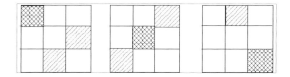

Figure 10.3   Maximal simple $3 \times 3 \times 3$ core. Cross-hatched boxes indicate superdiagonal elements.

Table 10.1   Henrion-Andersson diagonality criteria

Superdiagonality criterion	$\max \sum_{s=1}^{S} g_{sss}^2$
Slice diagonal criterion	$\max \sum_{p=1}^{P} \sum_{k=1}^{K} g_{ppk}^2$
Maximum variance criterion	$\max \sum_{p=1}^{P} \sum_{q=1}^{Q} \sum_{r=1}^{R} \left(g_{pqr}^2 - \bar{g}\right)$

$\bar{g} = PQR^{-1} \sum_{p=1} \sum_{q=1} \sum_{r-1} g_{pqr}^2.$

or a Parafac model with one or two orthonormal components. After finding the appropriate transformation matrices, the inverses of these transformation matrices are then used to adjust the components (see also Brouwer & Kroonenberg, 1991). Andersson and Henrion (1999) also proposed a procedure for optimizing slice diagonality, be it that they clearly had a full core array in mind.

Kiers (1992a) proposed both orthonormal and nonsingular transformations of the full core array. He reported results from simulation studies that showed that his orthonormal transformation procedures could not recover superdiagonality adequately, but that nonsingular transformations could do so.

Henrion and Andersson (1999) and Andersson and Henrion (1999) discussed three general criteria for simplifying the core array via orthonormal rotations: (1) maximizing the superdiagonal entries of the core array, (2) maximizing slice diagonality, and (3) maximizing the variance of the squared core entries. The last criterion, however, is not a target criterion (see Table 10.1). Each of the criteria is based on squared core entries, (g_{pqr}^2). As explained in Andersson and Henrion (1999), the first two diagonality criteria can be normalized with respect to the total sum of the squared core entries, $\sum_{p,q,r} g_{pqr}^2$, and the maximum variance with respect to its theoretical maximum, $PQR(PQR-1)\bar{g}$. Therefore, the success of the criteria can be expressed as proportions of their norms.

In Andersson and Henrion (1999), an algorithm was presented that can compute each of the three criteria. The results of the Andersson and Henrion (1999) study show that their procedures succeed in improving (super)diagonality of arbitrary core arrays and in concentrating the variance in a few core entries. However, as they did not carry out simulations with known solutions, the recovery rate was not determined.

10.4.4   Rotating the core array to arbitrary simplicity

In order to rotate the core array to simplicity, Kiers (1997a) proposed to rotate the core along each of its three sides using orthonormal rotation matrices, following a suggestion by Kruskal (1988). Kiers (1998b) also proposed a method for oblique rotation (SIMPLIMAX) by minimizing the sum of the m smallest elements of the rotated core array. Furthermore, Kiers (1998b) outlined how his procedure can be used for theoretical work and be adapted in many ways to facilitate this. The procedure seems to be rather sensitive for suboptimal solutions and therefore requires many random starts to find a globally optimal solution. The orthonormal procedure for rotating the core is also part of the simultaneous rotation of core and components, and will be discussed further in Section 10.4.5. Examples of rotations and their effects are given in Section 10.4.7.

10.4.5   Simultaneous rotation of components and core array

Kiers's work on rotations in three-mode analysis culminated in a procedure for simultaneously rotating the components and the core array Kiers (1998d). The fundamental idea was to define a orthomax simplicity function that is the weighted sum of the orthomax criteria for all three component matrices and the core array. This criterion subsumes varimax:

$$f_{\text{corecomp}}(\mathbf{S}, \mathbf{T}, \mathbf{U}) = f_{\text{core}} + f_{\text{components}}, \tag{10.11}$$

where $\mathbf{S}$, $\mathbf{T}$, and $\mathbf{U}$ are the orthonormal rotation matrices for the first, second, and third modes, respectively. The criterion function for the core array consists of the weighted sum of the three orthomax rotations of each of the matricized forms of the core array:

$$f_{\text{core}} = \sum_{\ell=1}^{3} w_\ell f_{\text{OR}}(\tilde{\mathbf{G}}^\ell, \gamma_\ell), \tag{10.12}$$

where $\tilde{\mathbf{G}}^1, \tilde{\mathbf{G}}^2, \tilde{\mathbf{G}}^3$ indicate the matrices whose columns consist of the vectorized horizontal, lateral, and horizontal slices of the core array $\tilde{\mathbf{G}}$ rotated by $\mathbf{S}$, $\mathbf{T}$, and $\mathbf{U}$, respectively, and γ_ℓ is the orthomax parameter (e.g., $\gamma = 0$ is quartimax and $\gamma = 1$ is the varimax criterion). The w_ℓ are the weights indicating the importance of the rotation for each of the sides of the core array. The equivalent expression for the component matrices is

$$f_{\text{comp}} = w_A f_{\text{OR}}(\tilde{\mathbf{A}}, \gamma_a) + w_B f_{\text{OR}}(\tilde{\mathbf{B}}, \gamma_b) + w_C f_{\text{OR}}(\tilde{\mathbf{C}}, \gamma_c), \tag{10.13}$$

where the tildes signify the rotated versions of the component matrices, the γ's are the orthomax parameters, and the w's the weights for the importance of the simplicity of the component matrices.

Note that, although we rotate not only the core array along its three modes but also each of the component matrices, we do not have six, but only three orthonormal rotation matrices $\mathbf{S}$, $\mathbf{T}$, and $\mathbf{U}$.

Kiers (1998d) makes a number of suggestions for carrying out the analyses in an orderly fashion. First of all, probably on content-based arguments, one should decide what to do with each mode. This involves two choices: (1) whether the modes should have normalized or principal coordinates; (2) whether all modes should be simultaneously rotated or only some of them. Given the limited experience, no hard and fast rules can be given. It is known, however, that for a core to be a Murakami core (see Section 10.6) the mode with the highest number of components should be

Table 10.2   Coping data: Illustration of simultaneous rotation of core and components

	Relative weights			Varimax criterion values $- f$					
	A	**B**	**C**	Core	**A**	**B**	**C**	Sum	Total
1	—	—	—	*4.37*	*1.37*	*2.24*	*3.36*	*6.97*	*11.29*
2	0.0	0.0	0.0	**5.68**	1.28	2.83	3.30	7.41	13.09
3	0.1	0.1	0.1	5.64	1.32	3.06	3.84	8.22	13.87
4	0.5	0.5	0.5	5.28	1.41	3.49	4.82	9.71	14.99
5	1.0	1.0	1.0	4.93	1.50	3.64	5.06	10.20	15.13
6	2.0	2.0	2.0	3.77	1.92	3.81	5.25	10.97	14.74
7	3.0	3.0	3.0	2.26	2.31	3.95	5.35	11.61	13.87
8	100	100	100	2.11	**2.33**	**3.95**	**5.36**	11.63	13.74

The first line indicates the unrotated original solution. Mode A, Situations; Mode B, Scales; Mode C, Children. The values in **bold** indicate the highest possible value for the criterion in question. Line 5 is the solution with natural weights.

in principal coordinates. The best advice that can be given so far is to start with the simplest situation and adjust it if the results are not satisfactory. This may sound very arbitrary, but it should be remembered that rotations do not affect the solution itself, but are only used for improved interpretation. So far all published analyses have opted for using the varimax criterion, and Kiers (1997a) reported simulation results that showed that the choice of type of orthomax procedure is not very crucial. With respect to the weights to be chosen beforehand, Kiers suggests using fixed *natural weights* for the core array (for precise definitions, see Kiers, 1998d). The relative importance of simplicity in the various parts of the model is then regulated via the weights for the components. Details will be discussed in connection with the example.

10.4.6   Illustration of simultaneous rotations

For the illustration we will use the $3 \times 3 \times 3$ solution for the Coping data (Röder, 2000). In Table 10.2, we have listed a selection of solutions with different relative weights for the components, that is, relative to the natural weights.

The top line in Table 10.2 shows the simplicity measures for the original unrotated solution. The first thing to notice is that relative weights larger than the natural weights (line 5; 1.0, 1.0, 1.0) will produce a core array that is less simple than the original core array, which had a core simplicity of $f_{core} = 4.37$. This value is less than the maximum simplicity index of $f_{core} = 5.68$. In most analyses less simplicity than the original core array does not seem desirable. If we assume that the interpretation of

Table 10.3   Coping data: Effect of simultaneous rotations of core array and components on the values in the core array

	Original values			Core simplicity			Natural weights			Component simplicity		
	S1	S2	S3	S1	S2	S3	S1	S2	S3	S1	S2	S3
Child component 1												
S1	1.3	7.9			8.3		8.1	−1.0			3.9	4.8
S2	−2.2		3.9			3.7	1.3	4.9			−2.5	4.8
S3										1.1	−2.9	−2.7
Child component 2												
S1	1.7	1.2	−1.7			−1.8	1.1		−1.2		1.3	−1.2
S2	6.5	−2.0	1.0	7.2			7.0			6.7		−1.4
S3			2.2			2.1			1.5	1.9		
Child component 3												
S1	5.1			5.0			4.4		2.7	−2.5	2.6	−2.5
S2		1.7	−2.1		1.5	−2.8		−1.2	−2.6			
S3			1.0		1.1	1.2	1.5	1.4	−1.2	3.6		2.9

S1, S2, S3 — Scale components 1, 2, and 3; *S1, S2, S3* — Situation components 1, 2, and 3.

Large Core entry > |3.5|; *Medium* |2.0| < Core entry < |3.5|; values below |1.0| have been blanked out; *Original core:* 4 large, 5 medium; *Simplest core:* 4 large, 2 medium; *Natural weights:* 4 large, 2 medium; *Simplest components:* 5 large, 7 medium. The order of the scale components of the original solution has been changed to facilitate comparison with the other solutions.

the scales (Mode B) is of prime importance, we would like the scale mode to be quite simple and with the natural weights ($f_B = 3.64$) we get close to the maximum value ($f_B = 3.95$). The same is true for the other two modes, so that using the natural weights seem a fairly satisfactory compromise if one wants to achieve simplicity in all parts of the model.

To see what this entails for the parameters, we have presented the solutions of the original analysis and those of lines 2 (maximal core simplicity), 5 (natural weights) and 8 (maximal component simplicity) for the core array in Table 10.3, and for the scale components in Table 10.4.

Table 10.3 shows that simplifying only the components is not a good option for the core array, as many more medium values are introduced and the larger values are smaller than in the other rotations. It also shows that, from the point of view of simplicity of the core array, using natural weights seems a reasonable compromise.

Rotations clearly lead to cleaner components, but maximal core simplicity complicates the Mode B components, albeit not by much. The difference between maximal component simplicity and the natural weights is rather small so that one can easily settle for the natural weights. We will not show the other two component matrices as the basic principles remain the same. At least two other examples exist of a simi-

Table 10.4   Coping data: Effect of simultaneous rotation of core array and components on values of Mode B (Scales)

	Original values			Core simplicity			Natural weights			Component simplicity		
	S1	S2	S3	S1	S2	S3	S1	S2	S3	S1	S2	S3
Ang	**1.1**	*−0.3*	*−0.3*	**0.9**	**−0.4**	0.0	**1.0**	−0.1	0.0	**1.0**	0.0	−0.0
Ann	**0.4**	**0.5**	*−0.3*	*0.3*	**0.7**	−0.2	0.1	**0.7**	−0.1	0.0	**0.8**	0.0
Sad	0.2	**0.4**	*−0.3*	0.2	**0.6**	−0.2	0.1	**0.6**	−0.2	−0.0	**0.6**	−0.1
Afr	0.2	0.1	0.2	0.1	0.2	0.2	0.0	0.2	0.2	0.0	0.1	0.2
Sup	−0.0	0.1	−0.1	0.0	0.1	−0.1	0.0	0.1	−0.1	−0.0	0.1	−0.1
Agg	−0.2	0.1	−0.1	−0.1	0.1	−0.1	−0.1	0.1	−0.1	−0.1	0.1	−0.1
App	*0.3*	0.1	**0.8**	0.0	0.1	**0.7**	0.0	0.1	**0.7**	0.0	−0.1	**0.7**
Avo	*−0.3*	−0.2	**−0.7**	−0.0	*−0.3*	**−0.6**	0.1	*−0.3*	**−0.6**	0.1	−0.1	**−0.6**

S1, S2, S3 — Scale components 1, 2, and 3; **Large** coefficient > |0.35| *Medium* |0.25| < coefficient < |0.35|; *Original core:* 6 large, 6 medium; *Simplest core:* 6 large, 2 medium; *Natural weights:* 5 large, 1 medium; *Simplest components:* 5 large, 0 medium. Ang = Angry; Ann = Annoyed; Afr = Afraid; Sup = Support; Agg = Aggression; Appr = Approach; Avo = Avoidance.

larly reasoned selection of weights in a practical application. One is contained in the original paper proposing the technique (Kiers, 1998d), and the other is contained in Kiers and Van Mechelen (2001).

10.4.7   Illustration Henrion–Andersson criteria

The original 3×3×3 core array from the Coping study is presented in Table 10.5, together with core arrays resulting from applying the three simplicity criteria of Andersson and Henrion (1999), that is, a core array optimally rotated toward maximal superdiagonality, a core array with maximal slice diagonality, and a core array with maximum variance. In addition, we have taken the core arrays of three variants of the simultaneous core and component rotations: the maximally simple core array, the core array when the components are maximally simple, and the core array transformed by means of natural weights (see Section 10.4.6).

The core arrays in the table have on purpose not be arranged to show their maximum similarity. The values of the core arrays listed are those that were derived by the techniques in question. What this illustrates is that the core arrays are determined up to a permutation of the columns and rows. If one does not take this into account unrealistic results occur. For instance, the maximally simple core has a superdiagonality value of 0.8%, whereas after rearranging this value is 67.4%. From different starting positions the Henrion–Andersson algorithm gives very stable results, but without rearranging the values in the core arrays these results might not be very useful, as the

Table 10.5   Coping data: Core arrays and values of the Henrion–Andersson criteria

	C1			C2			C3		Super Diag.	Slice Diag.	Max Var.
S1	S2	S3	S1	S2	S3	S1	S2	S3			
Original											
1.3	**7.9**	0.9	1.7	1.2	−1.7	**5.1**	0.2	0.0			
−2.2	−0.4	**3.9**	**6.5**	−2.0	1.0	−0.9	1.7	−2.1	58.2	64.8	17.3
−0.2	−0.2	−0.3	0.2	0.7	2.2	0.9	0.7	1.0			
Maximal superdiagonality											
8.3	0.2	−0.1	0.0	−0.9	−5.0	−0.2	−1.7	0.3			
0.2	−0.0	0.1	−1.0	−1.6	0.6	−0.5	−2.0	−0.6	**67.7**	70.1	22.3
0.3	−3.7	−0.2	−1.6	2.7	0.5	−0.1	−0.1	7.2			
Maximal slice diagonality											
6.8	−1.3	−0.6	**1.4**	4.0	0.1	**1.9**	1.8	1.5			
0.1	−**2.4**	1.0	0.1	−**0.7**	−0.4	0.4	.6	−0.6	59.2	**70.4**	17.3
1.7	−1.5	**2.0**	−0.7	−0.3	−**7.8**	−4.7	−0.8	**2.1**			
Maximal variability											
0.1	**8.3**	−0.0	0.1	−0.3	−1.8	**5.0**	0.1	−0.2			
−0.4	0.1	**3.7**	**7.2**	0.0	0.0	0.0	1.5	−**2.8**	67.5	70.2	**22.6**
−0.0	0.1	−0.3	−0.0	−0.7	−2.1	−0.6	−1.0	−1.2			
Core maximally simple											
0.1	**8.3**	−0.0	0.1	−0.3	−1.8	**5.0**	0.1	−0.3			
−0.4	0.1	**3.7**	**7.2**	0.1	0.0	0.0	1.5	−**2.8**	67.4	70.1	22.5
0.0	−0.1	0.4	0.0	0.7	2.1	0.6	1.1	1.2			
Natural weights											
−0.1	**8.1**	−1.0	1.1	−0.8	−1.2	**4.4**	0.8	**2.7**			
0.4	1.3	**4.9**	**7.0**	−0.6	0.6	−0.9	−1.2	−**2.6**	60.9	63.4	18.4
−0.1	0.3	−0.3	−0.7	−0.2	1.5	1.5	1.4	−1.2			
Components maximally simple											
−0.5	**3.9**	**4.8**	−0.1	1.3	−1.2	−**2.5**	**2.6**	−**2.5**			
−0.6	−2.5	**4.8**	**6.7**	−0.7	−1.4	−0.7	−0.1	−0.3	37.5	47.2	8.4
1.1	−**2.9**	−**2.7**	1.9	0.7	0.4	**3.6**	−0.1	**2.9**			

S1, S2, S3 — Scale components 1, 2, and 3; C1, C2, C3 — Child components 1, 2, and 3.

The values for the Henrion–Andersson criteria are given as percentages with respect to their maximal value (see Section 10.4.3). The **bold** values in the second, third, and fourth core arrays indicate the type of the criterion, and the *italic* values indicate larger deviating values. In the other core arrays, **bold** values simply indicate the larger values. The **bold** values in the criterion section of the table indicate the maximum value that could be obtained on the criterion given the data.

above example shows. The criterion values in Table 10.5 have been calculated for the core array optimally permuted toward superdiagonality.

The middle part of Table 10.5 shows that the maximum values for the three Henrion–Andersson criteria are 67.7% for superdiagonality, 70.4% for slice diagonality, and 22.6% for maximum variance. Furthermore, the maximally simple core and the maximally superdiagonal core array have nearly maximum values on all three criteria. The core arrays rotated with natural weights and maximal slice diagonality do worse on all but the slice diagonality criterion. The natural weights rotated core array has acceptable, but somewhat lower, values on all three criteria, and finally the last core array shows what happens when only varimax rotations are applied to the components. The number of larger values decreases dramatically and there is a tendency to equalize the weight across the whole of the core array. This is also evident from all quality indicators. In other words, if one wants to have an interpretable core array, having very simple components all around will not help interpretation and might even make direct interpretation of the core array impossible, or very difficult indeed.

10.5   THEORETICAL SIMPLICITY OF CORE ARRAYS

So far we have primarily considered empirical attempts at simplifying core arrays and component matrices. However, intensive research has been carried out (as indicated previously) into theoretical simplicity of core arrays. One of the problems studied was how many zeroes a core array can have without loss of fit to the data, in other words, how the rotational freedom can be used to create as many zeroes in the core array as possible. Kiers, Ten Berge and their colleagues have derived several results on the maximum number of zeroes that core arrays of various sizes can have (Kiers, Ten Berge, & Rocci, 1997c; Kruskal, 1988; Murakami et al., 1998b; Rocci, 1992; Ten Berge & Kiers, 1999). In this section we will discuss some of the results, especially those that have or may have practical relevance.

10.5.1   Murakami core

Murakami et al. (1998b) discovered that an extremely simple core array exists when the product of the number of components of two modes minus one is equal to the number of components of the third mode; such a core is here referred to as a *Murakami core*. In particular, when $QR - 1 = P$ (or $PR - 1 = Q$ or $PQ - 1 = R$), and $Q \geq R$ (as can be done without loss of generality), the core array has exactly $R(Q + R - 2)$ nonzero elements provided the mode with the largest number of components is in principal coordinates. Unfortunately, in the range of core sizes that is generally considered, there are only very few core arrays that fulfill this condition. For example, if $1 + \max(P, Q, R)$ is prime a Murakami core does not exist. Half the numbers below 10 are primes, which means that of the smaller core arrays only five triplets of components are Murakami cores: (3,2,2) — 4 nonzero entries; (5,3,2) — 6 nonzero entries; (7,4,2) — 8 nonzero entries; (8,3,3) — 11 nonzero entries; (9,5,2) — 10 nonzero entries. Thus, in general, about 80% of the entries are 0. However, core arrays corresponding to the latter sizes are relatively uncommon.

In their paper, Murakami et al. (1998b) also mention that the case with $P = QR$ (and $Q = PR$ and $R = PQ$) can be "trivially" shown to have only P (Q,

Table 10.6 Coping data: Illustration Murakami core

	Original Core			Rotated Core		
	S1	*S2*	*S3*	*S1*	*S2*	*S3*
Child component 1						
S1	14.518	0.364	12.570	−0.002	**−7.821**	−0.001
S2	11.692	7.431	−11.534	**19.548**	0.000	0.000
Child component 2						
S1	−2.444	14.713	−1.240	0.000	0.000	**19.548**
S2	5.354	−10.502	−9.463	−0.001	**17.915**	0.000

S1, S2, S3 — Scale components 1, 2, and 3; *S1, S2, S3* — Situation components 1, 2, and 3.

R) nonzero elements after nonsingular transformation. All core arrays with one component in one of the modes fulfill this condition, for instance, $(P,P,1)$. If a mode only has one component, the core array consists of a single slice. Therefore, the P nonzero elements lie on the diagonal of that slice. When all modes have at least two components the condition cannot be fulfilled if P is prime, so that only the core arrays of the following triplets of numbers of components can be simplified: $(4,2,2)$ — 4 nonzero entries, $(6,3,2)$ — 6 nonzero entries, $(8,4,2)$ — 8 nonzero entries, and $(9,3,3)$ — 9 nonzero entries, given no mode has more than 9 components.

It is important to note that the simplicity of these core arrays exists independent of the data, so that no substantive claim can be derived from their simplicity. Another caveat is that an extremely simplified core array will not necessarily lead to a simple interpretation of the components.

Illustration of the Murakami core. Table 10.6 provides the rotated $2{\times}3{\times}2$ Murakami core for the Coping data. From a complex core array we can obtain an extremely simple one via nonsingular transformations. However, as a result of the transformations, the components are now very strongly correlated, where they were orthogonal before.

10.5.2 Some simplicity results for core arrays

Ten Berge and Kiers (1999) proved "that the $p{\times}p{\times}2$ arrays have rank p [that is, a Parafac solution with p components] or $p + 1$, almost surely" and "that almost every $p{\times}q{\times}2$ array has a CANCECOMP/PARAFAC decomposition in min[p,2q] dimensions." (p. 172). The interesting aspect of these observations for practical applications is that they also apply to core arrays. If a core array can be perfectly fitted by a Parafac model then the whole data array can, due to the uniqueness of the Parafac model itself (see Brouwer & Kroonenberg, 1991). This two-step procedure,

Table 10.7   Size of Parafac solutions for arrays with size $p \times q \times 2$

Size core array	Number of Parafac components	Size core array	Number of Parafac components
$3\times3\times2$	3,4	$2\times2\times2$	2,3
$4\times3\times2$	4	$3\times2\times2$	3
$5\times3\times2$	5	$4\times2\times2$	4
$6\times3\times2$	6	$p\times2\times2$	4
$p\times3\times2$	6		

The $3\times2\times2$ core is also a Murakami core (see Section 10.5.1).

that is, Tucker3 on the data followed by Parafac on the core, and directly applying Parafac on the data, is a topic for further investigation.

The implication of a Parafac solution for Tucker3 core arrays is that in several cases the Tucker3 model will be equivalent to a Parafac model. When the number of components is as in the models marked in bold in Table 10.7, it might be worthwhile to consider which model is the most advantageous in representing the same explained variability.

10.6 CONCLUSIONS

In this chapter we have demonstrated a number of ways in which the basic form of, the Tucker models can be subjected to transformations, of the components, the core array, or both. In all cases, the aim is an easier interpretation of the results. We also demonstrated that there can be a potential conflict between simple components and simple core arrays, and thus researchers should be very clear at the outset what aspect of the solution they would like to be simple. On the other hand, there is no harm in being eclectic, the more so because rotations do not affect the model fit but only provide different ways to look at the same variability.

CHAPTER 11

GRAPHICAL DISPLAYS FOR COMPONENTS

A picture is worth a thousand words.[1]

11.1 INTRODUCTION

Various kinds of supplementary information are necessary for an in-depth interpretation of results from a multiway principal component analysis. In this chapter we will discuss various types of plots useful in multiway component analyses, review different methods to plot components, and pay attention to plots of single modes as well as plots that portray more than one mode at a time. Elsewhere in this book other types of plots in use in multiway analyses are discussed, such as plots to assess model

---

[1]"One look is worth a thousand words" was coined by Fred R. Barnard in *Printers' Ink*, 8 December 1921, p. 96. He changed it to "One picture is worth a thousand words" in *Printers' Ink*, 10 March 1927, p. 114 and called it "a Chinese Proverb so that people would take it seriously." *http://www2.cs.uregina.ca/~hepting/research/web/words/index.html*. Accessed May 2007.

fit (see Section 8.5, p. 179), sums-of-squares plots to assess the fit of levels of modes (see Section 12.6.2, p. 290), and various residual plots (see Section 12.7, p. 292).

11.2   CHAPTER PREVIEW

The types of plots considered are scatter plots of coordinate spaces involving one or more modes. Considerations for constructing such plots are discussed throughout the chapter and examples are provided. First, we will discuss various options for displaying component spaces for a single mode, then, a variety of plots for more than one set of components in a single space, such as line plots, joint biplots, nested-mode biplots, and nested-mode per-component plots.

The major sources for the material presented in this chapter are Kroonenberg (1983c, Chapter 6), Kroonenberg (1987b), and Geladi, Manley, and Lestander (2003), but especially Kiers (2000a).

11.2.1   Data sets for the examples

Most plots will be demonstrated using the Sempé girls' growth curves data, but in addition a plot from Andersson's spectrometry data is included, one from a publication by Dai (1982), and one from a publication by Nakamura and Sinclair (1995).

Illustrative data set: Girls' growth curves data.   Thirty girls (first mode) were selected from the French auxiological study (1953 to 1975) conducted under supervision of Michel Sempé (Sempé, 1987). Between the ages of 4 and 15 they were measured yearly (third mode) on a number of variables, eight of which are included here (second mode): weight, length, crown–coccyx length (*crrump*), chest circumference (*chest*), left upper-arm circumference (*arm*), left calf circumference (*calf*), maximal pelvic width (*pelvis*), and head circumference (*head*)[2]. The data set is thus a three-way block of size 30 (girls) × 8 (variables) × 12 (time points.) Before the three-mode analyses, the data were profile preprocessed; see Section 6.6.1, p. 130. One of the consequences is that all scores are in deviation of the average girl's profile. After preprocessing, the average girl's profile contains zeroes on all variables at all ages, and therefore in many plots she is located at the origin of these plots. Figure 11.1 gives an impression of the growth curves of this average French girl. The variables weight, length, and crown–coccyx length have been divided by 10 to equalize the ranges somewhat; no further processing was done.

The data were fitted with a Tucker3 model with three components for the first mode (girls), three components for the second mode (variables), and two components for

---

[2]Earlier versions of the graphs can be found in Kroonenberg (1987b); 1987 ©Plenum Press; reproduced and adapted with kind permission of Springer Science and Business Media. The data can be obtained from the data set section of the website of The Three-Mode Company; *http://three-mode.leidenuniv.nl*. Accessed May 2007.

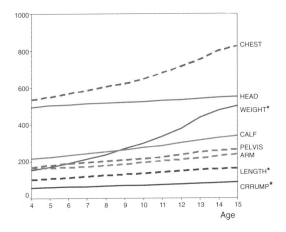

Figure 11.1 Girls' growth curves data: Growth curves of the average girl. The starred variables have been divided by 10 to reduce their ranges.

the third mode (years of age). This solution had a fit of 77% and was able to describe the most important aspects of the physical development of the 30 French girls.

11.3 GENERAL CONSIDERATIONS

As preparation for the material of this chapter, Chapters 1 and 2 of Krzanowski (2000) are warmly recommended. In those chapters concepts such as distances, inner products, direction cosines, subspaces, projections, and graphical interpretation of principal components are treated in great detail, especially from a user's point of view; also Appendix B provides some guidance for these concepts. In this chapter, we will not provide such a systematic introduction but explain various aspects when they arise.

11.3.1 Representations of components

As explained in more detail in Section 9.3.1, there are two versions of representing principal component analysis. We have referred to these as the *covariance form*, in which the variables are in principal coordinates and the subjects in standard coordinates, and the *distance form*, in which the subjects are in principal coordinates and the variables in normalized coordinates. Because in the distance form the Euclidean distances between the subjects are preserved as well as possible given the reduced dimensionality of the space, it is the form of choice when one wants to portray the entities of a single mode. Detailed explanations of the correctness of plotting the distance form are presented in Legendre and Legendre (1998) and Kiers (2000a). The distance form is the following

$$x_{ij} = \sum_s g_{ss}(a_{is}b_{js}) \sum_s [g_{ss}a_{is}][b_{js}] = \sum_s a_{is}^* b_{js}, \qquad (11.1)$$

where $\mathbf{B}$ is orthonormal, $\mathbf{B'B} = \mathbf{I}_s$, and $\mathbf{A}^*$ is orthogonal, $\mathbf{A}^{*\prime}\mathbf{A}^* = \mathbf{G}^2$, with $\mathbf{G}^2$ a diagonal matrix with the squares of the singular values g_{ss}.

The next question is what happens to the other mode, which is in normalized coordinates, in particular, the variable mode in the distance form. In the full-dimensional space the original variables are the (almost certainly nonorthogonal) axes. After calculating the principal components in the full space, these components are the orthonormal coordinate axes that span the same J-dimensional full space as the variables. The variables are directions or variable axes in this space. When we restrict ourselves to the S major principal components, the variable axes are also projected onto this S-dimensional space. To investigate them, we can plot the variable axes in the reduced space using the rows of $\mathbf{B}$, that is, $(b_{j1}, \ldots, b_{jS})$ (see Kiers, 2000a, p. 159).

11.4 PLOTTING SINGLE MODES

In multiway analysis, either one treats the component matrices evenhandedly and plots all of them in principal coordinates, or one assigns the modes different roles and plots them in different metrics. In the former option, high-dimensional distances are correctly represented in the low-dimensional plots within the accuracy of the approximation. In the latter case, the variables are generally portrayed in principal coordinates to represent their correlations as well as possible and the subjects in normalized coordinates. If plotted together variables are generally represented as arrows and subjects as points.

11.4.1 Paired-components plots

With the term *paired-components plot* we refer to plots in which the components of a single mode are plotted against each other in two- or higher-dimensional scatter plots. Such plots should have aspect ratios of 1, so that distances in the plots are properly portrayed.

Normalized versus principal coordinates.   When the explained variabilities are almost equal, there will be few visual differences between the plots in normalized and principal coordinates, but differences can be considerable when the explained variabilities are very different. Note, however, that the principal coordinates are scaled by the square roots of the explained variabilities which makes the differences less dramatic. In Fig. 11.2, the proportional explained variabilities, which were 0.56 and 0.14 (a factor of 4), become 0.74 and 0.37 at the root scale (a factor of 2).

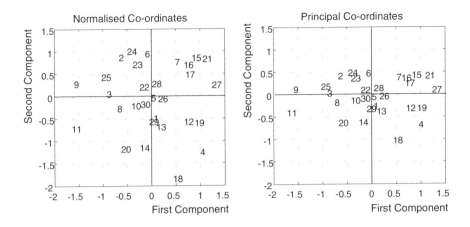

Figure 11.2   Girls' growth curves data: Difference between normalized and principal coordinates for the first mode (girls).   Square roots of explained variability of the two components which serve a scaling constants are 0.74 and 0.37, respectively. For comparability the normalized components have been scaled so that the lengths of the coordinate *axes* in both plots are equal.

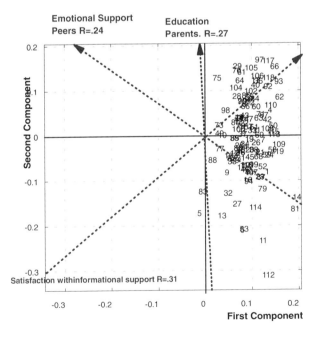

Figure 11.3   Coping data: Demonstration of the inclusion of external variables to enhance the interpretation of the subject space.

Adding explanatory information.   It always pays to make the plots as informative as possible. Especially when the subject or object space is being displayed, the figure typically looks like an unstructured set of points. There are several ways to enhance such plots, in particular, by using non-three-mode (background) information. If a cluster analysis has been carried out the subjects can be labeled according to these clusters, or they may be labeled by another grouping variable. External variables may be plotted in the space, such that they can be used for interpretation by establishing which subjects have high and which subjects have low values on these external variables. Assuming the subject space is centered and the axes are indicated by a_1 and a_2, one then regresses a standardized external variable z_Y on the axes, that is, $z_Y = \beta_1 a_1 + \beta_2 a_2 + e$. In the plot, the axis for $\hat{z}_Y$ is drawn through the points $(0,0)$ and (β_1, β_2); in addition the squared multiple correlation indicates how much of the variable Y is explained by the axes. Figure 11.3 gives an example of adding external information for the full Röder (2000) Coping data (for an explanation of the data and the variables, see Section 14.4, p. 354).

An imaginative use of auxiliary information is contained in a plot published in a paper by Dai (1982) on the industrial design of chairs (Fig. 11.4).

11.4.2   Higher-dimensional spaces

When we have higher-dimensional spaces, a clear presentation on paper becomes more difficult and many attempts have been made to cope with this problem. The most effective solutions consist of dynamic graphics in which one can spin the configuration in higher-dimensional, effectively three-dimensional, space along the axes.

Paired-components plots.   Failing that, one may inspect the plots by pairs via a *paired-components plot* to create a mental three-dimensional picture. In Fig. 11.5 the variables of the girls' growth curves data have been plotted as vectors. Because the variables are in principal coordinates and the appropriate input preprocessing has been applied (see Section 9.4, p. 218), the angles between the vectors are approximations to the correlations. If such conditions are not fulfilled, the angles can be interpreted as measures of association, the smaller the angles the higher the variables are associated. The circle in the plot is the *equilibrium circle* (see Legendre & Legendre, 1998, p. 398ff.), which can be used to assess how well variables are presented by the low-dimensional space given normalized coordinates. The closer the end points of the arrows are to the circle, the better the variables are represented in the space.

Minimum spanning tree.   One problem with three-dimensional graphs is that it is not always clear whether two vectors that are close together in the graph are also close together in high-dimensional space. Both in a high-dimensional and a low-dimensional space, vectors or arrows from the origin are often only represented by their end points, and thus reference is often made to "points" rather than "vectors". This usage is sometimes followed here as well. The end points of vectors might end

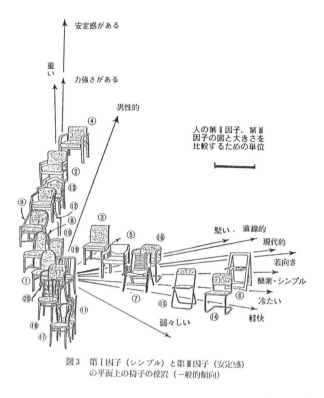

図3　第Ⅰ因子（シンプル）と第Ⅲ因子（安定感）
の平面上の椅子の位置（一般的傾向）

Figure 11.4　Chair-styles data: The inclusion of external information to enhance an object space. The Japanese text next to the vectors describes the adjectives used to characterize chairs. Source: Dai (1982); reproduced with kind permission from the Japanese Psychological Review.

up close together in a low-dimensional space because their projections just happen to lie close together. To solve that issue, a *minimum spanning tree* may calculated. Minimum spanning trees are generally based on distances in the full-dimensional space, that is, for each pair of variables j and j'. First, such distances are calculated over all subjects and conditions: $d_{jj'} = \left(\sum_i \sum_k (x_{ijk} - x_{ij'k})^2 \right)^{1/2}$. Then the minimum spanning tree is constructed in such a way that the sum of all the paths connecting the points (without loops) is the shortest possible. As closest neighbors in the full-dimensional space are connected, one must doubt the closeness of two points in a low-dimensional space when they are not. By drawing lines between closest points in the graph, one can judge more easily whether points that seem close in the low-dimensional space are actually close or not. The minimum spanning tree is equally useful for assessing the closeness of the individuals as portrayed in Fig. 11.2.

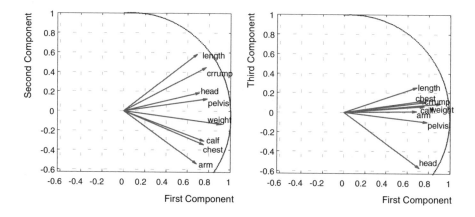

Figure 11.5   Girls' growth curves data: Side-by-side paired components plots of the second mode (variables). The variables have been connected to the origin and scaled so that the angles indicate the similarities between the variables.

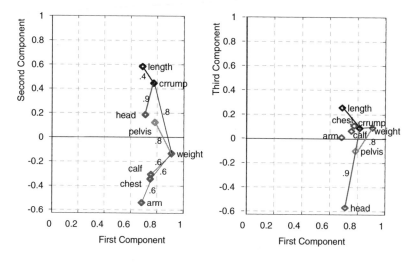

Figure 11.6   Girls' growth curves data: Side-by-side paired-components plots of the variable space. The variables have been connected by a minimum spanning tree based on the three-dimensional configuration. The lengths of the connecting paths are indicated as well.

In this case the distances $d_{ii'}$ are computed between pairs of individuals across all variables and conditions.

A poor man's version for calculating distances between points is to calculate distances in the projected space if it has a higher dimension than two. In that way, one

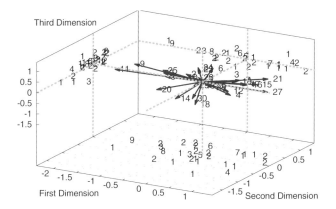

Figure 11.7   Girls' growth curves data: Three-dimensional representation of the subject space anchored by the coordinate axes and the projections onto the 1-2 plane, 1-3 plane, and the 2-3 plane. The figure shows the position of the girls in the three-dimensional space by vectors which are labeled with the girls' sequence numbers. Points on the walls and the floor are the projections onto the 1-2 plane, 1-3 plane, and the 2-3 plane, respectively, and they are labeled with the first character of their labels.

can judge whether in the two-dimensional plane points are also close in the three- or higher-dimensional space. In Fig. 11.6 the minimum spanning tree is drawn for the variables of the girls' growth curves data based on a three-dimensional configuration. From the first two dimensions one might be under the mistaken impression that head circumference is fairly close to pelvis, while it is not.

Three-dimensional graphs.   Ideally we would like to look at the three-dimensional space itself, and a figure such as Figure 11.7 is an example of how it can be constructed. The three-dimensional impression is created by boxing-in the points and projecting their values on the walls and the floor. The walls and the floor are in fact the three paired-components plots. The coordinate axes are included as well as their projections on the walls and the floor. For such a plot to work there should not be too many points or it becomes too messy to interpret.

Size as a third dimension.   Another alternative is to use different sizes of points to indicate their values in the third dimension (see Fig. 11.8). This option can be combined with the anchoring of the points in some plane, which can be either the base of the plot or the plane of the first two dimensions (see Fig. 11.9, taken from Nakamura and Sinclair, 1995, p. 104). For all representations, it is true that they only really work for a limited number of points with a reasonably clear structure.

Stereo-pairs and anaglyphs.   A further approach, which has seen very little application but should be useful for publications, is the use of stereo vision. Examples are

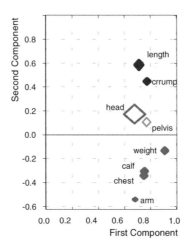

Figure 11.8   Girls' growth curves data: Two-dimensional representation of the three-dimensional variable space. The sizes of the points indicate their values on the third axis. An open diamond indicates a negative value; a closed diamond a positive value.

stereo-pairs which consist of two graphs of the same space constructed in such a way that side-by-side they provide a stereo view of the space. Unfortunately, there are very few examples of the use of stereo-pairs, and not many standard computer programs produce such plots. For a general introduction see Holbrook (1997)[3]. Comparable variants are *anaglyphs*, which are two graphs, one in blue and the other in red, but with a slightly shifted picture, which when viewed through a pair of spectacles with blue glass for one eye and red glass for the other eye give the illusion of depth. To be able to employ these one needs programs[4] to produce the graphs and, depending on one's training, optical aids to see proper depth in the pictures. Three-dimensional spinning of the coordinate space generally has to precede the making of a stereo picture to find the most informative point of view to inspect the three-dimensional graph. For further information on three-dimensional stereo graphs see Huber (1987).

11.4.3   All-components plots

When there is a natural order in the levels of a mode, it can be advantageous not to plot the components against each other, but to plot them against the levels of the mode. Depending on the spacing of these levels, this plotting can be done against the level number or against the real values of the levels. As an example, Fig. 11.10 depicts the two components from the age mode of the girls' growth curves data.

[3] *http://oxygen.vancouver.wsu.edu/amsrev/theory/holbrook11-97.html*. Accessed May 2007.
[4] For an example see *http://www.stereoeye.jp/software/index_e.html*. Accessed May 2007.

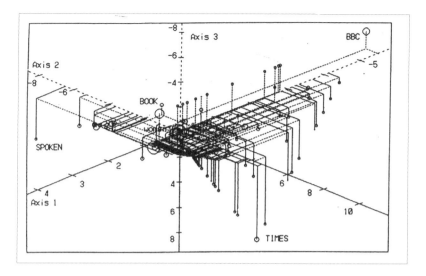

Figure 11.9   World of woman data: Three-dimensional representation of a component space with anchoring of the points on the 1-2 plane. The picture displays the collection of words closely associated with the word *woman* as can be found in the "Bank of English", the corpus of current English compiled by Cobuild. The corpus contains words appearing in the English language in sources like the BBC, books, the spoken word and The Times. Source: Nakamura & Sinclair (1995), Fig. 2, p. 104. 1995 ©Oxford University Press. Reproduced with kind permission from the Oxford University Press.

In Fig. 11.10, the principal coordinates have been used to emphasize the relative importance of the two components. Given the patterns in the two components, to facilitate interpretation, one should consider a rotation toward a target of orthonormal polynomials to get the first component to portray the constant term, and the second one the deviation from this overall level. This results in Fig 11.11, but note that in that figure the components are normalized coordinates.

All-components plots are almost standard in chemical applications using the Parafac model, because there a component very often represents the spectrum of a specific analyte. This spectrum can be compared with the spectra of known analytes for identification. Figure 11.12 gives an example of the results from a four-component Parafac solution. It displays the input spectrum of excitation frequencies applied to a set of sugar samples[5]. Experimental details for this research, as well as full analyses for this type of data, can be found be in Bro (1999a). The graph show a number of theoretically unacceptable anomalies such as the rising pattern at the end of one of the

[5]This picture was made using unpublished data collected by Claus Andersson, KVL, Copenhagen, Denmark.

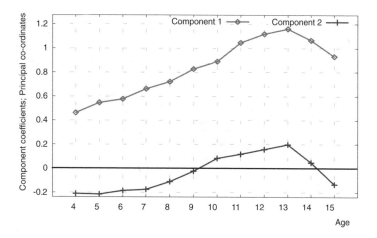

Figure 11.10   Girls' growth curves data: All-components plot. The horizontal ages consists of the age of the girls and the vertical axis displays the component coefficients in principal coordinates.

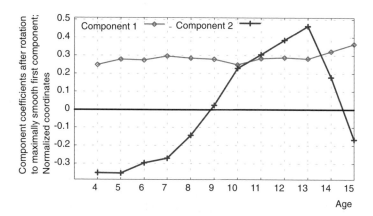

Figure 11.11   Girls' growth curves data: All-components plot. The horizontal ages consists of the age of the girls and the vertical axis displays the component coefficients in normalized coordinates after rotating the age mode to an maximal smooth first component.

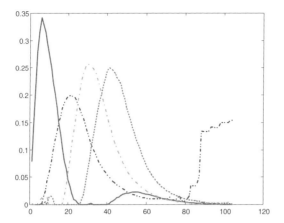

Figure 11.12   Andersson sugar data: Plot of Parafac components for the excitation mode in a fluorescence spectroscopy experiment on sugar. The irregularities in the spectra can be suppressed by introducing further restrictions on the components such as single-peakedness.

components. In his paper, Bro discusses such anomalies and how Parafac analyses can be improved via constraints on the components to regularize the results.

11.4.4   Oblique-components plots

Component spaces from a Parafac analysis are seldom orthogonal which poses some problems in plotting the components. In Section 11.5.1 we will look at portraying the components across modes, but here our aim is to discuss plots of the components within a mode. Plotting nonorthogonal components as rectangular axes destroys any distance or correlation interpretation of the plot. Therefore Kiers (2000a) proposed computing an auxiliary orthonormal basis for the component spaces and projecting unit-length Parafac components (or any nonorthonormal axes for that matter) onto such a plot.

The following steps are necessary to construct a plot of the loadings of the variables in a two-mode situation in which the unit-length columns of $\mathbf{A}$ are nonorthogonal. First, one searches for a transformation matrix for the component matrix $\mathbf{A}$, such that $\mathbf{A}^* = \mathbf{A}\mathbf{T}$ is column-wise orthonormal; then the variable coefficients $\mathbf{B}$ are transformed by the inverse of $\mathbf{T}$, that is, $\mathbf{B}^* = \mathbf{B}(\mathbf{T}')^{-1}$, which are the coefficients to be plotted.

Let us first write the Parafac model in its two-mode form $\mathbf{X} = \mathbf{B}\mathbf{F}'$, with $\mathbf{F} = (\mathbf{A} \otimes \mathbf{C})\mathbf{G}'$, with $\mathbf{G}$ the matricized form of the $S \times S \times S$ superdiagonal core array $\mathcal{G}$ with the g_{sss} on the superdiagonal. To plot the variables in $\mathbf{B}$, this matrix needs to be in principal coordinates; thus, $\mathbf{B}\mathbf{G}$ is plotted with $\mathbf{G}$ an $S \times S$ diagonal matrix with the g_{sss} on the diagonal. A transformation matrix $\mathbf{Q}$ is derived to find the

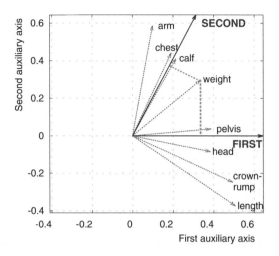

Figure 11.13   Girls' growth curves data: Variable mode from a two-dimensional Parafac analysis. The Parafac components are plotted in the two-dimensional space spanned by orthonormal "auxiliary axes".

auxiliary orthonormal axes for $\mathbf{F}$, such that $\mathbf{F}^* = \mathbf{FQ}$, and $\mathbf{B}^* = \mathbf{B}(\mathbf{Q}')^{-1}$ can be plotted. However, because the Parafac axes are unique, it is not enough to plot the variables with respect to the auxiliary axes, we also need to know the directions for the original Parafac axes in that space. As shown in Kiers (2000a, p. 159), the rows of $(\mathbf{Q}')^{-1}$ are the required axes. The coordinates of the variables on the Parafac axes are found by projecting them perpendicularly onto these axes as indicated in the plot (see Fig. 11.13).

11.5   PLOTTING DIFFERENT MODES TOGETHER

The basic aim of carrying out a three-mode analysis is generally to investigate the relationships between the elements of different modes. To assist in evaluating this aim, it is desirable to have plots across modes. Given the unique orientation of the components of the Parafac model in combination with the parallel proportional profiles property, it seems sensible not to make spatial plots of these components, but to portray the sth components of the three modes in a single plot, a *per-component plot* or *line plot*[6].

---

[6]In the statistical literature there are other definitions for the term "line plot", but in this book we will only use it in the sense defined here.

11.5.1   Per-component plots or line plots

To construct per-component plots, each term of the model $\mathbf{a}_s \otimes \mathbf{b}_s \otimes \mathbf{c}_s$ is separately portrayed by plotting the coefficients for all modes along a single line. The argument for using principal coordinates is not applicable here, as we are not concerned with spatial representations but only across-modes comparisons. Therefore, one should use either the orthonormalized components or, even better, unit mean-square scaling of components. The advantage of the latter scaling is that it compensates for the differences in number of levels in the components of the three modes; see Section 9.4, p. 218.

Per-component plots can also be useful in the Tucker3 model when it is possible to interpret all components of all modes (quite a task), and when it makes sense to interpret each and every one of the terms of the model, that is, all $\mathbf{a}_p \otimes \mathbf{b}_q \otimes \mathbf{c}_r$ that have sizeable core elements, g_{pqr}. Especially after core rotations that have achieved great simplicity in the core array coupled with (high) nonorthogonality of the components, per-component plots may be very useful. In this case, too, components scaled to unit mean-squares are indicated (for examples in the Tucker3 case see Kroonenberg & Van der Voort, 1987e).

Example per-component plot: Girls' growth curves data.   Crucial in the interpretation of the per-component plot in Fig. 11.14 is that the mean scores per variable have been removed at each time point. This means that the girl with a zero score on the girl components has average growth curves for all variables and the scores portrayed here are deviations from the average girl's growth curves. Thus, if the product of the three terms is positive for a girl, she is growing faster than the average girl; if it is negative, the girl is lagging behind in growth. As all variables and all years have positive coefficients, whether a girl grows faster or slower than average is entirely determined by the coefficients on the girl component. Differences in body length are the most dramatic; those in arm circumference are much less. Differences with the average girl peak around the 13th year and diminish after that. The reason for this is that some girls have their growth spurts earlier or later than the average girl. It would have been really nice if the measurements had continued for another three years to see whether the differences level off, as one would expect.

11.5.2   Two-mode biplots

A standard (two-mode) biplot is a low-dimensional graph in which the rows (say, subjects) and the columns (say, variables) are displayed in a single plot. It is constructed by performing a singular value decomposition on the data matrix: $\mathbf{X} = \mathbf{A}\Lambda\mathbf{B}'$, where Λ is the diagonal matrix with singular values, which are the square roots of the eigenvalues. The biplot technique often uses one of two asymmetric mappings of the markers. For example, in a *row-metric preserving* two-dimensional biplot (Gabriel & Odoroff, 1990), the row markers have *principal coordinates* $(\lambda_1 a_{i1}, \lambda_2 a_{i2})$ and the column markers have *normalized coordinates* (b_{j1}, b_{j2}) (Greenacre, 1993, Chapter 4).

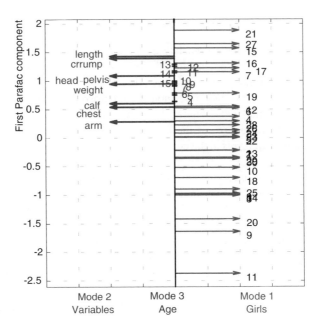

Figure 11.14   Girls' growth curves data: Per-component plot from a two-dimensional Parafac analysis with first component of each of the three modes;. Left: variables; Middle: Age; Right: Girls. Abbreviations: CrRump = Crown–coccyx length; Pelvis = maximum pelvic width. All other measures, except weight, are circumferences. The mean squared lengths of all components of all modes are 1.

In row-isometric biplots, the distances between the row markers are faithfully represented but those between the columns are not, with the reverse for column-isometric biplots (Gower, 1984). An alternative is to divide λ equally between the rows and the columns, that is, the rows are displayed as $\mathbf{A}^* = \mathbf{A}\mathbf{\Lambda}^{1/2}$ and the variables as $\mathbf{B}^* = \mathbf{B}\mathbf{\Lambda}^{1/2}$, where we assume that $\mathbf{A}^*$ and $\mathbf{B}^*$ consist only of the first few (mostly two) dimensions.

Commonly, the subjects are displayed as points and referred to a row markers. The variables are then displayed as arrows or vectors and referred to as column markers. In the plot, the relative order of the subjects on each of the variables can be derived from their projections on the variable vectors, and the angles between the variable vectors indicate to what extent these orders are similar for the variables (see Fig. 13.10 for an example with detailed interpretation). The extent to which a biplot is successful in representing these aspects of the data depends on the quality of the low-dimensional (mostly two-dimensional) approximation to the original data.

Gabriel (1971) introduced the word biplot and developed further interpretational procedures with his colleagues (e.g., Gabriel & Odoroff, 1990). The idea goes further back, at least to Tucker (1960). Based on Kroonenberg (1995c), a more detailed introduction to the biplot, its relation with the singular value decomposition, interpretational rules, and the effect of preprocessing on a biplot are presented in Appendix B.

11.5.3 Joint biplots

A *joint biplot* in three-mode analysis is like a standard biplot and all the interpretational principles of standard biplots can be used. What is special is that one constructs a biplot of the components of two modes (the *display modes*) given a component of the third or *reference mode*. Each joint biplot is constructed using a different slice of the core array. The slicing is done for each component of the reference mode. Each slice contains the strength of the links or weights for the components of the display modes. The coefficients in the associated component of the reference mode weight the entire joint biplot by their values, so that joint biplots are small for subjects with small values on the component and large for those with large coefficients.

The starting point for constructing a joint biplot after a Tucker3 analysis is the $I{\times}J$ matrix $\Delta = \mathbf{A}\mathbf{G}_r\mathbf{B}' = \mathbf{A}_r^*\mathbf{B}_r^{*\prime}$ or the $I{\times}J$ matrix $\Delta_k = \mathbf{A}\mathbf{H}_k\mathbf{B}' = \mathbf{A}_k^*\mathbf{B}_k^{*\prime}$ after a Tucker2 analysis. For each core slice, $\mathbf{G}_r$ (or $\mathbf{H}_k$), a joint biplot for the $I{\times}P$ component matrix $\mathbf{A}^*$ and the $J{\times}Q$ component matrix $\mathbf{B}^*$ needs to be constructed such that the P columns of $\mathbf{A}^*$ and the Q columns of $\mathbf{B}^*$ are as close to each other as possible; see Kroonenberg and De Leeuw (1977). Closeness is measured as the sum of all $P{\times}Q$ squared distances $\delta^2(\mathbf{a}_i^*, \mathbf{b}_j^*)$, for all i and j.

The construction of a joint biplot is as follows (see also Kroonenberg, 1994, pp. 83–84). The $P{\times}Q$ core slice $\mathbf{G}_r$ is decomposed via a singular value decomposition into

$$\mathbf{G}_r = \mathbf{U}_r\Lambda_r\mathbf{V}'_r,$$

and the orthonormal left singular vectors $\mathbf{U}_r$ and the orthonormal right singular vectors $\mathbf{V}_r$ are combined with $\mathbf{A}$ and $\mathbf{B}$, respectively, and the diagonal matrix Λ_r with the singular values is divided between them in such a way that

$$\mathbf{A}_r^* = (I/J)^{1/4}\,\mathbf{A}\mathbf{U}_r\Lambda_r^{1/2} \text{ and } \mathbf{B}_r^* = (J/I)^{1/4}\,\mathbf{B}\mathbf{V}_r\Lambda_r^{1/2}, \qquad (11.2)$$

where the fourth-root fractions take care of different numbers of levels in the two component matrices. The columns of the adjusted (asterisked) component matrices are referred to as the *joint biplot axes*. When the $\mathbf{G}_r$ ($\mathbf{H}_k$) are not square, their rank is equal to $M = \mathbf{min}(P, Q)$, and only M joint biplot axes can be displayed. The complete procedure comes down to rotating each of the component matrices by an orthonormal matrix, followed by a stretching (or shrinking) of the rotated components. The size of the stretching or shrinking of the axes is regulated by the square roots of λ_{mm}^r and the (inverse) fourth root of J/I. Note that even if there is a large difference in explained variability of the axes, that is, between $(\lambda_{mm}^r)^2$ and $(\lambda_{m'm'}^r)^2$, there can be a sizeable visual spread in the plot as the component coefficients are multiplied by the $(\lambda_{mm}^r)^{1/2}$ which makes their values much closer.

As an example with explained variabilities equal to 0.30 and 0.02 (ratio 15:1), the respective components are multiplied by their fourth roots or 0.74 and 0.38 (ratio 2:1).

As $\mathbf{A}_r^* \mathbf{B}_r^{*\prime} = \mathbf{\Delta}_r$, each element δ_{ij}^r is equal to the inner product $\mathbf{a}_i^* \mathbf{b}_j^{*\prime}$, and it provides the strength of the link between i and j in as far as it is contained in the rth core slice. By simultaneously displaying the two modes in one plot, visual inferences can be made about their relationships. The spacing and the order of the subjects' projections on a variable correspond to the sizes of their inner products, and thus to the relative importance of that variable to the subjects.

One of the advantages of the joint biplot is that the interpretation of the relationships of variables and subjects can be made directly, without involving component axes or their labels. Another feature of the joint biplots is that via the core slice $\mathbf{G}_r$ ($\mathbf{H}_k$) the coordinate axes of the joint biplots are scaled according to their relative importance, so that visually a correct impression of the spread of the components is created. However, in the symmetric scaling of the components as described here, the distances between the subjects are not approximations to their Euclidean distances (i.e., they are not *isometric*), nor are the angles between the variables (approximate) correlations. The joint biplot for the Tucker3 model is explicitly meant to investigate the subjects with respect to the variables, given a component of the third mode. For the Tucker2 model, the joint biplot provides information on the relationships between subjects and variables given a level of the third mode.

In practice, joint biplots have proved a powerful tool for disentangling complex relationships between levels of two modes, and many applications make use of them, as do several examples in this book.

Example: Girls' growth curves data — Joint biplots.

Tucker3 joint biplots.   To illustrate a joint biplot, we again use the girls' growth curves data. The time mode was chosen to be the reference mode and its components are shown in Fig. 11.10. The first time component shows a steady increase with a peak at 13 years of age. The last two years show a fairly steep decline. The primary interpretation of the first time component is that it indicates overall variability at each of the years. Thus, there was increasing differentiation between the girls up to year 13 and this difference decreased for the last two years. Figure 11.15 shows the joint biplot with girls and variables as display modes associated with the first time component. The joint biplot portrays the first two (varimax-rotated) axes of a three-dimensional plot belonging to a $3\times3\times2$ solution. By and large, the variables fall in two main groups, the length (or skeletal) variables (Length and crown–coccyx length) and the circumference (or soft tissue) variables (Arm, Calf, Chest). Pelvic circumference is associated with both the skeletal variables and the soft-tissue variables. The third axis (not shown) provides a contrast with the head circumference, and chest and arm.

Girls 19 and 27 have sizeable projections on all variables and are relatively tall and sturdy, while 9 and 11 are short and petite. In contrast, girl 4 has average length but

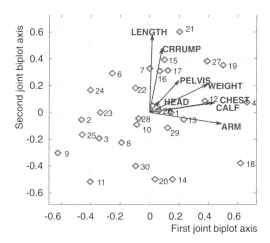

Figure 11.15   Girls' growth curves data: Joint plot from $3\times3\times2$-Tucker3 analysis for first component of age mode. Abbreviations: CrRump = Crown–coccyx length; Pelvis = maximum pelvic width. All other measures, except weight, are circumferences.

is rather hefty; girl 18 is shorter but has a similar amount of soft tissue. There are no really tall and slim girls (6 and 24 come closest), which seems reasonable since part of someone's weight is due to bones, but short girls can get heavy due to increase in soft tissue.

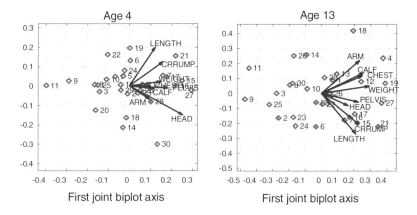

Figure 11.16   Girls' growth curves data: Joint plots from a 3×3-Tucker2 analysis at 4 years old and at 13 years old. Abbreviations: CrRump = Crown–coccyx length; Pelvis = maximum pelvic width. All other measures, except weight, are circumferences. Note that the scales for the two plots are different, and that the second axes are more or less mirrored.

Table 11.1   Girls' growth curves data: Tucker2 core slices at 4 and 13 years of age

Girl components	Variable components 4-years-old			Variable components 13-years-old		
	V1	V2	V3	V1	V2	V3
G1	**0.32**	−0.03	−0.04	**0.83**	0.02	0.03
G2	0.11	−0.12	−0.02	−0.06	**−0.40**	−0.00
G3	−0.10	−0.04	0.17	0.13	−0.01	**0.21**
Weight axes	0.20	0.04	0.01	1.08	0.24	0.06

The weight of the axes are those of the joint biplots for each age group.

Tucker2 joint biplots.   Figure 11.16 shows two joint biplots from a 3×3-Tucker2 analysis. The left-hand plot shows the relative positions of the girls on the variables at the start of the observational period. Young girls clearly show less variability or deviation from the average girl, as can be seen from the smaller scale of the coordinate axes. Moreover, two axes suffice for this age group, showing that there is less differentiation between the girls. Note that at this age girls 9 and 11 were already comparatively small, and 21 was already rather tall. At the age of 13, when the differences between the girls were at their largest, the variables fan out more, indicating that the girls are different from each other on different variables. The Tucker models consist of a single space for the girls and a single one for the variables for all time points but the core slices combine these common components in different ways over the years as can be seen in Table 11.1. At age 4, the relationship between girl and variable components is very diffuse, whereas at age 13 there are clear links between them, indicated by the sizeable core elements.

11.5.4   Nested-mode biplots

The alternative to joint biplots, which display two modes given a component of the third or reference mode, is to construct a matrix in which the rows consist of the fully crossed levels of two of the modes (also called *interactive coding*), and columns consist of the levels of the remaining mode. The coordinates for such plots can be constructed starting with the basic form for the Tucker3 model,

$$x_{ijk} \approx \sum_{p=1}^{P}\sum_{q=1}^{Q}\sum_{r=1}^{R} g_{pqr}\left(c_{kr}a_{ip}b_{jq}\right) = \sum_{r}\left(\sum_{p}\sum_{q} g_{pqr}c_{kr}a_{ip}\right)b_{jq}$$

$$\approx \sum_{r} f_{(ik)r}b_{jq} \tag{11.3}$$

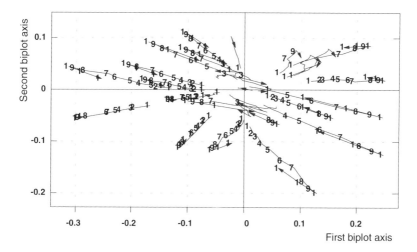

Figure 11.17   Girls' growth curves data: Nested-mode biplot for the normalized component space of the variables. Each trajectory represents a girl's development over time with respect to the variables with high values on the first variable component. Numbers are sequence numbers; 10, 11, 12 are indicated with a 1; real ages are sequence numbers + 3.

with $f_{(ik)r} = \sum_p \sum_q (g_{pqr} c_{kr} a_{ip})$ the interactively coded scores, so that $\mathbf{F}$ is a tall combination-mode matrix. A content-based decision has to be made whether the distances between the $f_{(ik)r}$ have to be displayed as (approximations to the) distances in high-dimensional space, which requires them to be in principal coordinates. In this case, the rows of $\mathbf{B}$ with normalized coordinates are displayed as axes (distance form). Alternatively, if we want the correlations between the variables in Mode B to be correctly displayed, they have to be in principal coordinates (covariance form). When plotting both of them in a single biplot, the second alternative would be the most attractive one, but it all depends on the purpose of the plot.

Because the interpretation of components is generally carried out via the variables, it often makes sense to have the subjects×conditions in the rows and the variables in the columns. Either the subjects are nested in the conditions, or vice versa, and because of this nesting, these plots are referred to as *nested-mode biplots*. When, in addition, one of the modes in the rows has a natural order, its coefficients can be fruitfully connected for each level of the other mode. For instance, conditions or time points can be nested within each subject. The resulting trajectories in variable space show how the values of a subject change over time or over the conditions. When the variables have been centered per subject–occasion combination, the origin of the plot represents the profile of the subject with average values on each variable on all occasions, and the patterns in the plot represent the deviations from the average subject's profile.

An interesting aspect of the subject–condition coefficients on the combination-mode components is that per variable component they are the inner products between the variables and the occasions and thus express the closeness of the elements from the two modes in the joint biplot. For further technical details, see Kiers (2000a) and Kroonenberg (1983c, pp. 164–166)

Example: Girls' growth curves data — Nested-mode biplots.   Figure 11.17 shows the trajectories of the girls in the normalized variable space. The general movement is outward from the origin, indicating that the girls in question deviate more and more from the average profile, but it is also clear that this deviation is decreasing by the age of 15. Given the lack of data beyond age 15, it is not clear whether most of them will regain their initial position with respect to the profile of the average girl.

In Fig. 11.18 the average position of the girls on their growth curves is indicated by their sequence number. In addition, the projection of the variable axes are included. They are plotted as the rows of the orthonormal matrix **B**. As all variables point to the right, it means that girls at the right-hand side of the origin, such as 24, increase their values with respect to the average girl while the girls on the left-hand side, such as 9 and 11, lag behind and until their 12th year lag further and further behind. Girls in the southeast corner, such as 18, gain especially more in body mass as is evident from their elevated increases in the soft-tissue variables, while girls like 21 increase especially in skeletal width.

11.5.5   Nested-mode per-component plots

Sometimes it is not very useful to display the interactively coded scores, $f_{(ik)r}$, for different components r and r' in one plot. Sometimes it is clearer to plot the scores of the subjects–condition scores for each of the components separately. Such plots are sometimes easier to use, explain, or present than the nested-mode biplots in which one has to inspect projections on vectors.

In case of a good approximation of the model to the data, the interactively coded component scores as described above will resemble the component scores from a standard principal component analysis on a data matrix in which the columns are variables, and the rows the subject–condition combinations. Other writers (e.g., Hohn, 1979) have also suggested using such component scores.

For the Tucker2 model the component scores may be derived by rewriting the basic form (Eq. (4.11), p. 53) as follows:

$$x_{ijk} = \sum_{p=1}^{P} a_{ip}d_{pjk} + e_{ijk} \text{ with } d_{pjk} = \sum_{q=1}^{Q} b_{jq}h_{pqk}. \tag{11.4}$$

Example: Girls' growth curves data — Nested-mode per-component plots.   The trajectories of the girls for the first variable component are depicted in a nested-mode per-component plot (Fig. 11.19). All variables have positive values on this

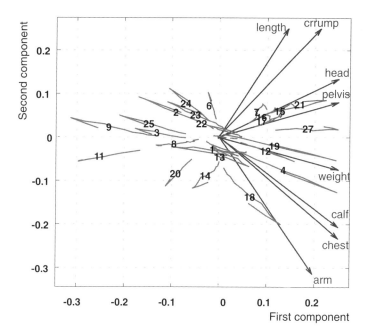

Figure 11.18   Girls' growth curves data: Nested-mode biplot for the normalized component space of the variables. Numbers indicate girls; if they are located close to the origin they have been deleted. The trajectories end at age 15. Arrows indicate the projections of the *variable axes* in the low-dimensional space. Angles do not correspond to correlations.

component, so that the trajectories show the girls' relative growth with respect to the growth of the average girl. A number of patterns can be discerned with respect to these curves: (1) all but one girl cross the horizontal line, indicating that smaller (larger) than average at 4 means smaller (larger) than average all the way through puberty; (2) only a handful of girls have growth curves parallel to that of the average girl, showing that they have the same growth rate; and (3) the measurements stopped a couple of years too early to establish where the girls ended up with respect to the average girl.

11.6   CONCLUSIONS

Given the complexity of multiway analysis, it is clear that interpretation becomes more challenging when "multi" means "greater than three". However, well-chosen plots can be of great help in getting an overview of the patterns in the data, and in making sense of them. As Kiers (2000a, p. 169) remarked, the plots used in multiway analysis "rely on two-way (PCA) models, obtained after rewriting the three-

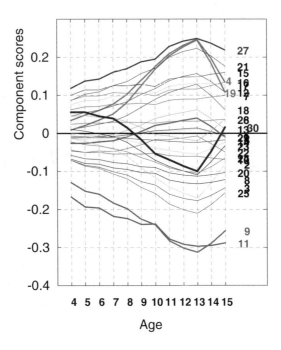

Figure 11.19 Girls' growth curves data: Nested-mode per-component plot for the first component of the variables. Scores on the first component of the girl–time combinations. Numbers at the right are the identification numbers of the girls, corresponding to those of the earlier plots.

way models at hand." Therefore, they do not capture all information and should be used in conjunction with the numerical results of the multiway analyses. Nevertheless, without graphing the results in one way or another, it seems that no multiway analysis can succeed in conveying what is present in the data without taking resort to one or more graphical displays.

CHAPTER 12

RESIDUALS, OUTLIERS, AND ROBUSTNESS

12.1 INTRODUCTION

In most disciplines, multiway component analysis is and has been primarily a vehicle for *summarization*, that is, describing a large body of data by means of a small(er) number of more basic components. More and more, however, it has also been used for *parameter estimation* in those cases where multiway models are based on *a priori* substantive, theoretical information. The Beer–Lambert law in spectroscopy is an example of this (see Bro, 1998c). Apart from these purposes, the method should also be useful for *exposure*, that is, detecting not only the anticipated, but also unanticipated characteristics of the data (see Gnanadesikan & Kettenring, 1972, p. 81). In fact, Gnanadesikan and Kettenring argue "from a data-analytic standpoint the goals of summarization and exposure ought to be viewed as two sides of the same coin.".

With respect to exposure, there are at least two important types of special information that need to be exposed. First, outliers need to be detected, in particular, unusual data that reveal themselves in the residuals and that may indicate special features of some data that the model cannot handle. Second, influential points also need to be

Applied Multiway Data Analysis. By Pieter M. Kroonenberg
Copyright © 2007 John Wiley & Sons, Inc.

detected because they may have a direct effect on the outcomes or estimation of the model itself. The latter may make the model itself invalid, or at least highly inappropriate, for the larger part of the data. In many cases, the distinction is not always clear-cut. For instance, influential points often are outliers as well, and outlying points may have no effect on the outcome.

Two approaches toward handling unusual observations may be distinguished. The first approach seeks to identify outliers and influential points with the aim of removing them from the data before analysis. In particular, unusual observations are investigated by residual plots and they are tested with measures for extremeness and for their influence on the estimates of the parameters of the model (*testing for discordance*). A second approach is to use *accommodation procedures* such as robust methods, which attempt to estimate the model parameters "correctly" in spite of the presence of unusual observations. There are several procedures for robustifying two-mode principal component analysis, but for the multiway versions there have been only few detailed investigations: Pravdova, Estienne, Walczak, and Massart (2001), Engelen et al. (2007a), and Engelen and Hubert (2007b).

12.2 CHAPTER PREVIEW

In this chapter we will deal with both "standard" analysis of residuals and with robust methods for multiway analysis. However, as the robustness research in the multiway area is of recent origin, the discussion will be more of a research program than a report about the practicality of carrying multiway analysis in a robust way. First, the general (theoretical) framework for the analysis of residuals will be discussed. Then we will discuss existing proposals for two-mode analysis and comment on their multiway generalizations. When discussing robustness, some general principles will be outlined as well as the existing proposals for two-mode PCA. Subsequently, some general points will be presented that influence the generalization to three-mode component models, and an attempt will be made to outline how one may proceed to make advances in the field. Several aspects will be illustrated by way of a moderately sized data set originating from the OECD about trends in the manufacturing industry of computer related products.

12.2.1 Example: Electronics industries

The Organization for Economic Co-operation and Development (OECD) publishes comparative statistics of the export size of various sectors of the electronics industry: information science, telecommunication products, radio and television equipment, components and parts, electromedical equipment, and scientific equipment. The specialization indices of these electronics industries are available for 23 countries for the years 1973–1979. The specialization index is defined as the proportion of the monetary value of an electronic industry compared to the total export value of manufactured goods of a country compared to the similar proportion for the world as a whole (see D'Ambra, 1985, p. 249). Profile preprocessing was applied to the data (see Section 6.6.1, p. 130, for details about this type of preprocessing). Given the age of the data, it is not our intention to give a detailed explanation of the data. They primarily serve as illustrative material.

12.3 GOALS

In analysis of interdependence, such as principal component analysis and multidimensional scaling, the fit of the model is generally inspected, but, in contrast with analyses of dependence such as regression analysis, residuals are examined only infrequently. Note that also in multiway analysis there may be a good fit of the model to the data but a distorted configuration in the space defined by the first principal components due to some isolated points. The examination of such isolated outlying points is of interest in practical work and may provide an interpretation of the nature of the heterogeneity in the data (see Rao, 1964, p. 334).

Residuals are generally examined in univariate and multivariate analysis of dependence, be it that in the latter case their analysis is inherently more complex. Not only are the models themselves complex, but there are many more ways in which individual data points can deviate from the model."Consequently, it is all the more essential to have informal, informative summarization and exposure procedures" (Gnanadesikan & Kettenring, 1972, p. 82).

An informal analysis of residuals of multiway analysis, however, is not without hazards. The specific structure of the data (i.e., the multiway design and the initial scaling) may introduce constraints on residuals or on subsets of residuals. Furthermore, the presence of outliers, in particular, outlier interactions among the modes, may affect more than one summary measure of the residuals and thereby distort conclusions drawn from them. In short, all the woes of the regular analysis also pertain to the analysis of residuals, which is performed to detect inadequacies in the former. An additional complexity is that multiway data are almost always preprocessed before the multiway analysis itself, so that due to the presence of outlying points the analysis may already be flawed before one has begun. A final point is that virtually all multiway component models are estimated by least-squares procedures, which are very sensitive to outlying and influential points.

There are three major goals for the *examination of residuals*, which are relevant for multiway analysis (see Fig. 12.1):

1. *Detection of outliers.* Detection of points that appear to deviate from the other members of the sample with respect to the chosen model.

2. *Detection of influential points.* Detection of points that deviate within the model and that for a large part determine its solution.

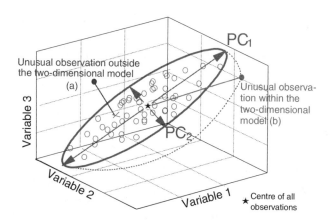

Figure 12.1   Three-dimensional variable space with a two-dimensional principal component space with an outlier (a) and an influential point (b).

3. *Detection of unmodeled systematic trends.* Detection of trends that are not (yet) fitted by the model, because not enough components have been included, or because they are not in accordance with the model itself.

Not all of these goals can be analyzed by only examining the residuals. The major goals for the *investigation of robustness* are:

1. *Accommodation.* Estimation of the parameters in spite of the presence of outliers and influential points.

2. *Identification.* Identifying outlying and influential points.

Whereas residuals are examined after the standard models and procedures have been applied to the data, robust procedures affect the model and analysis procedures themselves and are therefore much more "invasive" as well as mathematically more complex.

12.4   PROCEDURES FOR ANALYZING RESIDUALS

12.4.1   Principal component residuals

Given that S principal component are necessary to describe a particular data set adequately, the projections of the J-dimensional data on the last, that is, $(J - S)$ non-fitted principal components will be relevant for assessing the deviation of a subject i from the s-dimensional fitted subspace. Rao (1964, p. 334) suggested using the length of the perpendicular of a subject i onto the best-fitting space, d_i, in order to detect its lack of fit to the low-dimensional space defined by the S fitted components.

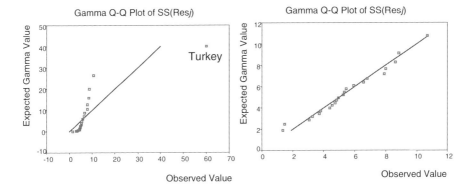

Figure 12.2 OECD data: Q-Q plot for residual sums of squares of countries. Left: Gamma distribution based on all SS(Residual); Right: Gamma distribution based all countries except Turkey. These and subsequent Q-Q plots were generated with SPSS (SPSS Inc., 2000).

This length, here referred to as the *Rao distance*, is equal to the square root of the residual sum of squares, or for three-way data $d_i = \sqrt{\sum_{j,k} e_{ijk}^2} = \sqrt{SS(Res)_i}$.
Because of the least-squares fitting function, we know that the total sum of squares may be partitioned into a fitted and a residual sum of squares,

$$\sum_{i=1}^{I}\sum_{j=1}^{J}\sum_{k=1}^{K}(x_{ijk} - \hat{x}_{ijk})^2 = \sum_{i=1}^{I}\sum_{j=1}^{J}\sum_{k=1}^{K}x_{ijk}^2 - \sum_{i=1}^{I}\sum_{j=1}^{J}\sum_{k=1}^{K}\hat{x}_{ijk}^2, \qquad (12.1)$$

or SS(Residual) = SS(Total) − SS(Fit). Similarly, as has been shown by Ten Berge et al. (1987), we may partition the total sum of squares of a subject i as SS(Total)$_i$ = SS(Fit)$_i$ + SS(Residual)$_i$, and similarly for the other modes. The Rao distance for assessing the lack of fit of each subject i may thus be written as

$$d_i = \sqrt{\sum_{j}\sum_{k}(x_{ijk} - \hat{x}_{ijk})^2} = \sqrt{SS(Total)_i - SS(Fit)_i}. \qquad (12.2)$$

Because the fitting of multiway component models is carried out via least-squares methods, procedures for assessing residuals from least-squares procedures such as regression analysis, should in principle also be valid for multiway models. However, in principal component analysis the predictors are latent variables, while in regression the predictors are measured quantities. The similarity of the two kinds of methods implies that proposals put forward for least-squares residuals should also be applicable for principal component residuals.

Gnanadesikan and Kettenring (1972, p. 98) supplemented Rao's proposal by suggesting a gamma probability plot for the squared Rao distances, d_i^2, in order to search for aberrant points. In Fig. 12.2 we have presented such a plot for the OECD data. In the left-hand plot, the gamma distribution is based on all countries; in the right-hand plot Turkey is excluded from the calculations. Turkey's anomalous data are evident.

Gnanadesikan and Kettenring (1972, p. 99) also suggested using various other plots involving the last few principal components, but we will not go into these proposals because such components are not available for almost all multiway methods.

12.4.2   Least-squares residuals

As noted previously, many proposals exist for investigating least-squares residuals for (multivariate) linear models with measured predictors. Here, we will only discuss those proposals that are relevant to linear models with latent predictors, such as principal component analysis.

There seem to be two basic ways of looking at the residuals, $e_{ijk} = x_{ijk} - \hat{x}_{ijk}$, with $\hat{x}_{ijk}$ the structural image of the data point x_{ijk}. The first is to treat them as an *unstructured sample* and employ techniques for investigating unstructured samples, such as found in Gnanadesikan (1977, p. 265).

1. Plotting the residuals against certain external variables, components, time (if relevant), or predicted (= fitted) values ($\hat{x}_{ijk}$). Such plots might highlight remaining systematic relationships [Goal (3.)]. They may also point to unusually small residuals relative to their fitted values indicating that such residuals are associated with overly dominant points, or combination of points [Goal (2.)].

2. Producing one-dimensional probability plots of the residuals, for example, full normal plots for the residuals, or χ_1^2 plots for squared residuals. According to Gnanadesikan (1977, p. 265), "[r]esiduals from least-squares procedures with measured predictors seem to tend to be 'supernormal' or at least more normally distributed than original data". The probability plots may be useful in detecting outliers [Goal (1.)] or other peculiarities of the data, such as heteroscedasticity.

The second overall approach to residuals is to take advantage of the structured situation from which the residuals arose, that is, from a design with specific meanings for the rows and columns, in other words, to treat them as a *structured sample*. Exploiting the specific design properties may introduce problems, exactly because of the design, and because the constraints put on the input data, which may influence the residuals.

Multiway residuals are more complex than two-way residuals but the unstructured approach remains essentially the same. It might, however, in certain cases be useful to consider several unstructured samples, for instance, one for each condition. This might be especially appropriate in the case of multiset data (see Section 3.7.2, p. 39), that is, data in which the measures for each k-mode element have been generated independently. For the structured approach it seems most useful to look at a multiway partitioning of the residuals into sums of squares for the elements of each mode separately (using (squared) Rao distances) (see Table 12.1).

For all proposals for carrying out a residual analysis it should be kept in mind that we are interested in rough, or first-order, results. Attempting to perfect such analyses by adding more subtlety carries the danger of attacking random variation. After all, we are only dealing with measures from which, in principle, the main sources of variation have already been removed.

12.5 DECISION SCHEMES FOR ANALYZING MULTIWAY RESIDUALS

In this section we will present two decision schemes for the analysis of residuals from a multiway component analysis. By following this procedure, a reasonable insight may be gained into the nature of the residuals, so that decisions can be made about the quality of the multiway solution obtained and about the need for further analysis. The use of the scheme will be further explained and illustrated in subsequent sections using the OECD data.

A distinction must be made between analyzing regular residuals and squared residuals. Table 12.1 presents the proposed scheme for the analysis of structured multiway residuals and Table 12.2 presents the proposed scheme for the analysis of unstructured multiway residuals. For the first step the use of squared residuals via the squared Rao distance is preferred, because fewer numbers have to be investigated, and use is made of the structure present in the residuals. Moreover, the resulting numbers can be interpreted as variation accounted for, and the sums of squares can be directly compared with the overall (average) fitted and residual sums of squares. Finally, any irregularity is enhanced by the squaring, and no cancelation due to opposite signs occurs during summing. For the second step the regular residuals are preferred, especially because at the individual level the signs are important to assess the direction of the deviation and to discover trends.

The problem of examining the individual residuals can, however, be considerable, because the number of points to be examined is exactly as large as before the multiway analysis: no reduction has taken place in the number of points to look at. What makes it easier is that supposedly the structure has already been removed by the multiway analysis.

12.6 STRUCTURED SQUARED RESIDUALS

Next to looking at the Rao distances themselves, the major tool for evaluating structured squared residuals is the *sums-of-squares plot* or *level-fit* plot for each mode. It has for each level f the $SS(Res)_f$ on the vertical axis and the $SS(Fit)_f$ on the horizontal axis.

Table 12.1   Investigation of squared residuals as a structured sample

---

Analysis

A. Investigate for each mode the *squared Rao distances* per element.   For example for the first mode of a three-way array this becomes,

- $d_i = \text{SS(Res)}_i = \sum_{j,k} e_{ijk}^2$ with $e_{ijk}^2 = (x_{ijk} - \hat{x}_{ijk})^2$.

B. Inspect the *distributions* of the

- SS(Total)$_f$ of the elements per mode to detect elements with very large SS(Total)$_f$ which might have had a large influence on the overall solution, and elements with a very small SS(Total)$_f$, which did not play a role in the solution;

- SS(Residual)$_f$ of the elements per mode to detect ill-fitting and well-fitting points;

- Relative SS(Residual)$_f$ to detect relative differences between residual sums of squares; that is, investigate Rel SS(Total)$_f$ = SS(Residual)$_f$/SS(Total)$_f$ ($f = i$, j, k, ...) by using histograms, stem-and-leaf displays, probability or quantile plots, and so on.

C. Use *sums-of-squares plots* to identify well-fitting and ill-fitting points.

Suggested action

1. IF no irregularities AND acceptable fit STOP, OR for surety GOTO Step 2.

2. IF no irregularities AND unacceptable fit GOTO examination of unstructured residuals AND/OR increase number of components AND redo the analysis.

3. IF one level f of each mode, thus $f = i', j', k'$, respectively, in the three-mode case, BOTH fits badly AND has a very large SS(Total)$_f$ for each $f = i', j', k'$, check for a clerical error at data point (i', j', k').

4. IF some elements of any mode fit badly GOTO examination of unstructured residuals.

5. IF one element f of a mode has a very large SS(Total)$_f$, AND a very small SS(Res)$_f$, mark this element as missing AND redo the analysis to assess the influence of this element on the overall solution, OR rescale input, especially equalize variation in that mode, AND redo the analysis.

Table 12.2   Investigation of residuals as an unstructured sample

---

Analysis

Investigate the unstructured residuals $e_{ijk} = x_{ijk} - \hat{x}_{ijk}$.

A. Examine the distribution of the residuals via a normal probability (or quantile) plot.

B. Examine plots of the residuals, e_{ijk}, versus

- fitted values, x_{ijk}, for trends, systematic patterns, or unusual points;

- data for remaining trends;

- external variables for identification of systematic patterns in the residuals.

Suggested action

- IF trend, or systematic patterns have been found THEN reexamine the appropriateness of the model AND STOP, OR describe these patterns separately AND STOP, OR increase the number of components AND redo the analysis.

- IF a few large residuals are present AND no systematic pattern is evident THEN check the appropriate data points, AND/OR STOP.

- IF no large residuals or trends are present STOP.

12.6.1 Sums-of-squares plots

To assess the quality of the fit of the levels of a mode, it is useful to look at their residual sums of squares in conjunction with their total sums of squares, in particular, to investigate per level f the *relative residual sums of squares*, Rel.SS(Residual)$_f$ = SS(Residual)$_f$/SS(Total)$_f$. If a level has a high relative residual, a limited amount of the data are fitted by the model. If the total sum of squares is small, it does not matter that a level is not fitted well. However, if the SS(Total)$_f$ is large, there is a large amount of variability that could not be modeled. In other words, either there is a larger random error, or there is systematic variability that is not commensurate with the model, or both. One could also present this information via the *fit/residual ratio* per level, that is, SS(Fit)$_f$/SS(Residual)$_f$, from which the relative performance of an element can be gauged. In particular, large values indicate that the level contains

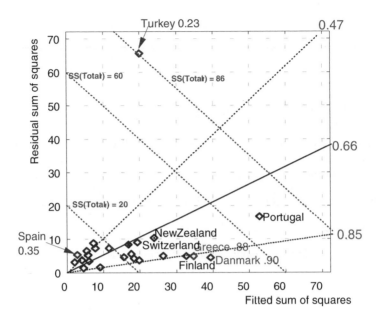

Figure 12.3   OECD data: Sums-of-squares plot of countries. Lines with constant fit/residual ratios start at the origin, and lines with constant SS(Total) under an angle of −45°. The solid line starting at the origin runs through the point with average relative fit (0.66). Specially marked are the worst-fitting countries, Turkey and Spain, and best-fitting countries, Denmark and Greece.

more fit than error information, and vice versa. The SS(Residual)$_f$ and the SS(Fit)$_f$ as well as their relationships can be shown directly in a sums-of-squares plot.

12.6.2   Illustration and explanation

The OECD data (Section 12.2.1) will be used to illustrate the sums-of-squares plots. Figure 12.3 shows the major aspects of the sums-of-squares plot. By plotting the sums of squares, rather than the relative sums of squares, the total sums of squares are also contained in the plot. In particular, one may draw lines of equal total sums of squares. Because the axes represent sums of squares, the total sums of squares are obtained by directly adding the x-value, SS(Fit), and the y-value, SS(Residual). The further the levels of a mode, here countries, are out toward the northeast corner of the plot, the higher their total sum of squares. Thus, countries with large amounts of variability can be immediately spotted from their location in the plot. In Fig. 12.3, Turkey is the country with the largest total sum of squares (85.6), and Germany (not marked) has the smallest (5.2). If a country has a very small sum of squares, this generally means that the country has average scores on all variables given that in most

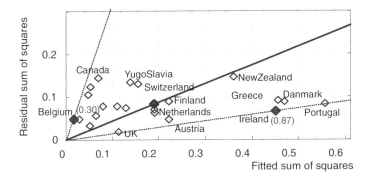

Figure 12.4   OECD data: Sums-of-squares plot of countries after deleting Turkey from the analysis.

profile data sets variables have been fiber-centered in some way (see Section 6.6.1, p. 130).

12.6.3   Influential levels

The plot also shows whether there is a relationship between the $SS(Fit)_f$ and the $SS(Total)_f$, which is common in least-squares procedures. Levels with large variabilities will in general be better fitted than those with small $SS(Total)_f$. If a level has such a comparatively large relative sum of squares, it will be located in the southeast corner.

For the electronics industries data, we see that this tendency shows up for the countries ($r = 0.76$). At the same time, we see that Turkey is quite different from the other countries. It couples a very high total sum of squares (85.6) with a comparatively low relative fit (0.23). In other words, a large part of its total sum of squares is not fitted by the model. Such phenomena require further investigation, as should be done for the Turkish data. Spain couples a low total sum of squares (8.2) with a relative fit of 0.35, indicating that it has generally average values on the variables, and that we are mostly looking at random fluctuations around that average. The best-fitting countries are Denmark (Rel. fit = 0.90) and Greece (Rel. fit = 0.88).

To establish the influence of Turkey on the analysis, the analysis was rerun after deleting the Turkish data, resulting in a much more balanced sums-of-squares plot (see Fig. 12.4).

Figure 12.5 provides a view on the relative fit of the industries in the analysis. What is evident from this plot is that, in particular, the components industries and to a lesser extent the electronics industries making scientific equipment are ill represented in the solution. Thus, only very partial statements can be made about the development of these industries over the years, as the model has only a fit of 0.17 and 0.29 to their data. In fact, all other industries have a fit of 0.75 or better.

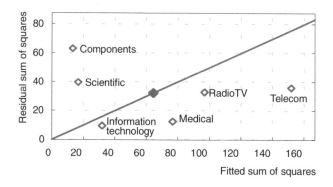

Figure 12.5   OECD data: Sums-of-squares plot of the electronics industries.

12.6.4   Distinguishing between levels with equal fit

Another interesting feature of sums-of-squares plots is that they show which levels have equal residual sums of squares but different total sums of squares. The plot shows which levels have large residual sums of squares but do not fit well, and which levels have a large SS(Residual) combined with a large total sum of squares. Without a residual analysis, it is uncertain whether a point in the middle of a configuration on the first principal components is an ill-fitting point or just a point with little overall variation. To assist in investigating this, lines of equal relative fit can be drawn in a sums-of-squares plot. Such lines all start at the origin and fan out over the plot. One of these line functions as the standard reference line because it runs from the origin through the point with coordinates (SS(Fit), SS(Res)), which is both the line of the average fit and that of the overall relative fit. In Fig. 12.3 the solid standard reference line runs from the origin through the point with average relative fit of 0.66. Countries above this line fit worse than average and countries below the line fit better than average. In the figure, lines have also been drawn with a relative fit of ±1 standard deviation (0.47 and 0.85).

12.6.5   Effect of normalization

When a mode has been normalized, that is, the variations (variances or total sums of squares) of the levels have been equalized, this is directly evident from the arrangement of the levels on a line at an angle of $-45°$ with the positive x-axis given an aspect ratio of one (see Fig. 12.6 for an example).

12.7   UNSTRUCTURED RESIDUALS

Besides analyzing whether certain levels of the modes do not fit well or fit too well, it is often necessary to find out the reason for this unusual fit, and to this end the individual

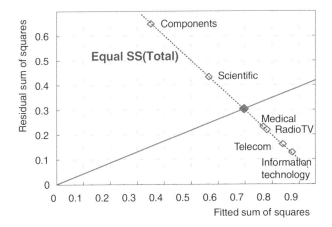

Figure 12.6   OECD Data: Sums-of-squares plot with normalized variances. The closed marker indicates the point with coordinates (SS(Fit), SS(Res)) and it has an average relative fit.

residuals should be examined. Taking our lead from regression diagnostics, a residual plot of the standardized residuals versus the (standardized) predicted values can be used to spot unusual observations, especially those outside the model (i.e., outliers). For unusual observations within the model (i.e., influential points), the residual plot may be less useful exactly because of the influence they have already exerted on the model. Because there are more observations than in two-mode analysis, the hope is that it will be more difficult for individual data points to have a large influence on the model.

In order to retain an overview, it can be helpful to examine the residuals per slice rather than all $I \times J \times K$ residuals in a single plot. This will allow both a better overview and an assessment of any level that was found to be unusual in terms of it residual sum of squares. The residual plot of the OECD data (Fig. 12.7) shows that the unusual position of Turkey occurs for more than one industry, and that a detailed examination of its entire data set is called for.

If Gnanadesikan's remark (1977, p. 265) that residuals tend to be "supernormal", or at least more normal than the data, is true, the supernormality should show in a histogram in which the standardized residuals are compared to the normal distribution. Figure 12.8 shows that there is evidence for supernormality in the residuals of this data set. Note the Turkish outlying observations, which also were evident in the residual plot (Fig. 12.7).

In summary the analysis of the residual sums of squares and the residuals themselves provides insight into the quality of a solution found for the data.

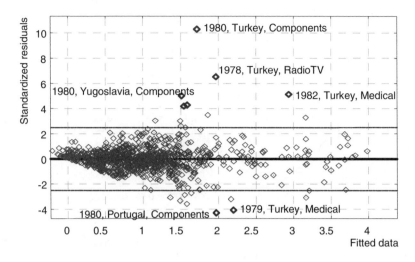

Figure 12.7 OECD Data: Plot of standardized residuals versus fitted values. The central line marks the standardized residual of 0; the upper and lower lines indicate standardized residuals of ±2.5.

12.8 ROBUSTNESS: BASICS

The development of robust methods for three-mode components is still in its infancy, but the research area seems to be gaining in importance. So far (i.e., mid-2007)

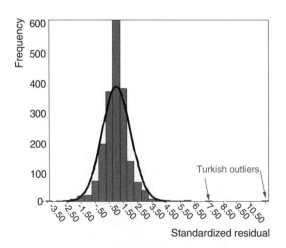

Figure 12.8 OECD data: Histogram of the standardized residuals.

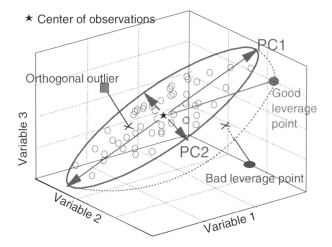

Figure 12.9   Four types of (un)usual points: regular points (open circles); good leverage points (closed circle); bad leverage points (closed ellipse); and orthogonal outliers (closed square).

only a few papers have appeared explicitly dealing with robustness in this context (Pravdova et al., 2001; Vorobyov, Rong, Sidiropoulos, & Gershman, 2005; Engelen et al., 2007a; Engelen & Hubert, 2007b). The multiway developments are inspired by the considerable efforts taking place toward robustifying two-mode principal component analysis, and considerable progress has been made in the developments of good algorithms (see Hubert, Rousseeuw, & Vanden Branden, 2005; De la Torre & Black, 2003, for recent contributions). An overview can be found in Rousseeuw, Debruyne, Engelen, and Hubert (2006).

12.8.1   Types of unusual points

On the basis of Hubert et al. (2005), one can be a bit more precise than in Section 12.3 about the different kinds of unusual points that may be encountered in a data set (see Fig. 12.9). Most of the points will be (1) *regular points* lying more or less within the component space. The outlying points can be (2) *orthogonal outliers* residing far outside the component space but projecting within the boundary of the regular points; (3) *good leverage points* far away from the regular points but more or less inside the component space; or (4) *bad leverage points* that lie far outside the component space and project onto the component space outside the boundary of the regular points.

Within multiway analysis, a further distinction must be made between (1) *outlying individual data*, for example, the value for the size of the ICT industry in Turkey in 1978; (2) *outlying fibers*, that is, levels of modes within other modes, for example,

Figure 12.10   Outliers in three-mode analysis: each of the four panels is a slice of the data array: First: no outliers; Second: individual outliers; Third: outlying fiber (top). Fourth: slice outliers: a slice with women's faces has already been deleted. Source: De la Torre & Black (2003), Fig. 1, p. 118; 2003 ©Springer and Kluwer Academic Publishers. Reproduced with kind permission from the author and Springer Science and Business Media.

the data of all industries of Turkey in 1978, or (3) *outlying slices*, that is, levels of a mode across the other two modes, for example, the data of Turkey as a whole.

Figure 12.10 contained in the De la Torre and Black (2003) study on learning problems in computer vision may serve as a further illustration, where each face constitutes an individual data point, and the four segments are slices of the $2 \times 6 \times 4$ data array. The various objects in the second slice such as the manekinekko (the signalling cat) are individual outliers (type 1), the row with spotted individuals (in the third slice) is a fiber outlier (type 2), and the fourth slice of female faces constitutes a slice outlier that has already been removed.

How to deal with these different types of outliers may be different for different circumstances. For instance, one may change a single outlier to a missing data point, but a complete outlying slice may better be completely deleted or temporarily removed from the analysis.

12.8.2   Modes and outliers

The basic assumption in two-mode robustness is that methods have to be robust with respect to outlying observations (subjects, objects, etc.) or data generators, but less is said about unruly variables. In principal component analysis, variables that do not have much in common with the other variables simply do not show up in the first components, and their influence on the solution can be reduced by equalizing variances, for instance.

In the standard setup, for multiway analysis one can have outlying subjects, outlying variables, outlying conditions, outlying time points, and so on. As indicated previously, if one restricts oneself to subjects there are two possibilities, corresponding to the two common data arrangements. To detect outlying slices, the data are viewed according to the I by $J \times K$ arrangement (a wide combination-mode matrix), that is, the subjects have scores on $J \times K$ variables. In some robust procedures the robust estimates for the parameters are based on a subset of slices so as to eliminate the influence of outlying slices on the parameter estimates. To detect outlying fibers we need an $I \times K$ by J arrangement (a tall combination-mode matrix) with the robust estimates based on a subset of the $I \times K$ fibers (i.e., rows). In the latter case an imbalance occurs in the sense that some subjects will have no observations in certain conditions. This can only be handled in a multiway analysis by designating these fibers as missing in order to make the data set a "complete" multiway array for analysis.

12.9   ROBUST METHODS OF MULTIWAY ANALYSIS

One may distinguish a number of approaches toward robustifying principal component analysis, which are candidates for extension to multiway analysis. Some of these methods will be briefly discussed in this section, such as prior cleaning of the data before a multiway analysis, robust preprocessing, multiway analysis on robust covariance matrices, projection pursuit methods, robust regression, and using penalty functions.

12.9.1   Selecting uncontaminated levels of one or more modes

Pravdova et al. (2001), Engelen et al. (2007a), and Engelen and Hubert (2007b) proposed to clean multiway data from contamination by outliers or inappropriate points before the multiway analysis proper; Pravdova et al. (2001) used the Tucker3 model, and Engelen et al. (2007a) and Engelen and Hubert (2007b) the Parafac model for their exposition. As their robust procedures are basically two-mode procedures (such as robust PCA), they are carried out per slice or matrix of the multiway array. These slices are cleaned of offending data points and are thus decontaminated. During the three-mode analysis proper, a standard alternating least-squares algorithm is used and the contaminated levels are treated as missing or are not used in the estimation of their mode. Via a residual analysis per mode, contaminated levels are removed before a final three-mode analysis. Even though not mentioned explicitly, these removed levels can be treated as supplementary levels and estimated again after the final analysis. Note that this approach is based on outlying fibers rather than individually deviating data points. From the description, the impression is that the procedure is rather involved and does not involve minimizing loss functions, which makes the results sometimes difficult to evaluate. In addition, the analysis is still performed with least-squares procedures, which are notoriously sensitive to outlying observation, so that a further robustifying could take place here, but it is not clear whether this is necessary.

12.9.2 Robust preprocessing

The next approach to robustification of a multiway analysis is to apply robust methods for preprocessing, that is, robust centering using the L_1-median estimator (i.e., the spatial median or median center) and robust normalization using the Q_n-estimator (for details, see Stanimirova, Walczak, Massart, & Simeonov, 2004, who also give the appropriate references). An alternative is to use the minimum covariance determinant estimator (MCD) (Rousseeuw, 1984; Rousseeuw & Van Driessen, 1999), which, however, can only be used in situations where the number of rows is larger than the product of the columns and tubes and their permutations; see Croux and Haesbrocck (2000).

After robust preprocessing, a standard multiway algorithms are applied. This, however, may not be the best solution, because the estimation is still done with least-squares loss functions. However, Engelen et al. (2007a) and Engelen and Hubert (2007b) have obtained promising results with the least-squares loss function after robustification of the input. A potential worry is that outlying observations become even more extreme after robust measures for location and scale have been used. These measures ignore the outlying points during calculations so that the variances are smaller than the raw variances, which means that the deviations from the center will tend to be larger than the original ones in both an absolute and a relative sense.

A complication is that in standard multiway preprocessing centering and normalization take place over different parts of the data; that is, centering is mostly per fiber and normalization is mostly per slice (see Section 6.9, p. 141). For instance, this is in contrast with the present all-in-one MCD procedure in which first a I by $J \times K$ wide combination-mode matrix is created and the minimum covariance determinant procedure is applied after that.

12.9.3 Robust estimation of multimode covariance matrices

A third approach is to first calculate a robust multimode covariance matrix using the most appropriate robust covariance procedure. Then, the resulting covariance matrix is analyzed with a robust procedure for PCA mentioned in the next section. Such a procedure for three-way data is based on the I by $J \times K$ wide combination-mode matrix. However, a quite likely situation in three-mode analysis is that $I < J \times K$, which will lead to problems in estimation procedures that need full-rank covariance matrices. Croux and Haesbroeck (2000) show that the MCD estimator and the S-estimators of location and shape are well suited for analyzing covariance matrices, provided that in the raw data $I > J \times K$.

Kiers and Krijnen (1991b) and Kiers et al. (1992b) present a version of the standard algorithm for multimode covariance matrices and show that its results are identical to those from the standard alternating least algorithms for both the Parafac and Tucker3 models. These procedures are attractive as they do not require full-rank covariance matrices. In other words, one could first robustify the covariance matrix via an appropriate robust estimation procedure and then apply the algorithms to the robustified covariance matrix.

The MCD procedure is one of complete accommodation as first the robust estimates are derived and individual observations no longer play a role. However, the robustification only seems possible if there are more subjects than variables×conditions. For instance, it is not possible to analyze the full OECD data in this way because there are 22 countries and 42 (6×7) "variables". In contrast, applying standard multiway methods to robust covariance matrices are not influenced by such restrictions.

The great advantage of procedures based on robust multimode covariance matrices, given the present state of affairs, is that they can be used to robustify multiway analysis without much additional statistical development. However, since there is little experience with the nonrobust, nonstochastic techniques for multimode covariance matrices, other than the two examples presented in Kroonenberg and Oort (2003a), it is difficult to gauge the usefulness of in connection with robust covariances. The same applies to the stochastic versions discussed by Bentler and Lee (1978b, 1979), Bentler et al. (1988), and Oort (1999).

12.9.4   Robust solutions via projection pursuit methods

A fourth approach is to extend proposals for two-mode PCA using robust methods based on projection pursuit methods, which originated with Li and Chen (1985) and were further developed by Croux and Ruiz-Gazen (1996) and Hubert, Rousseeuw, and Verboven (2002). The term *projection pursuit* refers to the search for that projection in the variable space on which the projections of the subjects have maximum spread. In regular PCA, directions successively are sought which maximize the variance in the variable space, provided the directions are pair-wise orthogonal. To robustify this, a robust version of the variance is used. A recent program to carry out the projection pursuit component analysis, RaPCA, was developed by the Antwerpen Group on Robust and Applied Statistics (Agoras) and is contained in their MATLAB robustness Toolbox LIBRA (Verboven & Hubert, 2005).

At present, the most sophisticated version of robust PCA based on projection pursuit was developed by Hubert et al. (2005), who combined a projection pursuit procedure with subspace reduction and robust covariance estimation. Their procedure carries the acronym RobPCA and is also included in the LIBRA robustness MATLAB toolbox. In chemical applications in which there are often more variables than subjects and many components are required, the projection pursuit approach seems to be very effective.

The question of how to generalize this approach to multiway data is far from straightforward. One option might be to replace each step in the standard algorithm for the multiway models by some projection pursuit algorithm, but acceleration procedures may be necessary to make this feasible.

12.9.5   Robust solutions via robust regression

A fifth approach to solving the estimation of multiway component models is to apply a robust procedure to each step of the alternating least-squares procedures. In particular, this seems feasible when one uses a regression step for the estimation as is standard in the Parafac model and was proposed by Weesie and Van Houwelingen (1983) for the Tucker3 model. In their paper, they already suggested replacing the least-squares regression by a robust variant. However, whether the convergence of such algorithms is assured, is not known. Vorobyov et al. (2005) developed two iterative algorithms for the least absolute error fitting of general multilinear models. The first was based on efficient interior point methods for linear programming, employed in an alternating fashion. The second was based on a weighted median filtering iteration. Croux et al. (2003) proposed to fit multiplicative two-mode models with robust alternating regression procedures using a weighted L^1 regression estimator, thereby extending the Gabriel and Zamir (1979) crisscross algorithm; see Section 7.3.1, p. 148. It will be interesting to see their work extended into the multiway realm.

12.9.6   Robust solutions using penalty functions

A fifth approach has been formulated within the area of computer vision by De la Torre and Black (2003). The approach is based on adding a penalty function to the basic two-mode PCA minimization function. Their procedure consists of simultaneously estimating location and scale parameters, and components. The standard least-squares discrepancy or loss function is replaced by a discrepancy function from a robust class of such functions. Full details can be found in their original publications, which include earlier references.

12.9.7   Short-term and quick-and-dirty solutions

Given the present state of knowledge about robustness in multiway component models, one might consider using primitive forms of robust PCA to be able to have some control over unusual points, fibers, or slices in three-way data. Some procedures come to mind. (1) Perform some robust procedure on the slices of a mode, identify the outlying points, deal with them in some coherent fashion, and then fall back on the standard estimation procedure. The papers of Pravdova et al. (2001) and Engelen et al. (2007a), and Engelen and Hubert (2007b) fall more or less in this category; (2)

Use robust procedures on the multimode covariance matrix. (3) Check in some way or another for outliers in the residuals, designate them as missing data points, and continue with the standard algorithms. (4) Matricize the multiway array and apply robust two-mode procedures.

12.10 EXAMPLES

In this section we take a brief look at robust information that can be acquired from two-mode PCA and that can be useful for multiway analysis. This (unsophisticated) excursion will provide some further indication of the possibilities of including robust procedures in multiway analysis. The discussion consists of three parts: (1) Per-slice analysis; (2) Subjects by Variables×Occasions (Wide combination-mode matrix); (3) Subjects×Occasions by Variables (Tall combination-mode matrix).

12.10.1 Per-slice analysis

In this subsection we will look at the kind of information one may acquire from looking at single slices. We will use just two years of the OECD data , the first and the last, in order to have some indication of the possibilities and difficulties of this kind of approach (see Section 12.2.1 for a description of the data).

Data for 1978 without Poland.   Poland had no valid scores in 1978, so it will not be included in the analysis. First, we tried to evaluate what the minimum covariance determinant estimator (MCD) – see Section 12.9.2 – had to offer and how sensitive it is to the number of (outlying) points that are excluded from the calculations of the means and covariances. The results are summarized in Fig. 12.11, it shows the sizes of the distances versus the number of points included in the estimation. When only two countries are excluded from the calculations, those two are considered outliers (Switzerland and Finland), but when eight countries are excluded from the calculation (as is suggested by the default settings of the FastMCD program[1]) these same eight countries were marked as serious outliers. The disconcerting aspect for this example is that it looks as if the robust methods created their own outliers. Clearly, with respect to multiway methods, this needs further investigation.

To get an impression of the nature of the extremeness of the excluded countries, we have also carried out a classical PCA of the 1978 data. Figure 12.12 shows that the countries excluded are at the rim of the configuration, but visually there would be no serious reason to discard them. In other words, the robust analysis had pointed us to possible anomalies, which should be investigated but which might not necessarily be as bad as they seem. Note that this is not necessarily a good procedure, as in the robustness literature it is argued that the classical methods are inadequate to detect outliers and thus robust methods cannot be evaluated by the classical approach.

[1]- Fortran version 1997*http://wis.kuleuven.be/stat/robust/*. Accessed May 2007.

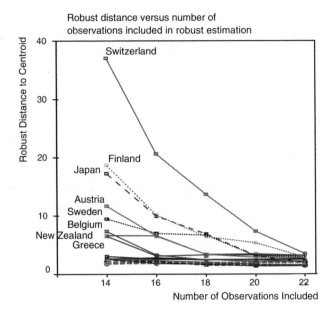

Figure 12.11   OECD data: Effect of number of observations selected for the MCD estimation on the identification of outliers. Total number of countries is 23.

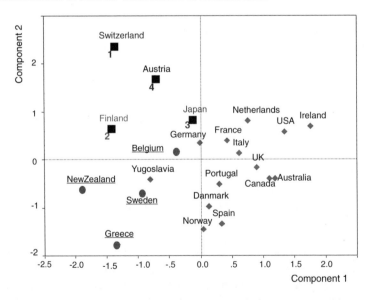

Figure 12.12   OECD data: Loadings from a classical PCA with the excluded countries shown as squares, numbered according to their diagnosed degree of outlying; the underlined countries with round markers are the ones with the next largest robust distances.

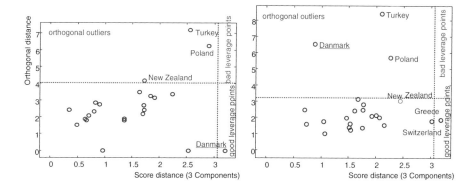

Figure 12.13   OECD data: Diagnostic distance plots for RaPCA (left) and RobPCA (right); three-component solutions in both cases. The lines indicate a 95% boundary based on the χ^2-distribution.

We have only shown the results of the robust estimation of the location and scale parameters. The procedures used later for the whole data set can also be applied to the separate slices, but for this quick overview these results have been omitted in order to avoid redundancies.

12.10.2   Countries by industries × years (Wide combination-mode matrix)

By arranging the complete data set as a wide combination-mode matrix of Countries by Industries × Years matrix, we are ignoring the fact that the industries are the same across years and vice versa. Therefore, both modes could possibly be modeled more parsimoniously by their own components.

The OECD data set was subject to both the pure projection pursuit method RaPCA (Hubert et al., 2002) and the combined projection pursuit and MCD estimation procedure, RobPCA (Hubert et al., 2005). The corresponding scree plots (not shown) are rather different in that RobPCA shows a clear knee at three components, but for RaPCA the decline is nearly linear for the first five components. It should be remarked that for RobPCA the solutions are not nested (eigenvalues: 7.9, 4.8, 1.6; three-component solution; eigenvalues: 7.4, 6.4, 1.9, 1.2; four-component solution), while they are for RaPCA: 7.9, 6.0, 3.7, 2.0. One of the most important diagnostics for the solutions is the diagnostic distance plot (Rousseeuw & Van Zomeren, 2000), which portrays the distance of the levels of a mode to the center in the component space (score distance) against the orthogonal distance to the component plane (see also Fig. 12.1).

Both techniques indicate that Turkey and Poland have large orthogonal distances and that they just fall short of being bad leverage cases (see Fig. 12.13). However, the results from both techniques differ widely with respect to Denmark. In RobPCA it has a large orthogonal distance, but in RaPCA it is a good leverage point. This difference can be traced back to the fact that the third component of the RaPCA solution is dominated by Denmark, drawing it into the component space, but it has low scores on all three components in the RobPCA solution, thus giving it a large distance from the component space. It is not our intention to go into this problem much deeper. It is solely presented to show that both techniques may provide a good feel for the nature of the (slice) outliers. On the basis of this information, further investigations and subsequent decisions can be made.

12.10.3   Countries × years by industries (Tall combination-mode matrix)

The arrangement as a tall combination-mode matrix with countries × years by variables can be used to assess whether certain country–year combinations are outlying relative to the other combinations. What it cannot do is assess whether certain country–year combinations are outlying within a country, that is, whether certain values for the industry as a whole are vastly different from one year to the next. It is clear that proper robust multiway methods should cater for such evaluations.

To give an impression of the kind of results one may obtain from the existing robust methods, both robust PCAs discussed earlier have been applied to the tall combination-mode matrix of countries × years by variables matrix. In particular, we would like to get some information on the dimensionality of the component space in this arrangement, which incidentally is the basis for calculating starting values for the standard TUCKALS algorithms so that the dimensionality of this arrangement is directly relevant to the multiway analysis at hand.

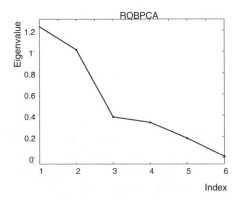

Figure 12.14   OECD data: Scree plot for both RaPCA and RobPCA.

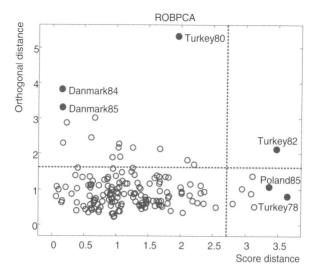

Figure 12.15   OECD data: Diagnostic distance plot for RobPCA on the countries×years by industries matrix.

From Fig. 12.14 it can be seen that one has to choose either two or four components because the third and fourth eigenvalues are almost equal. For the present demonstration, the two-component solution was selected. It can be shown that in the tall combination-mode matrix with more rows than columns, the two techniques produce the same solution if all components are derived. Moreover, the RaPCA solutions are nested, while the RobPCA ones are not (see Engelen, Hubert, & Vanden Branden, 2005, p. 119).

The next step is to carry out the two-dimensional solutions and identify the unusual fibers (country–year combinations). This can be done by a diagnostic distance plot, which is shown here for the RobPCA solution (Fig. 12.15). Again, Turkey supplies a number of deviating fibers (or rows in this case); the 1980 data produce an orthogonal outlier, the 1982 data a bad leverage point, and the 1978 data a good leverage point. An additional confirmation of the outlying position of the extreme points can be obtained by looking at the Q-Q plots for both the squared orthogonal distance and the squared score distance.

Figure 12.16 shows the squared orthogonal distance against the cumulative gamma distribution, both for all points and when the five most extreme points are deleted, while Fig. 12.17 shows the squared distances both for all points and when the seven most extreme points are deleted. The conclusion from these figures is that by deleting the most extreme points the remaining points more or less follow a gamma distribution. Thus, using arguments from Gnanadesikan and Kettenring (1972), we doubt whether

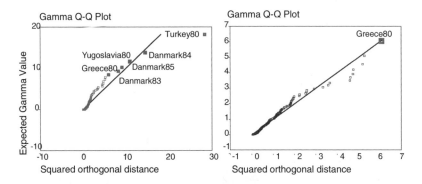

Figure 12.16 OECD data: Gamma Q-Q plot for squared orthogonal distances - two-dimensional RobPCA on the countries×years by industries matrix. All rows included (left), all but the five most extreme rows included.

these extreme points are in line with the remaining ones. A clear disadvantage is that with the Q-Q plots the orthogonal and score distances are not investigated jointly.

To finish off the discussion of the example, we present Fig. 12.18, showing the two-dimensional score space for the OECD data with RobPCA (based on 145 of 161 of the rows or fibers). In the figure the trajectory for Turkey is shown, and its erratic behavior is emphasized in this way. Most countries form more or less tight clusters, indicating that the electronic industries were relatively stable over the period under investigation. It would need a much more detailed investigation into the data and the availability of additional information to explain possible anomalies, and decide

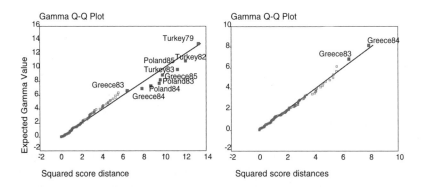

Figure 12.17 OECD data: Gamma Q-Q plot for squared score distances – two-dimensional RobPCA on the countries×years by industries matrix. All rows included (left); all rows except for the seven most extreme rows (right).

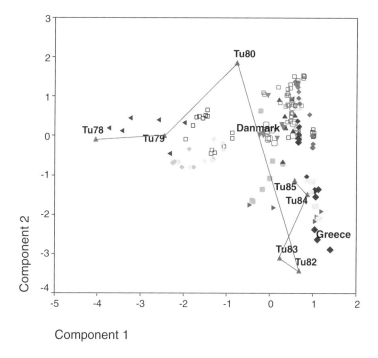

Figure 12.18   OECD data: Scores plot - two-dimensional RobPCA on the countries×years by industries matrix.

whether these are due to the opening of a new factory, inadequate recording, or simply to missing data.

12.11   CONCLUSIONS

A detailed analysis of the quality of solutions from multiway analyses can be performed by a straightforward but limited investigation of the residuals, as was done in the first part of this chapter. Some indication can be had of whether the deviations occur within or outside the model.

Robust methods that have been developed and are being developed for two-mode principal component analysis will become included in the three-mode framework, but much research is necessary to determine how to do this in an optimal manner. Several typical multiway problems arise which are not present in two-mode analysis. Furthermore, the handling of missing data in robustness is also under full development in two-mode analysis, which bears promise for multiway analysis. Because of the size of multiway data and their repeated measures characteristics, missing data are generally more a rule than an exception.

A difficult point at present is the question of what should be done about bad leverage fibers. Most multiway models start from a fully crossed design, and eliminating fibers can only be done by declaring them missing. In such cases, one might actually examine the fibers in more detail to find whether specific values in the fiber are out of line so that these individual points rather than whole fibers can be declared missing. Procedures for downweighting points in multiway analysis could also be developed, but little work has been done so far in this area.

PART III

MULTIWAY DATA AND THEIR ANALYSIS

In Part III the emphasis is on actually carrying out multiway component analyses using the procedures described in Part II. The properties of the multiway data at hand are crucial to how an analysis should proceed, so that the following chapters have been organized around specific data formats, such as profile data, rating scale data, longitudinal data, contingency tables, and binary data. In addition, multiway clustering is discussed, partly as a way to enhance multiway component analyses. The specific examples are all, except in the last chapter, based on three-way data. The majority of the data for the detailed analyses come from the social and behavioral sciences, but data sets from environmental studies and growth studies are analyzed as well. In addition, applications from (analytical) chemistry and agriculture are frequently mentioned.

CHAPTER 13

MODELING MULTIWAY PROFILE DATA

13.1 INTRODUCTION[1]

A common research goal in nonexperimental research is to improve the understanding of a conceptual domain by studying a number of variables measuring aspects of that domain. However, most conceptual domains, especially in the social and behavioral sciences, are intrinsically multivariate. For such investigations, subjects (or equivalent entities) are commonly considered random samples from some population, even though in reality they seldom are. Given a lack of randomness, during an analysis one should not only investigate the structures derived from the variable correlations but also pay attention to individual differences or group differences, because correlations between variables may be the result of group differences rather than being the "true" correlations among the variables derived from random samples. A balanced interest in subjects and variables is called for.

[1]This chapter is based on joint work with Richard Harshman, University of Western Ontario, Canada and Takashi Murakami, Chukyo University, Toyota, Japan.

When the data are more complex than individuals by variables, for instance, when the individuals are measured under several conditions and/or at more than one time point, multiway methods are required for their analysis. Common types are *multiway profile data*, which consist of subjects having scores on variables under several conditions and/or occasions, and *multiway rating scale data*, which consist of situations that are judged on a number of scales by several individuals, possibly at several occasions - see Chapter 14. More-than-three-way variants of both these data types occur although far less frequently — see Chapter 19. The structure of such data are analyzed by the generalizations of two-mode procedures as discussed in detail in Chapter 4. The present chapter illustrates the two major types of multiway component models, that is the Tucker and Parafac models, within a single framework, so as to provide insight into each model, their mutual relationships, and the data themselves.

Multiway profile data for numerical variables often have a multivariate repeated measures design. The major differences between the standard multivariate analysis and treatment with multiway models is that in the former the primary attention is often on mean performance levels rather than on individual differences. Furthermore, multiway methods generally lack an explicit stochastic framework. If one wants to perform tests especially of the statistical stability of the results, one can use the bootstrap procedures outlined in Sections 8.8 and 9.8.2. In addition, there are a limited number of stochastic multiway component or factor approaches for multiway profile data (Bentler & Lee, 1979; Oort, 1999).

Because multiway profile data often have different measurement scales and ranges, which means that their values are not necessarily comparable, some form of pre-processing is generally required (see Chapter 6). Given their nature, multiway profile data will nearly always be centered across subjects, and rarely across conditions. This is in contrast with multiway rating scale data (see Chapter 14), where often the data are not centered across the subjects but across concepts, situations, and so on. Depending on the comparability of the type of variables and their measurement levels, multiway profile data are usually normalized over all (centered) values of a variable. In other word, usually profile preprocessing is applied to multiway profile data.

The major purpose of analyzing multiway profile data by means of multiway models is to investigate the relationships between the three modes with proper attention to individual differences. When such differences are truly present and of intrinsic interest, multiway analysis is particularly useful. When individuals are purely seen as a random sample from one or more populations, generally the means and the co-variance matrices are sufficient, and one may restrict oneself to analyzing them rather than the raw data. The value of a multiway analysis of raw data is enhanced if additional information is available about the subjects (see Section 9.8.3, p. 234). If it is not, one can still fruitfully describe the differences between the subjects in terms of the multiway data themselves, but not in terms of other characteristics of the subjects.

13.2 CHAPTER PREVIEW

The focus of this chapter is on illustrating component analysis of multiway profile data using the two most common types of multiway models — Parafac models and Tucker models. The data on judgments of parental behavior toward their sons (Kojima, 1975) will be used throughout the chapter to illustrate these models and their analysis. We will first discuss the analysis of the data by a model, and then discuss various aspects of applying this model in general. Finally, we will briefly compare the two models and the results from the analyses, ending with a number of recommendations.

13.3 EXAMPLE: JUDGING PARENTS' BEHAVIOR

13.3.1 Stage 1: Objectives of the analysis

The perception of parental behavior by parents and their children was the central concern of the study from which the illustrative data in this chapter have been drawn (Kojima, 1975). Kojima wanted to validate the component (or factorial) structure of the questionnaires used to measure parental behavior. He argued that this could be done by comparing the reactions of parents and their children to parallel versions of the same instrument, more specifically he used two versions of the Child's Report of Parent Behavior Inventory (CRPBI) (Schaefer, 1965), one for the parents and one for the children.

Kojima performed separate component analyses for each of the judgment conditions and evaluated the similarities of the components using congruence coefficients. He also used Tucker's inter-battery method of factor analysis (Tucker, 1958). Using the Parafac model a single set of parallel proportional components for the scales will be derived, which are valid in all conditions simultaneously. The parallel proportional profile principle also applies to the boys whose behavior is assessed. As indicated earlier, the model does not provide a way to represent different correlations between the components across conditions or subjects, since constancy of correlations is a natural consequence of factors being proportional across related parts of the data.

The three-component Parafac solution presented in the next section should be seen as an example of the kind of answers that can be obtained with the model. We do not claim that this model solution is necessarily the best or most detailed Parafac analysis that can be obtained from the data. In later sections, we will discuss such issues in more detail.

13.3.2 Stage 2: Parental behavior data and design

The Japanese version of the CRPBI was used for this study. In order to have parents judge their own behavior, Kojima (1975) developed a strictly parallel version of this inventory for them (PR-PBI). The CRPBI is a three-point Likert-type questionnaire designed to assess children's perceptions with respect to parental acceptance, permitted psychological autonomy, and level of parental control. The scales were unidirectional and the respondents were asked whether each statement was "Not like", "Somewhat like", and "Like" the parental behavior. The original English versions are nearly identical in indicating behaviors of mothers and fathers, but the structure of the Japanese language made it possible to make a single version suitable for both parents. The substantive questions to be addressed in our analysis are (1) to what extent the structures of the questionnaire subscales are independent of who judges parental behavior, (2) whether individual differences between judges exist, and (3) how such differences, if they exist, can be modeled and presented.

The data are ratings expressing the judgments of parents with respect to their own behavior toward their sons, and the judgments of their sons with respect to their parents. Thus, there are four conditions: both parents assessing their own behavior with respect to their sons — Father-Own behavior (F-F), Mother-Own behavior (M-M); and the sons' judgment of their parents' behavior — Son-Father (B-F), Son-Mother (B-M). The judgments involved 150 middle-class Japanese eighth-grade boys on the 18 subscales of the inventory (see Table 13.1). Thus, the three-way profile data consist of a 150 (boys)×18 (scales)×4 (judgment combinations) data array.

13.3.3   Stage 3: Model and dimensionality selection

Before the actual Parafac analysis, profile preprocessing was applied to the data, that is, the condition means per scale were removed and the centered scales were normalized; see Section 6.6.1, p. 130. Given the unidirectionality of the scales, another possibility is to treat the data as three-way rating scale data, so that an additional centering across scales can be considered. Such a centering was not pursued here. Removing the boy-by-judge means leads for each child to removal of the mean differences between the judges in scale usage. This results in entirely different analyses from those in the original publication.

The first objective of a Parafac analysis is to establish how many reliable parallel proportional components can be sustained by the data. This issue is dealt with in detail in Section 13.5.3, where we investigate to what extent inappropriate or degenerate solutions exist, and to what extent the components cross-validate in split-half samples.

As shown in detail in Section 13.5.3, the fitted sums of squares for the one-, two-, and three-components unrestricted solutions were 18%, 31%, and 42%, respectively, while the reliability of the four-component solution was questionable. Therefore, we chose to present the three-component solution with a fit of 42%, which is probably reasonable for this type of three-way data. Moreover, it was advantageous to impose an orthogonality restriction on the subject components (see Section 13.5.3). This prevented the high correlation observed between the subjects components, which probably indicated a bounded degenerate solution, or at least a difficult solution to interpret (Harshman, 2004). As shown in Table 13.3 the fit for the restricted solution remained 42%. It should be pointed out that, as in two-mode analysis, it would be a mistake to look for a specific amount of fit. The purpose of component analysis is to find an adequate representation of the major structural patterns in the data. Whether these structures explain a large part of the variability is also dependent on aspects such as their reliability and the homogeneity of the population from which the sample was drawn. From an inspection of the residual sums of squares per subject, per scale, and per condition, it could be seen that there were no levels that fitted substantially better than other levels, and there were no large groups of levels that did not fit at all. This gave us sufficient confidence to proceed with this solution.

13.3.4   Stage 4: Results and their interpretation

As the Parafac model specifies that in each mode per component s the same source of variation underlies the variability, a component rather than a subspace interpretation is called for (see also Section 9.2.3, p. 211). To construct such a component inter-pretation for the Parafac results, three aspects are involved for each component: (1) the coefficients in each of the modes, that is, a_s, b_s, and c_s, (2) the way in which they combine to generate their contribution to the observed score $(a_s \times b_s \times c_s)$, that is, how they contribute to the structural image of the data (see also Section 2.3.1, p. 18), and (3) the weight of the component (i.e., g_{sss} — the larger the weight, the more important the component (see also Section 9.2.3, p. 211). Each component s corresponds to a *source of variation*, which contributes to the observed scores. The impact of a source across the levels of each mode is different for each of the mea-surement combinations. For example, if *parental love* is the source in question, then (1) scales measuring parental love should have high coefficients on the parental love component – their coefficients b_{js} are large; (2) children judging that they receive a lot of parental love should also have high coefficients on that component – their a_{is} are large, and conditions in which parental love is particularly evident should also have high coefficients on the component in question – the c_{ks} are large. Thus, the source parental love is represented by a component that is present in all modes. To deduce the nature of the source from the components one has to consider the sizes of the coefficients and their patterns on the components. The actual interpretation of a Parafac model for multiway profile data usually starts with the variables (scales in the present example), but the interpretation of the component or the source of the variation in the data should be in line with the observed variability in all modes.

Table 13.1 shows the coordinates of the scales both on the components scaled to unit mean-squared lengths — left-hand panel (as used in Figs. 13.1 – 13.3), and as the variable–component correlations (i.e., the structure matrix) – right-hand panel. The scales have been arranged to correspond roughly with their spatial arrangement in three-dimensional space. Several of the commonly found dimensions underlying the CRPBI here return as groups of scales, which need not necessarily align with the parallel proportional components themselves.

Table 13.1 Parental behavior data: Scale coordinates

Scale	CRPBI	Unit mean-square lengths			Component–scale correlations		
		1	2	3	1	2	3
Parental support or Acceptance - 1							
Acceptance	AC	1.28	0.92	1.29	**0.49**	0.34	**0.47**
Acceptance of individuation	AC	1.27	0.50	1.03	**0.49**	0.18	**0.38**
Child centeredness	AC	1.06	0.90	1.23	**0.41**	0.33	**0.45**
Positive involvement	AC	0.95	1.15	1.44	**0.37**	**0.42**	**0.53**
Psychological control - 2a							
Instilling persistent anxiety	PC	−0.69	1.44	1.15	−0.26	**0.53**	**0.42**
Hostile control	PC	−1.03	1.39	1.07	**−0.40**	**0.51**	**0.39**
Control through guilt	PC	−1.11	1.33	1.01	**−0.43**	**0.49**	**0.37**
Enforcement	FC	−0.95	1.29	0.98	**−0.36**	**0.47**	**0.36**
Behavioral control - 2b							
Intrusiveness	PC	−0.11	1.46	1.21	−0.04	**0.54**	**0.44**
Control	FC	−0.40	1.45	1.15	−0.15	**0.53**	**0.42**
Possessiveness	PC	−0.20	1.38	1.21	−0.08	**0.51**	**0.44**
Rejection - 3							
Inconsistent discipline	PC	−1.19	0.34	0.82	**−0.46**	0.13	0.30
Withdrawal of relations	PC	−1.37	0.51	0.76	**−0.53**	0.19	0.28
Hostile detachment	AC	−1.61	0.01	0.48	**−0.62**	0.00	0.18
Rejection	AC	−1.72	0.23	0.65	**−0.66**	0.09	0.24
Lax discipline - 4							
Nonenforcement	FC	−0.20	−0.69	0.34	−0.08	−0.25	0.12
Extreme autonomy	FC	0.01	−0.58	0.37	0.00	−0.21	0.14
Lax discipline	FC	−0.15	−0.12	0.94	−0.06	−0.05	0.34
Standardized weights (g_{sss})		0.148	0.135	0.134			

Unit mean-squares entries are regression coefficients. *Scale groups*: 1 = Parental support or Acceptance; 2a = Psychological control; 2b = Behavioral control; 3 = Rejection; 4 = Lax discipline.
Official CRBPI scales: PC = Psychological control; FC = Firm control; AC = Acceptance.

Figures 13.1, 13.2, and 13.3 show the per-component plots for components 1 to 3, respectively. As Kojima's interest centered around the similarity in structure of the scale space, we take this as our starting point for the interpretation.

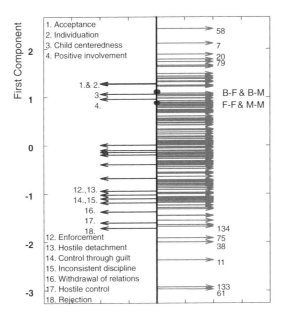

Figure 13.1 Parental behavior data: First components of the Son mode, the Scale mode, and the Condition mode. The arrows on the left-hand side represent the scales, the arrows on the right-hand side the sons, and the four marks on the vertical line indicate the conditions; that is, B-F: Boy judges his father's behavior toward him; B-M: same with respect to his mother; F-F: the father judges his own behavior toward his son; M-M: same for the mother. The standardized weight for this component is 0.15 or 15%. The numbers for scales in this plot only serve for identification, and do not relate to the order of the scales in Table 13.1.

Acceptance–Rejection. The first scale component representing the Acceptance–Rejection dimension occurs frequently in studies using the CRPBI, even though in Kojima's data Psychological control is weakly associated with Rejection. All judging conditions had comparable positive coefficients with marginally higher values for the boys' judgments. In other words, there was no judge×scale interaction on the first component: all judges agreed about whether there was an accepting or rejecting regime in a family. There were considerable differences between the regimes the boys experienced. For instance, in the case of boys 58, 7, 20, and 79, parents and son agreed that both parents' behavior could be characterized more by Acceptance than by Rejection. The reverse can be said of the behavioral regime for boys 61, 133, 11, and 38. Recall that the final effect of the component on the structural image is determined by the product of all the coefficients on the first component, $a_{i1}b_{j1}c_{k1}$, and that a positive product increases the score of the (i, j, k)-score, $\hat{x}_{ijk}$, and a negative product decreases it (see also Section 9.5.3, p. 230).

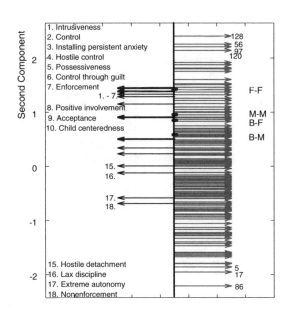

Figure 13.2 Parental behavior data: Second components of the Son mode, the Scale mode, and the Condition mode. The arrows on the left-hand side refer to the scales, the arrows on the right-hand side to the sons, and the four black marks on the vertical line to the conditions; that is, B-F: Boy judges his father's behavior toward him; B-M: same with respect to his mother; F-F: the father judges his own behavior toward his son; M-M: same for the mother. The standardized weight for this component is 0.14 or 14%. The numbers for scales in this plot only serve for identification, and do not relate to the order of the scales in Table 13.1.

Control. This component is characterized by the controlling influence of the parents, with Acceptance siding slightly with Control except for Acceptance of individuation. Nonenforcement and extreme autonomy are only very faintly seen as the opposite of control. Rejection is unrelated to the amount of control that is being exercised. Parents judged themselves more controlling than their sons did, but both parents and boys considered the father more controlling than the mother. Again the regimes were very different in the families, but there was no qualitative disagreement about the kind of control in families; only about its quantity. The parents of boys 123, 56, 97, and 120 were seen as very controlling compared to the average, while the reverse was true for the parents of boys 86, 17, and 5.

Response style. On the third component all scales have comparable positive coefficients, parents' judgments hover around zero and those of the boys are negative. For the interpretation it is easier to reverse both the signs of the boys' coefficients and those of the judgment conditions, so that boys 32, 121, ... have positive coefficients and boys 92, 144, ... negative ones. This leaves the overall sign of the product term,

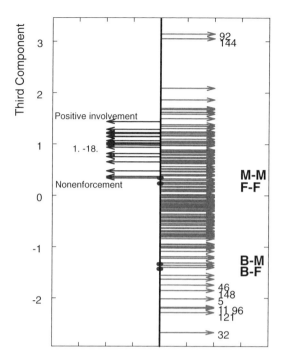

Figure 13.3   Parental behavior data: Third components of the Son mode, the Scale mode, and the Condition mode. The solid arrows on the left-hand side refer to the scales, the dashed arrows on the right to the sons, and the four black marks on the vertical line to the conditions; that is, B-F: Boy judges his father's behavior toward him; B-M: same with respect to his mother; F-F: the father judges his own behavior toward his son; M-M: same for the mother. The standardized weight for this component is 0.14 or 14%.

$a_{i3}b_{j3}c_{k3}$, unchanged[2]. A likely interpretation of this component is that it represents a response style. Some boys, notably 32 and 121, tended to judge higher on all scales, while others such as 92 and 144 tended to judge systematically lower on all scales. As mentioned before, this component carried no weight in the judgments of the parents as their coefficients are virtually zero on this component.

13.3.5   Summary of the analysis

There is major agreement between parents and their sons about the parents' behavior, as is evident from their coefficients always being on the same side of zero. The differences have to do with differences in judgment about the amount of control

---

[2]Such sign reversals do not change the results of the analysis as long as it is done consistently. The sign of the complete term $a_{is}b_{js}c_{ks}$ must remain the same.

exerted by the parents. The boys did not feel to the same extent controlled as the parents considered themselves controlling. Fathers and boys agreed that fathers were somewhat more controlling than mothers. Finally, the boys seemed to have more of a response style than parents did. This could indicate that parents were more discriminating in their use of the scales.

An interesting question arises from the analysis using the Parafac model. If we follow Cattell and Harshman that components derived by means of Parafac have more real meaning than (rotated) maximum-variance components, one may wonder about the meanings of the Parafac components in the present study. It is clear that the variable–component correlations (see Table 13.1) do not follow the "official" grouping of the scales observed by the originators of the instrument. If these patterns can also be found in other studies, some rethinking about what exactly the scales measure seems to be in order. An entirely different view of the study could be that the differences observed are of a cross-cultural nature; there are differences between Japanese and American parents and their sons.

13.4 MULTIWAY PROFILE DATA: GENERAL ISSUES

This section deals with practical issues that need to be addressed when analyzing multiway profile data. First, general issues will be discussed independent of the model in question; after that specifics of each model will be taken up.

13.4.1 Stage 1: Objectives

The major aim of applying multiway analysis to multiway profile data is to unravel complex patterns of dependencies between the observations, variables and conditions. We have seen that in the Parafac model the components are weighted by the g_{sss}, while in the Tucker models all combinations of the components (or elements of the core array), that is, g_{pqr} for the Tucker3 model and h_{pqk} for the Tucker2 model, are present. This highlights the essential difference between the models: in the Parafac model there are only S weights for the component combinations, while in the Tucker models there are $P \times Q \times R$ or $P \times Q \times K$ component combinations, respectively. The triple-indexed core arrays recreate the complexity of the original data, but on a smaller scale, because there are fewer levels in the components than in the original modes. In addition, the components themselves are not observed, but they are latent quantities, and therefore are more difficult to interpret. *A fortiori* the core containing combinations of these latent quantities, is even more difficult to interpret.

The Tucker and Parafac classes of models also differ in that a common interpretation across modes is sought for the components in the Parafac model, while separate interpretations have to be found for the components within a mode in the Tucker models (component interpretation). It is only after this has been done that the links between combinations of components is addressed. Another important difference in interpretation is that, as the Tucker models are projection models, it is often not the maximum variance components that are interpreted but the relationships between the levels of a mode (subspace interpretation). In such cases, the components are only seen as the *supports* for the lower-dimensional spaces rather than entities that carry intrinsic meaning. More details can be found in Chapter 9.

13.4.2   Stage 2: Data description and design

Types of data and their requirements.   The multiway profile models discussed here require data that can be conceived of as interval data, and this is the only type of data that will be considered in this chapter. There exists a procedure for the Parafac model called ALSCOMP3, developed by Sands and Young (1980), to handle other types of measurement levels via optimal scaling, but no further developments have been made since its inception, and no working version of their program seems to be available.

Preprocessing.   As mentioned earlier, profile preprocessing is commonly applied to multiway profile data, that is, the preprocessed data, z_{ijk}, are centered per variable–condition combination (jk) in the three-way case, and are normalized per variable slice j (see Section 6.6.1, p. 130).

Fiber centering is recommended because then the scores on each variable at each occasion are deviation scores with respect to the average subject's profile, so that all preprocessed scores z_{ijk} are in deviation of the variable means. The suggested centering has the additional advantage that if there is a real meaningful, but unknown, zero point of the scale, it is automatically removed from the scores; for a detailed discussion of this point see Section 6.4.1 (p. 117). A further effect of this centering is that the component coefficients of the subjects are in deviation of their means. The slice normalizations over all values of each of the variables mean that the normalized deviation scores carry the same meaning for all occasions. Thus, a standard deviation has the same size in the original scale for each occasion. The effect of the profile preprocessing is that the structure matrix of variable–component correlations can be computed (Table 13.1). The recommended preprocessing is the only way to achieve this interpretation.

13.4.3   Stage 3: Model and dimensionality selection

As model selection is specific to each class of multiway models, we will defer discussion to the sections on the models themselves.

13.4.4   Stage 4: Results and their interpretation

There are essentially four different ways in which the basic results of a multiway analysis can be presented: (1) *tables* of the coefficients or loadings for each mode – rotated or not (see Table 13.1); (2) *paired-components plots*: separate pair-wise graphs of the components per mode (see Fig. 13.6); (3) *all-components plots* showing all components of a mode in a single plot with the sequence numbers (or time points) on the horizontal axis (see Fig. 11.10, p. 268); and (4) *per-component plots*: plots with one component from each mode in a single plot, as was done for the Parafac model in the earlier example (see Figs. 13.1, 13.2, and 13.3). This last type of plot conforms most closely to the spirit of the Parafac model, because it allows for a component interpretation, by way of a simultaneous inspection of the coefficients of the three modes per component. The paired-components plots make more sense in the subspace interpretation when the spatial characteristics of the solution are to be examined. This is more in line with the characteristics of the Tucker models. Further details will be discussed in Sections 13.5.4 and 13.6.4.

13.4.5   Stage 5: Validation

In validating multiway results, one may look toward the analysis itself (internal vali-dation) and investigate the statistical stability of the solution via bootstrap analyses, which will not be attempted in this chapter (see, however, Section 8.8.1, p. 188, and Section 9.8.2, p. 233). One may also look at the residuals to see which parts of the data have not been fit well and to what extent there is systematic information left. Another approach to validation is to look at information external to the data set to see whether they can help to shed further light on the patterns found (external validation). Unfortunately, no external information for the Kojima data was available, so that the validation procedures described in Section 9.8 (p. 233) cannot be demonstrated with this data set.

13.5   MULTIWAY PROFILE DATA: PARAFAC IN PRACTICE

The primary aim of the analyses in this section is to discuss the practical issues in connection with a Parafac analysis on multiway profile data. This will be illustrated for the three-way case. For the exposition we will lean heavily on the *magnum opus* of Parafac's godfather, Richard Harshman (Harshman & DeSarbo, 1984b; Harshman & Lundy, 1984c, 1984d), who produced a complete catalogue described in Harshman (1984a) under the title "How can I know if it's 'real'?".

13.5.1   Stage 1: Objectives

The aim of most analyses carried out with the Parafac model is to uncover the exis-tence of components that show parallel proportional profiles, and if possible to identify these components as real ones that carry true substantive meaning. In many analytical chemistry applications this is not a problem, as the components often correspond to physical or chemical properties of the substances under investigation. Harshman, Bro, and co-workers have shown several analyses in which the uncovered components could be given such a status (e.g., Harshman, 1994a; Harshman, Ladefoged, & Goldstein, 1977; Bro, 1998c; Smilde et al., 2004). However, in the example of the previous section, the correspondence between the components and the theory of parental behavior is far more difficult to establish. On the other hand, the fact that stable three-component solutions could be found should give valuable information to the substantive researchers. Thus, given that stable Parafac components have emerged, it should spur on researchers to make sense of these parallel proportional profiles.

13.5.2   Stage 2: Data description and design

Because the Parafac model is based on parallel proportional profiles and is therefore sensitive to violation of this principle in the data, it is important that attention be paid to the possibility that such profiles might not be present in the data. In two-component Parafac analyses, the solution is almost always identical to a $2\times2\times2$-Tucker3 solution. In such cases, there are nonsingular transformations that rotate the Tucker3 core array to superdiagonality. Often, the only way to check the existence of parallel proportional profiles is via an analysis with the Parafac model itself.

13.5.3   Stage 3: Model and dimensionality selection

The choice of the "best" or most appropriate solution of the Parafac model is not an easy one and the procedure to arrive at such a solution requires considerable attention to detail. Because the model is a restrictive one, it may not fit due to the lack of parallel profiles in the data, so that several analyses for each plausible number of components are necessary to find an appropriate solution.

Uniqueness.   Harshman (1970) showed that a necessary condition for the uniqueness of a solution is that no component matrix has proportional columns. One may also distinguish between weak and strong uniqueness (Krijnen, 1993, Chapter 2). A strong unique solution is one for which there are no other solutions that fit almost as well. If there are, the solution is only weakly unique. Krijnen suggested to check for weak uniqueness by comparing the fit of a regular solution of the Parafac with a *parallel solution*, that is a Parafac solution with two proportional columns in any of the component matrices. If the difference is small, the solution is considered weakly unique (see Section 5.6.4, p. 88).

For the Kojima data, Table 13.2 shows that equating the last two components, which have a congruence of -0.87 in the standard solution, leads to a loss of fit of 1.3%, while keeping the subject components orthogonal leads to a negligible loss of 0.3%. All solutions have virtually identical second and third components. However, the first component of the orthonormal solution is not quite the same as that of the standard solution with a congruence of 0.64, but it is identical to that of the parallel solution. Moreover, the first components of the orthogonal and parallel solutions are less similar to their second and third components than is the case in the standard solution. From

Table 13.2 Congruence coefficients for the scales of the three three-component Parafac solutions (standard, with two parallel scale components, and with orthonormal subject components)

Analysis		PF3			Parallel			Orthonormal		
		1	2	3	1	2	3	1	2	3
Unrestr.	1	1.00								
Parafac	2	−0.82	1.00							
3 comp.	3	0.64	−0.87	1.00						
Two	1	**0.85**	−0.40	0.19	1.00					
parallel	2	0.73	**−0.95**	**0.97**	0.28	1.00				
comp.	3	0.73	**−0.95**	**0.97**	0.28	**1.00**†	1.00			
Ortho.	1	0.83	−0.37	0.17	**0.99**	0.26	0.26	1.00		
subject	2	−0.70	**0.98**	**−0.87**	−0.23	**−0.94**	**−0.94**	−0.20	1.00	
comp.	3	0.64	**−0.86**	**0.99**	0.19	**0.97**	**0.97**	0.17	**−0.86**	1.00

† Congruence is equal to 1.00 because components have parallel profiles by design. Fit three-component Parafac = 0.421; Fit Parafac with two parallel scale components = 0.408; Fit Parafac with orthogonal subject components = 0.418.

this information, one may conclude that the case for a strong uniqueness for the Kojima data is not very strong, and that one may opt for interpreting one of the restricted solutions. In fact, combining the two restrictions, thus having an orthonormality restriction on the components of the subjects and a parallel restriction on the scales, does not lead to an additional loss compared to the parallel solution, so that imposing both restrictions at the same time might be even more advantageous with respect to interpretation. In the analysis in Section 13.3 we have only included the orthogonality restriction on the subject components.

Multiple solutions. Given that the Parafac model for fallible data only supports parallel proportional profiles and is not a complete decomposition, one has to run analyses with different numbers of components to determine how many parallel proportional profiles the data contain. In other words, we are really concerned with model fitting and not so much with data approximation. This also means that both using too few and using too many components can lead to unsatisfactory solutions (see also Table 13.3). In general, it turns out that specifying too many components is more problematic than specifying too few. However, when two components are close together in terms of their contributions to the fit in the case of a too small model, different analyses may pick up different components. Only when the number of components is increased can stable solutions be found that represent both these components. Whether this occurs is very much data dependent; such a situation was, for instance, encountered during the analyses in Section 8.9.3 (p. 196) (see also Murakami & Kroonenberg, 2003).

Split-half procedures.   One solution to get insight on the stability of the components is to split the data in half and perform a separate analysis on both parts. If there is a true underlying solution, it should show up in both analyses. Important restrictions to this procedure are that there must be a random mode that can be split. In multiway profile data the subject mode can generally be used for the splitting. It is necessary that there are sufficient "subjects" in order to minimize the influence of the idiosyncrasies of specific individuals. How much is enough is difficult to say in general, because much depends on the noise level of the data and the clarity of the underlying structure (see also Section 8.6.1, p. 184). For the Kojima data we will follow Harshman and DeSarbo's (1984) suggestion to make two (or more) orthogonal splits. This entails randomly splitting up the data into four more or less equal parts, say A, B, C, and D, after which four new data sets can be created, to wit A+B and C+D, and A+C and B+D.

Degeneracy.   In Section 5.6.5 (p. 89) the problem of nonconvergent solutions was discussed in which two components tend to become perfectly negatively correlated and the g_{sss} are increasing without bound. Such solutions from a Parafac analysis are called *degenerate*, and we will encounter such solutions during the model selection below.

The basic problem causing the degeneracy is that algorithms to compute the model parameters cannot cope with data that do not conform to the Parafac model and therefore produce degenerate, uninterpretable solutions. It should be noted, however, that in some cases, depending on the starting values, an algorithm may or may not end up in a degenerate solution. Walking across the terrain of a former mining operation, one does not always fall into a mine shaft. To avoid such pitfalls, several different starting values should always be used when computing Parafac solutions. Because it is not always obvious whether one has a degenerate or only a slowly converging solution, seemingly aberrant solutions should always be examined in detail and should be tested further.

Investigating convergence.   The first objective in getting acceptable solutions is to have convergent solutions. Table 13.3 shows the results of the analyses for both the complete data set and four orthogonal splits, arranged by number of components. From this table it is evident that two components can be reliably obtained without much difficulty, be it that one of the split-half data sets (S4) had considerable difficulty in converging to a solution. For the full data it was possible to find the same three-component solution from each starting position, but not for all split-halves. Finally, a four-component solution is also possible, but its recurrence in the split-halves is uncertain. Because there are only four levels in the third mode, going beyond four components does not make much practical sense. Note that having convergent

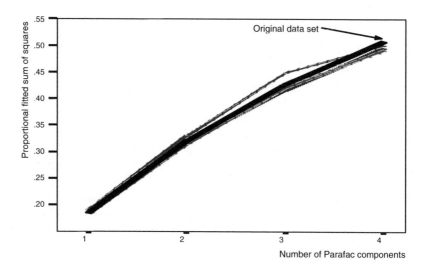

Figure 13.4   Parental behavior data: Parafac scree plots for the original data set and the four orthogonal split-half solutions. The solid thick line represents the solution of the complete data set.

solutions does not necessarily mean that the solutions are the same, a matter taken up in the next section.

The proportional fitted sums of squares for the four solutions of the full data set were 0.18, 0.31, 0.42, and 0.49, showing that nearly 50% of the variability in the data could be modeled by a parallel proportional profiles model. An (inverted) scree plot of the proportional fitted sums of squares (Fig. 13.4) is only mildly informative as to the selection of the number of components, because of the small number of possible components, but this will often be the case. More effective is the plot in evaluating the effect of using split-half solutions. For all numbers of components, each of these solutions is very close to the solution for the original complete data set, so that overall stability is not an issue. However, because of this the plot cannot be used in selecting a particular dimensionality.

Examining convergent solutions.   Detailed analyses of convergent solutions via component comparisons are necessary to establish component stability. This can most easily be done using congruence coefficients, ϕ (see Section 8.9.3, p. 196). We will use the two-component solutions of the overall analysis and the corresponding split-half ones as an example. These comparisons may also be used as an additional way to assess the uniqueness of the solution. Assuming there is a unique solution in the population, the same components should recur in the split-half solutions, barring sampling errors and instability due to insufficient subjects. It is an empirical question whether 75 subjects in the split-half samples are enough to establish stability. In

Table 13.3 Fit measures for Parafac solutions with 1 through 4 components

Data set	Prop. fit	No. it.	Standardized component weights				Degeneracy indicators			Status 5 sols.
							TC	EV	CN	
PF1										
All	0.183	45	0.18							5(+)
S1	0.189	42	0.19							5(+)
S2	0.189	44	0.19							5(+)
S3	0.186	44	0.19							5(+)
S4	0.185	44	0.19							5(+)
PF2										
All	0.315	51	0.17	0.16			−0.1	1.0	1.0	5(+)
S1	0.331	51	0.31	0.27			−0.4	0.8	1.4	5(+)
S2	0.319	146	0.23	0.17			−0.2	0.9	1.1	5(+)
S3	0.321	57	0.16	0.16			0.0	1.0	1.0	5(+)
S4	0.321	835	0.73	0.67			−0.8	0.3	3.9	5(+)
PF3										
All	0.421	650	0.39	0.35	0.18		−0.6	0.6	1.9	5(9th)
S1	0.449	1000	8.31	7.87	0.31		−1.0	0.0	46.7	5(6th)
S2	0.416	174	0.17	0.16	0.13		−0.1	0.9	1.3	2 sols (7th)
S3	0.424	282	0.23	0.22	0.16		−0.7	0.3	3.8	5(8th)
S4	0.427	381	1.08	0.49	0.43		−0.6	0.3	4.2	4(+); 1(8th)
PF4										
All	0.492	216	0.26	0.24	0.16	0.15	−0.3	0.7	1.6	1(OK); 4(nc)
S1	0.502	400	0.23	0.16	0.15	0.14	−0.2	0.9	1.3	2 sols.
S2	0.496	588	0.94	0.94	0.56	0.22	−0.7	0.3	3.8	3(nc); 2(8th)
S3	0.491	1000	7.88	7.73	7.38	7.03	−1.0	0.0	77.1	5(nc)
S4	0.508	365	0.27	0.22	0.20	0.11	−0.3	0.8	1.4	1(nc); 3(+); 1(6th)
Limits	1.000	1000	< 1	< 1	< 1	< 1	≫ −1	> .5	> 5	

Prop. fit = proportional fit; No. it = number of iterations; Degeneracy indicators: TC = triple cosine; EV = smallest eigenvalue; CN = condition number triple cosine matrix. 5(+) = 5 identical fitted sums of squares up to the 10th decimal; $j(i\text{th}) = j$ solutions with equal fitted sums of squares up to ith decimal; nc = not converged; i sols = i different solutions.

The S4 PF3 solution is problematic, with a standardized fit for its first component > 1 and two cosines > −0.50. Decreasing the convergence criterion, using different starting points, and increasing the number of iterations did not have an effect. The normalized core consistency attained extremely low values. There are two virtually identical solutions with a fit difference in the 8th decimal. Further investigation seems called for.

Table 13.4 Congruences between the components of two-component Parafac models

Data set	Comp. no.	ALL 1	S1 2	S2 1	S3 1	S4 1	ALL 2	S1 1	S2 2	S3 2	S4 2
ALL	1	1.00									
S1	2	0.89	1.00								
S2	1	0.96	*0.74*	1.00							
S3	1	0.95	0.97	0.85	1.00						
S4	1	*0.84*	0.98	*0.68*	0.93	1.00					
ALL	2	**0.37**	0.72		0.56	0.80	1.00				
S1	1	0.52	**0.80**	0.32	0.65	0.86	0.96	1.00			
S2	2	0.30	0.66	**0.08**	0.48	0.75	0.99	0.96	1.00		
S3	2		0.64		**0.49**	0.71	0.95	0.85	0.93	1.00	
S4	2	0.54	0.58	0.32	0.71	**0.90**	0.98	0.97	0.95	0.92	1.00

ALL = all 150 subjects; Si = ith orthogonal split; values below 0.30 not shown. **Bold** numbers represent the congruence coefficients between first and second components of the same analysis. *Italic* coefficients indicate values lower than 0.85 for the same components.

an empirical study, Lorenzo-Seva and Ten Berge (2006) found that subjects judged components to have a fair similarity when ϕ was in the range 0.85–0.94, while two components with a ϕ higher than 0.95 were considered to be identical for the purpose of interpretation.

Table 13.4 shows that coefficients for the scales of the four split-half samples are in fair to excellent agreement with the main first component, with lower values for S4 and S1 with $\phi = 0.84$ and $\phi = 0.89$, respectively. S4 was also the split-half analysis with problematic convergence, probably due to a ϕ of 0.90 between its components. The congruence coefficients for the second components with the main second component are all equal to or above 0.95. On the basis of these coefficients, we may conclude that the same components were present in the overall solutions and the split-halves. However, a complication is that among the split-halves there are ϕ's as low as 0.68. There is an additional caveat, because the ϕ's between the first and second components within analyses are vastly different, ranging from 0.08 to 0.90 (bold in Table 13.4). From this, we have to conclude that the congruence coefficients within analyses are highly unstable, and without going much deeper into the analyses it is difficult to assess the importance of this instability.

Assessing trilinearity. As discussed in Section 8.6.3 (p. 185), (normalized) *core consistency* can be used as a measure for evaluating Parafac models. The principle is based on the distance between the core array derived from the Parafac components and a superdiagonal core array $\mathcal{G}$. A core array is said to be *superdiagonal* if for, say, a three-component model only the g_{111}, g_{222}, and g_{333} have sizeable values, and all other core elements are 0 (see Fig. 4.10). Bro (1998c, pp. 113–122) proposed constructing a *core consistency plot* that has the core values on the vertical and the core elements on the horizontal axis with the superdiagonal elements plotted first. For the Kojima data, core consistency plots for the unrestricted three-dimensional solution, the three-dimensional solution with orthogonal subject components, and the convergent four-dimensional solutions are presented in Fig. 13.5.

The two-component solution has a core consistency of 100%, indicating a perfect superdiagonal core array. The unrestricted three-component solution (Fig. 13.5(a)) with a normalized core consistency of 51% shows that there are many not-quite-zero elements in the core array, but the three superdiagonal elements are the largest ones, be it only just. In other words, there is some Tucker structure: there are some clearly nonzero off-diagonal elements in the core array. However, imposing orthogonality on the subject components (as was done in the example) raises the normalized core consistency to 75% with a much clearer superdiagonal structure (Fig. 13.5(b)). Finally, the four-component solution (Fig. 13.5(c)), even though it has converged, shows a complete lack of normalized core consistency (-0.9 %), supporting the choice of the three-component solution in the example.

13.5.4   Stage 4: Results and their interpretation

Earlier we illustrated the presentation of the results with a table containing the scale coefficients (Table 13.1), accompanied by per-component plots showing the components one by one but with all three modes in the same plot (see Figs. 13.1, 13.2, 13.3). If one wants to make paired-components plots, it is not the correlated components that should form the coordinate axes, but orthonormalized versions of them, in other words we need oblique-components plots. This is necessary to obtain the correct interpretation in these plots; see Section 11.4.4 (p. 269) and Kiers (2000a). Figure 13.6 shows such a plot with auxiliary orthogonal axes for the unrestricted solution. The Parafac components are shown as dashed lines labeled with capital letters for their component numbers.

Parafac core arrays.   The core consistency plot (Fig. 13.5(a)) allows us to discover Tucker structure, but one has to look at the Parafac core array itself to evaluate this structure; see Section 9.7.1, p. 232. Table 13.5 shows the Parafac core arrays. We see that the superdiagonal elements are the largest ones apart from g_{211}, which is 1.80, while $g_{333} = 1.61$, providing the detailed evidence of a Tucker structure. When imposing orthogonality on the subject components, the Parafac core array has a much more superdiagonal structure than the model without the constraint. Its largest off-diagonal element (0.55) is less than half as small as the smallest superdiagonal element (1.31).

One of the advantages of the Parafac interpretation of the core array over the Tucker model's core array is that the interpretation is still based on parallel proportional profiles. At the same time, the components cannot be rotated to facilitate their interpretation, because then they lose their proportional profiles. Probably the best

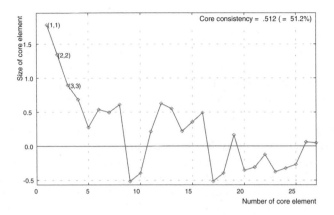

(a) Three-component solution: Unrestricted

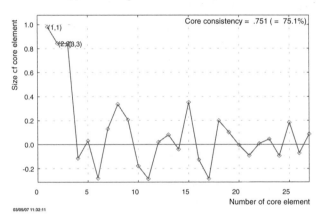

(b) Three-component solution: Orthogonal subject components

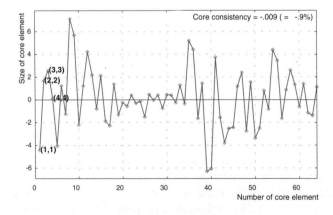

(c) Four-component solution

Figure 13.5 Core consistency plots for the Kojima data.

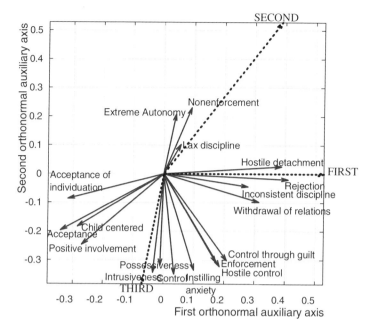

Figure 13.6 Parental behavior data: Oblique-component plot with the first two auxiliary orthonormalized axes for the unrestricted three-component Parafac solution. Angles in three-dimensional space between the Parafac components: $\theta_{(1,2)} = 145°$, $\theta_{(1,3)} = 50°$, $\theta_{(2,3)} = 151°$.

thing is to be eclectic and decide empirically which components are easier to interpret when no *a priori* components are available, as is so often the case in the social and behavioral sciences.

13.6   MULTIWAY PROFILE DATA: TUCKER ANALYSES IN PRACTICE

The primary aim of the analyses in this section is to discuss the practical issues carrying out a Tucker analysis on multiway profile data.

13.6.1   Stage 1: Objectives

The Tucker models are primarily used to find a limited set of components with which the most important part of the variability in the data can be described. These models are thus especially useful for data reduction and for exploration of variability. The main reason that the Tucker model is not as immediately useful as the Parafac model in the search for developing or identifying general principles or rules that underlie patterns in the variables is its rotational freedom. The basic results of applying the

Table 13.5   Parafac core arrays

	Conditions 1			Conditions 2			Conditions 3		
	S1	S2	S3	S1	S2	S3	S1	S2	S3
Unconstraint solution									
B1	**4.70**	1.42	−1.36	1.59	0.90	−1.31	−0.64	−0.68	0.11
B2	*1.80*	1.31	−1.05	1.39	**3.39**	−1.00	−0.56	−0.58	0.08
B3	0.73	1.61	0.56	0.56	1.24	0.41	−0.22	−0.49	**1.61**
Orthogonal subject components									
B1	**1.60**	−0.46	0.34	0.03	0.55	−0.44	−0.01	0.07	−0.11
B2	−0.19	0.21	−0.29	0.13	**1.32**	0.31	−0.14	−0.15	0.14
B3	0.05	0.55	−0.47	−0.06	−0.20	0.16	0.01	0.29	**1.31**

The numbers in **bold** indicate the elements of the superdiagonal of the core array. The *italic* entry in the top panel is larger than the smallest superdiagonal element. S1, S2, S3 indicate the scale components; B1, B2, B3 the boys components. The differences in the sizes of the numbers are caused by the correlations among the components in the unrestricted solution. The differences in sizes of the core elements are the consequence of the high correlation between the subject components of the unconstraint solution.

Tucker models to multiway data are maximum variance subspaces in which in each component space any orientation of the axes is as good as any other in terms of fit of the solution to the data. If a subspace interpretation is desired, that is descriptions are to be made in terms of the original variables, subjects, and conditions rather than latent entities, the rotational freedom is extremely helpful in unraveling these patterns.

In multiway profile data the data are mostly normalized deviations from the variable means at each condition and/or occasion, due to profile preprocessing (see Section 13.4.2). A large part of the variability in the original data, in particular, the variability between the means per condition, is not contained in the multiway analysis. This should always be borne in mind when interpreting results. However, what facilitates the interpretation is that the removed means represent the average subject's profile, that is, the person who has an average score on each variable under each condition. In many applications, a variable×condition matrix of means contains important information in its own right and should be carefully analyzed, for instance, via a biplot (Gabriel, 1971); see also Appendix B.

13.6.2   Stage 2: Data description and design

Preprocessing, sample size considerations, and data requirements have already been discussed in Section 13.4.2.

Types of variability.   In contrast with the Parafac model, the Tucker models can handle both system variation and object variation, and therefore can be used to fit data in which the correlations between the components change over time (see Section 4.6.1,

p. 60). It is especially the simultaneous handling of these types of variations which makes the Tucker models extremely useful for the investigation of the patterns within and between modes in exploratory situations where there is little *a priori* knowledge. Alternatively, the models can be used when one does not want to prejudice the analysis by specifying restrictive models.

13.6.3   Stage 3: Model and dimensionality selection

When confronted with multiway data choosing the most appropriate Tucker model is not always easy, because the numbers of components can vary independently for each mode. In this section we will try to find a reasonably fitting Tucker3 model, and similarly a Tucker2 model, by first evaluating a series of models with varying numbers of components in the three modes. The choice between settling for a Tucker2 model or a Tucker3 model depends on whether we want to reduce two or three of the modes to their components. When we have a mode with very few, very different levels, it is not always useful to reduce it to its components. On the other hand, when there is little difference between the levels in the way they relate to the components of the other two modes, using one or two components might be a very effective way of simplifying the interpretation. This is especially true when one of the modes is essentially unidimensional. The single core slice corresponding to that component is necessarily diagonal, and the coefficients in the single-dimensional mode serve to expand or contract the common configuration of the elements of the other two modes (Ten Berge et al., 1987). In some cases these coefficients may be interpreted as response styles (Van der Kloot, Kroonenberg, & Bakker, 1985b).

As explained in Section 8.5 (p. 179), two types of plots are useful for selecting models to be interpreted: (1) the deviance plot (Fig. 13.7) displaying the residual sums of squares against the degrees of freedom, and (2) the multiway scree plot (Fig. 13.8) showing the residual sums of squares against the total number of components, that is, $N_3 = P + Q + R$ for the Tucker3 model and $N_2 = P + Q$ for the Tucker2 model, respectively. The models with the best SS(Residual)/df ratios and those with the best SS(Residual)/N ratios lie on a convex hull as indicated in these figures.

Balancing these ratios, the $3 \times 2 \times 2$-Tucker3 model and the 3×2-Tucker2 model seem acceptable candidates for interpretation. More detailed investigations can be undertaken by comparing the fit of the levels of the modes, the fit of the components for each mode and the interpretability of the solutions.

Unlike the Parafac model, the Tucker models present no serious problems with multiple solutions, nonconvergence, and similar algorithmic difficulties. Timmerman and Kiers (2000) conducted an extensive simulation study and only encountered problems when the numbers of components deviated much from the actual number of components analyzed, and even then primarily when a random start was chosen rather than a rational one.

For the present data, and this is not very unusual, the scales components and the subjects components for the Tucker2 model and the Tucker3 model are very similar.

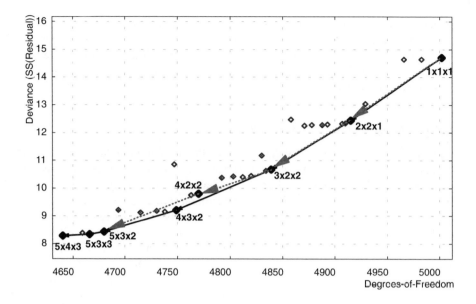

Figure 13.7   Parental behavior data: Deviance plot for all Tucker3 models $P \leq 5$ and $Q \leq 4$ and $R \leq 3$. The solid line connects optimal choices based on the deviance convex hull. The dashed lines connect the optimal choices resulting from the multiway scree plot (see Fig. 13.8).

Therefore, we will primarily limit our discussion to the latter and will only refer to the specific features of the former.

It is useful to compare the fit of the Tucker models with those of comparable Parafac ones. For the two-component models the proportional fitted sums of squares are 0.3145 (Parafac), 0.3145 ($2\times2\times2$-Tucker3), and 0.3152 (2×2-Tucker2). For the three-component models the proportional fitted sums of squares are: 0.4208 (Parafac), 0.4220 ($3\times3\times3$-Tucker3), and 0.4224 (3×3-Tucker2). Clearly, on the basis of fit alone, there is not much to choose between comparable models. In fact, the two-component Parafac and the $2\times2\times2$-Tucker3 model are identical up to nonsingular transformations (see Section 8.9.3, p. 196). Depending on the aims of the analysis, one might prefer the orthonormalized subject-component variant of the Parafac analysis with its smaller number of parameters and its unique solution.

Evaluating fit.

Fit of components.   The overall fit of both the $3\times2\times2$-Tucker3 and the 3×2-Tucker2 models is 41% (Table 13.6), which is a fairly common value, considering the size of the data and the fact that the fit refers to individual data points rather than to covariances. There are no real differences between the Tucker3 and Tucker2 models

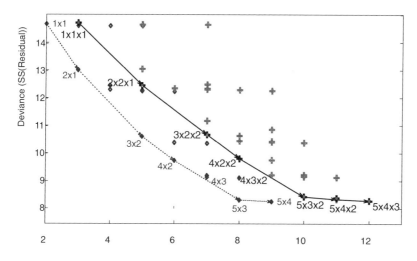

Figure 13.8 Parental behavior data: Multiway scree plot for all models with $P \leq 5$ and $Q \leq 4$ and $R \leq 3$ – Tucker3 model; and $P \leq 5$ and $Q \leq 4$ – Tucker2 model.

in terms of the fit per component for the boys and the scale. Moreover, the additional parameters in the Tucker2 model for the conditions are not really necessary. The conditions can be fit well by two components.

Fit of subjects. The basic form of the Tucker models makes it possible to assess the fit or lack of it for each level of each mode separately. The sums-of-squares plot, Fig. 13.9(a) (see Section 12.6.1, p. 289), shows that there are several boys who have large total sums of squares (e.g., 11 and 16) so that they have scores far removed from the average on one or more variables. The boys located near the origin (e.g., 28 and 63) have small residual and small fitted sums of squares and therefore have scores

Table 13.6 Proportional fit of the components of the three modes for the $3 \times 2 \times 2$-Tucker3 and 3×2-Tucker2 models. Per mode the proportional fits sum to the total fit = 0.41

		Tucker3 model			Tucker2 model		
	Mode	1	2	3	1	2	3
1	Boys	0.18	0.13	0.10	0.19	0.13	0.10
2	Scales	0.27	0.13		0.28	0.14	
3	Conditions	0.31	0.10				

that are close to the means of the variables. On the whole, the spread looks very even and there are no particular outlying or suspicious values.

Fit of scales.   Figure 13.9(b) shows that Extreme autonomy, Nonenforcement, and Lax discipline are the worst-fitting scales in this solution, while Positive involvement, Hostile control, and Acceptance fit best. Whether a scale fits better than another one is clearly not related to the type of disciplining involved. The normalizations ensures that all variables have an equal total sum of squares, which aligns the scales on a line of equal total sum of squares in the graph (see also Section 12.6.2, p. 290).

Fit of judgment conditions.   As there are only four conditions, constructing a plot is superfluous. The Boy–Father and Boy–Mother conditions have proportional fits of 0.49 and 0.56, respectively. The Father–Father and Mother–Mother conditions have better fits, 0.62 and 0.73, respectively. The (modest) difference is probably due to more inconsistency in the boys' judgments over the scales.

Fit of combinations of components.   Besides the question of the number of components and their contribution to the overall fit, one may also investigate the contributions of the component combinations to the fit of the model to the data. As long as the components are orthogonal within the modes, the core array can be partitioned into the explained variabilities for combinations of components by squaring the core elements and dividing them by the total sum of squares (see Section 9.5.1, p. 225). Thus, explained variability of combination (p,q,r) is $g_{pqr}^2/\mathrm{SS}(\mathrm{Total})$, where $\mathrm{SS}(\mathrm{Total}) = \sum_i \sum_j \sum_k z_{ijk}^2$ with z_{ijk} the preprocessed data. Details are given in the next Section.

13.6.4   Stage 4: Results and their interpretation

In this section we will look at the ways in which the basic results from a multiway profile analysis can be enhanced for interpretation. In particular, we will discuss components, the core array and plots of components per mode, rotations, both of component space and/or the core array, joint representations of the modes, and using constraints.

Displaying components and core array.   As mentioned in Section 13.4.4, there are essentially four different ways in which the components of each mode can be presented: tables of the coefficients for each mode, separate paired-components plots, per-component plots, or all-components plots showing the components against their sequence numbers or time points. Rather than presenting graphs for the boys components and for the scale components separately, we present them together in joint biplots (see Section 11.5.3, p. 273).

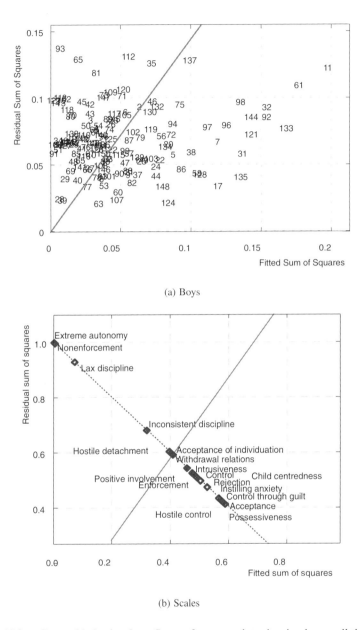

(a) Boys

(b) Scales

Figure 13.9   Parental behavior data: Sums-of-squares plots showing how well the model fits the data of each boy (a) and each scale (b). The solid line indicates the locus of all boys (scales) with the same fit/residual ratio. Boys (scales) above the average line have a worse fit/residual ratio than average while the boys (scales) below the line have a better fit/residual ratio. The scales lie on a line of equal total sums of squares, because they were normalized.

Table 13.7   Unrotated and rotated condition components and their fit

Condition	Abbrev.	Unrotated		Rotated		Proportional SS(Residual)
		1	2	1	2	
Boy judging Father behavior	(B-F)	**0.63**	−0.37	**0.73**	0.03	0.49
Boy judging Mother behavior	(B-M)	**0.56**	−0.39	**0.68**	−0.03	0.56
Father judging his behavior	(F-F)	0.42	**0.66**	0.01	**0.79**	0.62
Mother judging her behavior	(M-M)	0.33	**0.53**	−0.01	**0.62**	0.73
Proportional fit of components		0.31	0.10	0.25	0.16	

Unit length components; varimax rotation on the orthonormal components.

Judgment components.   The mode of the judgment conditions is more conveniently presented in a table, because there are so few levels (Table 13.7). From this table, it follows that the primary contrast is between the boys' judgments and those of their parents. The major contrast, although small, is that the coefficients related to the father behavior are a bit higher than those related to mother behavior. This was also evident in the Parafac analysis (see Fig. 13.2), but the reliability of this difference can only be established via bootstrap analyses (see Sections 8.8.1 and 9.8.2) or a new study. The model selection graphs in Section 13.6.3 show that adding condition components will not really increase the fit, so that no further important distinctions between the conditions can be found.

Core array.   The core array (Table 13.8) contains the weights the combinations of components, one from each mode. Given that we are dealing with principal components, in almost all Tucker3 analyses the core element g_{111}, weighting the first components from all the modes in the structural image, is the largest element of the core array, and so it is for the Kojima data: $g_{111} = 1.75$. The second largest element in the core array is $g_{221} = 1.42$ combining the second boys component, the second scale component, and the first condition component. This indicates that the largest linkage is between the scales and the scores the boys' parents get on these scales. In other words, the regimes in different families are very different, in the eyes of both the parents themselves and their sons.

The third largest element in the core array, $g_{312} = 1.24$, represents the linkage of the third boy component, the first scale component, and the second condition component. This Tucker structure also showed up in the Parafac core array (see Table 13.5). Note in the third panel of Table 13.8 that the adjusted core array after rotation of the condition components has many more mid-sized elements than the original core array. This general effect (i.e., distribution of variability) is the price to pay for having a simpler component structure. Kiers (1998d) has devised procedures

Table 13.8   Tucker3 core array

| | Unrotated core | | | | % Explained | | | |
| | C1 | | C2 | | C1 | | C2 | |
	S1	S2	S1	S2	S1	S2	S1	S2
B1	**1.75**	−0.47	−0.07	−0.19	**16.9**	1.2	0.0	0.2
B2	0.36	**1.42**	−0.42	−0.05	0.7	**11.1**	1.0	0.0
B3	0.16	0.33	**1.24**	0.18	0.1	0.6	**8.6**	00.2

| | Rotated core | | | | Optimally rotated core | | | |
| | C1 | | C2 | | C1 | | C2 | |
	S1	S2	S1	S2	S1	S2	S1	S2
B1	**1.51**	−0.30	0.88	−0.41	**1.76**	0.01	0.28	−0.28
B2	0.53	**1.23**	−0.16	0.71	0.13	**1.48**	0.28	−0.13
B3	−0.53	0.19	**1.13**	0.33	0.05	−0.07	**1.20**	0.57

The numbers in **bold** indicate the larger elements of the core array. C stands for condition (judge) components; S for scale components; B for boys components. The rotated core is the result of a counterrotation due to the varimax rotation of the condition components (see right-hand panel of Table 13.7), and the optimally rotated core is the result of a simultaneous varimax rotation of the three modes of the core array (Kiers, 1997a).

to simultaneously simplify the components and the core array (see Section 10.4.5, p. 249).

The extended core array (Table 13.9) shows very clearly the similarity and differences between the boys' judgments and the parents'. The top 2×2-part of the slices

Table 13.9   Tucker2 core slices

| | Boys judge Father behavior | | Boys judge Mother behavior | | Fathers judge Father behavior | | Mothers judge Mother behavior | |
	S1	S2	S1	S2	S1	S2	S1	S2
B1	**1.15**	0.15	**0.98**	0.28	**0.72**	0.27	**0.51**	0.34
B2	0.39	**−0.90**	0.35	**−0.83**	−0.02	**−0.54**	−0.22	**−0.47**
B3	0.35	0.17	0.42	0.08	**−0.93**	0.27	**−0.64**	0.18

The core elements in **bold** indicate the larger elements of the core array. S stands for scale components; B for boys components.

of the extended Tucker2 core array are very similar for both boys' and parents' judgments, even though the values are higher for the boys than for their parents, but the core elements h_{31k} for $k = 1, \ldots, 4$ have positive signs and lower values for the boys and negative signs and higher values for the parents. When compared with those of the full core array (Table 13.8), we see that the first Tucker3 core slice corresponds to the consensus between the boys and their parents, while the second Tucker3 core slice corresponds to their differences.

The values on the first scale component are nearly all positive, so that the contrast between the judgments is present on all scales. Thus, boys with positive values on the third boys component tend to give their parents higher values on all scales than their parents do, while boys on the negative side of the third boys component tend to do the reverse. As before, this suggests a response style for the boys but less so for their parents.

Joint biplots for components from two modes given the third.   A large part of the analysis of multiway data concentrates on the links between the modes contained in the core array, that is, the g_{pqr} in the three-way case. The interpretation of these links can be hard when no clear meaning can be attached to the components themselves. After all, they represent a direction in the component space of maximal variance, which is not necessarily the direction of maximal interpretability. Not being able to assign a substantive meaning to a component restricts the possibility of interpreting combinations of such components. To get around this, one may construct a *joint biplot* of the components of two modes (the *display modes*) given a component of the third (the *reference mode*).

A joint biplot of the display modes in three-mode analysis is like a standard biplot, and all the interpretational principles of standard biplots can be used (see Appendix B). The specific features are that each joint biplot is constructed using a different slice of the core array and that the levels of the reference mode weight the entire joint biplot by their component coefficients. In Section 11.5.3 (p. 273) a detailed explanation is provided for joint biplots.

Joint biplots for the Kojima boys.   First a decision has to be made which modes will be the display modes and which the reference mode. Judging from Tables 13.7 and 13.8 it seems to make the most sense to choose the rotated condition mode as the reference mode, because there is a clear interpretation for each of the condition components: that is, the first condition mode represents the boys' judgment of their parents' behavior and the second one the parents' judgments of their own behavior toward their sons. Given that the focus of the research is how the boys' and parents' judgments differ, the judgment mode and the scales mode should be used for the display modes. For this demonstration we have therefore chosen to concentrate on the latter option, but for other analyses the other perspective might be more fruitful.

The joint biplots will show how the judgments differ for each of the three boys' components, and they will represent the structures of the boys with positive scores

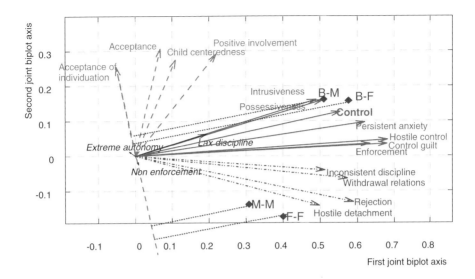

Figure 13.10 Parental behavior data: Joint biplot of the scales and judgment conditions for the first boys component. Solid arrows: Control variables; Dashed arrows: Parental support variables; Dash-dot arrows: Rejection variables; Lax discipline variables do not have sizeable representation in the joint biplot. Judgment conditions shown with projections on the Acceptance of individuation variable.

on their components. For boys with near-zero values on a component, the corresponding joint biplot does not apply, while for boys with negative values on the boys components, one of the displayed modes should be mirrored with respect to the origin.

As an example we will present here a joint biplot of the scales and judgment conditions for the first component of the boys, rather than present all three of them. The joint biplot is based on the first core slice associated with the first boys component, which is given in the top left panel in Table 13.8 as the first row; thus, on $g_{111} = 1.75$, $g_{121} = -0.47$, $g_{112} = -0.07$, $g_{122} = -0.19$. In this slice of the core array, the combination of the first components of the three modes (S1,B1,C1), or $g_{111} = 1.75$, is by far the dominant value, so that this combination of components dominates the patterns in the joint biplot. This means that the arrangement of the scales and the judgment conditions on the first axis is almost the same as their arrangement on their first components. The proportional explained values for the two joint biplot axes are 0.18 and 0.003. Even though the second joint biplot axis seems negligible, it still contains recognizable information about the differences between the judgments of the boys and their parents, respectively. It is, however, clear that validation of such patterns is highly desirable through cross-validation or replication.

The joint biplot (Fig. 13.10) shows the four groups of scales, with the "Psychological and behavioral control" group (Intrusiveness, Possessiveness, Control, Installing persistent anxiety, Hostile control, Enforcement, and Control through guilt) and the "Rejection" group (Inconsistent discipline, Withdrawal of relations, Rejection and Hostile detachment) as the most important ones. The Lax discipline scales (Nonenforcement, Lax discipline, and Extreme autonomy) do not play a role here, as they have small near-zero vectors. The difference between the boys and the parents is in the scores on the "Control" group of variables; the boys' scores for their parents are much higher than their parents' own scores, as is evident from their projections on the scale vectors of this group. The projections on the scales of the "Rejection" group are almost equal for boys and parents, which parallels the Parafac interpretation. The most distinguishing feature between the boys and their parents is the judgment on the Parental support scales. To illustrate this, the projections on Acceptance of individuation are shown in the plot. Parents have near zero projections on this variable and boys small positive values. This indicates that boys evaluate Parental support more favorably that their parents do, even though this direction, corresponding to the second axis, is far less important than that of the first biplot axis (see also Table 13.1 for the Parafac coordinates).

13.6.5   Stage 5: Validation

Internal validation.   In Section 13.3.3 we looked at the relative fit of the components and the combinations of the components via the core array, but we have not yet investigated the residuals from the model at the level of the individual data points. In the same way as other least-squares procedures we may construct a residual plot of the standardized residuals versus the standardized fitted values (Fig. 13.11).

From this plot we see that there are some larger standardized residuals, but given that we have 10,800 residuals that is only to be expected. The general conclusion from the residual plot is that the residuals are well behaved.

A further look at the adequacy of the solution can be had by looking at the distribution of the standardized residuals presented in Fig. 13.12. There are only a limited number (14) of observations with standardized residuals $> |3.0|$, so that there is no reason to be seriously worried about them. Moreover, the errors seem to have a nearly normal distribution.

13.7   CONCLUSIONS

In this chapter a large number of the general procedures presented in Part II were applied to multiway profile data using one particular three-way data set. Applying the common three-mode models to the same data not only provided some insight into their use in practice but also explained how they work vis-à-vis each other.

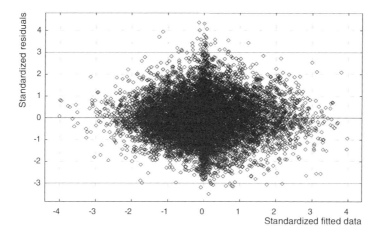

Figure 13.11   Parental behavior data: Residual plot of standardized residuals versus standardized fitted values. The horizontal lines define the interval between +3.0 and −3.0, which may serve to evaluate the size of the standardized residuals.

```
Distribution of standardized residuals
---------------------------------------
     f Stem  Leaves
    10 -3.0  *
    47 -2.5  **
   168 -2.0  ****
   445 -1.5  **********
   977 -1.0  ***********************
  1699 -0.5  **************************************
  2175 -0.0  ************************************************
  2037  0.0  **********************************************
  1584  0.5  ************************************
   935  1.0  *********************
   438  1.5  **********
   166  2.0  ****
    79  2.5  **
    26  3.0  *
    10  3.5  *
     4  4.0  *

One star represents   44 cases, except for the rightmost one,
which may represent less.

The smallest standardized residuals are   -3.48   -3.18
The largest  standardized residuals are    4.37    4.32
```

Figure 13.12   Parental behavior data: Distribution of the standardized residuals between 5.0 and −5.0.

For this data set the Tucker and Parafac models gave comparable results, so that the nature of the data and the requirements and preferences of the researcher should play an important role in model and dimensionality selection. However, in analytical chemistry and also for EEG and fMRI data, there are explicit models that have the form of one of the models (see Section 19.3.1, p. 472), especially the Parafac model, so that the theory dictates the model choice. On the other hand, strong theory is often not available in the social and behavioral sciences, so that representativeness and ease of interpretation should guide the choice. In practice, it is often advantageous to use both types of models to get a good understanding of what the data have to tell. Eclecticism in model choice often leads to more insight than does dogmatism.

CHAPTER 14

MODELING MULTIWAY RATING SCALE DATA

14.1 INTRODUCTION[1]

Multiway rating scale data typically consist of scores of subjects on a set of rating scales, which are used to evaluate a number of concepts, where scales and concepts should be seen as generic ones. Such data become four-way when they are collected at several time points. Very often in psychology such data can be characterized as stimulus–response data, where the concepts or situations are the stimuli and the answers to the rating scales the responses. To get higher-way rating scale data, one could imagine that the judgments are collected under different conditions such as different time regimes. However, rating scale data that are more than three-way are exceedingly rare, and a web search conducted in May 2007 did not come up with any

---

[1]This chapter is based on joint work with Takashi Murakami, Chukyo University, Toyota, Japan; in particular, Murakami and Kroonenberg (2003); parts are reproduced and adapted with kind permission from Taylor & Francis, *www.informaworld.com*.

concrete example. Therefore, we will restrict ourselves in this chapter to three-way rating scales.

Rating scale data originate from subjects having evaluated a set of concepts or situations, and judged them to have particular values on each of the rating scales. One could thus conceive of these values as the ratings of the strength of the relationship (or similarity) between scales and concepts. The complete data take the form of a three-way array of concepts by scales by subjects, where, typically, the subjects will be placed in the third mode (scaling arrangement, see Section 3.6.2, p. 35). Specific to three-way rating scale data is the circumstance that all scales have the same range and the actual values are comparable across scales. Prime examples of rating scales are bipolar semantic differential scales and monopolar Likert scales, . In Chapter 1 we introduced an example of students judging Chopin's preludes on a series of semantic differential scales. Another example treated in this chapter is the Coping data set (Röder, 2000), in which children indicated on Likert scales how strongly they felt certain emotions in various situations related to their daily (school) life. A discussion in the context of marketing of multiway rating scale data (referred to as multiattribute rating data) can be found in Dillon, Frederick, and Tangpanichdee (1985).

Because the individuals play such an important role it is necessary to pay attention to the kinds of individual differences that may occur and how these can be modeled. Individual differences can be handled both outside and inside the framework of the three-mode analysis as will be demonstrated in this chapter.

14.2 CHAPTER PREVIEW

In this chapter we will first discuss different types of individual differences and the way they enter into three-way rating scale data. Not only will we present ways to analyze the data themselves, but we will also look at differences in results from different three-mode models. In particular, we will look at response styles and how they enter into the analyses, and we will give suggestions as to how they may be recognized in three-mode analyses. In addition, issues of centering and normalizing for three-way rating scale data will be treated (Section 14.3.1). A subset of the Coping data (Röder, 2000) will form the first example, followed by a discussion of practical issues related to the analysis of rating scale data. The Multiple personality data (Osgood & Luria, 1954), consisting of semantic differential judgments by a woman with multiple personalities, will provide a more detailed insight into the analysis of three-way rating scale data.

14.3 THREE-WAY RATING SCALE DATA: THEORY

Three-way rating scale data will be represented as an $I \times J \times K$ three-way data array $\mathcal{X} = (x_{ijk})$ of I concepts by J scales by K subjects. Thus, the subjects are seen as judges who express themselves about the relationships between the concepts and scales.

14.3.1 Sources of individual differences

Random individual differences. Suppose $\mathbf{X}_k$ is the $I \times J$ (concept by scale) data matrix of subject k, that is, the kth slice of the data array $\mathcal{X}$. If all individual differences can be regarded as random variations around common values, the data may be modeled by expressing their scores as deviations around the common scale means:

$$x_{ij(k)} - \overline{x}_{ij} = \sum_{p}^{P} a_{ip} b_{jp} + e_{ij(k)}, \tag{14.1}$$

and in matrix notation,

$$\mathbf{X}_k - \mathbf{1}'\overline{\mathbf{x}} = \mathbf{A}\mathbf{B}' + \mathbf{E}_k, \tag{14.2}$$

where $\mathbf{1}$ is an I-dimensional column vector of ones, $\mathbf{A} = (a_{ip})$ is an $I \times P$ matrix of concept scores, $\mathbf{B} = (b_{jq})$ is a $J \times Q$ matrix of scale loadings, and the $\mathbf{E}_k$ are $I \times J$ matrices of residuals. For the situation at hand, P is necessarily equal to Q. The J-dimensional column vector of score means $\overline{\mathbf{x}}$ is supposed to be the same for all subjects. The left-hand side of Eq. (14.2) thus represents the per subject column-wise centering with respect to the overall scale mean. If we assume that no residual matrix $\mathbf{E}_k$ has any significant systematic pattern, the individual differences are essentially random fluctuations. In this case, one obtains the same results for concepts and scales by analyzing the concept by scale table averaged over individuals. Such an analysis makes sense if the number of concepts is sufficiently large, and it has been the standard procedure in most semantic differential studies (e.g., see Heise, 1969, for an overview of semantic differential research).

Response styles. Response styles or biases during judgments of scales are a well-known source for individual differences, one type of which is reflected in the use of the scales themselves. Some subjects express themselves boldly and use all scale points including the extremes, while others only use a few scale points around the mean. Such response styles can often be seen directly in the raw data. Van der Kloot and Kroonenberg (1982, 1985a) discuss this type of response style and its effect on the results of three-way analyses. Another type of response style is due to social desirability. Unfortunately, such a response style is much harder to model. One possible way to do this is to assume that the scale means express social desirability and the sizes of the standard deviations express individuals' scale usage. This implies that the model equation (Eq. (14.2)) now reads as

$$\mathbf{X}_k - \mathbf{1}'\overline{\mathbf{x}}_k = s_k \mathbf{A}\mathbf{B}' + \mathbf{E}_k, \tag{14.3}$$

where $\mathbf{1}$, $\mathbf{A}$, $\mathbf{B}$, and $\mathbf{E}_k$ are defined as before. The J-dimensional column vector of score means $\bar{\mathbf{x}}_k$ is now specific to each individual k, and s_k is a constant for subject k reflecting its scale usage. Given that these individual differences in response styles have been modeled by these parameters, investigations into more fundamental individual differences can proceed by first eliminating the means and standard deviations from the data[2]; thus, we proceed with the centered and normalized data $\mathbf{Z}_k$, with $\mathbf{Z}_k = (\mathbf{x}_k - \mathbf{1}'\bar{\mathbf{x}}_k)/s_k$, and $s_k = \sqrt{\sum_{i,j}(\mathbf{X}_k - \mathbf{1}'\bar{\mathbf{x}}_k)^2_{ij}}$.

Differences in scale and concept usage.   Often, more fundamental types of individual differences, in particular, individual differences in concept scores and those in scale loadings, are the focus of studies using three-way rating scale data. Such differences can be modeled by defining for each subject separate matrices $\mathbf{A}_k$ and $\mathbf{B}_k$ for concept scores and scale loadings, respectively:

$$\mathbf{Z}_k = \mathbf{A}_k\mathbf{B}'_k + \mathbf{E}_k. \tag{14.4}$$

In Eq. (14.4) the columns of each $\mathbf{Z}_k$ are centered but they are not normalized. Instead, the data array is normalized per subject slice, so that the differences between relative sizes of scale variances remain in the $\mathbf{Z}_k$. There are two important reasons for handling the data in this manner. First, the column-wise (or fiber) normalization of an individual table is often impossible because zero variances occur frequently, that is, subjects use the same scale value for each concept. Second, the differences in the scale variances can be informative in understanding the characteristics of the rating behavior. A disadvantage of this type of normalization is that the coefficients of the scale components (the columns of $\mathbf{B}_k$) are not correlations between the original scales and the components.

Another reason to be careful with separate analyses is that separate analyses are only justifiable under the condition that the reliability of individual data is very good, or that the number of concepts to be rated is very large. In real situations, the reliability of ratings is generally not high enough, and the number of concepts is seldom large enough to obtain stable estimates of an individual loading matrix, $\mathbf{B}_k$. Therefore, we advocate the use of three-way methods to analyze all individual data matrices simultaneously.

Differences in scale loadings and concept scores.   Even though Eq. (14.4) looks very simple, its implication is very general and in fact too general to be useful in practice. To clarify this, we have to make further specifications of possible individual differences. To facilitate the discussion, we will assume that the ideal conditions as mentioned above apply, so that individual data matrices can be analyzed separately and for each subject we obtain a loading matrix for scales and a score array for

---

[2]Simultaneously handling the estimation of a multiway model, the means, and standard deviations is only necessary in the case of missing data (see Section 7.3.7, p. 153).

concepts. Such individual matrices may differ both in content and in number of components.

Differences in number of components. If subjects rate all concepts belonging to one domain as also belonging to another domain, the components of these domains coincide, and the number of components is reduced by one. If some subjects do this but others do not, different numbers of components will be observed for different subjects. This will necessarily also affect the scale loadings and the concept scores.

Differences in scale loadings. More subtle types of individual differences may occur as well. For instance, if one subject uses a single scale from one domain in conjunction with all scales of another domain rather than its own, this scale has a salient loading on the "wrong" dimension. In such a case this scale is used differently across subjects.

Differences in concept scores. Concept scores produced by two subjects may be different to such an extent that they are almost independent. For example, suppose students are asked to rate their classmates. If one student gives evaluative ratings of the others on the basis of their academic achievements, and another student does the same on the basis of physical abilities, their scores may be uncorrelated. One may also imagine that a subject rates a set of concepts in nearly reverse order on evaluation scales, compared to most other subjects. In that case, this person's concept scores on the evaluation dimension are negatively correlated with those of other subjects.

Discussion. It is important to realize that the differences in concept scores and scale loadings can occur independently. On the one hand, even if the meaning of a scale is completely misunderstood, locations of concepts can be almost the same as in the normal case. On the other hand, if the correlational structure between scales of a subject is exactly the same as for most other subjects, it is still possible that the subject in question rates concepts differently from the other subjects.

Our three major classes of individual differences — in the number of components, in component scores, and in component loadings seem to correspond to the three types of individual differences described by Snyder and Wiggins (1970): (a) the basic underlying dimensions are not general; (b) different scales function differently as markers of the basic dimensions for different individuals; and (c) individuals utilize the basic dimensions differently in making scalar judgments across classes of concepts.

As discussed in Murakami and Kroonenberg (2003), it is possible to evaluate these three kinds of individual differences using three-mode methods. In the following we will present the theoretical basis for this contention and discuss the relative merits of the various three-mode models in this respect.

Extent of individual differences. Because, in general, people who share the same language can communicate with each other at a high level of understanding, it seems reasonable to assume that the overall level of individual differences to be found in rating scale data is not overly large. As a matter of fact, in most studies using three-mode component analysis on semantic differential scales, the most important component of the subject mode considerably dominates the remaining components (e.g., Levin, 1965; Snyder & Wiggins, 1970; Takeuchi, Kroonenberg, Taya, & Miyano, 1986). One of the few clear exceptions is the reanalysis by Kroonenberg (1983c, 1985a) of the Multiple personality data described by Osgood and Luria (1954), but this is a rather unusual case (see also Sections 14.6 and 19.5 for analyses of these data, both as a three-way and as a four-way data set).

The implication of the overall consensus is that we have to take into account the possibility that the existence of individual differences may be primarily confined to specific scales and concepts rather than that they are manifest throughout the entire data set. In semantic rating scales, it seems not unreasonable to think that more subjective judgments, such as ATTRACTIVE–UNATTRACTIVE and LIKABLE–UNLIKABLE, are more likely to be a source of individual differences than more objective judgments such as HARD–SOFT. Moreover, the ratings of ambiguous concepts might be another source of individual differences.

14.3.2   Individual differences and three-mode models

In this section we will show how different three-way models represent different kinds of assumptions about individual differences in concept scores and scale loadings. The existence of different models makes it possible to evaluate which individual differences are most prominent, and how they can be described in a parsimonious way.

Models with one type of individual differences.   If there are few individual differences in the scale loadings, we may replace $\mathbf{B}_k$ in Eq. (14.4) by $\mathbf{B}$, the common loading matrix shared by all subjects. This gives us the Tucker1-B model:

$$\mathbf{Z}_k = \mathbf{A}_k \mathbf{B}' + \mathbf{E}_k. \qquad (14.5)$$

On the other hand, if there are few individual differences in the concept scores, we may replace $\mathbf{A}_k$ in Eq. (14.4) by $\mathbf{A}$, the common score array shared by all the subjects. This leads to the Tucker1-A model:

$$\mathbf{Z}_k = \mathbf{A}\mathbf{B}'_k + \mathbf{E}_k. \qquad (14.6)$$

When there are no individual differences for either mode, we are back at an ordinary component model that is valid for each subject k:

$$\mathbf{Z}_k = \mathbf{A}\mathbf{B}' + \mathbf{E}_k. \qquad (14.7)$$

Which of these models fit the data is impossible to tell in advance; it is an empirical matter that has to be settled via comparison of model fits. Some of the subtleties in individual differences mentioned above cannot be properly modeled in this way. Therefore, we will reconsider the general model of Eq. (14.4).

Models with several types of individual differences.   The individual differences models discussed in this section require three-way methods for the estimation of their parameters.

Tucker2 model.   Suppose that there is a complete catalogue of components from which each subject makes a selection in order to judge concepts with a given set of scales, and a similar component catalogue for concepts. We assume that these catalogues are not overly large, in line with our earlier assumptions about the common basis of understanding between people. We further assume that the component models of almost all the subjects are constructed by their choice of columns from the catalogues of concept scores and scale loadings, and by the linear combinations of these columns.

This situation can be expressed formally as follows within the context of the Tucker2 model (see Section 4.10, p. 53). Suppose that $\mathbf{A}$ is a matrix of concept scores consisting of columns in the catalogue explained above, and $\mathbf{B}$ is a matrix of scale loadings constructed similarly. The data matrix of each individual k can then be expressed as

$$\mathbf{Z}_k = \mathbf{A}\mathbf{H}_k\mathbf{B}' + \mathbf{E}_k, \qquad (14.8)$$

where $\mathbf{H}_k$ is the matrix representing the strengths of the links between the chosen columns of $\mathbf{A}$ and $\mathbf{B}$. For example, if the data matrix of subject k is approximately represented by the sum of the products of the 2nd column of $\mathbf{A}$ and the 5th column of $\mathbf{B}$, and the 3rd column of $\mathbf{A}$ and the 1st column of $\mathbf{B}$, then h_{25k} and h_{31k} are 1 and the remaining elements of $\mathbf{H}_k$ are 0.

Various kinds of individual differences can be modeled via special structures in the extended core array $\mathcal{H} = (\mathbf{H}_k)$. Moreover, procedures such as cluster analysis on the frontal slices of the extended core array may assist in grouping types of individual scale structures and semantic concept spaces. When there are only individual differences of degree but not in kind, the subject space will be approximately one-dimensional so that each $\mathbf{H}_k$ is approximately diagonal – Fig. 14.1(a) (see also Section 4.5.3, p. 56). If two scale components, for example, $\mathbf{b}_1$ and $\mathbf{b}_2$, are fused into one component for several subjects, their $\mathbf{H}_k$ are no longer diagonal, and two off-diagonal elements, for example, h_{12k} and h_{21k}, will be large – see Fig. 14.1(b).

In order to accommodate a specific usage of a particular scale, the number of columns of the scale matrix $\mathbf{B}$ is increased by a separate component for that scale, and the corresponding element in the relevant component is decreased – Fig. 14.1(c). Each frontal slice of the extended core array is no longer square but has an extra column. If a subject k uses the scale in question in the usual way as a scale of its component, the element h_{14k} is salient as well as h_{11k}. If another subject, say, k', uses this scale in a different way, $h_{24k'}$ rather than $h_{14k'}$ might be salient – Fig. 14.1(d).

In the case where subjects rate concepts on two entirely unrelated dimensions, as in the previous example of students who used the same scales to rate their classmates on

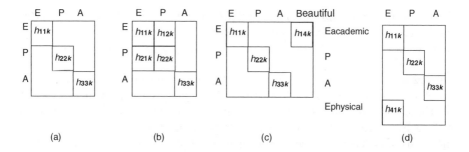

Figure 14.1   Illustration of the effect of individual differences on Tucker2 core slices. (a) Basic starting point: three scale domains and three concept domains – three components each. (b) Subject does not distinguish between scale domains E and P – components are fused for this subject. (c) The Beautiful scale is used differently by one or more subjects – additional column; (d) Scale component is used for different concept domains by different subjects – additional row.

either academic achievement or physical ability, the number of columns of **A**, rather than **B**, will be increased. The concept scores for the ratings on academic achievement will appear in, say, the first column and the concept scores for the ratings on physical ability in the second column. In the frontal slices for students who rated according to academic achievement, h_{11k}, h_{32k}, and h_{43k} are salient, while those of students who rated according to physical ability, h_{21k} rather than h_{14k} is salient. When several subjects rated concepts on a dimension in a completely reversed order compared to other subjects, it is not necessary to increase the number of columns. Their value of h_{14k} is simply negative.

In the analysis of real data there could be more complex combinations of several types of individual differences, so that patterns emerging in the extended core array may be far more messy. However, the basic patterns just described can be helpful in disentangling the more complex ones.

Tucker3 model.   The Tucker3 model is a more parsimonious model, as components are also computed for the subjects, and it can be conceived of as a model in which the extended core array is modeled with several more basic core slices:

$$\mathbf{Z}_k = \mathbf{A}\mathbf{H}_k\mathbf{B}' + \mathbf{E}_k = \mathbf{A}\left(\sum_r c_{kr}\mathbf{G}_r\right)\mathbf{B}' + \mathbf{E}_k, \qquad (14.9)$$

but there is no necessity to see it as such, as follows from its sum notation version:

$$z_{ijk} = \sum_{p=1}^{P}\sum_{q=1}^{Q}\sum_{r=1}^{R} g_{pqr}(c_{kr}a_{ip}b_{jq}) + e_{ijk}. \qquad (14.10)$$

Even though the model gives more a parsimonious representation of three-way data than the Tucker2 model, it seems less straightforward to use it to evaluate some of the individual differences. The cause may be the orthogonality of $\mathbf{C}$, the subject loading matrix. Because the centering in the scaling arrangement is across concepts and not across subjects, and because of the common base of understanding among people mentioned previously, the first component usually has positive scores for all subjects. The orthogonality of $\mathbf{C}$ forces the other components to have about half the signs of the scores negative. Although this may correspond well in the case where half of the subjects rate concepts in the opposite way to the other half on the specified dimension, it may not work in the case where there are two or more independent dimensions of the concepts but a subject only uses one of them.

Parafac model. Both the Tucker2 and Tucker3 models have advantages in analyzing rating scale data over their competitor, the Parafac model:

$$\mathbf{Z}_k = \mathbf{A}\mathbf{D}_k\mathbf{B} + \mathbf{E}_k. \tag{14.11}$$

As the Parafac model assumes differences between levels in all dimensions, it can be given similar interpretations as the Tucker2 model. However, there are no orthonormality constraints on the components, which might make it easier in some cases. The restriction of the Parafac model is the diagonality of the $\mathbf{D}_k$. The diagonality means that for every subject the links between the concept and scale domains are the same, and (individual) differences can only exist in the strengths of these links (i.e., in the d_{ss}^k). Thus, if differences exist as depicted in Fig. 14.1, they cannot be modeled with Parafac due to the Tucker structure.

A further complication is that the model will tend to produce degenerate results from many rating scale data due to the limited individual differences existing in at least one, usually the first, subject dimension. Even if this is not the case, highly correlated dimensions may occur, which could be difficult to interpret. A further drawback is that the number of components in all modes is necessarily the same, while in the Tucker models, one may specify different numbers of components in different modes. Therefore, the latter allow for discrimination between dimensions with considerable individual differences and those with small individual differences. In addition in the Tucker models, the core array can reveal the different patterns that arise from the individual differences due to the concepts and due to the scores. These characteristics seem to tip the advantage toward the Tucker models when analyzing rating scale data. One way to eliminate some of the problems of the Parafac model is to remove further means from the data, for instance, the consensus solution corresponding to the first subject component, by centering the subjects as well.

In the following, we will use a reduced real data set to illustrate the handling of individual differences using a Tucker3 analysis. Second, with another data set we will show how to assess empirically the relative values of the Tucker2, the Tucker3, and the Parafac models for three-way rating scale data.

14.4   EXAMPLE: COPING AT SCHOOL

14.4.1   Stage 1: Objectives of the analysis

The purpose of this example is to demonstrate how one may proceed with an analysis of three-way rating scale data. For clarity of exposition, a small, specially selected, sample of children from the Coping data is analyzed without going into too much technical detail. In this example only the Tucker3 model will be used, but in the next example a full model comparison is made.

14.4.2   Stage 2: Data description and design

The Coping data consist of the ratings of 14 Dutch primary school children (mean age 9.8 years). The children were especially selected on the basis of a preliminary three-mode analysis on a larger group of 390 children. The basis for selection was that there were maximal differences between them and together provided a relatively clear structure in the analysis (for a full account of the data, see Röder, 2000).

The children were presented with a questionnaire describing six common situations: restricted in class by the teacher (*Teacher No*), restricted at home by the mother (*Mother No*), too much work in class (*Workload*), class work was too difficult (*Too difficult*), being bullied at school (*Bullied*), and not being allowed to play with the other children (*Not Participate*). For each of these situations, the children had to imagine themselves being in such a situation and they were requested to indicate what they would feel (*Emotions*: Sad, Angry, Annoyed, Afraid) and how they generally dealt with such situations (*Strategies*: Avoidance coping, Approach coping, Seeking social support, Aggression). Thus, the data set has the following form: 6 situations by 8 emotions and strategies by 13 children.

The specific interest of the present study was to find out to what extent the interrelations between situations and emotions and strategies were different for different children. In the analysis, we assume that there is a single configuration of situations and a single configuration of emotions and strategies, but not all children use the same strategies and do not have the same emotions when confronted with the various situations. In terms of the analysis model, we assume that the children may rotate and stretch or shrink the two configurations before they are combined. In other words, the combined configuration of the situations and the emotions–strategies may look different from one child to another but they are constructed from a common perspective on the relationships between both the situations and the emotions–strategies.

14.4.3   Stage 3: Model and dimensionality selection

As explained in Chapter 8, a model can be selected in several ways, but here we will only show the deviance plot (see Section 8.5.2, p. 181). The model chosen for this example was the $2 \times 3 \times 3$ model with 2 situation components, 3 emotion–strategy

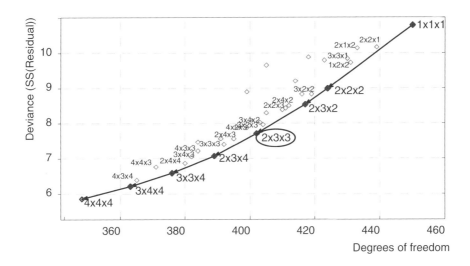

Figure 14.2   Coping data: Deviance plot. Models on the convex hull are shown in ellipses.

components, and 3 child components with a relative fit of 0.45. This model lies on the convex hull of the deviance plot (Fig. 14.2) and on that of the multiway scree plot (not shown). The Ceulemans–Kiers st-criterion (see Section 8.5.3, p. 182) is not much help given the near-linearity of the convex hull, so that more subjective considerations were used. The components of the modes account for 0.29 and 0.16 (1st mode), 0.26, 0.12, and 0.07 (2nd mode), and 0.27, 0.11, and 0.07 (3rd mode). All situations fit reasonably well except for Restricted by the mother (Mother No, relative fit = 0.18). The standardized residuals of the emotions and strategies were about equal, but the variability in Afraid, Social support and Aggression was not well fitted (0.17, 0.10, and 0.19, respectively). The relative fit of most children was around the average, except that children 6 and 14 fitted rather badly (0.14 and 0.08, respectively).

14.4.4   Stage 4: Interpretation of the results

Individual differences between children.   The 14 children were selected to show individual differences and these differences are evident from the paired-component plots of the subject space: Component 1 versus 2 (Fig. 14.3(a)) and Component 1 versus 3 (Fig. 14.3(b)). To get a handle on the children's characteristics in relation to their scores on the questionnaire, their coefficients on the components were correlated with the available background variables (external validation, see Section 9.8.3, p. 234). The highest correlations with the first component coefficients were with Happy in School, Quality of relationship with the Teacher, Having a Good Time at School (average about 0.65). On the whole, for most children, except for 1 and 6 who have

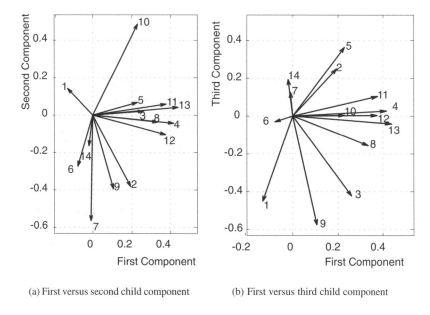

(a) First versus second child component   (b) First versus third child component

Figure 14.3 Coping data: Three-dimensional child space — paired-components plot.

negative values on the first component, their scores on the first component go together with positive scores on a general satisfaction with school. The second component correlated with Not Ill versus Ill (however, 10 and 1 were the only ill children) and Emotional support by the teacher (about 0.55 with low scores for 3, 6, 7 and high ones for 1, 8, 13). Finally, the third component correlated 0.70 with Internalizing problem behavior (a CBCL scale — see Achenbach & Edelbrock, 1983), but only eight of the fourteen children had valid scores on this variable, so that this cannot be taken too seriously. In principle the correlations are sufficiently high to assist in interpreting the joint biplots, however, correlations based on fourteen scores, some of which are missing, do not make for very strong statements.

Even though the correlations are all significant, one should not read too much into them with respect to generalizing to the sample (and population) from which they were drawn, as the children were a highly selected set from the total group, but it serves to illustrate that appropriate background information can be used both to enhance the interpretability of the subject space and its validity.

How children react differently in different situations. The whole purpose of the analysis is to see whether and how the children use different coping strategies and have different emotions in the situations presented to them. To evaluate this, we may look at joint biplots (see Section 11.5.3, p. 273). After evaluating several possibilities, it was decided to use the Situation mode as reference mode, so that the Child mode

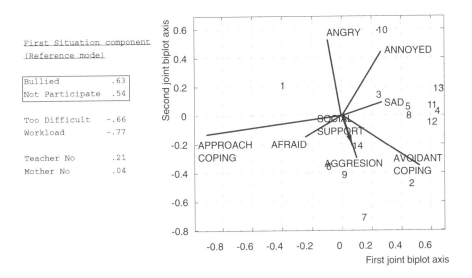

First Situation component
(Reference mode)

Bullied	.63
Not Participate	.54
Too Difficult	-.66
Workload	-.77
Teacher No	.21
Mother No	.04

Figure 14.4   Coping data: Joint biplot for being bullied (Bullied) and not being allowed to participate (Not Participate).

and the Emotion–strategy mode appear in the joint biplot as display modes. For the first situation component the three joint biplot axes account for 21.8%, 7.2%, and 0.4%, respectively, so that only the first two biplot axes need to be portrayed (see, however, comments in Section 11.5.3, p. 273, on the relative sizes of the axes in a joint biplot). The same was true for the second situation component, where the three joint biplot dimensions accounted for 10.3%, 5.2%, and 0.0001%, respectively.

To facilitate interpretation, two joint biplots are presented here for the first situation component: One for the situations loading positively on the component (Fig. 14.4), and one for the situations loading negatively on the component (Fig. 14.5). As explained in Section 11.5.3 (p. 273) this can be achieved by mirroring one of the modes around the origin. Here, this was done for the emotions and strategies. When interpreting these figures it is important to realize that they do not contain information on the absolute scores but only the relative information on the relationship between the three modes, because we have removed all main effects and two of three two-way interactions (see Section 6.6.4, p. 132, for a more detailed discussion on preprocessing rating scale data). In particular, in this graph the origin is the estimated mean score for all emotions and strategies, so that a zero value on the component of the situations represents the estimated mean of that situation. The interpretation of the joint biplot itself uses the projections of children on the emotions and strategies, with high positive values indicating that the child has comparatively high scores for such an emotion or strategy. The situations weight the values of these projections, so that it can be

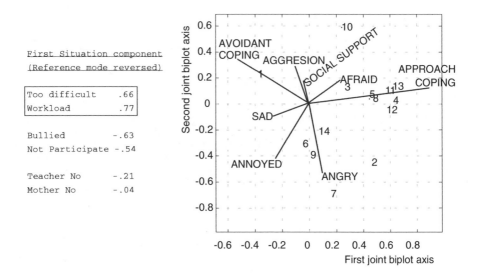

First Situation component
(Reference mode reversed)

Too difficult	.66
Workload	.77
Bullied	-.63
Not Participate	-.54
Teacher No	-.21
Mother No	-.04

Figure 14.5   Coping data: Joint biplot for class work too difficult (Too difficult) and too much work in class (Workload).

established whether the child uses a particular emotion in a particular situation relative to other emotions and situations (see Appendix B for a more precise explanation).

Bullying and not being allowed to participate.   Most children, especially 5, 8, 9, 11, 12, and 13, use an avoidant coping strategy comparatively more often than an approach coping strategy when bullied or being left out (Fig. 14.4). From the external variables we know that these are typically the children who are happy at school and who do well. Only child 1 (who is not so happy at school and does not do so well) seems to do the reverse, that is, using an approach rather than an avoidant coping strategy. Child 10 who according to the external variables is more ill than the others, is particularly angry and annoyed in such situations, while 2 and 7 (relatively robust children) resort toward aggressive behavior. Large differences with respect to social support are not evident.

Class work too difficult and too much work in class.   When faced with too difficult or too much class work, the well-adjusted children, especially 5, 8, 9, 11, 12, and 13, use an approach coping strategy comparatively more often than an avoidant coping strategy (Fig. 14.5). Only child 1 seems to do the reverse, that is, using an avoidant rather than an approach coping strategy. Child 10, who is more ill than the others, is somewhat aggressive in such situations, while the more robust children 2 and 7

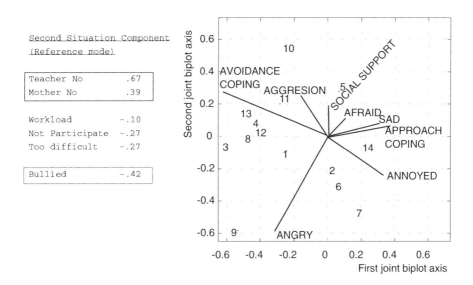

Figure 14.6   Coping data: Joint biplot for restricted by the teacher (Teacher No) and restricted by the mother (Mother No). The graph is valid for being bullied (Bullied) if the emotions–strategies are mirrored around the origin.

are particularly angry and rather annoyed. Large differences with respect to social support are not evident.

Restricted by the teacher and the mother.   For the children who are happy at school and do well (i.e., 3, 4, 8, 12, 13), being restricted by the teacher and to a lesser extent by the mother is handled especially by avoidance coping, children 10 and 11 are comparatively angry as well, child 5 seeks some social support and is relatively angry, afraid and sad (Fig. 14.6). Child 14 is fairly unique, in that he uses more approach coping but it is comparatively sad and annoyed as well. Children 2, 6, and 7 are primarily more angry and annoyed but do not favor one particular strategy over an another, and finally the reaction of child 9 is primarily one of anger more than any other emotion.

Note that being bullied also has a sizeable (negative) loading on the second situation dimension, indicating that being bullied is more complex than is shown in Fig. 14.4. For bullying, the reverse pattern from the one discussed for being restricted by the teacher and the mother is true. So, to get a complete picture for bullying, the information of the two joint biplots should be combined, which is not easy to do. In a paper especially devoted to the substantive interpretation of the data, one would probably search for a rotation of the situation space, so that bullying loads only on one component, and interpret the associated joint biplot especially for bullying. In this chapter, we will not pursue this further.

14.5   ANALYZING THREE-WAY RATING SCALES: PRACTICE

Several aspects of handling three-way rating scale data with the three main three-mode models are explained in great detail in Part II and the issues involved in model selection are the same as for three-way profile data (Chapter 13). Therefore, only a few specific aspects will be discussed in detail here.

14.5.1   Stage 1: Objectives

The major purpose of analyzing three-way rating scales data with three-mode models is to investigate individual differences. If there are no individual differences, the averages across subjects can be computed and the concepts by scales matrix with averages can be analyzed by, for instance, principal component analysis and drawn in a biplot. It is only when there are truly interesting individual differences that it is useful to resort to three-mode analysis. The value of three-mode analysis is enhanced if additional information is available about the subjects, as was shown in the previous example. If such information is not available, all one may say is that there are certain (groups of) subjects who rate the concepts in a different way, and these differences can be described in detail, but it is not possible to indicate which types of subjects make which types of judgments.

14.5.2   Stage 2: Data description and design

Types of data and their requirements.   Prime examples of three-way rating data are those arising from semantic differential research in which subjects have to rate a number of concepts on a set of rating scales. One specific characteristic of semantic differential data is that the scales are *bipolar*, that is, the anchors at the end points of the scales are two adjectives that are assumed to be antonyms, like GOOD and BAD or BEAUTIFUL and UGLY. Another common form of rating scales are Likert scales, where only one adjective determines the scale and the anchors at the end points indicate a superlative degree of the adjective in the positive and negative sense, such as NOT SAD AT ALL and EXTREMELY SAD. Such scales are referred to as *monopolar*.

The distinction between monopolar scales and bipolar scales is important, because in analyzing the latter one cannot center across the scales. The reason is that outcomes of centering depend on the orientation of the scales, and for bipolar scales there is no *a priori* orientation. On the other hand, a monopolar scale for an adjective is unidirectional because it runs from *not at all* (adjective) to *extremely* (adjective). Thus, centering across scales can be appropriate for monopolar scales but is not permissible for bipolar scales. The scales of the Coping data are an example of monopolar scales, while the semantic differential scales in the Multiple personality data (see Section 14.6) are an example of bipolar scales.

Handling many subjects.   One of the difficulties with individual differences analysis is that when the number of subjects becomes very large, it will be difficult (and boring) to discuss and explain the particulars of each individual without losing sight of the overall picture. To avoid this, one may either employ cluster analysis (see Chapter 16 for three-mode clustering) or regression analysis. In the latter case, one uses the background variables as criterion variables and the components as predictors to explain the axes in the subjects space. In Section 14.4.4 we presented the correlations of the external variables with the components for the Coping data. The predicted scores on the criterion variables can also be projected into the subjects space, and this was done for the full set of Coping data in Section 11.4.1 (p. 262). In multidimensional scaling this procedure is known as preference mapping (see Carroll, 1972a; Meulman, Heiser, & Carroll, 1986). A final option would be to use the background variables as instrumental variables by including them in the estimation procedure.

14.6   EXAMPLE: DIFFERENCES WITHIN A MULTIPLE PERSONALITY

14.6.1   Stage 1: Objectives

The aim of the analyses in this section is to present a detailed analysis of a set of bipolar rating scale data. In particular, we will present a model comparison of the three most common three-mode models applied to three-way rating scale data. We will also use constraints on the component spaces and perform rotations of the components and the core array. For this purpose, we will use data from probably the most famous case of a multiple personality (Osgood & Luria, 1954). In each personality, Eve White, Eve Black, and Jane produced scores on 10 semantic differential scales (see Table 14.2 for the scales used) about concepts related to her present life and mental state (see Table 14.8 for the concepts used). Previous three-mode analyses of these data have appeared in Kroonenberg (1983c, 1985a). The present analysis is not a rehash of the previous ones but introduces several more recent aspects that were not present in the former analyses. Notwithstanding, the presentation is not meant as a serious contribution to the understanding of the multiple personality, but will largely remain a demonstration of the analysis of three-way rating scale data. A four-way analysis of the same data can be found in Section 19.5 (p. 476).

14.6.2   Stage 2: Multiple personality data: Data and design

Osgood and Luria (1954) published the scores on semantic differential scales for each of the three personalities Eve White, Eve Black, and Jane, measured at two occasions (Testings I and II). In essence, the data are a four-way data set: concepts by scales by personalities by testings. We will, however, treat them here as a three-way set with

15 concepts $\times$ 10 scales $\times$ 6 administrations, but return to the data as a four-way set in Section 19.5 (p. 476). As semantic differential scales are bipolar scales, the data are centered per scale across concepts. Normalization is not considered necessary because all rating scales have the same range, and we want to have the scales with large variation to influence the analyses more. Thus, the data to be analyzed are $\tilde{z}_{ijk} = x_{ijk} - \bar{x}_{.jk}$, where the first mode consists of concepts, the second of scales, and the third of administrations. For vividness, we will refer to "personalities" rather than the unimaginative "administrations".

The major substantive question is what the similarities and differences are between the three personalities in their vision of their daily life, their family life, and their illness. Moreover, there is the question of whether there is a change in this vision during the period of the three months between measurements. In itself this is not very likely, so that the question should be rephrased in terms of stability of the judgments in the sense of a test–retest reliability.

14.6.3 Stage 3: Model and dimensionality selection

When choosing an appropriate model we will look at a deviance plot (Fig. 14.7) for all three-mode models with three components or less, that is, all models smaller than or equal to the $3 \times 3 \times 3$-Tucker3 model, smaller than or equal to the 3×3-Tucker2 model, and Parafac models with one, two, and three components.

Between-model and within-model selection.   Necessarily, all models with one component in all modes are equal, and as is generally but not always the case, the Parafac model with two components and the $2 \times 2 \times 2$-Tucker3 model are equal as well. The Parafac model with three components shows signs of degeneracy (see Section 8.6.2, p. 185), given that the smallest triple cosine is -0.78, the smallest eigenvalue of the triple-cosine matrix is 0.31, and its condition number is 4.0. However, increasing the number of iterations and decreasing the threshold for convergence to 10^{-10} leads to the same solution, and what is more, from ten different starting points, the proportional fits for all analyses are the same within ten decimals. Furthermore, the normalized core consistency is only 22%, with the Bro core consistency -9.5, and a superdiagonality of 15% (see Section 8.6.3, p. 185). To date, not many detailed analyses have been published with noisy data, convergent models with nearly equal components, and messy Parafac core arrays. The impression is that this is an example of what has been called *bounded degeneracy* by Harshman (2004).

From the deviance plot (Fig. 14.7) it may be concluded that the concepts by scales by administrations $3 \times 2 \times 2$-Tucker3 model, and the concepts by scales 3×2-Tucker2 model are the preferred models. These models are also on the convex hull in the multiway scree plots. The Ceulemans–Kiers st-criterion (see Section 8.5.3, p. 182), which selects the model where the convex hull has its smallest angle, results in our opinion in the too simple $2 \times 1 \times 2$-Tucker3 and the 2×1-Tucker2 models. This choice implies that there is only one scale component. A stability analysis using the bootstrap

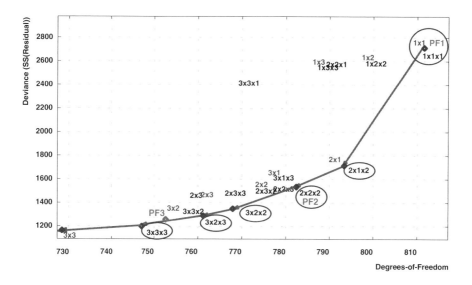

Figure 14.7   Multiple personality data: Deviance plot for Tucker3 ($p \times q \times r$), Tucker2 ($p \times q$), and Parafac (PFs) models with three components or less.

could probably determine whether the additional scale components are stable enough for interpretation, but we will provide further information on the usefulness of all components in the model. Whether a three-component Parafac model would be useful is a matter for further investigation, but it will not be pursued here.

Evaluating the $3 \times 2 \times 2$-Tucker3 model and the 3×2-Tucker2 model.   The next step is to investigate the contributions of the components themselves to see whether indeed all components of all modes provide sizeable contributions. Similarly, one may investigate whether all elements of all modes are sufficiently well represented by the model. In Table 14.1 this information is summarized.

From this table it is clear that in all modes all components contribute to the solution, even though the third component of the concepts and the second of the scales have rather smaller contributions in both models. It is their smaller sizes that makes the *st*-criterion point to the simpler models. Therefore, in the further analysis, we have to pay attention to whether these small components provide sufficient additional information to make it worthwhile to keep them in the model. With respect to the fit of the personalities in the Tucker2 model, all personalities fit well with an exceptionally high fit for Eve Black II. The scales showed more variability with the lowest fit for Fast (fit = 38%, T2; 45%, T3), and the highest for Worthless (88%, T2 & T3). With respect to the concepts, especially Mother and Child had very low fit (26% T2, 37% T3), while Peace of Mind, Hatred, Doctor, and Fraud had fit values of over 80%. It should be borne in mind that data analyzed have been centered in such a way that for

Table 14.1 Multiple personality data: Evaluation of fit of the $3 \times 2 \times 2$-Tucker3 model and 3×2-Tucker2 model

Mode	Sum	Component		
		1	2	3
Tucker3				
Concepts	0.68	0.37	0.26	0.05
Scales	0.68	0.59	0.09	
Personalities	0.68	0.38	0.30	
Tucker2				
Concepts	0.70	0.38	0.26	0.06
Scales	0.70	0.61	0.09	

Tucker 2: Fit of Personalities						
Personality	White I	White II	Jane I	Jane II	Black I	Black II
Fit	0.61	0.69	0.61	0.70	0.66	0.86

each personality and each scale they represent deviations from the mean judgment of that personality on the scale. Thus, concepts which have scores deviating considerably from each personality's mean are the ones with high variability and they typically fit well. The reverse side of this is that concepts that receive generally mean scores have low variability and fit badly. In conclusion, we can say that the chosen solutions seem adequate for interpretation, provided proper information can be derived from the last scale and concept components.

14.6.4 Stage 4: Results and their interpretation

Tucker3 analysis. After an initial inspection it was decided to perform varimax rotations for the personalities and for the scales, but not the concepts.

Concepts. Due to the centering of the data, the concept components are centered as well, so that in the paired-components plots the origin represents the hypothetical concept that has an average score on all the scales. As the concepts are in deviation from their means, they spread out through the entire space (Fig. 14.8). Several groups of concepts can be discerned: Hatred, Fraud, and Me form a group, as do Love, Sex, Spouse, and Job, as well as Doctor, Father, and Mother, while Peace of Mind and Confusion take diametrically opposite positions in space.

Scales. In the two-dimensional scale space (Fig. 14.9) it seems possible to make an interpretation of the two axes as Evaluation and Activity, two of three EPA-dimensions designated by Osgood as characterizing scale spaces in most situations. As all scales

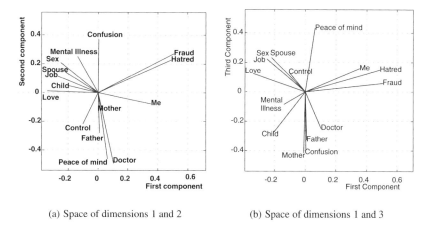

(a) Space of dimensions 1 and 2   (b) Space of dimensions 1 and 3

Figure 14.8   Multiple personality data: Three-dimensional concept space after varimax rotation.

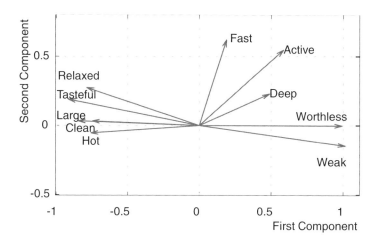

Figure 14.9   Multiple personality data: Scale space.

are bipolar, we were able to reverse the scales Worthless and Weak into Valuable and Strong, so that they side with the other positive evaluative scales[3].

To explore to what extent it was possible to label the axes as we did, first a varimax rotation was applied to the space, to align the scales as much as possible with the axes. As a next step, the scales were given fixed values on only one of the axes, such

---

[3]We have relabeled the end point "Tasty" as "Tasteful" as this seems a better antonym for "Distasteful" in current parlance.

Table 14.2   Multiple personality data: Scale space — Unrotated, varimax rotated, and constraint to equal values

Scale	Unrotated		Rotated		Constraint	
	1	2	1	2	*Evaluative*	*Active*
Hot	**0.311**	−0.059	0.277	−0.153	**0.378**	0.000
Relaxed	**0.323**	0.292	**0.397**	0.177	**0.378**	0.000
Large	**0.307**	0.035	**0.302**	−0.063	**0.378**	0.000
Clean	**0.348**	0.035	**0.341**	−0.075	**0.378**	0.000
Tasteful	**0.375**	0.203	**0.420**	0.076	**0.378**	0.000
Worthless	**−0.407**	−0.002	**−0.387**	0.125	**−0.378**	0.000
Weak	**−0.416**	−0.159	**−0.445**	−0.022	**−0.378**	0.000
Deep	−0.201	0.245	−0.115	0.295	0.000	**0.577**
Active	−0.240	**0.585**	−0.046	**0.631**	0.000	**0.577**
Fast	−0.078	**0.663**	0.133	**0.655**	0.000	**0.577**
Fit solution	68.0%		68.0%		65.2%	

All components are in normalized coordinates, and the varimax rotation was applied to the normalized components (see also Section 10.3.3, p. 243).

that all values on a component were equal. The size of each value on a component was such that the length of the component stayed equal to one (rightmost panel of Table 14.2). The resulting component matrix was kept fixed during the analysis (see also Section 4.10.7, p. 72). This leads to a 2.8% loss of fit, which in this example, given the already high fit, seems an acceptable decrease.

Personalities.   The personality space (Fig. 14.10) is unusual in that the first dimension, which represents the largest variability, shows a contrast between the personalities, while only the second dimension shows that the personalities also have much in common. The conclusion from this figure is that there is a great consistency in the judgments by the personalities over time (i.e., three months) and that Eve White and Jane are virtually indistinguishable, even though Eve White II is more like Eve White I than either of the Janes, and Jane II is more like Jane I than either of the Eve Whites.

Here we also investigate the space further, first by a varimax rotation that indeed nicely aligns the personalities with the rotated axes, and second by forcing the personalities on the axes with equal coefficients (see Table 14.3). The latter setup led to a negligible loss of fit of 0.5%, so that we have no problems accepting such a constraint.

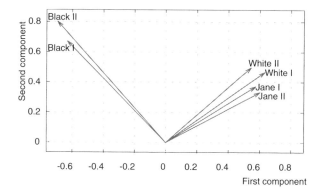

Figure 14.10   Multiple personality data: Personality space.

Table 14.3   Multiple personality data: Personality space. Unrotated, varimax rotated, and constraint to equal values.

	Unrotated		Rotated		Constraint	
Personality	1	2	1	2	1	2
White I	0.632	0.468	**0.545**	0.010	**0.500**	0.000
White II	0.545	0.498	**0.514**	0.064	**0.500**	0.000
Jane I	0.573	0.372	**0.470**	−0.021	**0.500**	0.000
Jane II	0.601	0.333	**0.466**	−0.055	**0.500**	0.000
Black I	−0.632	0.671	−0.017	**0.652**	0.000	**0.707**
Black II	−0.695	0.801	0.011	**0.753**	0.000	**0.707**
Fit solution	68.0%		68.0%		67.5%	

All components are in normalized coordinates, and the varimax rotation was applied to the normalized components (see also Section 10.3.3, p. 243).

Concept space after simplification.   Given that the scale space and the personality spaces could be considerably simplified, the basic analysis was rerun with both constraints in place. This resulted in the concept space of Table 14.4, which was first subjected to a varimax rotation on the normalized coordinates. This equivalent with the Harris–Kaiser independent cluster rotation (see Section 10.3.3, p. 243). The basic groups of items found before (Fig. 14.8) are also easily identified in this simplified solution.

Core array after simplification.   The core array in Table 14.5 illustrates one of the difficulties in interpreting a three-mode analysis. In one interpretation, the core

Table 14.4 Multiple personality data: Concept space after simplification of both scale and personality spaces. Fit of solution is 63%

	Varimax rotated		
Concept	1	2	3
Hatred	**0.57**	0.01	−0.03
Fraud	**0.56**	0.04	−0.08
Me	**0.32**	0.01	**0.33**
Doctor	−0.18	**0.55**	0.17
Mother	−0.13	**0.32**	−0.19
Father	−0.18	**0.31**	0.02
Love	−0.26	−0.25	−0.00
Spouse	−0.11	−0.28	−0.01
Sex	−0.10	**−0.37**	−0.04
Job	−0.12	**−0.42**	0.08
Peace of mind	−0.06	0.02	**0.60**
Confusion	0.07	0.09	**−0.54**
Child	−0.22	0.10	−0.27
Mental illness	−0.01	−0.17	−0.23
Self control	−0.15	0.02	0.18
Percentage explained	33%	16%	15%

Components are in normalized coordinates.

array contains the weights for the structural image, that is, g_{pqr} is the weight for $(\mathbf{a}_p \otimes \mathbf{b}_q \otimes \mathbf{c}_r)$. If g_{pqr} is small, the (p, q, r) combination is irrelevant; if it is larger, the combination is important, and how important can be gauged from the amount of explained variability based on the core element (g_{pqr}^2/SS(Total)). This technical explanation is, however, only useful if we have relatively straightforward interpretations of the components as well as a restricted number of large core elements. In the present case, it was possible to derive really simple components for the scales (Evaluation and Activity) and for the personalities (Eve Black and Eve White/Jane), but labels for the components for the concepts are far less obvious. Moreover, the simplicity has been bought by having a considerable number of component combinations, which need to be interpreted, because the core array is far from simple. In Section 10.4.5 (p. 249) Kiers's (1998d) procedures are discussed for simplifying both the core array and the components at the same time. However, due to the introduction of the fixed constraints, these procedures can only be applied to the concept mode, which is not going to help much.

Table 14.5 Multiple personality data: Core array after simplification of both scale and personality spaces. Fit of solution is 63%

Concept components	No.	Unstandardized core Evaluative	Active	Explained variability Evaluative	Active
Eve Black					
Love, Child vs. Hatred/Fraud	1	−15.6	−9.6	0.058	0.022
Doctor/Father/Mother vs. Sex,Job	2	19.5	15.0	0.090	0.053
Peace of mind vs. Confusion	3	18.7	7.9	0.083	0.015
Eve White & Jane					
Love, Child vs. Hatred/Fraud	1	32.6	2.8	0.252	0.002
Doctor/Father/Mother vs. Sex,Job	2	3.5	8.4	0.003	0.017
Peace of mind vs. Confusion	3	12.6	−7.5	0.038	0.013

The solution, as in many other Tucker3 analyses, is to abandon the idea of trying to find nameable components for all modes that is finding a component interpretation. It seems better to strive for a subspace interpretation by examining joint biplots of two display modes given one component of the remaining reference mode (see Section 11.5.3, p. 273).

Graphical representations of Eve Black and Eve White/Jane. The analysis of the Multiple personality data shows that there is a near-perfect split between Eve Black and Eve White/Jane. Thus, as a result of the initial analysis, one might consider averaging over each of the personalities and analyzing them separately, but in this way one loses the coherence and comparability between solutions. The joint biplots of scales and concepts as display modes and with the personalities as reference mode are a nice compromise, because both Eve Black (Fig. 14.11(a)) and Eve White/Jane (Fig. 14.11(b)) can have their own joint biplot. Because the core slices from Table 14.5 have order 3 by 2, only two-dimensional joint biplots are possible (see Section 11.5.3, p. 273).

Tucker2 analysis. Given the comparable fit of the $3 \times 2 \times 2$ Tucker3 model and the 3×2 Tucker2 model (68.0% versus 69.9%), it comes as no surprise that the concept spaces and the scale space are virtually identical. The only difference in the two models lies in the $3 \times 2 \times 2$ core array of the Tucker3 model and the $3 \times 2 \times 6$ extended core array of the Tucker2 model with a core slice for each personality (see Table 14.6).

Conclusions of the Tucker analyses. From the results of the two Tucker analyses, we may conclude that there are vast differences between the personalities. However,

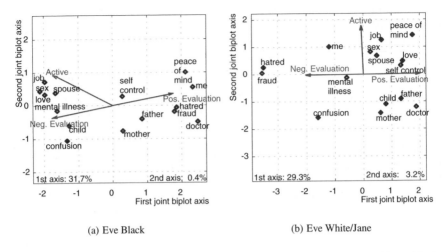

(a) Eve Black (b) Eve White/Jane

Figure 14.11 Multiple personality data: Joint biplots for each personality with the relationships between concepts and scales.

Jane is virtually indistinguishable from Eve White. Apart from that, there are also a number of concepts on which the personalities agree.

- Black and White/Jane agree on

    - Equally positive evaluation of Peace of Mind/Me and negative about Confusion;
    - Neutral (White & Jane) to Positive Evaluation (Black) of Doctor/Father/Mother versus Sex, Job, and Spouse;
    - Fast/Deep/Active of Sex/Job/Spouse, and not Doctor/Father/Mother.

- Black and White/Jane disagree on

    - Evaluation of Love/Child: White & Jane — positive; Black — negative;
    - Evaluation of Hatred/Fraud/Me: White & Jane — negative; Black — positive.

14.7 CONCLUSIONS

This chapter provided an overview of the way one might go about analyzing three-way rating scale data. The discussion touched upon the differences between three-way profile data and three-way rating-scales data. These differences primarily reside in the way the data are conceptualized, either as scores acquired by subjects on variables or

Table 14.6 Frontal slices of the Tucker2 extended core array

Concept components	Unrotated scale components				Constraint scale components			
	I		II		I		II	
	S1	S2	S1	S2	S1	S2	S1	S2
Eve White								
C1	**12.3**	−1.0	**9.9**	−3.3	**15.9**	3.3	**16.0**	−0.8
C3	1.8	7.3	7.3	5.1	4.1	4.4	**9.9**	5.8
C2	**14.1**	−5.4	**14.4**	−4.8	**8.8**	−4.7	**5.4**	−3.8
Jane								
C1	**11.9**	−4.3	**13.5**	−3.8	**14.8**	0.2	**16.4**	0.5
C3	−0.4	8.6	−2.1	4.3	−0.3	5.4	−0.5	2.1
C2	**9.9**	−2.7	**10.8**	−2.5	**6.3**	−2.5	**7.5**	−0.9
Eve Black								
C1	**−19.7**	−6.5	**−22.6**	−4.2	**−12.0**	**−9.7**	**−14.2**	−6.8
C3	−3.2	2.6	0.6	4.4	8.5	6.8	**12.4**	**10.8**
C2	**12.1**	3.0	**15.4**	3.5	**13.5**	7.1	**15.2**	5.9

Core elements with more than 2% explained variability in bold.

scores assigned by subjects in their role as judges of the similarity between a concept and a scale. The distinction between monopolar and bipolar scales was pointed out, as well as their preferred centerings and normalizations.

The modest example of the Coping data served to illustrate some of the considerations going into analyzing monopolar scales, while the bipolar semantic differential judgments of Osgood's Multiple personality data were analyzed in much greater detail using constraints and transformations.

CHAPTER 15

EXPLORATORY MULTIVARIATE LONGITUDINAL ANALYSIS

15.1 INTRODUCTION[1]

15.1.1 What are multivariate longitudinal data?

Multivariate longitudinal data are nearly by definition three-way data, as they consist of scores of observational units (subjects, objects, etc.) on variables measured at various time points (occasions); that is, they form a fully crossed design. Higher-way data can be imagined in which the fourth mode consists of different conditions under which the longitudinal data have been collected. However, such data sets are rare. Many longitudinal data sets have only one or two variables, and observations consist of single measurements of these variables over a long series of measurements. In such cases, there are really only two modes, variables and points in time. Alternatively, one or two subjects are measured on a single variable for a long period of time, in

---

[1]Parts of the text in this chapter have been taken from Kroonenberg, Lammers, and Stoop (1985b). Reproduced and adapted with kind permission from Sage Publications.

Applied Multiway Data Analysis. By Pieter M. Kroonenberg   **373**
Copyright © 2007 John Wiley & Sons, Inc.

which case we again have two modes, subjects and points in time. The interest in this chapter is in the situation when there are so many subjects, variables, and points in time that some type of condensation of at least two modes is required in order to describe or model the patterns present in the data. In this way, the type of questions that can be asked mostly have the form: Which groups of subjects have for which variables different patterns of development over time?

15.1.2 Types of models

A useful way of ordering techniques for multivariate longitudinal data is to look at the different roles time can play in the analysis of longitudinal data. First, time can be included in the definition of stochastic models. Second, the ordered aspect of time can be included in descriptive, nonstochastic models by, for instance, assuming smooth changes from one time point to the next. Finally, time can be used as a post-hoc interpretational device in otherwise unrestricted modeling.

Multivariate longitudinal data and multiway models *Stochastic modeling*

- *without latent structures*: general linear model [repeated measures analysis of variances, autoregressive models, multivariate time series models];

- *with latent structures*: structural equation models; latent growth curves

  - non-multiway: subjects are exchangeable [multivariable–multioccasion or multimode models];

  - multiway: information on subjects is present in the model via components and/or factors [multimode models: multiway variants]

Descriptive modeling

- component models with functional restrictions on the time modes, for example, smooth functions, explicit time-related functions (growth functions).

- component models with constraints on the components, for example, smoothness constraints

Time as interpretational device

- Tucker1 model [only components for variables]

- Tucker2 model with uncondensed time mode [no time components, only components for variables and subjects]

- Tucker3 model [components for each of the three modes, including the time mode]

- exploratory multimode structural equation models

- STATIS

- using rotations to a time structure (e.g., orthogonal polynomials) or toward smooth functions

---

15.2 CHAPTER PREVIEW

The core of this chapter is a detailed exposition of exploratory three-mode models, which can be used for longitudinal analysis with a special emphasis on three-mode models for continuous variables. We will concentrate on data consisting of more than a few subjects, variables, and time points, and designs and research questions taking individual differences into account. Moreover, we will look at truly longitudinal data in which the same subjects are measured over time, rather than cross-sectional data consisting of several samples of subjects.

To illustrate the way three-mode component models can provide useful descriptions of multivariate longitudinal data, we will present detailed analyses of two data sets. In particular, one example dealing with the growth of 188 Dutch hospitals over time Kroonenberg et al. (1985b), and one example about the morphological growth of 30 French girls (Kroonenberg, 1987a, 1987b).

15.3 OVERVIEW OF LONGITUDINAL MODELING

15.3.1 Stochastic modeling: Using explicit distributional assumptions

Doubly multivariate repeated measures.   A standard, mainly experimental, setup consists of a single dependent variable measured several times with a between-subject design. The aim of such studies is to test differences over time between means of groups that have undergone different (levels of) treatments. Typically, such designs are analyzed by repeated measures analysis of variance. The data are mostly analyzed using multivariate analysis of variance, where multivariate refers to the repeated measurement of the same variable. An extension is the much less common doubly-multivariate repeated measures analysis of variance design, where several variables are measured over time. In fact, the basic data for such a design are three-way data with a between-subject design.

Such designs make assumptions about the distributions of the variables, which are then used to construct significance tests as to whether the means or trends in means are significantly different from each other. Thus, the observations on the subjects are seen as the result of repeated sampling from specific distributions, but the subjects are not really seen as individuals. Anybody else drawn randomly from the same distribution would have served as well. What is lacking in such an approach is attention to the individuals themselves; moreover, the possibly evolving structure of the variables is not a central part of the investigation.

Structural equation models. As part of the solution, specific models have been devised, which on the one hand specify how the latent variables are related to the observed measurement (the measurement model), and on the other hand specify the relationships between the latent variables (the structural model). The complete model is now generally referred to as a *structural equation model* (SEM). A detailed, both conceptual and technical, introduction is contained in the book by Bollen (1989), and more recent books dealing explicitly with longitudinal designs for continuous data are Verbeke and Molenberghs (2000) and Fitzmaurice et al. (2004).

The development, analysis, and testing of structural equation models is generally carried out using (means and) covariance matrices as a starting point and hypothesized models are used to calculate estimates for these means and covariances. The adequacy of the fit between the observed values and those reconstructed on the base of the model is assessed with various fit indices; see Hu and Bentler (1999) for an overview. Generally, goodness-of-fit testing relies on distributional assumptions about multivariate normality, but some of the procedures have been shown to be rather robust given large samples. Nonnormal procedures and tests have been developed as well; see Boomsma and Hoogland (2001) for an overview.

Structural equation models have frequently been used for multivariate longitudinal data, especially in panel data with a limited number of variables and a small to medium number of time points. The analyzed covariance matrices have the form of multivariable–multioccasion covariance matrices, or multimode covariance matrices. An example of a structural equation model is shown in Fig. 15.1.

In non-three-mode covariance models, the individuals themselves do not really play a role and they are exchangeable in the sense that any other subject from the same population would have done as well. Within this framework, autoregressive models and dynamic factor models have been proposed. For an overview of these multivariate longitudinal models and multiway modeling see Timmerman (2001).

Latent growth-curve models. Another type of data often analyzed with structural equation models are latent growth-curve models in which not only the (co)variances but also the (structured) means of the latent variables are estimated (e.g., see Bijleveld, Mooijaart, Van der Kamp, & Van der Kloot, 1998, for an overview of the use of structural equation models for longitudinal data) and an in-depth treatment of the whole field Verbeke and Molenberghs (2000) and Fitzmaurice et al. (2004).

Three-mode covariance models. Three-mode covariance models are characterized by the inclusion of information on subjects through components or factors for the subjects (i.e., component scores). The basic data used for modeling are still the multimode covariance matrices, but the latent models can be formulated to include either the component scores themselves, or, more commonly, only their (co)variances or correlations. Oort (1999) provides the most comprehensive overview of multimode common factor models, while Oort (2001) deals entirely with longitudinal models in this context. Finally, Kroonenberg and Oort (2003a) discuss the relative advantages and disadvantages of the descriptive and the confirmatory approach toward three-mode covariance modeling; one of their examples deals with longitudinal data.

15.3.2 Descriptive modeling

As the major part of this chapter is concerned with descriptive, exploratory modeling, we will be touch upon some general principles that may be used in component models. It is assumed that the researcher intends to investigate the changes over time between the variables in some condensed form such as linear combinations, using either time itself or composite functions that describe the time development of the scores. There are basically two approaches: either one uses predetermined functions and estimates their parameters alongside other parameters of the component models, or one imposes general restrictions on the components such as smoothness and monotonicity.

Component models with functional restrictions on the time modes.   In order to use specific functions, one needs to have a good grasp of the underlying processes. Latent growth modeling is an example in which this is explicitly the case and in which time-related functions can be used. For instance, in Bus's Learning-to-read study discussed in Chapter 5 of Timmerman's (2001) monograph, the psychological and experimental information is that initially the subjects do not have the accomplishments to fulfill the task (baseline). As they are learning how to master the task, their performance improves until the task is more or less fully mastered. Thus, some kind of S-shaped curve will describe the performance over time. Following Browne (1993) and Browne and Toit (1991), Timmerman (2001, Chapter 5) used a Gompertz function to model latent growth curves. In other situations, other curves, such as sinusoidal curves in periodic phenomena, will be more appropriate.

Component models with general constraints on the components.   Timmerman (2001, Chapter 4) discusses a number of possibilities for including smoothness constraints in three-mode component models. In particular, she outlines the use of B-splines for general smoothness and I-splines for monotonic increasing functionsI-splines,monotonicity. Note that this approach is not limited to longitudinal designs. The assumption in this context is that from one observation to the next, regular changes can be expected and approximated by smooth curves. This implies that there is sufficient autoregressiveness in the process itself. For example, yearly measurements of yield in an agricultural experiment do not fulfill this condition, because weather conditions are not a very continuous phenomenon from one year to the next and certainly not an autoregressive one.

Time as an interpretational device.   Whereas in the studies discussed in the previous sections time was included in the design and analysis, general descriptive models do not always have facilities for including time explicitly. The analysis is performed as if the time order does not exist, and only afterwards, during the interpretation, do we call in the help of the ordered aspect of time to evaluate the solution. Excluding an essential part of the design has it drawbacks, but on the other hand not making *a priori* assumptions has its advantages as well.

Several devices are available to use time in a post-hoc manner. One can make trajectories in dimensional graphs connecting time points and inspect to what extent the time order is indeed reproduced in the analysis. Moreover, one might get an idea how to include time in a more sophisticated statistical model. A further possibility lies in optimally rotating time components toward ordered functions, such as higher-order polynomials or otherwise smooth functions.

15.4   LONGITUDINAL THREE-MODE MODELING

In this section we explicitly discuss the analysis of multivariate longitudinal data with three-mode component models. Whereas in multiway profile data the main dependence is between variables, in longitudinal data the serial dependence or autocorrelation between observations on different occasions is important as well. Interactions between the two kinds of dependence introduce further complications.

15.4.1   Scope of three-mode analysis for longitudinal data

The promise of three-mode principal component analysis and its analogues in longitudinal multivariate data analysis lies in the simultaneous treatment of serial and variable dependence. In the Tucker3 model the serial dependence can be assessed from the component analysis of the time mode, the variable dependence from the variable mode, and their interaction from the core array or from the latent covariance matrix (see Section 15.4.2). Using standard principal component analysis to analyze the data either arranged as a tall combination-mode matrix of subjects × occasions by variables, or arranged as a wide combination-mode matrix of subjects by variables × occasions, the variable and serial dependence, and their interactions become confounded.

15.4.2   Analysis of multivariate autoregressive processes

Lohmöller (1978a, 1978b, 1989) made several proposals toward the interpretation of serial and variable dependence via three-mode principal component analysis of multivariate longitudinal data. His contribution will be discussed in this section.

Introduction.   Basic to Lohmöller's approach is the assumption that the changes in the variables and the scores of the subjects on these variables can be modeled by multivariate autoregressive processes. Given that the assumption is tenable, he shows how one can get an indication of the size of the parameters of the assumed autoregressive process from the results of the three-mode analysis. In other words, once a three-mode analysis has been performed, an interpretation of its results can be given in terms of the parameters of the autoregressive process.

Lohmöller's procedure to arrive at these indicators or "estimators" is rather indirect. Some 625 data sets were first generated according to specific autoregressive processes, then analyzed with three-mode principal component analysis. Empirical "estimating" equations were derived for the parameters of the autoregression processes by regressing them on the parameters in the three-mode analyses. Lohmöller himself recognized that the procedure has serious drawbacks, as for each new kind or size of data set new simulation studies have to be performed. On the other hand, via his simulation studies he was able to investigate which results in a three-mode analysis are particularly sensitive to changes in specific parameters in the autoregressive models, and which results reflect the general characteristics of the autoregressive models. Lohmöller (1978a, 1978b) pointed out that, in fact, three-mode path models are to be preferred, because in those the autoregressive processes can be modeled directly. Some data sets, however, are too large to be handled by such an approach, because the $JK \times JK$ multimode covariance matrix can become too large.

Component analysis of time modes.   One of the problems in three-mode analysis of longitudinal data is the interpretation for the decomposition of the time mode, as its correlation matrix is more often than not a (quasi-)simplex. Entries of a simplex are such that the correlations between two time points are a decreasing function of their difference in rank order. Typically the bottom-left and top-right corners of the correlation matrix have the lowest entries (for examples see Table 15.2). As Guttman (1954) has shown, simplexes have standard principal components. In particular, the components of equidistant simplexes can be rotated in such a way that the loadings on the first component are equal, those on the second component are a linear function of the ranks of the time points, those on the third component are a quadratic function of the ranks, and so on. After extraction of the first two principal components, all variables have roughly the same communalities, and the configuration of time points resembles a horseshoe or "U", the opening of which increases with the relative size of the first eigenvalue.

The problem is not so much the standard solution as the fact that there are many different processes that could have generated the simplex. In other words, certain processes are sufficient for the appearance of a simplex in the correlation matrix, but they are not necessary. It is therefore very difficult to "prove" the existence of such processes from the appearance of a simplex in the time mode without a substantive theory why the processes should be present.

One kind of change that produces correlation matrices with a (Markov) simplex structure (e.g., see Jöreskog, 1970) is a first-order autoregressive processes, to be discussed later. Thus, a simplex-like correlation matrix may be explained by such

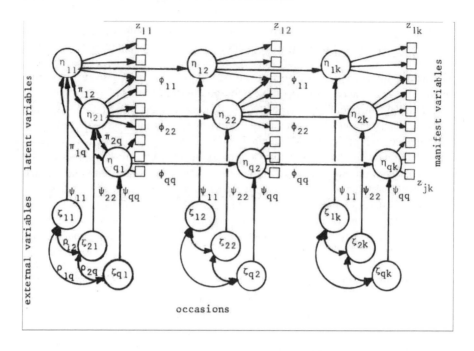

Figure 15.1   A three-wave multivariate autoregressive model with per phase three correlated latent variables (η), nine observed variables (z), and three correlated external variables (ζ)

an autoregressive process. Furthermore, a process of steady growth with a level and a gain component produces a correlation matrix with a (Wiener) simplex form (e.g., see Jöreskog, 1970). Thus, one may describe the underlying change process, at least approximately, by a level and a gain component. In the case that one has a simplex-like correlation matrix, the results of a component analysis on this correlation matrix may therefore be interpreted both in terms of parameters of the autoregressive process and in terms of a level and a gain component, provided the subject matter warrants the use of such models.

Autoregressive processes.   In Fig. 15.1 an example is given of the kind of autoregressive models one might consider. The model was suggested by Lohmöller (1978a, 1978b) in this context, and we will follow his description.

The structural part of the model (see also Section 15.3.1) has the form of a multivariate regression equation; it is assumed that the state of the latent variables η_k at occasion k depends on only two influences: the state of the variables at occasion $k - 1$ (i.e., it is a first-order process), and the state of the external variables ζ_k at occasion k,

$$\eta_k = \Phi\eta_{k-1} + \Psi\zeta_k, \quad k = 1, \ldots, K, \tag{15.1}$$

where η_k and η_{k-1} are the vectors of all Q latent variables. In the model it is assumed that the external variables ζ_k and ζ'_k ($k \neq k'$) are uncorrelated, which is almost always an oversimplification as it implies that all time-dependent influences are included in the model. We have written the external variables, possibly slightly incorrectly, as latent rather than manifest ones. Finally, it is assumed in this model that the latent and external variables are normalized.

The matrix Φ, called the transition matrix, describes how strongly the latent variables at occasion $k - 1$ influence those at occasion k. When Φ is diagonal, as is the case for the model in Fig. 15.1, the latent variables only influence themselves, and no other latent variables (i.e., $\phi_{qq'} = 0, q \neq q'$). When Φ is diagonal, the changes in the component structure of the variables are entirely due to external influences.

The matrix Ψ describes the influences of external variables on occasion k on the latent variables on occasion k. The external variables represent the entire influence of the environment on the latent variables. When Ψ is diagonal, as in the model in Fig. 15.1, each external variable only influences one latent variable (i.e., $\psi_{qq'} = 0, q \neq q'$). Note that here the matrices Ψ and Φ are assumed to be independent of k, so the model assumes that the first-order influences remain identical over time. Differences in influence over time of both latent and external variables cannot be accounted for by this particular autoregressive process, and as such it is almost always an oversimplification of reality. The structural part has been entirely discussed until now in terms of latent variables, and therefore we also need a measurement model to link the data with the structural model. In the present case, the measurement model is simply the three-mode principal component model itself, in which the components are the latent variables of the autoregressive model.

Latent covariance matrix.   Before entering into a discussion of the role autoregressive models can play in three-mode analysis, it is necessary to look at what we will call the *latent covariance matrix* $S = (\sigma_{qr,q'r'})$, called *core covariance matrix* by Lohmöller (1978a, 1978b). The $\sigma_{qr,q'r'}$ are the inner products of the elements of the Tucker2 core array

$$\sigma_{(qr),(q'r')} = \sum_{i=1}^{I} h_{i(qr)} h_{i(q'r')}, \tag{15.2}$$

where we assume that the I observational units or subjects constitute the unreduced first mode in a Tucker2 model with a second mode of variables, and a third mode of occasions (see Fig. 15.2). Depending on the scaling, the covariances may be cross products, 'real' covariances, or correlations. A similar covariance matrix could be defined for the Tucker3 model, but condensing a mode first and then computing the covariance matrix for it does not seem sensible, and no experience has been gained with a similar matrix for the Parafac model.

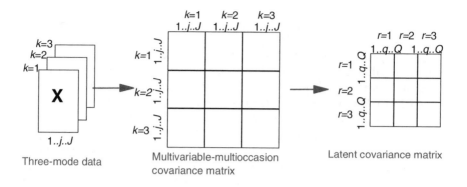

Figure 15.2 From $I \times J \times K$ three-way data to the $JK \times JK$ multimode covariance matrix (multivariable–multioccasion matrix) for the I subject scores on the JK variable–occasion combinations to the latent covariance matrix of the I component score on the combinations of Q variable components and R occasion components.

The core elements $h_{i(qr)}$ may be interpreted as scores of the observational units on the $Q \times R$ *structured components* of the second and third modes. In this context the αth structured component is a $J \times K$ vector $\boldsymbol{\xi}_\alpha$ with elements ξ_{jk}^α, and $\xi_{jk}^\alpha = b_{jq}c_{kr}$, with $jk = 1, \ldots, JK$, and $\alpha = 1, \ldots, QR$. In the example in Section 15.5, one of the structured components, for instance, is labeled as *gain in degree of specialization*. In that case an h_{iqr} represents the gain in degree of specialization of the ith hospital.

The value of $\sigma_{\alpha\alpha'} = (\sigma_{(qr),(q'r')})$ thus indicates the covariance of the αth and α'th structured components. Within structural equation modeling where the mode of observational units is stochastic, the latent covariance matrix arises in a natural way. If we follow Bloxom's (1968) formulation, the Tucker2 model can be written as

$$\mathbf{x} = (\mathbf{C} \otimes \mathbf{B})\boldsymbol{\xi} + \epsilon, \tag{15.3}$$

where $\boldsymbol{\xi} = (\boldsymbol{\xi}_\alpha)$ is the random vector of unobserved scores on the QR structured components, $\mathbf{x}$ the random vector of observations on the $J \times K$ variables, and ϵ the random vector of unobserved residuals. If we indicate the $JK \times JK$ residual covariance matrix by $\boldsymbol{\Theta}$, then the $JK \times JK$ multimode covariance matrix $\boldsymbol{\Sigma}$ is modeled as

$$\boldsymbol{\Sigma} = (\mathbf{C} \otimes \mathbf{B})\mathbf{S}(\mathbf{C}' \otimes \mathbf{B}') + \boldsymbol{\Theta}, \tag{15.4}$$

where $\boldsymbol{\Theta}$ contains the residual variances on the diagonal and nonzero off-diagonal elements in the case of correlated residuals. Loosely speaking, one may say that the latent covariance matrix underlies the observed multimode covariance matrix and embodies the basic covariances present in the data. This interpretation is based on the fact that it contains the covariances between the unobserved scores on the structured components.

Normalization of S.   The options for normalization of the latent covariance matrix parallel those in ordinary PCA. The component scores may be set equal to the lengths of the eigenvalues (principal coordinates), that is the distance form, or they may have length one (normalized coordinates), that is the covariance form (see Section 9.3.1, p. 215). Using normalized score coordinates, the size of the data is transferred to the variable components **B** and/or the occasion components **C**. This corresponds with the usage in, for instance, psychology, where the variables are in principal coordinates. Lohmöller generally scaled the latent covariance matrix such that score components are in normalized coordinates, and we have done the same in the example in Section 15.5 for comparability with his results.

The major purpose in discussing the latent covariance matrix from a Tucker2 model is that its structure can be used to investigate the parameters of postulated autoregressive processes underlying the observations. Lohmöller (1978a, 1978b, 1989) derived general relationships between latent covariances and general characteristics of the autoregressive processes based on a series of simulation studies.

Apart from this way of interpreting the latent covariance matrix, one may interpret the variances and covariances directly, provided the score components are in principal coordinates. The latent variances of the scores on the structured components, σ_{aa}, may be divided by the total variation present in the data – SS(Total) – so that they can be interpreted as the proportions variance accounted for by the structured components. After all

$$\sigma_{\alpha\alpha} = \sigma_{(qr)(qr)} = \sum_{i=1}^{I} h^2_{i(qr)},$$

and the squared elements of the core array can be interpreted as explained variances. The covariances $\sigma_{\alpha\alpha'}$ may be transformed into direction cosines between the structured components α and α':

$$\sigma^*_{aa'} = \sigma_{\alpha\alpha'}/\sigma_{\alpha\alpha}^{1/2}\sigma_{\alpha'\alpha'}^{1/2}, \tag{15.5}$$

Interpreting latent covariances in this way is more direct and has a wider applicability than the interpretation via parameters of autoregressive processes. On the other hand, the latter interpretation gives more specific and more substantive information because of the postulated model.

Linking autoregressive parameters to three-mode results.   In the introduction to this chapter we referred to two major sources of dependence in multivariate longitudinal data: variable and serial dependence. It is of interest to know whether these kinds of dependence influence each other. One may have a structure between variables (*variable dependence*) that is not changing over time. Furthermore, subjects may maintain their relative positions on the variables irrespective of the structure of the variables; that is, there is stability of the variables or high autocorrelations (*serial dependence*). Finally, variables and subjects may change simultaneously, in which case there is neither stability nor stationarity.

A set of latent variables is said to be *stationary* when the same component structure is present on all occasions. Furthermore, the set is *homogeneous* when the variables are highly correlated so that they can be represented in a low-dimensional space by a few components. In autoregressive processes the homogeneity is indicated by the covariances of the latent variables at each time point (the π_{iqq} in Fig. 15.1).

In three-mode component analysis we derive one set of orthonormal variable components over all occasions simultaneously, and the dimensionality of the component space is an indication of the overall homogeneity. As there is only one component matrix for the variables, one could get the impression that the model does not allow for nonstationarity. This, for instance, is indeed the case in a model without a core array like the Parafac model (see Section 4.6.1, p. 57), but not in the Tucker2 model, in which the deviations from stationarity show up in the core array and the latent covariance matrix (see Section 15.4.2). Lohmöller's contribution is that he attempted to investigate what kind of stationarity could be inferred from the latent covariance matrix. He claimed that for an autoregressive process as shown in Fig. 15.1, increasing and decreasing homogeneity can be gleaned, *ceteris paribus*, from the size and the signs of the covariances between the latent variables in the latent covariance matrix. We will return to this point in some detail when discussing the example in Section 15.5.6.

A (latent) variable will be called *stable* when the relative positions of the observational units on that (latent) variable stay the same in time. The stability of a (latent) variable may be judged from the covariances of the latent variables on different occasions. A set of variables will be called stable when all variables are stable. In autoregressive processes the stability of a latent variable η_q is given by ϕ_{qq} (Fig. 15.1); ϕ_{qq}^2 indicates to what extent the latent variable is determined by its predecessor. Stable variables are sometimes called *trait-like*, that is, mainly determined by the defining construct or trait, and unstable variables are sometimes called *state-like*, that is, mainly determined by the moment at which they are measured (see, Cattell, 1966, p. 357). The size of the covariances of a latent variable between time points is thus an indication of the stability of a variable. High values indicate a trait-like and low values a state-like variable. The overall stability of a set of variables, $\bar{\phi}$, may be determined from the first eigenvalue of the time mode, given that first-order autoregressive processes underlie the data (see Lohmöller, 1978a, p. 29). When a second-order autoregressive model applies, the stability will most likely be overestimated using the first eigenvalue.

Lohmöller also investigated the structure of the latent covariance matrix for autoregressive models like those in Fig. 15.1 in case of different stabilities of the latent variables, and in case of equal stabilities on changing dimensions. However, we will not go into that part of his study.

Finally, it is interesting to consider the situation in which all latent variables have equal stability (i.e., uniform autocorrelations) so that no partial cross-lag correlations exist. For this situation, Lohmöller (1978a, p. 4) showed that the latent covariance matrix is a $(QR \times QR)$ identity matrix, $\mathcal{J}$ or a diagonal matrix $\mathcal{D}$, depending on the particular scaling of the components. This means that the three-mode model for the observed covariance matrix Eq. (15.4) reduces to

$$\Sigma = (\mathbf{C} \otimes \mathbf{B})\mathcal{D}(\mathbf{C} \otimes \mathbf{B}) + \Theta. \tag{15.6}$$

This model, which was also described by Bentler and Lee (1978b), may be used as a kind of "null-hypothesis" against which to evaluate latent covariance matrices.

Discussion.   The above approach to evaluating change phenomena in multivariate longitudinal data very much depends on the appropriateness of the multivariate autoregressive models. It is also still very sketchy from a mathematical point of view and therefore requires further investigation. Further practical experience is also necessary to assess its potential. It seems that in some cases the assumption of an underlying autoregressive process is not unreasonable, as in the example in Section 15.4.2. In connection with this example we will discuss rough and ready ways to assess whether the assumption of the autoregressive model is tenable.

Lohmöller's major contribution has been that he provides a framework for the interpretation of multivariate longitudinal data, which cannot easily be handled directly by causal modeling or time series analysis. One of the problems in that area is that the present estimation procedures for multimode covariance matrices may have difficulty handling the size of covariance matrices considered here, apart from the fact that there are often not enough observations to make maximum likelihood and generalized least-squares estimation feasible. The advantage of such approaches, however, is that very refined modeling is possible within a hypothesis-testing framework.

Oort (2001) discussed confirmatory longitudinal three-mode common factor analysis within the structural equation framework and Kroonenberg and Oort (2003a) presented a comparison between two approaches of handling multimode covariance matrices with three-mode models, but the latter authors did not discuss Lohmöller's approach.

15.5 EXAMPLE: ORGANIZATIONAL CHANGES IN DUTCH HOSPITALS

15.5.1 Stage 1: Objectives

In order to gain some insight into the growth and development of large organizations, Lammers (1974) collected data on 22 organizational characteristics (variables) of 188 hospitals in The Netherlands ("subjects") from the annual reports of 1956–1966 (time). His main questions with respect to these data were (1) whether the organizational structure as defined by the 22 variables was changing over time, and (2) whether

Table 15.1   Hospital study: Description and mnemonics of the variables

	Variables	
Var. No.	Label	Description
1	Training	Training capacity
2	Resear	Research capacity
3	FinDir	Financial director
4	Facili	Facility index
5	QExtern	Ratio of qualified nurses in outside wards
6	QRatio	Ratio of qualified nurses/total number of nurses
7	Functi	Number of functions
8	Staff	Total staff
9	RushIn	Rushing index
10	ExStaf	Executive (managerial and supervising) staff
11	NMedProf	Nonmedical professionals
12	Admin	Administrative (i.e., clerical) staff
13	ParaMd	Paramedical staff
14	NonMed	Other nonmedical staff
15	Nurses	Total number of nurses
16	Beds	Total number of beds
17	Patien	Total number of patients
18	Opennns	Openness
19	ClinSp	Main clinical specialties
20	OutPSp	Main outpatient specialties
21	ClinSub	Clinical subspecialties
22	OutPSub	Outpatient subspecialties

there were different kinds of hospitals with different organizational structures and/or different trends in their structures. These two questions will be taken up in this section (for an earlier analysis see Kroonenberg et al., 1985b). In the next section we will try to assess (3) the stability of the latent variables, (4) the stationarity of the latent-variable domain and (5) the interaction between serial and variable dependencies. In particular, we will try to assess the parameters of a possibly underlying autoregressive process.

15.5.2   Stage 2: Data description and design

Prior to the three-mode analysis, the majority of the variables (see Table 15.1) were categorized for practical reasons into roughly ten intervals of increasing length in order to remove skewness of the counted variables, ease visual inspection, and prepare the data for other analyses. The variables were slice centered and slice normalized per variable in order to remove incomparable standard deviations, while maintaining the trends over the years in the years. As discussed in Section 6.5.2, p. 129, fiber centering

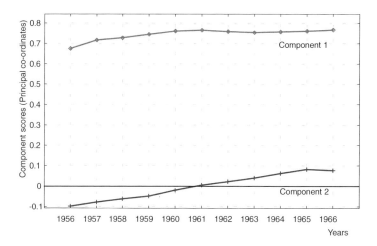

Figure 15.3 Hospital study: Time components in principal coordinates (all-components plot). Explained variability: first component 55%; second component 0.4%.

is another, possibly better, option, but this would have removed the time trends from the data.

15.5.3 Stage 3: Model and dimensionality selection

The hospital data were analyzed both with a $2 \times 2 \times 2$-Tucker3 model with two components each for the hospitals, the variables, and the years and with a 2×2-Tucker2 model. Details of the dimensionality selection can be found in the original publication (Kroonenberg et al., 1985b).

15.5.4 Stage 4: Results and their interpretation

Time trends. For the inspection of the time-mode components (Fig. 15.3), it is advantageous to present the components in principal coordinates (standardized component weights: $\nu_1 = 0.55$; $\nu_2 = 0.004$), because an assessment of their relative importance is crucial. The strong, stable time trend dominating the figure shows that the overall structural organization remains the same, except for a slight increase in the first years (say, 1956–1961). The second trend, gain, shows a very steady increase but is relatively unimportant. From Section 15.4.2 we know that we may expect such components from longitudinal data to show a simplex structure in the time mode. Table 15.2 shows the correlation matrix of the time mode (top) and of two of the variables (number of beds and main outpatient specialties).

Some authors (Van de Geer, 1974; Lohmöller, 1978a) suggest that it is advantageous to rotate the components from a simplex to orthogonal polynomials, leading to

Table 15.2   Hospital study: Correlations between years

time mode (based on 188×22 observations)

	1	2	3	4	5	6	7	8	9	10	11
1	100										
2	96	100									
3	94	97	100								
4	93	95	98	100							
5	89	92	94	95	100						
6	87	90	92	93	97	100					
7	86	88	90	91	94	95	100				
8	85	88	90	91	94	95	97	100			
9	83	85	87	89	91	93	94	96	100		
10	81	84	86	87	90	91	93	95	97	100	
11	80	82	85	86	89	90	92	94	95	97	100

number of beds

	1	2	3	4	5	6	7	8	9	10	11
1	100										
2	97	100									
3	97	98	100								
4	96	98	99	100							
5	95	96	98	98	100						
6	94	96	97	98	99	100					
7	94	95	97	97	98	99	100				
8	93	95	96	97	98	98	99	100			
9	92	94	95	95	97	97	98	99	100		
10	92	93	94	94	96	96	96	98	99	100	
11	91	92	93	93	95	95	95	97	97	99	100

main outpatient specialties

	1	2	3	4	5	6	7	8	9	10	11
1	100										
2	94	100									
3	90	92	100								
4	87	90	98	100							
5	79	82	90	90	100						
6*	**70**	**72**	**79**	**80**	**89**	100					
7*	72	74	82	83	86	**79**	100				
8	74	77	84	86	89	**83**	93	100			
9	70	72	79	80	82	**76**	88	93	100		
10	66	68	77	78	80	**75**	86	90	96	100	
11	65	66	75	77	79	**72**	84	89	95	98	100

Note the curious break in the simplex of the main outpatient specialties at the years 6, 7, and 8. The correlations rise after year 6, and only systematically fall off again after year 8.

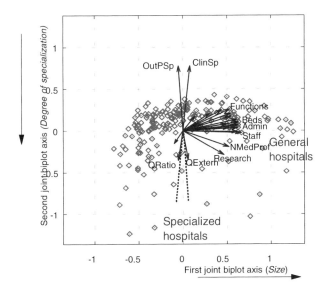

Figure 15.4   Hospital study: Joint biplot of variables and hospitals (display modes) associated with the first time component (reference mode).

a first component that has more or less equal entries, a second component with entries increasing linearly over time when the time points are equidistant, and a third component that shows a quadratic function of time, that is, first an acceleration and then a deceleration, or vice versa. For the present data, it was attempted to rotate the time mode to such a matrix of orthogonal polynomials, but the rotation matrix was practically an identity matrix ($r_{11} = 0.9996; r_{12} = r_{21} = 0.0294; r_{22} = 0.9996$). Not surprisingly, it only transferred a very small amount of the growth in overall level from the first to the second component. We will therefore continue to show the unrotated time components.

Hospitals and variables.   To answer the first question with respect to the changes in organizational structure, we will inspect a joint biplot with the hospitals and variables as display modes and time as the reference mode (see Section 11.5.3, p. 273). Figure 15.4 is the joint biplot associated with the first time component, which reflects the overall stable characteristics of the variable and hospital domains.

The joint biplot shows that in terms of the variables the axes may be interpreted as *size* and *degree of specialization*, with the former component not only representative of the variables intended to measure size (see Table 15.1), but also of most of the other variables. The degree of specialization is primarily indicated by a deficit of main specialties (OutPSp and ClinSp), a somewhat larger research capacity (Research), greater proportions of qualified nurses (QRatio), and more qualified nurses outside the wards (QExtern). In terms of the hospitals the axes can be interpreted as

Table 15.3 Hospital study: Full core array with relationships between the components of the three modes

	First hospital type			Second hospital type	
	Size	Degree of specialization		Size	Degree of specialization
Raw values of full core array					
Level	150	1		−1	53
Gain	2	−7		10	−5
Standardized contribution to SS(Fit)					
Level	0.490	0.000		0.000	0.060
Gain	0.000	0.001		0.002	0.000

general hospitals and *specialized hospitals*, respectively. From the relative sizes of the standardized component weights (48% and 8%, respectively) we may conclude that the first components of the hospitals and the variables are by far the most important ones. The second component is essentially determined by the fact that some 15–20 hospitals lack a considerable number of main specialties compared to the other hospitals; that is, they are more specialized. Incidentally, the sharp boundary of the hospitals on the positive Y-axis in Fig. 15.4 is caused by ceiling effects due to the fact that a large number of hospitals have all the main specialties a hospital can have. From Fig. 15.3 in combination with Fig. 15.4, we can deduce that the answer to the *first research question* is that the overall organizational structure was stable; that is, the relative position of the hospitals remained unaltered, but there is a steady but small increase or decrease in overall level or size, depending on the signs of the loadings on other components (see Section 9.5.3, p. 230 for a discussion on keeping track of signs for components and core arrays).

Interaction between hospitals and variables. To answer the *second research question*, a decomposition in terms of components alone does not suffice, and the full core array must be inspected as well. First of all, Table 15.3 confirms the answer to the first research question. The combination of the first components of all three modes (general hospitals, size, and level), $g_{111} = 150$, explains most of the fitted variation (Proportion of SS(Fit) due to g_{111}/ SS(Fit overall) $= 0.49/0.56 = 0.88$. The gain in size of the general hospitals, $g_{112} = 2$, is negligible over and above the increase already contained in the level component.

The second important combination ($g_{221} = 53$; Prop. SS(Fit) $= 0.06$) indicates that the specialized hospitals also maintain their overall level of specialization. There is a slight tendency ($g_{222} = -5$) to become less specialized, and to grow in overall size ($g_{212} = 10$). Similarly the large general hospitals tend to become somewhat less specialized ($g_{221} = -7$). The standardized contributions to the SS(Fit) show that

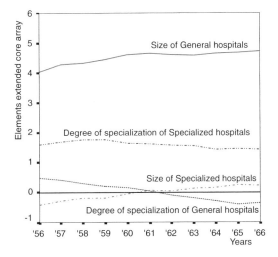

Figure 15.5   Hospital study: Trends of hospital–variable combinations. Expressed via the core elements, $h_{q,r}$ of the extended core array of the Tucker2 model.

these effects are very small, leading to the conclusion that the specialized hospitals do not have a very different growth pattern from that of the other hospitals.

A more detailed inspection of the time trends is given in Fig. 15.5, where the elements of the extended core array with years as unreduced third mode of a Tucker2 analysis have been plotted against time. The patterns, of course, are in accordance with the Tucker3 analysis presented earlier, but the development of the relations between the hospital and variable components over time is shown more explicitly.

15.5.5   Checking the order of autoregressive process

Before attempting to estimate the parameters of an autoregressive process, it should be established whether it is reasonable to postulate such a process for the data, and if so, whether it is of the right order. There seem to be a number of ways to do this.

First, check whether the time mode is a simplex, as we know that autoregressive processes generate simplexes. Inspection of the hospital data shows the correlation matrix to be a simplex (see Table 15.2), and moreover the points in time (years) are equidistant.

Second, perform a multiple regression of $\mathbf{x}_k$ for the data at time k on the earlier observations on the same variables, $\mathbf{x}_{k-1}, \mathbf{x}_{k-2}, \ldots$, where $\mathbf{x}_k$ is the ($IJ \times 1$) vector for the kth occasion. When the autoregressive process is first-order, only $\mathbf{x}_{k-1}$ should have a sizeable regression coefficient. Although ordinary least-squares estimation will lead to incorrect standard errors for the estimators of the regression coefficients, they are in general unbiased (e.g., see Visser, 1985, p. 71). Table 15.4 shows the relevant

Table 15.4 Hospital study: standardized regression coefficients for predicting an occasion from earlier occasions

	Predictors									
Criteria	$t-1$	$t-2$	$t-3$	$t-4$	$t-5$	$t-6$	$t-7$	$t-8$	$t-9$	$t-10$
2	96	—								
3	79	19	—							
4	85	14		—						
5	75	14	7		—					
6	87	10				—				
7	63	19	10				—			
8	70	16	8	5				—		
9	75	13	12				2		—	
10	76	15	4	4						—
11	75	11	8							

All values have been multiplied by 100.

standardized regression coefficients, and the hospital data seem to follow at least a second-order autoregressive process with a dominant first order. This implies that the lag-one correlations will overestimate the overall stability of the process.

Combining the above information, the assumption of a second-order autoregressive process with a rather strong first-order seems plausible. However, as pointed out in Section 15.4.2, Lohmöller's procedures were only developed for first-order autoregressive processes, because higher-order autoregressive processes turned out to be unmanageable. We will proceed as if the autoregressive process is a first-order one, keeping in mind this is only an approximation and might lead to an overestimation of the stability.

15.5.6 Assessment of change phenomena

In this section we will apply Lohmöller's proposals for assessing change phenomena to the hospital study. In order to remain compatible with his discussion, we will again use the results from the fiber-standardized data (i.e., standardized per variable on each occasion), instead of the recommended profile preprocessing of fiber centering the data per variable–occasion combination and slice normalization of variables over all occasions together. Our major tool will be the latent covariance matrix, discussed in Section 15.4.2.

The overall stability of the variable domain, $\overline{\phi}$, may be assessed in two different ways. First, Lohmöller gives tables linking the overall stability to the eigenvalues of the first two components from the time mode. From these tables, the overall stability is estimated as between 0.85 and 0.95. This estimate may be compared with the correlations between adjacent occasions (i.e., $r_{k,k-1}$ of the $(K \times K)$ correlation matrix $\mathbf{R}$ of the time mode in Table 15.2). The comparison of the lag-one correlations in Table 15.4 shows good agreement. In addition, it should be observed that the lag-one correlations do not vary much. This leads us to accept the assumption that $\bar{\phi}$ is independent of time, and that by and large the hospitals maintain their overall rank order of the variables over the years.

With a high overall stability, the variable components should be very stable as well. Following Lohmöller's guidelines we may infer from $\sigma_{11,11} = 1.71$, and $\sigma_{21,21} = 1.70$ of the latent covariance matrix (Table 15.5A) that both latent variables are equally stable and trait-like. One may seek confirmation for this by inspecting the cross-lag correlations for representative variables (see Table 15.2). Taking the variable Beds as indicator for size, the stability is obvious; taking the variable Main outpatient specialties to indicate degree of specialization, the stability is still clearly visible but rather irregular between years 6 and 8. The cause of the latter is a matter for separate investigation. One might speculate that the definition of what constitutes a specialty changed at the time.

The zero value of the covariance between size and degree of specialization for level, $\sigma_{21,11}$, indicates that no cross-lag covariances exist between the latent variables ($\phi_{pp'} = 0, p \neq p'$); that is, $\mathbf{\Phi}$ is diagonal. The interaction between level and gain of the two latent variables ($\sigma_{22,11} \neq 0; \sigma_{21,12} \neq 0$) shows that the set of variables is not stationary. There is a negative covariance between level of size and gain in degree of specialization ($\sigma_{22,11} = -2.57$), which indicates that hospitals that are large through the years tend to lose or at least not gain in degree of specialization, and vice versa. We suspect this might be the ceiling effect: large hospitals already had all the specialties they could have. Furthermore, there is a positive covariance between gain in size and degree of specialization ($\sigma_{21,12} = 1.57$): very specialized hospitals tend to become larger, and vice versa.

The importance of the deviations from stationarity are, however, relatively small. This can be assessed from the direction cosines between the structured components; see Eq. (15.5). From Table 15.5B it follows that the direction cosine between level of size and gain in degree of specialization is -0.02 ($\theta = 91°$), and the direction cosine between level of degree of specialization and gain in size is 0.01 ($\theta = 89°$). In other words, the deviations from stationarity do not succeed in introducing substantial nonorthogonalities between the structured components. These small direction cosines may seem strange when considering the sizes of the elements in Table 15.5A. It should, however, be realized that the normalizations of the elements of the latent covariance matrix have eliminated the dependence of the elements on the importance of the components to which they refer. This means that all components are treated on the same level. This normalization was chosen partly to fit in with the assumption that all latent variables were normalized per occasion, that is are fiber normalized. It has some advantage in highlighting the interactions, but at the same time gives a rather incorrect impression of their sizes.

Table 15.5   Hospital study: latent covariance matrix

| | | Size | | Degree of specialization | |
		Level	Gain	Level	Gain
A: Lohmöller scaling					
Size	Level	**1.71**	0.01	0.00	−2.57
	Gain	0.01	**1.44**	1.57	−0.10
Degree of	Level	0.00	1.57	**1.70**	−0.09
specialization	Gain	−2.57	−0.10	−0.09	**3.86**
B. Standardized					
Size	Level	**0.50**	0.00	0.00	−0.02
	Gain	0.00	0.00	0.01	0.00
Degree of	Level	0.00	0.01	**0.06**	0.00
specialization	Gain	−0.02	0.00	0.00	0.00
C. Labeling of elements					
Size	Level	$\sigma_{11,11}$	$\sigma_{11,12}$	$\sigma_{11,21}$	$\sigma_{11,22}$
	Gain	$\sigma_{12,11}$	$\sigma_{12,12}$	$\sigma_{12,21}$	$\sigma_{12,22}$
Degree of	Level	$\sigma_{21,11}$	$\sigma_{21,12}$	$\sigma_{21,21}$	$\sigma_{21,22}$
specialization	Gain	$\sigma_{22,11}$	$\sigma_{22,12}$	$\sigma_{22,21}$	$\sigma_{22,22}$

15.6   EXAMPLE: MORPHOLOGICAL DEVELOPMENT OF FRENCH GIRLS[3]

15.6.1   Stage 1: Objectives

The original study from which the present data were taken was initiated in order to get insight into the physical growth patterns of children from ages four to fifteen. Details about the study together with several analyses of the data for 30 normal French girls can be found in the volume *Data analysis. The ins and outs of solving real problems* (Janssen et al., 1987).

15.6.2   Stage 2: Data description and design

The thirty girls set were chosen from a larger data base because they had measurements available for all variables for each of the twelve years. The eight variables under consideration were the following: Weight, Length, Crown–coccyx length (CrownRump),

[3]This section is partly based Kroonenberg (1987b); 1987 ©Plenum Press; reproduced and adapted with kind permission of Springer Science and Business Media. The data can be obtained from the data set section of the website of The Three-Mode Company; *http://three-mode.leidenuniv.nl*. Accessed May 2007.

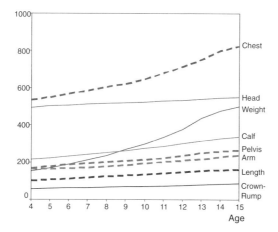

Figure 15.6   Girls' growth curves data: Mean curves of the eight variables under consideration, representing the average girl's growth curves. The variables have not yet been normalized to equal sum of squares; therefore, no scale has been attached to the vertical axis.

Chest circumference (Chest), Left upper-arm circumference (Arm), Left calf circumference (Calf), Maximum pelvic width (Pelvis), and Head circumference (Head); all measured in millimeters. Thus, the data form a $30 \times 8 \times 12$ longitudinal data block of girls by variables by years; see Chapter 11 for further analyses of these data.

15.6.3   Stage 3: Model and dimensionality selection

Preprocessing and mean curves.   Before the analysis proper, profile preprocessing was applied to the data; that is, the fiber means of each variable at each time point, $\bar{x}_{.jk}$, were removed and the variables were slice-normalized (i.e., divided by $s_{.j.}$), which removed the unequal influences of different measurement scales and variances. The effect of this is that the mean growth curve is removed from the analysis and the scores analyzed are those in normalized deviations from the average girl's growth curves; see Section 6.6.1 (p. 130).

From the growth curves of the average girl (Fig. 15.6), it can be seen that she has a nearly linear growth in most variables, except for chest circumference, which has a change in slope around the tenth year, while weight has a change in slope around the eighth year. Both variables level off around fourteen years of age.

Dimensionality selection.   In this section we will present the results from a Tucker2 analysis with 3 components for the girls mode and the variable mode, as well as a Tucker3 analysis in which, in addition, the time mode is represented by two components. The number of components for the analysis was determined by examining deviance plots (see Section 8.5.2, p. 181). In this plot for the Tucker3 model

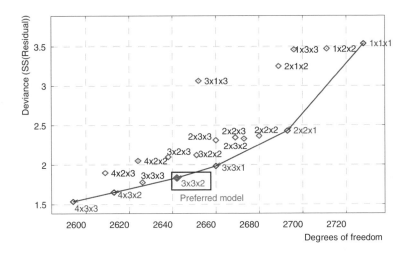

Figure 15.7   Girls' growth curves data: Deviance plot for the Tucker3 model. On the vertical axis the residual sum of squares is indicated and on the horizontal axis the associated degrees of freedom. The preferred models lie on the convex hull. The 3 (girl components)×3 (variable components)×2 (time components) model is the one presented in this chapter.

(Fig. 15.7) we settled for a $3\times3\times2$-Tucker3 model and a 3×3-Tucker2 model, because both the $2\times2\times1$-Tucker3 model favored by the Ceulemans–Kiers st-criterion (see Section 8.5.3, p. 182) and the $3\times3\times1$-Tucker3 model were deemed too simple as they only had one time component.

In addition to the overall fit, the fit of the components for each of the modes was examined as well as that of their combinations (see Table 15.6). The small difference in fit between the two Tucker models indicates that condensing the time mode to two components does not affect the results much, the more so because the fit of the components for the variables and the girls is essentially similar for both analyses. We will therefore switch between the two models whenever expedient.

15.6.4   Stage 4: Results and their interpretation

Variables.   The three components of the variables with normalized coordinates are portrayed together with the *equilibrium circle* Fig. 15.8 (see Legendre & Legendre, 1998, p. 398ff.). The figure shows that all variables fit equally well and that a third dimension is present because of the deviating pattern for the growth of the head. Its similarity to pelvis and crown–coccyx in two dimensions is somewhat deceptive. This becomes clearer when we portray the variable components in principal coordinates and connecting them in accordance with their minimum spanning tree as was done in Fig. 11.6 (see Section 11.4.1, p. 262). The figure shows that Head dips deep under the 1-2 plane and Pelvis less so, while Length is above the plane. A varimax rotation

Table 15.6 Fit for components of the $3\times3\times2$-Tucker3 model and the 3×3-Tucker2 model; Fit expressed as proportion of total sum of squares.

	Tucker3 model			Tucker2 model		
Mode	1	2	3	1	2	3
Proportional fit per component						
Girls	0.56	0.14	0.07	0.56	0.15	0.07
Variables	0.58	0.14	0.06	0.58	0.14	0.06
Years	0.75	0.02				

Nonnegligible elements of core arrays				
Tucker3 full core array		Tucker2 extended core array	
G1,V1,Y1	0.56	G1,V1	0.56
G2,V2,Y1	0.14	G2,V2	0.14
G3,V3,Y1	0.05	G3,V3	0.06
G3,V1,Y2	0.01	G3,V1	0.01
Overall fit	0.77	0.78	

Gp,Vq,Yr indicates the core element that is the combination of the pth girl component, the qth variable component, and the rth time component; there are no time components for the Tucker2 model.

Table 15.7 Varimax component solution for variables

		Components		
Variables	Abbreviation	1	2	3
Left upper-arm circumference	Arm	**0.51**	0.01	0.16
Chest circumference	Chest	**0.47**	0.16	0.12
Left calf circumference	Calf	**0.45**	0.17	0.18
Weight	Weight	**0.45**	0.32	0.24
Length	Length	0.08	**0.62**	0.14
Crown–coccyx length	CrownRump	0.16	**0.55**	0.28
Maximum pelvic width	Pelvis	0.27	0.34	**0.40**
Head circumference	Head	0.16	0.21	**0.78**

Component 1: *Soft tissue*; Component 2: *Skeletal length*; Component 3: *Skeletal width*.

of the component space (Table 15.7) shows that the three rotated components can be characterized by the *Soft tissue* variables, the *Skeletal length* variables, and the *Skeletal width* variables, be it that the pelvic width has noticeable loadings on all components. We will use these components to further examine the growth patterns.

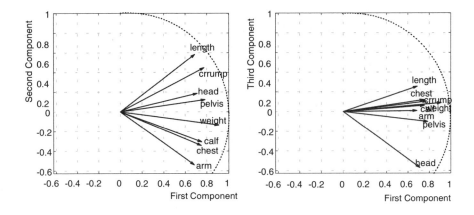

Figure 15.8   Girls' growth curves data: Components for the variables in normalized coordinates with the equilibrium circle added (see in Section 11.4.1, p. 262).

Girls.   We will not portray the component space of the girls as it is not very revealing except that, due to the centering employed, nearly all combinations of positive and negative coefficients on the three components occur. In accordance with the relative importance of the components (see Table 15.6), the variability in the third dimension is small and primarily caused by a few girls with larger coefficients.

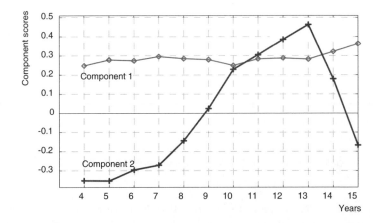

Figure 15.9   Girls' growth curves data: All-components plots of the Time components after rotation to an optimally constant first component. Normalized coordinates.

Years.   Figure 15.9 shows that the first time component after being optimally rotated toward a constant value is indeed nearly constant, indicating the overall highly constant rank-order of the girls as expressed by their positive correlation matrix. The second component reflects the changes in variability in growth rates, which is gradually and from age 8 more markedly increasing until age 13, when the variability decreases sharply, indicating that by then all the girls have had their growth spurts and their differences level off and stabilize. Note that the normalized coordinates preclude the assessment of the relative importance of the components from the plot.

15.6.5   Trends over time for individual girls

To investigate the trends in growth of the individual girls over time, we can construct nested-mode biplots (also called an interactive biplots) (see Section 11.5.4, p. 276) or nested-mode per-component plots. The row coefficients in nested-mode plots represent the structured component scores of the 30×12 girl–year combinations on each of the components of the variables. These coefficients of nested-mode plots are comparable with the coordinates of the interstructure in STATIS (see Section 5.8, p. 105). For each variable component, we can portray the girl$\times$age coefficients to examine the growth curves in a nested-mode per-component plot, and we can examine the trajectories of the girls over time in a nested-mode biplot (see Section 11.5.3, p. 273).

Individual growth curves per variable component.   Here, we will only show the growth curves corresponding to the first rotated component of the reference mode, that is, variables representing Soft tissue. In Fig. 15.6 we saw that the average girls had fairly similar growth curves for most variables, except for a faster increase in weight and chest circumference and a very slow increase in head circumference. The growth curves observed in the nested-mode per-component plot (Fig. 15.10) show to what extent individual curves followed those of the average girl with respect to the Soft tissue variables. The general pattern is one of increasing variability until 13 years of age and a decreasing variability after that. This reflects the situation that some girls grew more and/or earlier than others, and that around the 13th year the fast growers stopped and the average girl caught up. At the same time the late growers started to catch up with the average girl. At the end of the measuring period the differences between the girls had increased considerably compared to when they were 4, but the relative rank order at the beginning and at the end is generally maintained. Once a girl is smaller than her peers, she is very unlikely to catch up in puberty, even though she might temporarily do so if her growth spurt comes early. This follows from the fact that very few growth lines cross that of the average girl. Nearly all early growers fall back in the end to the relative position they had at the beginning. The message from the other components, Skeletal growth and Skeletal width, is similar.

Growth curves as trajectories in nested-mode biplots.   The same girl–year coordinates can also be plotted in the component space of the variables, rather than with

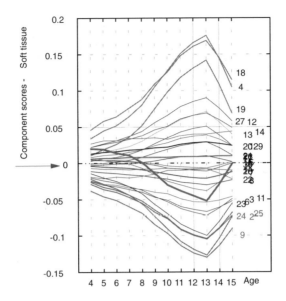

Figure 15.10   Girls' growth curves data: Individual growth curves on Soft tissue for the 30 girls; scores are deviations from the average growth curve (represented by the horizontal line at 0).

respect to each component separately. By connecting the scores over the years for each girl we get trajectories in the component space as portrayed in Fig. 15.11 for the Soft tissue–Skeletal length space.

The origin of the space consists of the coordinates of the 12 years of the average girl. All trajectories start in the neighborhood of the origin, move away from it, and more or less return to it. The star-like pattern indicates that there are many different combinations of growth in Soft tissue and Skeletal length, something that is difficult to observe from the per-component growth curve plots. It can also be observed that the largest distance from the origin occurs around 13 years of age, after which the curve returns to the origin, indicating that the relative difference from the average girl is diminishing. There is also one girl who is close to the origin all the time, indicating that her growth pattern mirrors that of the average girl as depicted in Fig. 15.6.

15.7   FURTHER READING

Timmerman (2001) presents one of the few extended treatments of longitudinal data within the three-mode context, treating simultaneous component analysis, the application of functional constraints on components, modeling longitudinal data of different people measured at different times (cross-sectional data and multiset), as

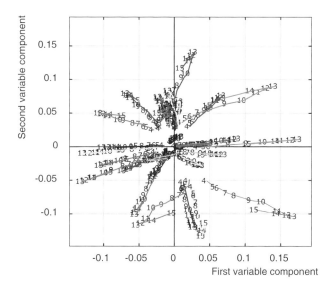

Figure 15.11   Girls' growth curves data: Individual growth curves for the 30 girls presented in the component space of Soft tissue (horizontal) and Skeletal length (vertical).

well as dynamic factor models; subjects not treated in this chapter. Unfortunately, two important topics could not be included here. In particular, Murakami (1983) uses the Tucker2 model in an innovative way to assess the changes in the components for variables and in the components for occasions both separately and in conjunction, and Meredith and Millsap (1982) employed a variant of canonical correlation analysis for a similar purpose. The book by Lohmöller (1989) is also an invaluable source for studying three-mode methods and longitudinal data, including his latent three-mode path models.

15.8 CONCLUSIONS

Insight into developmental processes in multivariate longitudinal data can be acquired by inspecting relationships between the components of the three modes: observational units, variables, and points in time. A detailed analysis of the latent covariance matrix can supply information on differential growth patterns if they exist, and the extended core array can help to inspect the changes in the interrelationships between the components over time. In fact, the entries in the extended core array can be seen as the scores of observational units on structured components of the latent variables and trends. A further level of detail is introduced by using the scores on the structured components of subjects and time to portray individual growth curves per variable component in a nested-mode per-component plot, and by plotting these same scores in a nested-mode biplot so that their trajectories could be examined and interpreted.

A description of data in terms of an autoregressive model has been shown to be acceptable for certain data sets, but first-order processes might be difficult to find. Our example and most of Lohmöller's needed a second-order term. Furthermore, the theory is still very much underdeveloped; only in specific cases are detailed statements possible. At present the most useful aspect is the interpretational framework that an autoregressive model can provide for three-mode analysis of multivariate longitudinal data.

CHAPTER 16

THREE-MODE CLUSTERING

16.1 INTRODUCTION[1]

16.1.1 Multiway profile data

As discussed in Chapter 13, multiway profile data typically consist in the three-way case of scores of individuals on a set of variables that are measured under a number of conditions; in agriculture one could think of realizations of varieties on which a number of attributes are measured in different environments. Generally, the variables or attributes have different measurement scales and the values across variables are not necessarily comparable. Even though we do not necessarily assume that there is a specific time structure or order relationship between the measurement conditions, the data generally conform to a multivariate repeated measures design. In connection

[1]This chapter is based on joint work with Kaye Basford, University of Queensland, Brisbane, Australia and Paul Gemperline, East Carolina University, Greenville, NC, USA; parts have been published in Kroonenberg, Basford, and Gemperline (2004).2004 ©John Wiley & Sons Limited; reproduced and adapted with kind permission.

with clustering very few, if any, techniques deal explicitly with four-way and higher-way data, so that we will necessarily restrict ourselves in this chapter to three-way data and cluster techniques to analyze them[2].

16.1.2   Ordination and clustering

In Chapter 13 the emphasis was on analyzing three-way profile data by developing dimensional models for the three modes of the data matrix. This approach is often referred to (especially in the biological and life sciences) as ordination. In the present chapter we address a different research question for the same type of data. In particular, we want to know whether it is possible to simplify the discussion of the properties of the individuals (subjects, objects, etc.) by grouping them according to common patterns of responses. The classification of individuals into groups is mostly done by cluster analysis; for two-mode profile data many such procedures exist, but for three-way profile data these procedures are few and far between. In the most common approach, clustering is not carried out on the profile data themselves but on similarity measures derived from the profile data, such as in the INDCLUS method (Carroll & Arabie, 1983). For direct clustering of three-way profile data, a limited number of techniques are available, an overview of which is presented in Section 2.4.6 (p. 23).

The three-mode mixture method of clustering discussed in this chapter is a direct generalization of the two-mode variant developed by Wolfe (1970), and the two methods are equivalent for a single condition. The major difference between most clustering techniques and the mixture method of clustering is that the latter is based on a statistical model with explicit distributional assumptions. Furthermore, the mixture method of clustering belongs to classical multivariate (multinormal and/or multinomial) statistics, unlike most other techniques discussed in this book, except for the three-mode common factor analysis or structural equation models mentioned in Section 2.4.3, (p. 22). The basic idea behind the mixture method of clustering is that under each condition the data are a sample from a mixture of multivariate distributions. The mixture is the result of several groups having their own multivariate distributions, and the aim of the technique is to uncover these groups. The three-mode assumption is that even though the shape of the distribution for each group is the same under each condition, its location under each condition may be different.

Next to three-mode method of clustering, other developments in the area of clustering three-way data concentrate on combining data reduction via component methods and clustering of the subjects mostly using k-means procedures. A recent paper in this area is Vichi et al. (2007). These techniques will, however, not be treated in this book.

---

[2]The term "multiway clustering" is a very common one in the literature but generally refers to splitting single samples in more than way, and it is thus mostly unrelated to *multiway data*

16.2 CHAPTER PREVIEW

In this chapter we will address the question of how to group individuals for which we have measured three-mode profiles, that is, individuals having scores on variables under different conditions. The model assumes that the same groups exist in all conditions under consideration. After the theoretical introduction, the search for groups is illustrated by a small example taken from an environmental study of the effect of pollution on blue crabs (Gemperline et al., 1992). The subsequent sections will provide guidance for applying the clustering technique in practical problems followed by a full-fledged example about the attachment between mother and infant.

16.3 THREE-MODE CLUSTERING ANALYSIS: THEORY

In this section we will give an outline of the three-mode mixture method of clustering. A more detailed and general exposition can be found in the original publications, especially Basford and McLachlan (1985b) and Kroonenberg, Basford, and Van Dam (1995b), which contains an extended treatment of the main example in this chapter. To make the discussion more concrete, we will assume we have data from I individuals who are measured on J variables under K conditions, and that the individuals will be partitioned into G groups. The name *mixture* refers to the joint distribution of the G groups which is the result of the combination or mixture of the G separate distributions. In our discussion we will concentrate on the situation with normally distributed variables, but situations that include multinomially distributed variables may also be handled (see Hunt & Basford, 1999). A further development along these lines is proposed by Vermunt (2007), who adapted the multilevel latent class model for categorical responses to the three-way situation. He has also shown that the mixture methods discussed here can be seen as special cases of his proposals.

16.3.1 Parameters: Means and covariances

The clustering method is based on the assumption that each of the I individuals belongs to one of G possible groups, but it is unknown to which one. Therefore, the aim is to assign the individuals to the group to which they have the highest probability of belonging. In order to be able to do so, it is necessary to establish the characteristics of each group and the probability of each individual belonging to each group. Observations are available on J variables under K conditions, so that the data comprise K vectors of multivariate observations on each individual, one for each condition. For individual i ($i = 1, \ldots, I$) these are denoted by $\mathbf{x}_{i1}$ through $\mathbf{x}_{iK}$. The vector of all observations for individual i has $J \times K$ elements and is denoted by $\mathbf{x}_i$.

If we assume that there exists only one group, and that the data come from K multivariate normal distributions, one for each condition, then the model specifies

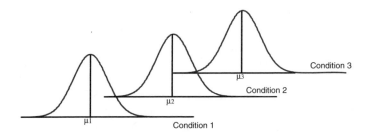

Figure 16.1   Univariate distributions of a single variable measured under three conditions: One group.

that the distributions have the same $J \times J$ covariance matrix, but each condition has its own mean vectors so that there are $J \times K$ means. In other words, all distributions have the same shape, but they differ in the location of their means. To illustrate the situation by a much simplified univariate example, Fig. 16.1 sketches the case of a single variable with one group and three conditions.

When there are more groups, each group has its own multivariate normal distribution with its own distributions of the variables in each group and these are allowed to have different means in each condition; thus, there are $J \times K \times G$ means, μ_{kg}. Figure 16.2 sketches the situation for one condition and a single variable with three groups having different means and variances. Different groups may have either the same covariances between the variables, that is, there is a *common (within-group) covariance matrix*, or each group may have a different or *group-specific (within-group) covariance matrix* (see Fig. 16.3).

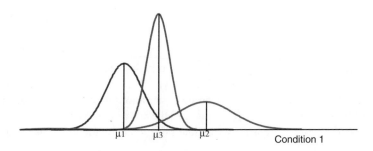

Figure 16.2   Distributions of a single variable measured in one condition: Three groups.

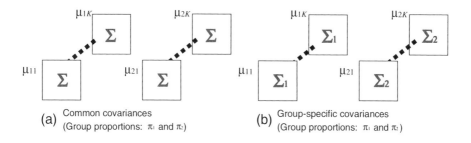

Figure 16.3   Group structure and parameters for the two-group situation: (a) Common covariance matrices, (b) Group-specific covariance matrices.

16.3.2   Parameters: Mixing proportions

Under the normal mixture model proposed by Basford and McLachlan (1985b), it is assumed that the relative sizes of the G groups are given by the *mixing proportions* $\pi_1, \ldots, \pi_G$ respectively, which are also unknown, so that they too have to be estimated.

16.3.3   Estimation and starting solutions

Because of the way in which the estimation of the parameters is carried out, one has to choose the number of groups before the start of the computational procedure and allocate the individuals to these groups. Together this determines the initial mixing proportions. Given a starting allocation, initial values for all parameters can be calculated, and they collectively are referred to as the starting solution.

The estimation of the parameters of the cluster model is carried out with maximum likelihood procedures, which aim to find those values of the parameters for which the observed values are as likely as possible. The estimation procedure is an iterative one, in which the likelihood is continuously increased. However, such an iterative procedure is only assured to converge to a local maximum, and thus only a suboptimal solution can be guaranteed. Which particular solution is obtained depends on the initial allocation of the individuals to the groups.

One can attempt to find an overall or globally best solution by using several different starting allocations (see Section 5.4, p. 81). Any initial partitioning of the individuals may be used, but the requirement is that the partitioning leads to high values for the likelihood function. Often these starting allocations are obtained by using the groupings from one or more two-mode hierarchical clustering methods. Starting allocations obtained via Ward's method (Ward, 1963) applied to some two-mode version of the data tend to be particularly useful, as they often lead to solutions with the highest log-likelihood due to the theoretical links between Ward's method and the mixture method of clustering (e.g., see Gordon, 1981, p. 50). Similarly, the allocations resulting from two-mode mixture analyses on the data of one or more of the conditions are candidates for providing high-quality solutions for the three-mode case.

16.3.4 Statistics: Posterior probabilities

Once a solution has been obtained, one can derive the *posterior probabilities* that an individual $i, (i = 1, \ldots, I)$, with observation vector $\mathbf{x}_i$, belongs to group $g(g = 1, \ldots, G)$. The probabilities are not themselves parameters because they have been derived from the actual parameters. On the basis of the posterior probabilities, each individual is assigned to the group for which it has the highest estimated posterior probability. In this way, the individuals are partitioned into a number of discrete, relatively homogeneous groups. It should be borne in mind that sometimes the allocation of an individual is more or less arbitrary, when the posterior probabilities are almost equal.

16.3.5 Independence of observations

The mixture model assumes that the measurements taken on individuals under separate conditions are independent of each other. There is no concern about the individuals being independent of each other, but the same individuals are measured under all conditions. Therefore, we are assuming independence of measurements on the same individuals over conditions. Independence is often a valid assumption for the agricultural data to which the model is frequently applied (the same genotypes, but separate plants, are grown in each location), but is more doubtful with other types of data. Treating the separate conditions as independent measurements is a compromise that enables some of the structure of the design to be accommodated, that is, the same variables are measured each time, but this approach does not do justice to the independence assumption. For the first example (see Section 16.4), the independence assumption might just be tenable, but for the main example (see Section 16.6), the independence across conditions is not a realistic assumption. The consistency of the clustering and the ordination results, however, suggests that the results do not seem to suffer unduly from this lack of independence. To our knowledge, no systematic study into the importance of the independence assumption exists for repeated measures three-way data.

Even though the mixture method of clustering is based on statistical and distributional assumptions, it is often used in three-mode analysis in an exploratory fashion. The outcomes of the three-mode mixture approach should be supported and validated by outcomes from complementary ordination procedures to overcome possible doubt about the violation of independence.

Figure 16.4   Blue crab (*Callinectes sapidus*).   Photograph by Mike Oesterling. Reproduced by courtesy of the Virginia Institute of Marine Science; taken from *http://www.vims.edu/env/research/crust.html*. Accessed May 2007.

16.4   EXAMPLE: IDENTIFYING GROUPS OF DISEASED BLUE CRABS

16.4.1   Stage 1: Objectives

Blue crabs (*Callinectes sapidus*) (Fig. 16.4) are of great commercial value in North Carolina so that the appearance of diseased crabs in 1986 in a specific region of the Pamlico River caused great concern. Gemperline et al. (1992) cited the hypothesis that "environmental stress weakens the organism so that its normal immunological response is unable to ward off opportunistic infection by chitinoclastic bacteria". These bacteria were considered to be the cause of penetration of the carapace causing lesions of 5–25 mm. The major objective of the study was to confirm that the disease of the crabs could directly be related to the trace elements that were found in their bodies. From simple observation and information on where they were caught, a grouping into three categories could be made: healthy crabs from the Albemarle Sound, healthy crabs from the Pamlico River, and diseased crabs from the Pamlico River.

In this small reanalysis, the concern was whether the three categories could be reliably distinguished by a three-mode cluster analysis on the amount of trace elements found in the crab tissues. A more extensive analysis combining three-mode clustering and three-mode component analysis was published in Kroonenberg et al. (2004); see also Kroonenberg, Basford, and Ebskamp (1995a). If the cluster analysis could not find the original groupings, and assuming the analysis did its work properly, other factors than damaging amounts of trace elements could be responsible. The original analyses of these data were carried out by Gemperline et al. (1992) using, among other techniques, a three-mode principal component analysis using the Tucker3 model.

16.4.2   Stage 2: Data description and design

Gemperline and his colleagues collected tissue samples of gill, hepatopancreas (the crab's organ which functions both as liver and pancreas), and muscle taken from blue crabs in each of three areas: Albemarle Sound, nondiseased Pamlico, and diseased Pamlico. Samples from three individual crabs were pooled, giving a total of 16 pooled tissue samples in each category and a grand total of 48 pooled tissue samples. Twenty-five trace elements were used for the main analysis, so that a 48 (tissue samples) by 25 (trace elements) by 3 (tissue types) three-mode data array was available for analysis. To facilitate the discussion, we will refer to an "individual" in the data set as a *blue crab* rather than "the combined tissue of three individual crabs collected from the same location".

16.4.3   Stage 3: Model and number-of-groups selection

After examining solutions using different numbers of groups and different starting values, the analysis with three groups using the original categories (Pamlico Diseased, Pamlico Healthy, and Albemarle) as starting allocation turned out to be the one with the highest value of the likelihood function (see Section 16.5.3 for further details on number-of-groups selection). Moreover, it turned out that the mixture method of clustering procedure placed each crab into its proper group with a posterior probability of 1. Thus, the clustering confirmed that crabs within their own category were more alike than crabs from other categories.

16.4.4   Stage 4: Results and their interpretation

Means.   For the blue crab data, there are 25 (variables)×3 (tissues)×3 (groups) means, which makes inspection rather complicated. In this instance, we will only show a few graphs that illustrate the means for those trace elements for which the diseased Pamlico crabs have elevated doses, because it is believed that too large doses of trace elements are the most likely cause of the disease in the crabs. The gill tissue of the diseased crabs contained very high doses of aluminium (Al), as well as a large number of trace elements, the hepatopancreas tissue had especially high doses of arsenic (As), and the muscle tissue seems to contain relatively more sodium (Na) and magnesium (Mg) (Fig. 16.5).

Common covariance matrix.   The 25×25 common covariance or correlation matrix of trace elements cannot easily be inspected. Therefore, a principal component analysis was run on the correlation matrix to summarize its patterns. Essentially, the trace elements uranium (U), aluminium (Al), iron (Fe), chromium (Cr), tin (Sn), vanadium (V), manganese (Mn), cobalt (Co) , arsenic (As), and silicon (Si) tended to be found together (correlations from 0.30 to 0.94); in addition, silver (Ag) and selenium (Se) tended to be found together in the crabs' body parts, and the same can

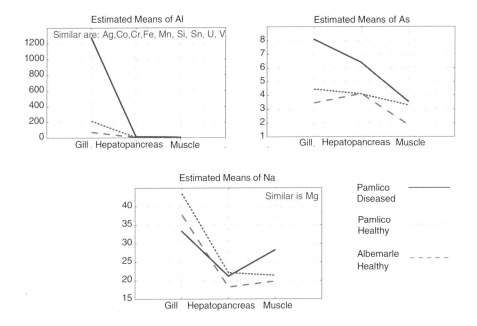

Figure 16.5 Blue crab data. Means plots showing elevated levels for the trace elements Al, As, and Na in the diseased Pamlico crabs: Gill (Al, As), Hepatopancreas (As), Muscle (Na).

be said for magnesium (Mg), calcium (Ca), and phosphorus (P). Using this information, a search can be initiated to find the source of these trace elements, such as a specific plant producing these chemicals. Some of these elements naturally occur in blue crabs but others should not be there.

It should be noted that there are some differences between the outcomes of the Gemperline et al. (1992) paper and the analysis presented here: the major reason for which seems to be that Gemperline and co-workers used a different kind of preprocessing and a larger number of components.

16.5 THREE-MODE CLUSTER ANALYSIS: PRACTICE

In this section, those theoretical and practical aspects of three-mode mixture method of clustering that have a direct bearing on choosing and carrying out a proper analysis will be addressed. Several of these points will be illustrated by the blue crab data.

16.5.1   Stage 1: Objectives

The major aim of any cluster analysis is to subdivide the individuals into groups, and, as we have seen, the major difference with most cluster analyses is that in three-mode clustering we have profiles for individuals on variables under several conditions. The practical advantage of the three-mode profiles is that more information on the differences between individuals is available than in two-mode clustering, so that one may expect that in reasonably structured data sets the clustering is more reliable and, hopefully, more stable.

A more general point is that irrespective of whether a theoretical construct is discrete or continuous, cluster methods will produce clusters, either by dissecting continuous dimensions or by seeking for natural clusters. Only afterwards can one discern the nature of the clusters, for instance, by using ordination methods. Results of the cluster analyses will therefore not provide definitive answers to questions regarding whether theoretical constructs are discrete or continuous. However, if the groupings derived by the cluster method correspond to other types of classifications, this will give further support to their typology. If not, one has at least gained further insight into the individual differences in the situation at hand. In the main example in Section 16.6 we will come back to this issue.

16.5.2   Stage 2: Data description and design

The main practical aspect in the design of the data which influences the analysis, is the number of individuals, especially compared to the number of variables and groups, and the question of whether one wants to assume group-specific or common covariance matrices.

Group-specific versus common covariance matrices.   In Section 16.3.1 it was explained that one may model either using a common covariance matrix for all groups or group-specific covariance matrices. A common covariance matrix has the advantage that far fewer parameters have to be estimated, especially when there are many groups. There are $\frac{1}{2}J(J + 1)$ parameters in the common covariance matrix and G times as many group-specific covariances. Solutions in the common covariance case tend to be more stable, but a tendency has been observed for the derived clusters to be of roughly equal size (e.g., see Gordon, 1981, p. 52).

In many practical situations, group-specific covariance matrices are more realistic, and this model is therefore to be preferred in those cases when it can be estimated. In cases when there are relatively few individuals compared to variables, it will often not be possible to estimate such group-specific covariance matrices, because of the larger number of parameters. If that is the case, programs will often report that the covariance matrix of a group cannot be inverted because the covariance matrix in that group has become singular, and the program will not proceed. The simplest way to still run a mixture method of clustering is to postulate that all groups have a common covariance matrix.

An intermediate alternative was suggested by Hunt and Basford (1999). They formulated intermediate models by employing the principle of local independence, which is the basis, for instance, of latent class modeling (e.g., Heinen, 1996). For the mixture method this principle implies that within the groups there is zero correlation between the variables. In other words, the observed correlations between variables for all observations result from the differences in means between the groups (Hunt & Basford, 1999, p. 290). Under the local independence assumption the group-specific covariance matrices are diagonal, which means that per group there are only J parameters rather than $\frac{1}{2}J(J+1)$. In this way, it might be possible to fit group-specific covariance matrices in some cases when otherwise this might be impossible. Hunt and Basford (1999) further suggested that if no good fit could be obtained, selected off-diagonal elements of the group-specific covariance matrices may be allowed to be estimated, so that an adequate fit could become possible.

Blue crabs: Group-specific versus common covariance matrices. The use of the mixture method of clustering for this data set containing $48 \times 25 \times 3$ observations was in great doubt from the beginning, due to the fairly large number of 25 variables compared to the small number of 48 individuals that in addition had to be divided into a number of groups. With three groups, the common covariance model has in total $G + G \times JK + G \times \frac{1}{2}J(J+1) = 563$ parameters, while the group-specific covariance model has an additional $(G-1) \times \frac{1}{2}J(J+1) = 650$ free parameters. In this case, there were several starting solutions from which no convergent solutions could be obtained, so that our strategy was to assume a common covariance structure across groups and only check afterwards whether a group-specific approach would work for the best solution.

Number of observations. Because mixture methods are based on distributional assumptions and the total sample is split into several groups, sample size can easily be a concern. The covariance matrices for groups with a small number of observations can be relatively ill-defined because of large standard errors. There is, however, one mitigating factor, namely, that not only the individuals allocated to a group contribute to the estimation of the parameters of a group's distribution, but all individuals that have a positive probability of belonging to this group do as well. The most extreme form of this is that for a group consisting of a single individual sometimes the (within-group) covariance matrix can be estimated because sufficient other individuals have a positive probability of belonging to this group. Nevertheless, one should not search for large numbers of groups if there are few individuals. Strict guidelines cannot be formulated because the strength of the grouping structure is also a determining factor. With respect to the means, groups with few individuals are likely to have large standard errors and thus will be difficult to distinguish from other groups.

Distributional assumptions. Formally, the mixture method employed here requires multivariate normality of the variables per condition in each group. One of the difficulties of checking the assumptions is that the composition of the groups is not

Table 16.1 Blue crab data: Skewness and kurtosis of the distributions of Al, As, and Na) for different tissue types

	Aluminum			Arsenic			Sodium		
	G	H	M	G	H	M	G	H	M
Skewness	3.1	13.1	8.4	7.6	−0.3	−0.7	−0.3	−0.6	0.8
Kurtosis	2.0	2.5	0.1	3.5	0.7	0.7	2.7	0.3	1.1

G = Gill tissue; H = Hepatopancreas tissue; M = Muscle tissue.

known beforehand. Due to the largely exploratory use of the technique not much formal checking has been done about the sensitivity to this assumption. It is probable that for large samples normality is not really an issue. In case of doubt, one could investigate both the univariate and multivariate kurtosis for the whole sample beforehand, and afterwards for the groups found (see Browne, 1982, section 1.5, for a multivariate test). If the populations are decidedly nonnormal, the mixture method of clustering might be far from optimal (e.g., see Everitt, 1980, section 5.2). Whether in real life this is a problem depends on the aim of the clustering, that is, whether one is seeking natural clusters or wants to dissect continuous observations. In the latter case, the multinormality assumption seems a reasonable one to make; in the former it is very much an empirical matter whether the method is appropriate. No real solution is available as yet for seriously skewed continuous data. In some cases transformations of the variables might be advisable. Alternatively, one might categorize the skewed distributions, so that the variable can be treated as a multinomial one. One can then use mixture methods developed by Hunt and Basford (1999) to handle both multinomial and normal variables at the same time.

Blue crab data: Distributions. Figure 16.6 shows that there are considerable differences between the distributions of some of the trace elements. Similar large differences and nonnormalities were present for many of the other trace elements. The considerable values for the kurtosis and the skewness also indicated large deviations from multivariate normality (Table 16.1). The deviations from normality were considerably smaller for the group distributions than for the overall distributions of the trace elements, so that the grouping itself caused some regularization toward normality. Some further regularization might have been possible by log-transformations provided it can be shown that severity of the disease is related to log-doses. No transformations were applied in the present case. Treating the distribution as multinomial was not investigated either. Notwithstanding the deviation from normality, the results of the analysis seem to be consistent, so that, at least in this analysis, the lack of multinormality did not appear to be important, but it is clear that considerable checking needs to be done in such cases.

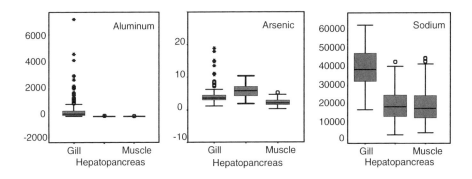

Figure 16.6   Blue crab data: Distributions for Al, As, and Na showing the considerable differences and nonnormality.

16.5.3   Stage 3: Model and number-of-groups selection

Starting allocations.   Crucial to the success of a three-mode mixture method of clustering is finding an optimal solution. As the likelihood function is only guaranteed to converge to a locally optimal solution, one has to run a number of analyses using different starting solutions to find a global optimum as was explained in Section 16.3.3.

Number of groups.   Finding either the correct number of clusters when natural clusters exist or an appropriate number of clusters when no natural clusters exist is clearly the first task in performing a cluster analysis. In the three-mode mixture method of clustering, several statistics and auxiliary information are available for decision making, but sometimes the criteria do not point unequivocally to one particular number of clusters.

Using the likelihood ratio test.   Testing for the number of components G in a mixture is an important but very difficult problem, which has not been completely resolved (McLachlan & Basford, 1988). An obvious way of approaching the problem is to use the likelihood function, which is maximized during the iterative procedure, to test for the smallest value of G compatible with the data. Many authors, including Wolfe (1970), have noted that, unfortunately, with mixture models regularity conditions do not hold for the tests based on the differences in the log-likelihoods between two nested models, so that the chi-square tests for these differences are not applicable. Notwithstanding, McLachlan and Basford (1988) suggest taking guidance from values of the likelihood ratio test statistic when comparing the cluster solution with g groups to the solution with $g - 1$ groups. In particular, they suggest basing decisions on the observation that although generally the likelihood function increases monotonically with the number of clusters (and thus parameters), much smaller gains are made with the addition of more clusters, so that the trade-off between fewer parameters to interpret and a higher value of the likelihood can be made. However, McLachlan and Basford (1988) recommend that the outcome of Wolfe's likelihood ratio test should not be rigidly interpreted, but rather used as a guide.

For all solutions with more than one cluster, the likelihood function very often has multiple maxima. In general there is no guarantee that the best maximum is also the global maximum, even though more confidence is inspired by a solution when the iterative procedure produces the same solution from different starting allocations.

Choosing the number of groups for the blue crab data.   To illustrate some of the above points, the decision procedure for finding an adequate number of groups and an adequate solution is presented for the blue crab data. In Table 16.2 the log-likelihood for several sets of starting values for two, three, and four groups are presented employing the common covariance method. The starting allocations of crabs to groups were based on the results from both a k-means and a Ward's cluster analysis on each of the tissues using SPSS (SPSS Inc., 2000).

The likelihood for a single group was -7427 estimating 401 parameters. For all numbers of clusters greater than 1, the starting values from the analysis of the gill provided the highest likelihood, and twice Ward's method performed better than k-means (Table 16.2). Using the starting solutions, increasing the number of clusters from one to four enlarged the likelihood each time using 401 extra parameters for each new cluster. Therefore, on the basis of the increases in likelihood, a four-cluster solution might be chosen. In particular, the increase in the likelihood in going from one group to two groups was 284, going from two to three groups 169, and going from three to four groups the increase was 65. Using the original grouping (Pamlico diseased or not diseased and Albemarle) as a starting allocation for a three-group solution, however, leads to the highest likelihood, even higher than the best four-group solution found on the basis of the two-mode starting solutions. For an optimal four-group solution, one needs to use the original grouping in three clusters and reallocate the seven crabs with the largest weights on the first dimension of a joint biplot (Fig. 16.8). It was not possible to improve that solution by allocating more or less crabs to the fourth group.

After the solutions were determined using the common covariance method, the best results based on the original groupings were used to investigate the existence of a solution with group-specific covariances. For the two-group and three-group cases, such solutions did indeed exist, but not for the four-group case. The likelihoods were much higher (-6013 and -4857, respectively), but the groupings of the crabs were identical to the common covariance case. Thus, the gain in likelihood was achieved by allowing for the group-specific covariances, and not from a superior grouping of the crabs. The identical partitionings also meant that the mean values of the groups were the same for the same number of clusters. Given the easier interpretability of the common group covariance case and the primary use of the means for deciding the influence of the trace elements on the crabs, it was decided to continue with the three-group solution based on the common covariance solution. However, for researchers

Table 16.2   Log-likelihood statistic for starting allocations based on k-means and Ward's clustering methods on each of the tissues for common covariance method and results for best solution with group-specific method

Cluster		Number of clusters			
method	Tissue	1	2	3	4
Number of parameters		401	802	1203	1604
		−7427			
K-Means	Gill		−7200	**−6974**	−6981
	Hepatopancreas		−7263	−7132	−7085
	Muscle		−7184	−7075	−7057
Ward	Gill		**−7143**	−7034	**−6909**
	Hepatopancreas		−7263	−7132	−7071
	Muscle		−7184	−7144	−7041
Original grouping				**−6878**	
Original grouping*					**−6878**
Likelihood differences			284	265	147
df differences			800	800	800
Group specific covariances		**−7427**	**−6013**	**−4857**	singular

Best solutions given the number of clusters are set in bold. The degrees of freedom for the test of the difference between (g) and $(g − 1) = 2×$ (number of parameters for (g) groups − number of parameters for $(g − 1)$ groups −1). A starting solution for four groups was created from the original grouping by allocating the seven most outlying crabs on the first dimension of Fig. 16.8 to a fourth group.

in the field it would be interesting to see whether there was any merit in using the group-specific covariance solution. Moreover, it might be useful for them to evaluate what the seven crabs, which formed the fourth group in the four-group solution, had in common in order to pinpoint further possible causes of the disease.

In a way, it was a bit disappointing and slightly worrying that the clustering technique did not succeed in retrieving this optimal partitioning into the original groups from other starting points. Most likely the response surface was so hilly that the algorithm got stuck at a local maximum every time (see also Section 5.4.2, p. 81). We suspect that the large number of variables and comparatively small number of individuals may have had something to do with this.

Using the size of within-group covariances.   In a one-cluster solution there are obviously no differences in means between groups, and all variability between individuals is contained in the covariance matrix of that one group. However, as the number of groups increases, part of the variability between individuals is transferred from the covariances to differences in means, so that the overall sizes of the (co)variances become smaller. In perfectly homogeneous clusters all the differences between groups are concentrated in the means, and the covariance matrices are diagonal; that is, within a group the variables are independent. This observation can be used to decide what is an appropriate number of groups by comparing the average absolute within-group covariances across solutions with different numbers of groups. If more than one group exists, then initially the average absolute within-group covariances will drop off relatively rapidly, but if an analysis is run with more groups than reliably exist in the data, the average absolute within-group covariances can be expected not to change much between different solutions. However, with the 25×25 covariance matrices for the blue crab data, it is not clear whether it is useful to check this, as additional statistical techniques are needed to compare several such matrices.

Using posterior probabilities.   McLachlan and Basford (1988) suggested that the estimates of posterior probabilities of group membership can be used in the choice of the number of groups G. In particular, one may use these probabilities for different models to assess the ambiguity of allocation of individuals to a particular group: the more ambiguous the allocations, the more uncertain the number of groups. However, the reverse need not be true. The data may allow perfectly acceptable groupings for different number of groups. For instance, both in the optimal two-group and the optimal three-group solutions of the blue crab data the posterior probabilities were equal to 1 for all crabs. This is a phenomenon often observed in three-mode mixture method of clustering. In essence this says that the groupings are nested in the sense that in the $g + 1$ grouping the new group is a subset of an already existing group.

Comparing partitionings.   By cross-classifying the different groupings in a contingency table, the statistically derived optimal classification can be compared to clusterings with other numbers of groups or with theoretically derived classifications. This makes it possible to assess the percentage agreement between the partitionings, and computing the concordance using measures such as the chance-corrected measure for agreement, Cohen's kappa (J. Cohen, 1960). Guidelines for interpreting the size of Cohen's kappa are found in Fleiss (1981). The same cross-tabulations may also be used to assess the stability of partitionings over different groups. Moreover, core individuals, who always belong to the same group, can be identified as well as individuals whose group membership is very unstable. Because the mixture method is not a nested method, individuals may be in different clusters depending on the number of groups in the solution. Thus, one of the strengths of the mixture method of clustering is that with an increasing (decreasing) numbers of groups, solutions form not necessarily a hierarchy, and it is an empirical rather than a method-dependent issue whether nesting takes place.

Table 16.3   Cross-classification of the original categories with the results of several cluster analyses

	Gill Ward		Gill k-means			Original grouping			Gill Ward				Original† grouping			
	1	2	1	2	3	1	2	3	1	2	3	4	1	2	3	4
Alb	16	0	16	0	0	16	0	0	16	0	0	0	16	0	0	0
PH	16	0	3	13	0	0	16	0	3	9	4	0	0	16	0	0
PD	0	16	1	0	15	0	0	16	0	0	8	8	0	0	11	7

Alb = Albemarle; PH = Pamlico healthy; PD = Pamlico diseased. †A starting solution for 4 groups was created from the original grouping by allocating the 7 most outlying crabs on the first dimension of Fig. 16.8 to a fourth group.

Comparing partitionings for the blue crab data.   The two-group solution created a clear split between the healthy and the diseased crabs, irrespective of their origin (Table 16.3). The three-group solution starting from the original grouping showed that the cluster analysis exactly retained the grouping. This best solution confirms that the original grouping based on collection area and disease state of the crab can be found directly from the trace elements. Note, by the way, that for the other three-group solution and the best four-group solution, the healthy–diseased split was (almost) perfectly maintained, showing that further differentiation is primarily within the two healthy states. The three healthy Pamlico crabs classified as Albemarle were in the same groups in the other three-group solutions and in the four-group solution.

Allocation of individuals to clusters.   Under the mixture approach to clustering, it is assumed that the data at hand can be considered as a sample from a mixture of several populations in various proportions. After estimating the parameters of the underlying distributions, the individuals can be allocated to these populations on the basis of their estimated posterior probabilities of group membership. In this way, individual observations can be partitioned into a number of discrete, relatively homogeneous groups. However, if the posterior probability is less than a specified value, the individuals concerned can remain unallocated but with known probabilities of belonging to the various groups. When the posterior probabilities are nearly equal, the allocation of an individual to a particular group is largely arbitrary and should be treated as such.

A more formal summary way to look at the size of the posterior probabilities is the *correct allocation rate*. The *correct allocation rate for each cluster* is defined as weighted sums of the *a posteriori* allocation probabilities of the individuals to the clusters, and the *overall correct allocation rate* is weighted over all clusters. However, in three-mode applications it has been observed that even in suboptimal solutions the posterior probabilities can be close to 0 or 1 for almost all individuals. This leads to correct allocation weights around one, so that these rates might not be very helpful in deciding which number of groups is preferable.

Posterior probabilities in the blue crab data.   For these data all posterior probabilities were either 0 or 1 in all solutions, except on one occasion where they were 0.9998 and 0.0002. So all correct allocation rates are 1 as well. In the two-mode mixture method of clustering, the posterior probabilities seem to be much more variable, and in that case these allocation rates seem to be more useful.

16.5.4   Stage 4: Results and their interpretation

The basic parameters of the mixture method of clustering are the group means for each condition, the covariance matrices, and the mixing proportions, and, derived from them, the posterior probabilities of an individual belonging to each of the groups. These quantities will be discussed in turn.

Means.   For each variable the clustering provides $J \times G$ means, which have to be interpreted. When the number of variables is not overly large (e.g., as is often the case in agriculture), the means can be displayed in a table showing per variable how the means of the groups differ over conditions. For the Strange Situation data (see Section 16.6.1 for a description) such a table is presented in Table 16.5. When there are many variables this may become unwieldy and some further analysis on the means might be advisable. In essence, the means form a $G \times J \times K$ three-way array. One option might be to do a three-mode ordination on the means to portray the relationships between the means, but this will not be pursued here.

Means plot.   Another way to present the means is to construct a *means plot* in which a plot per variable is made with the conditions on the horizontal axis and the condition means of the groups connected by lines in the plot. If available confidence intervals around the means should be added. Sometimes it takes a bit of ingenuity to find a proper order of the conditions on the horizontal axis: time, the clarity of the plot, or some other criterion can be used. In agriculture, the mean yield of the environments (conditions) is often used to order the conditions, as yield is the most import commercial variable. If there are a large number of groups, one might consider presenting them in two or more plots to ensure readability.

Examples of means plots for the blue crab data are presented in Fig. 16.5. Unless one is interested in small details, presenting means plots for each of the 25 variables does not provide immediate insight, but making a judicious selection of them will certainly help in presenting the results of the analyzed. The means plots of the blue crab data show that sizeable differences in the amounts of trace elements exist for different categories of crabs.

The means plot for the Strange Situation data (see Fig 16.9) is somewhat different because the variables rather than the conditions (i.e., episodes) have been placed on the horizontal axis, while both groups and conditions are represented as lines in the plot. This was possible because of the small number of groups and conditions and because all variables had the same range.

Standard errors for means.   Basford, Greenway, McLachlan, and Peel (1997) discussed an analytical and a bootstrap procedure to find estimates for the standard errors of the group means, but unfortunately could not find a straightforward and reliable procedure for calculating them. One easy heuristic, but rough and ready, procedure is to take the variances from the group-specific or common covariance matrices and use these to calculate the standard errors. If one wants to include the standard errors in the means plots, it may be argued that the interval defined by $\pm 1.5\times$ these standard errors can be used as a gauge to evaluate the differences between the portrayed means (see Basford & Tukey, 1999, pp. 266–267). Basford and Tukey also suggest this is to be preferred over a series of descriptive analyses of variance, because optimizing the mean differences through the cluster analysis precludes valid inferences of mean differences.

Correlations.   Even though it is the covariance matrix that is used in the estimation procedure, in general it will be much easier to interpret the correlation matrix because of different measurement scales of the variables.

Common within-group covariance and correlation matrix.   Given that an adequate solution has been found, the common correlation matrix can be inspected to evaluate the correlations among the variables. To assist in this, one might consider a standard principal component analysis to facilitate the interpretation, or to help order the correlation matrix so that the patterns present can easily be seen (see also p. 410). As an example, the common correlation matrix for the blue crab data is presented in a highly condensed and sorted form (Fig. 16.7).

Group-specific covariance and correlation matrices.   With large numbers of variables one will seldom have group-specific covariance matrices due to estimation problems, but also in interpretation one will need further statistical techniques like principal component analysis to make sense out of them. When there are more than a few groups, such an undertaking will require extensive analysis using techniques that are at least as complex, if not more so, than the three-mode clustering which produced them.

When the number of variables or attributes is not too large, visual inspection of the correlation matrices may suffice for interpreting the correlational structure of the variables for each group and to make a comparison between them. When the clustering has succeeded in finding very homogeneous groups, it is possible that all individuals will have the same values on one or more of the variables leading to zero variances. A comparison involving group-specific covariance matrices will be presented in the example of the Strange Situation data in Section 16.6.4.

	Al	Sn	U	Cr	V	Fe	Mn	Co	As	Si	Mo	Ni	Pb	Cu	P	Ca	Mg	Ag	Se
Al	1.0	.9	.9	.9	.8	.8	.7	.7	.5	.4	.3	.3	.3	.0	.0	-.0	.0	.0	-.0
Sn	.9	1.0	.8	.8	.8	.8	.7	.6	.5	.4	.3	.3	.3	.0	.0	-.0	.0	.1	-.0
U	.9	.8	1.0	.8	.8	.9	.7	.7	.6	.5	.3	.2	.4	.0	.0	-.0	-.0	.1	-.0
Cr	.9	.8	.8	1.0	.8	.8	.6	.6	.5	.5	.3	.4	.5	.1	.0	.0	.0	.0	-.0
V	.8	.8	.8	.8	1.0	.8	.6	.6	.5	.3	.3	.2	.3	.1	.0	.0	.0	.0	.0
Fe	.8	.8	.9	.8	.8	1.0	.7	.6	.5	.5	.2	.2	.5	.1	.0	-.0	-.0	.1	-.0
Mn	.7	.7	.7	.6	.6	.7	1.0	.7	.7	.2	.4	.3	.2	.0	.0	.0	.0	.0	-.0
Co	.7	.6	.7	.6	.6	.6	.7	1.0	.7	.3	.6	.2	.2	.0	-.1	-.1	-.0	.0	.0
As	.5	.5	.6	.5	.5	.5	.7	.7	1.0	.2	.4	.3	.2	.1	-.1	-.2	-.0	.1	.1
Si	.4	.4	.5	.5	.3	.5	.2	.3	.2	1.0	.1	.3	.6	.1	.0	.0	-.0	.2	.0
Mo	.3	.3	.3	.3	.3	.2	.4	.6	.4	.1	1.0	.2	.1	.0	-.2	-.1	-.0	.2	.3
Ni	.3	.3	.2	.4	.2	.2	.3	.2	.3	.3	.2	1.0	.3	.0	.0	.0	.2	.1	.0
Pb	.3	.3	.4	.5	.3	.5	.2	.2	.2	.6	.1	.3	1.0	.4	-.0	-.1	-.2	.2	.0
Cu	.0	.0	.0	.1	.1	.1	.0	.0	.1	.1	.0	.0	.4	1.0	-.0	-.0	-.0	.1	.2
P	.0	.0	.0	.0	.0	.0	.0	-.1	-.1	.0	-.2	.0	-.0	-.0	1.0	.8	.4	-.2	-.0
Ca	-.0	-.0	-.0	.0	.0	-.0	.0	-.1	-.2	.0	-.1	.0	-.1	-.0	.8	1.0	.4	-.2	-.0
Mg	.0	.0	-.0	.0	.0	-.0	.0	-.0	-.0	-.0	-.0	.2	-.2	-.0	.4	.4	1.0	.0	.0
Ag	.0	.1	.1	.0	.0	.1	.0	.0	.1	.2	.2	.1	.2	.1	-.2	-.2	.0	1.0	.5
Se	-.0	-.0	-.0	-.0	.0	.0	-.0	.0	.1	.0	.3	.0	.0	.2	-.0	-.0	.0	.5	1.0

Figure 16.7   Common within-group correlation matrix for blue crab data rounded to the first decimal.

16.5.5   Stage 5: Validation

Ordination.   Irrespective of whether there is a natural grouping of the individuals, cluster methods will produce groups because of the definition and basic assumptions of the methods. Ordination can often be useful to examine the nature of the clusters and shed some light on the possible existence of natural clusters. Because of the three-mode nature of the data, using a three-mode ordination seems an obvious choice, and three-mode principal component analysis has often been used for this purpose, especially in agriculture (see Basford, Kroonenberg, & Cooper, 1996, for an overview and reference to applications). An alternative is to ignore the three-mode structure and use all variables in a discriminant analysis with the newly derived grouping as a dependent variable.

Discriminant analysis.   Whereas cluster analysis is a technique to find clusters, discriminant analysis is a technique to predict known cluster membership on the basis of a set of predictors. In linear discriminant analysis, the assumption is that all groups have a common within-group covariance matrix; in quadratic discriminant analysis, the assumption is that each group has its own group-specific within-group covariance matrix, corresponding to the two options in the mixture method of clustering. Once a cluster analysis has been performed, one can use a discriminant analysis to validate the clusters, especially to investigate whether there is a clear distinct set of clusters or a split of one or more continuous dimensions. The type of discriminant analysis should mirror the assumptions about the covariance matrices in the cluster analysis. This ideal situation, however, can cause difficulties, as it is not uncommon to find singular covariance matrices for individual groups, mainly because one of the variables is constant for that group. In such a case, quadratic discriminant analysis will fail, and one might have to settle for a linear discriminant analysis instead.

Not in all cases will it be possible to carry out a discriminant analysis; it becomes especially difficult when the number of individuals is relatively small. Because the variables of all conditions are used together, this situation can arise easily. The blue crab data are a case in point, as with 48 crabs and 3×25 variables there is not much point in performing a discriminant analysis. That the mixture method of clustering can succeed in this situation is due to the efficiency of treating the same variables across conditions as the same entities, and because individuals can have positive probabilities in more than one group. A detailed example of a situation in which discriminant analysis was useful can be found in Kroonenberg et al. (1995b).

Validation of the blue crab clusters via ordination. To get an impression of the spatial organization of the clusters in the blue crab data and their relationships with the trace elements, here we present a joint biplot from a Tucker3 model with the trace elements and the blue crabs as display modes and the tissues as reference mode. Only the joint biplot for one of the components will be shown here, in particular, one that after rotation has been aligned with the gill tissue; see Chapter 13 for further details on Tucker3 analyses for profile data, Chapter 6 for preprocessing three-way data, and Section 11.5.3 (p. 273) for further details on joint biplots.

The three groups of crabs are clearly separated in the Gill joint biplot (Fig. 16.8), except that one healthy crab from one location on the Pamlico is very close to the Albemarle crabs. In the gill tissue of the diseased crabs, especially Al, As, Cr, Co, Fe, Mn, Sn, U, and V were found in comparatively large doses, while several healthy Pamlico crabs still had relatively large doses of Mg, Ni, and lead (Pb) and nearly all of them had relatively moderate levels of Na and titanium (Ti). In the gill tissue, larger amounts of copper (Cu) and to a lesser extent Se were found in the Albemarle crabs. The presence of especially copper is as it should be because the crabs' blood uses copper to bind oxygen rather than iron as in humans. Copper is, in fact, the source of the crabs' blue color.

From the point of view of the clustering, the separation found by the cluster method on the basis of the trace elements was confirmed in the ordination joint biplots, which also added information on the relative elevation, or lack of it of trace elements in various tissues.

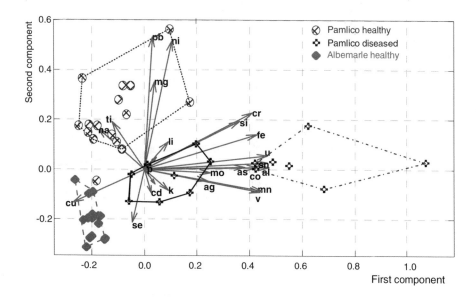

Figure 16.8   Blue crab data: Joint biplot for the Gill component. The outlines (convex hulls) indicate the four groups. One healthy Pamlico crab close to the Albemarle group has not been included in the healthy Pamlico group.

Comparison with theoretical groupings.   In those cases where on substantive or other grounds groupings are known *a priori*, it should be checked to what extent the grouping found by the cluster analysis concurs or is different from the substantive one. This can be done by cross-tabulating the two groupings, and for the blue crab data this was done in Table 16.3. In this particular case, the mixture method grouping coincided with the initial categorization of the crabs and the influence of the trace elements on the crabs was confirmed by the analyses. However, for the Strange Situation data, the clustering and theoretical partitioning were entirely different and more detailed analysis of this discrepancy gave further insight into the data (see Section 16.6.5).

16.6   EXAMPLE: BEHAVIOR OF CHILDREN IN THE STRANGE SITUATION[4]

The main practical considerations in carrying out a three-mode mixture method of clustering will be illustrated by an example from early childhood research. The data come from a research design in which infants experience a series of episodes with

---

[4]The content of this section is based on Kroonenberg et al. (1995b). Reproduced and adapted with kind permission from *British Journal of Psychology*; 1995 ©The British Psychological Society.

either the mother, a stranger, or both, intended to raise the stress on the infant in such a way that the infant's attachment behavior toward the mother can be assessed. This Strange Situation design is the mainstay in attachment research, but the present analysis method is all but a common one in this area.

16.6.1   Stage 1: Objectives

The Strange Situation procedure was designed to assess the quality of infant–mother attachment and is considered appropriate for infants between 12 and 18 months (for details, see Ainsworth, Blehar, Waters, & Wall, 1978). The procedure consists of seven 3-minute episodes arranged to continuously increase the stress , so that the infant's attachment system with respect to the mother is activated. The crucial episodes, the *reunion episodes* (R1 and R2), are those in which the mother returns after having been away, and the infant was left alone with a stranger. On the basis of the infant's behavior, especially during these episodes, the quality of the infant–mother attachment relationship is generally categorized as *insecure-avoidant* (A), *secure* (B), or *insecure-resistant* (C).

According to the classification instructions (see Ainsworth et al., 1978, pp. 59–63), the scores on five 7-point scales in the two reunion episodes play a crucial role in the clinical classification, that is, the Proximity Seeking (PS), Contact Maintaining (CM), Resistance (RS), Avoidance (AV), and Distance Interaction (DI) scales, where high scores on Avoidance are especially indicative for an A classification, and high scores on Resistance for a C classification.

The aim of a cluster analysis is to assess whether the clinical A,B,C typology corresponds to the statistical grouping derived via the mixture method of clustering. A full analysis of these data was given by Kroonenberg et al. (1995b).

16.6.2   Stage 2: Data description and design

A total of 326 infants, or rather infant–mother pairs, were included in our analyses. They originated from five different studies conducted at the Centre for Child and Family Studies of the Department of Education and Child Studies, Leiden University. A comprehensive description of the data can be found in Van Dam (1993). For the 326 infants, scores were available on the five variables mentioned above, measured during the two reunion episodes, so that the data set was a $326 \times 5 \times 2$ three-way data set.

16.6.3   Stage 3: Model and number-of-groups selection

Starting allocations.   Five starting allocations were used including those from two-mode clusterings and allocations resulting from two-mode mixture cluster analyses on the separate episodes. The single group solution had a log-likelihood of -6041. The five two-group analyses using group-specific covariance matrices had two different

Table 16.4   Cross-tabulation of the three-group solution versus the two-group solution both with group-specific covariance matrices (left), and the three-group solution with group-specific covariance matrices against the three-group solution with a common covariance matrix (right)

Group	Two groups: Group-specific covariances		Three groups: Common covariances			Total
	1	2	1	2	3	
1	**114**	0	**113**	1	0	114
2	0	**122**	48	**64**	10	122
3	0	**90**	0	11	**79**	90
Total	114	312	161	76	89	326

solutions with log-likelihood values of -959 (two times) and 336 (three times). For the three-group case with group-specific covariance matrices all analyses converged to the same three-mode cluster solution, independent of the starting allocation with a log-likelihood of 5811. Solutions with larger numbers of clusters only produced marginal increases in the likelihood. In comparison, the three-cluster solution for the common covariance matrix model had a likelihood of -5733. Thus, the three-cluster solution with group-specific covariance matrices was clearly the preferred solution. This solution was also satisfactory from the point of view of *correct allocation rates* as these were all equal to 1.00, indicating that all infants were allocated to groups with an a posteriori probability of 1.00 (see, however, comments in Section 16.5.3, p. 419).

Cross-tabulation of solutions.   Information on the stability of the division of the infants into groups was obtained by cross-tabulating the various partitionings (Table 16.4). It is evident that there was a nesting of three-cluster and two-cluster solutions for the group-specific covariance model. However, the common covariance model produced an entirely different solution. These results further support the three-cluster solution with group-specific covariance matrices.

16.6.4   Stage 4: Results and their interpretation

Group means.   The mean differences on Proximity Seeking (PS), Contact Maintaining (CM), and Distance Interaction (DI) contribute most to the distinction between groups (Table 16.5). Resistance (RS) and Avoidance (AV) are less important, even though they follow the pattern of PS and CM, and that of DI, respectively. These means show that the *first cluster* is characterized by high values in both episodes for Avoidance and Distance Interaction, and low values on Proximity, Resistance, and

Table 16.5   Strange Situation data: Estimated means for three-cluster solution

Group	Episode	Scales					N
		PS	CM	RS	AV	DI	
1	R1	2.1	1.0	1.5	**3.6**	**5.1**	114
	R2	1.9	1.0	2.0	**3.6**	**5.2**	
2	R1	3.0	1.8	2.0	3.3	**4.7**	122
	R2	**4.4**	**3.9**	2.5	2.6	3.3	
3	R1	**5.5**	**4.8**	2.8	1.8	1.0	90
	R2	**5.7**	**5.7**	3.3	1.7	1.0	

PS = Proximity Seeking; CM = Contact Maintaining; RS = Resistance; AV = Avoidance; DI = Distance Interaction.

Contact Maintaining. The *second cluster* shows a stable low RS, but increasing PS and CM coupled with decreasing AV and DI. Finally, the *third cluster* has consistently high PS and CM scores with low AV and DI coupled with a comparatively high level of RS. Overall, there seems to be a single (Proximity + Contact) versus (Avoidance + Distance) dimension that the cluster method uses to define groups.

The same information presented in Table 16.5 can also be represented in a variant of the means plot (Fig. 16.9). The Strange Situation data are somewhat special in the sense that there were only two conditions (episodes), all the variables had the same range, and there were only three groups, so that all variable means for all groups and all conditions can be portrayed in a single plot. The reverse pattern for Groups 1 and 3 can be clearly discerned, with Group 2 taking an intermediate position (see Section 16.6.5 for further interpretation of these patterns). Moreover, the plot shows that only in the second group substantial differences existed between the scores in the first and the second reunion episodes.

Group-specific correlation matrices.   The correlation matrices for the three groups are listed in Table 16.6. In both the first and the third group, all infants had the same score of 1.00 (the lowest possible) for one of the variables (Contact Maintaining and Distance Interaction, respectively), illustrating how the clustering technique converts covariances between variables into differences in means between the groups. These real differences in correlations in the groups explain why the common covariance model gave such a bad fit.

16.6.5   Stage 5: Validation

Spatial arrangement of clusters: Three-mode PCA.   To validate the results from the cluster analysis, a three-mode principal component analysis with a single component

Table 16.6 Group-specific correlation matrices per group

Scale	Group 1					Group 2				
	PS	CM	RS	AV	DI	PS	CM	RS	AV	DI
Proximity	1.00					1.00				
Contact	0.00	*0.00*				0.51	1.00			
Resistance	−0.08	0.00	1.00			0.10	0.13	1.00		
Avoidance	−0.23	0.00	0.33	1.00		−0.41	−0.25	0.11	1.00	
Distance	0.21	0.00	−0.11	−0.45	1.00	−0.28	−0.44	−0.28	0.01	1.00

	Group 3				
Proximity	1.00				
Contact	0.17	1.00			
Resistance	0.03	0.27	1.00		
Avoidance	−0.37	−0.17	0.10	1.00	
Distance	0.00	0.00	0.00	0.00	*0.00*

PS = Proximity Seeking; CM = Contact Maintaining; RS = Resistance; AV = Avoidance; DI = Distance Interaction. Italic *0.00* indicates that all infants in the group had the same value for this scale.

for the reunion or episode mode (see Section 4.5.3, p. 56) was performed on the two times five variables. Given that we intended to use this analysis as a validation for the cluster analysis, it was decided to limit the solution to a fairly easily presentable one with two components for infants and scales. The two-component solution accounted for 63% of the variability with a first component accounting for 47% and the second for 15%. The relative fits of the two reunion episodes were 0.66 and 0.75, indicating that the second episode carried a bit more weight in the solution. Figure 16.10 shows the two-dimensional joint biplot. Each of the infants is labeled according to its cluster membership, and the vectors represent the component loadings for the five variables.

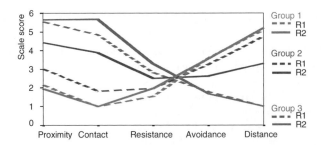

Figure 16.9 Strange Situation data: Groups means for scales for proximity seeking; contact maintaining; resistance; avoidance; distance interaction. R1 = First reunion Episode; R2 = Second reunion episode

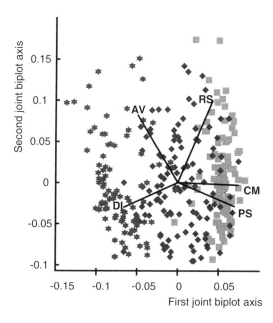

Figure 16.10   Strange Situation: Joint biplot using an one-dimensional reference mode (Episodes). Cluster identification: Stars, Group 1; Diamonds, Group 2; Squares, Group 3.

The first component can be interpreted as the extent to which infants primarily seek proximity and contact with the mother (i.e., use PS and CM), or primarily stay at a distance from her (i.e., use AV and DI). The coefficients for the five variables on the first component are 0.84 (CM), 0.77 (PS), 0.47 (RS), -0.55 (AV), and -0.74 (DI). This confirms the above statement about the importance of the PS and CM versus AV and DI contrast in the cluster analysis, supporting the adequacy of the three-group cluster solution.

Validation of clusters: Discriminant analysis.   To further support the adequacy of the cluster solution, a discriminant analysis was carried out with the groups from the three-mode cluster analysis as the dependent variable and the first component of the component analysis as predictor. This yielded a canonical correlation coefficient of 0.89. The first component provides for an 83% correct allocation of the infants to the groups found by the cluster analysis. (For an example using the original variables, see Kroonenberg et al. (1995b).)

Comparison with clinical classification.   In Section 16.6.1, the question was raised of whether the clinical classification, which was constructed using the guidelines set out by Ainsworth et al. (1978), would also be found in the empirical classification derived with the three-mode mixture method of clustering. To answer this question we

Table 16.7   Cross-tabulations of clinical classification versus joint solution for first+second reunion episodes (group-specific covariance matrices model)

Clinical	Cluster			Total
	1	2	3	
A	**52**	27	1	80
B	62	**78**	69	209
C	0	17	**20**	37
Total	114	122	90	326

have crossed the clinical classification with the clusters from the three-mode mixture method for the case of group-specific covariance matrices (Table 16.7). It is apparent that the clustering method and the clinical classification procedures lead to entirely different clusters. The percentage agreement between the two solutions is 46%, a very low figure indeed, while the chance corrected measure, Cohen's kappa was only 0.16, below any acceptable level for agreement. From the point of view of the cluster analysis, the clinical classification is clearly suboptimal. Even with the clinical classification as a starting allocation for the clustering, the algorithm converged to the solution already presented.

The cause of this difference lies in the different weighting of the variables. The clinical classification instructions emphasize the infants' scores Avoidance and Resistance in their allocation of infants to attachment groups, while downplaying the role of Distance Interaction. However, the clustering procedure weights all variables more or less equally and even somewhat disregards scores on Resistance, due to its limited variability. At the same time, the infants differ vastly in their scores on Distance Interaction and the clustering method is sensitive to that.

To bring this point home, we have relabeled the infants in the joint biplot Fig. 16.10 with their clinical classifications (Fig. 16.11), and comparing these figures, the different partitioning of the infants with the two types of classifications show up dramatically.

16.7   EXTENSIONS AND SPECIAL TOPICS

16.7.1   Multinomial three-mode cluster analysis

The most fundamental extension to the three-mode mixture method of clustering was the inclusion of categorical variables with multivariate multinomial distributions (Hunt & Basford, 1999, 2001). Moreover, it was shown that one can perform mixture method of clustering on mixtures of multinomial and normal distributions with and without missing data. A further extension was proposed by Vermunt (2007), who

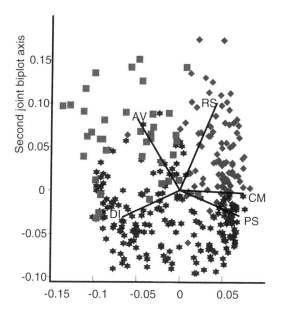

Figure 16.11   Strange Situation: Joint biplot with clinical classification. A: Squares, B: Stars, C: Diamonds

formulated a model that relaxes the assumption that all observational units are in the same cluster at all occasions.

16.7.2 Computer programs

The three-mode mixture method of clustering is implemented in the computer program called MixClus3 which is available both in a Fortran and an S-Plus version. This program can be obtained from Prof. K. E. Basford (as well as the two-mode version of the program, MixClus2); see the website of the Biometry section of the School of Land and Food Sciences, University of Queensland[5]. An earlier version of this program was published as an appendix in McLachlan and Basford (1988). Basford's program was restructured and amended to fit into the 3WayPack suite and can now also be used within that general framework; see the website of The Three-Mode Company[6]. Dr. L. Hunt (Waikato University, Waikato, New Zealand) has made her program MultiMix3 available on her website[7]. This program handles mixtures of multinomial and normal distributions with or without missing data, and also allows the imposition of special structures on the covariance matrices, such as local independence. The

[5] *http://www.uq.edu.au/lcafs/index.html?page=53572.* Accessed May 2007.
[6] *http://three-mode.leidenuniv.nl.* Accessed May 2007.
[7] *http://www.stats.waikato.ac.nz/Staff/lah.html.* Accessed May 2007.

models developed by Vermunt have been implemented in Latent Gold (Vermunt & Magidson, 2005)[8].

16.7.3   Further reading

Papers treating specific theoretical aspects of three-mode cluster analysis as presented in this chapter are Basford and McLachlan (1985a) and McLachlan and Basford (1988). Three-mode cluster analysis originated in agriculture, and most applications still come from this field. In particular, it has been useful to highlight differences between varieties in their phenotypic response to different environments over several years. Agricultural papers showing how three-mode cluster analysis works in practice are Basford and Kroonenberg (1989), Basford, Kroonenberg, and DeLacy (1991), and Chapman, Crossa, Basford, and Kroonenberg (1997).

16.8   CONCLUSIONS

In this chapter, the three-mode mixture method of clustering for continuous three-way profile data consisting of individuals by variables by conditions was explained and illustrated. The basic idea behind this method of clustering is that under each condition the data are a sample from a mixture of multivariate distributions. The mixture is the result of several groups having their own multivariate distribution, and the aim of the technique is to uncover these groups. The basic three-mode assumption for the techniques used here is that even though the shape of the multivariate normal distribution for each group is the same under each condition, its location may be different under each condition. Detailed examples were discussed, in particular, the blue crab data (Gemperline et al., 1992) and the Strange Situation data (Kroonenberg et al., 1995b). The results of the cluster analyses were validated with several ordination techniques, such as discriminant analysis and three-mode principal component analysis. The assumption of independence of the episode measurements in the mixture model does not appear to be so restrictive that it hinders obtaining interpretable results that show consistency over different analytical techniques used during the validation of the clusterings. Thus, on the basis of the evidence, we have reasonable confidence that three-mode mixture method of clustering can be a useful tool for summarization in terms of relatively homogeneous clusters, even if some of the assumptions of the technique are not met.

---

[8]*http://www.statisticalinnovations.com/products/latentgold.html.* Accessed May 2007.

CHAPTER 17

MULTIWAY CONTINGENCY TABLES

17.1 INTRODUCTION[1]

In all but two chapters in this book we treat the multiway analysis of continuous variables. This chapter and the next one are different in that the variables are categorical. It is not our intention in this chapter to give a full treatment of the possibilities of analyzing multiway categorical data, but we will only discuss the analysis of multiway categorical data in the form of three-way contingency tables. Moreover, only those methods are presented in this chapter which involve three-mode principal component analysis. This boils down to three-way correspondence analysis. There are several related topics that are still under development, in particular nonsymmetric correspondence analysis for multiway data (D'Ambra & Lombardo, 1993; Lombardo, Carlier,

---

[1]This chapter is based on joint work with André Carlier (†5-5-1999), Université Paul Sabatier, Toulouse, France ; portions of this chapter have been taken and adapted from Carlier and Kroonenberg (1996) with kind permission from the Psychometric Society.

Applied Multiway Data Analysis. By Pieter M. Kroonenberg
Copyright © 2007 John Wiley & Sons, Inc.

& D'Ambra, 1996), and multiway techniques for categorical data using optimal scaling (Sands & Young, 1980).

The standard approach for handling multiway contingency tables is to perform a loglinear analysis; for an solid treatment of loglinear models, see Agresti (1990). Using loglinear analysis one intends to find a model for (logarithms of) the observed frequencies. Such a model is linear for the logarithms of the expected values and resembles analysis of variance, in the sense that a full or saturated model has a constant term, main effect terms, and two-way and higher-way interaction terms. Parsimonious models are built by removing interaction terms starting with the highest interaction term(s). A model for a multiway contingency table with only main effects is the model of complete independence.

When the variables making up the contingency table have many categories, one will seldom be able to ignore interactions without a serious loss of fit. In this chapter we will concentrate on the analysis of those contingency tables for which the interactions are important and the complexity cannot be reduced by eliminating some of them. Multiway correspondence analysis discussed in this chapter aims to unravel multiway interactions rather eliminate them. Within loglinear modeling a similar purpose is served by association models; see Anderson (1996) and for an empirical comparison between the two approaches Kroonenberg and Anderson (2006).

So far multiway correspondence analysis has only been applied for three-way contingency tables. Such tables arise either from cross-classifying three categorical variables or from more than three categorical variables that have been cast into the three-way mold by interactively coding one or more of the variables.

17.2   CHAPTER PREVIEW

In this chapter we will present methods to analyze data that can be collected into a medium to large three-way contingency table. By medium to large we mean that the variables constituting the table have at least three categories, but typically at least one of them has (many) more categories. At the same time, there should be sufficient numbers of observations, both overall and in the cells, to make an in-depth analysis sensible.

The most basic model for a three-way contingency table is the model of complete independence. In this model, the proportions in the table can be reproduced from the marginal proportions of the rows, columns, and tubes. After removing from the table of that part which can be explained by the independence model, we are left with the interactions or the *total dependence* between the three variables. The major aim of the technique treated in this chapter, three-way correspondence analysis, is to acquire an understanding of this dependence, and the nature and structure of the mutual influence of the variables. The tools used for this purpose are (1) measuring the contributions of various parts of dependence to the total dependence via the inertia, (2) modeling the dependence via three-mode models, and (3) graphing the dependence via several types of biplots.

17.3 THREE-WAY CORRESPONDENCE ANALYSIS: THEORY

Three-way correspondence analysis is a generalization of ordinary or two-way correspondence analysis, and it shares many aspects with the simpler technique, so that it will be introduced first.

17.3.1 Two-way correspondence analysis

A two-way contingency table has a row variable with index $i = 1, \ldots, I$ and a column variable with index $j = 1, \ldots, J$. The contents of the cells in the table are generally frequencies f_{ij}, or relative frequencies or proportions p_{ij}. The differences between the observed proportions can be modeled with a part due to the model of independence between rows and columns and a part due to the dependence between rows and columns. The model of independence between row and column variables postulates that the p_{ij} can be modeled by the product of the marginal proportions $p_{i.}$ and $p_{.j}$, that is $p_{i.}p_{.j}$. What is left after the contribution of the independence model has been subtracted from each cell proportion is due to the *dependence*, $(p_{ij} - p_{i.}p_{.j})$. For the purpose of analysis, this dependence part is generally standardized by the square root of the expected value under the independence model $p_{i.}p_{.j}$. In other words, $(p_{ij} - p_{i.}p_{.j})/\sqrt{p_{i.}p_{.j}}$ is the standardized residual from the independence model. The parallel quantity X_{ij} expressed in frequencies rather than proportions, that is, $X_{ij} = (f_{ij} - f_{i.}f_{.j})/\sqrt{f_{i.}f_{.j}}$, is often called as a *chi-term*, because the sum of the squared X_{ij} make up the Pearson X^2 statistic, which is generally referred to as the chi-square statistic because of its asymptotic chi-square distribution.

Measuring dependence.   Let us consider an $I \times J$ contingency table with relative frequencies p_{ij}. There are two types of measures of dependence: the first measure, Π_{ij}, indicates the deviation from independence of each cell (i, j) in the contingency table; the second, Φ^2, indicates the total variance due to the dependence in the entire table. $\Phi^2 = X^2/n$, where X^2 is Pearson's chi-squared statistic and n is the total number of observations. Φ^2 is known as the *inertia* and as *Pearson's mean-square contingency coefficient*. We may write

$$\Phi^2 = \sum_{i,j} \frac{(p_{ij} - p_{i.}p_{.j})^2}{p_{i.}p_{.j}} = \sum_{i,j} p_{i.}p_{.j}\left(\frac{p_{ij} - p_{i.}p_{.j}}{p_{i.}p_{.j}}\right)^2 = \sum_{i,j} p_{i.}p_{.j}\Pi_{ij}^2, \quad (17.1)$$

where $p_{i.} = \sum_j (p_{ij})$, $p_{.j} = \sum_i (p_{ij})$. The measure of the deviation of dependence Π_{ij} may be rewritten as

$$\Pi_{ij} = \frac{p_{ij} - p_{i.}p_{.j}}{p_{i.}p_{.j}} = \frac{p_{ij}}{p_{i.}p_{.j}} - 1 = \frac{\Pr[i|j]}{\Pr[i]} - 1 = \frac{\Pr[j|i]}{\Pr[j]} - 1, \qquad (17.2)$$

showing that $1 + \Pi_{ij}$ is equal to the ratio of the probability of category i given category j, $\Pr[i|j]$, to the marginal probability of i, $\Pr[i]$, and to the ratio of $\Pr[j|i]$ to $\Pr[j]$. Therefore, Π_{ij} measures the *attraction* between categories i and j if $\Pi_{ij} > 0$ and the *repulsion* between the two categories if $\Pi_{ij} < 0$. It is the matrix containing the $\mathbf{\Pi} = (\Pi_{ij})$ which is decomposed into its components by correspondence analysis.

Modeling dependence. In order to inspect the structure of the dependence in a contingency table, we want to portray it in a graph. To do this, we need to find an optimal joint representation of the row and column categories in a low-dimensional space, which can be plotted for inspection. The appropriate tool for this is based on a generalization of the singular value decomposition (see also Section 4.3, p. 48).

If we indicate the number of independent components, or the rank, of the matrix $\mathbf{\Pi}$ by S_0, then the generalized singular value decomposition (GSVD)[2] of the matrix $\mathbf{\Pi}$ is defined as

$$\Pi_{ij} = \sum_{s=1}^{S_0} \lambda_s a_{is} b_{js}, \qquad (17.3)$$

where the scalars $\{\lambda_s\}$ are the singular values arranged in decreasing order of magnitude, and the a_{is} and b_{js} are the elements of the singular vectors or components $\{\mathbf{a}_s\}$ and $\{\mathbf{b}_s\}$, respectively. The components $\{\mathbf{a}_s\}$ are pairwise orthonormal with respect to the inner product weighted by $(p_{i.})$, and a similar property holds for $\{\mathbf{b}_s\}$ with respect to the weights $(p_{.j})$. This means that $\{\mathbf{a}_s\}$ and $\{\mathbf{a}'_s\}$ are orthonormal when weighted with $p_{i.}$ or $\sum_i p_{i.} a_{is} a_{is'} = 1$ if $s = s'$ and 0 otherwise.

If we want a small number, S, of components that explain most of the dependence and thus approximate the full solution as well as possible, we should choose them so that we get the best low-dimensional approximation to $\mathbf{\Pi}$. This means that we have to use $\widehat{\mathbf{\Pi}}^{(S)}$, which is the sum of the first S terms of Eq. (17.3) (Eckart & Young, 1936):

$$\Pi_{ij}^{(S)} = \sum_{s=1}^{S} \lambda_s a_{is} b_{js}. \qquad (17.4)$$

The result of this choice is that the total inertia Φ^2 can be split into a fitted part and a residual part. This can be used to assess the overall quality of a solution via the proportion explained Φ^2.

---

[2]This generalization of the SVD using weighted metrics is common in the context of correspondence analysis. However, the term "generalized singular value decomposition" is used in an entirely different sense in most of the linear algebra literature.

Plotting dependence: Biplots.   Given that we have decomposed the dependence, we want to use this decomposition to graph the dependence. The appropriate graph is the biplot (Tucker, 1960; Gabriel, 1971), Section 11.5.2 (p. 271), and Appendix B (p. 491). In two-way correspondence analysis the biplot displays in S dimensions the rows and columns of the matrix $\widehat{\Pi}$, collectively called *markers*. Using the positions of the markers of the ith row and the jth column, it becomes possible to estimate the value of the element $\widehat{\Pi}_{ij}$ of $\widehat{\Pi}$, and to interpret geometrically the directions of the markers in the plot. If $S = 2$, this can be done in one plot; if $S = 4$ two such plots are needed, and so on. Because it is difficult to evaluate markers in a three-dimensional space, the most commonly used biplots are two-dimensional.

17.3.2   Three-way correspondence analysis

Important properties of three-way correspondence analysis are that the dependence between the variables in a three-way table can be modeled and displayed. Moreover, it shares and extends many properties of two-way correspondence analysis.

Measuring dependence.   Whereas in two-way tables there is only one type of dependence, in three-way tables one can distinguish (1) *total dependence*, which is the deviation from the three-way independence model, (2) *marginal dependence*, which is the dependence due to the two-way interactions, (3) *three-way dependence*, which is due to the three-way interaction, and (4) *partial dependence*, which is the total dependence with one or more marginal dependence partialed out.

Measuring total dependence.   Three-way contingency tables have orders I, J, and K with relative frequencies p_{ijk}. Dependence in the table is again measured by the inertia Φ^2, which is defined analogous to the two-way case as in Eq. (17.2).

$$
\begin{aligned}
\Phi^2 &= \sum_{i,j,k} \frac{(p_{ijk} - p_{i..}p_{.j.}p_{..k})^2}{p_{i..}p_{.j.}p_{..k}} = \sum_{i,j,k} p_{i..}p_{.j.}p_{..k} \left[\frac{p_{ijk} - p_{i..}p_{.j.}p_{..k}}{p_{i..}p_{.j.}p_{..k}} \right]^2 \\
&= \sum_{i,j,k} p_{i..}p_{.j.}p_{..k}(\Pi_{ijk})^2.
\end{aligned} \tag{17.5}
$$

Φ^2 is based on the deviations from the three-way independence model, and contains all two-way interactions and the three-way interaction.

The measure for the dependence of cell (i, j, k), Π_{ijk}, may be rewritten as

$$
\Pi_{ijk} = \frac{\Pr[ij|k]}{\Pr[ij]} \cdot \frac{\Pr[ij]}{\Pr[i]\Pr[j]} - 1. \tag{17.6}
$$

The quantity $1 + \Pi_{ijk}$ is the product of (1) the ratio $\Pr[ij|k]/\Pr[ij]$, which measures the relative increase or decrease in the joint probability of the categories i and j given occasion k, and (2) the ratio $\Pr[ij]/\Pr[i]\Pr[j]$ which measures the relative increase or decrease in the deviation from the marginal independence. If the conditional probability for all k is equal, then $\Pr[ij|k] = \Pr[ij]$ and the first ratio is 1. Then $\Pi_{ijk} = \Pi_{ij.}$, and the three-way table could be analyzed with two-way correspondence analysis. The symmetric statement after permutation of the indices holds as well. Therefore, the Π_{ijk} measure the total dependence of the cell (i, j, k).

The elements of the two-way marginal totals are defined as weighted sums over the third index. Thus, for the $I \times J$ margins these elements are

$$\Pi_{ij.} = \sum_k p_{..k}\Pi_{ijk} = \sum_k p_{..k}\frac{p_{ijk} - p_{i..}p_{.j.}p_{..k}}{p_{i..}p_{.j.}p_{..k}} = \frac{p_{ij.} - p_{i..}p_{.j.}}{p_{i..}p_{.j.}}. \tag{17.7}$$

The elements of the other two-way margins, $\Pi_{i.k}$ and $\Pi_{.jk}$, are similarly defined. One-way marginal totals are summed over two indices and they are 0 due to the definition of Π_{ijk}; the overall total is 0 as well. For example, the one-way row margin i follows from (using Eq. (17.7))

$$
\begin{aligned}
\Pi_{i..} &= \sum_j \sum_k p_{.j.}p_{..k}\Pi_{ijk} = \sum_j p_{.j.}\frac{p_{ij.} - p_{i..}p_{.j.}}{p_{i..}p_{.j.}} \\
&= \sum_j \frac{p_{ij.}}{p_{i..}} - \sum_j \frac{p_{i..}p_{.j.}}{p_{i..}} = 1 - 1 = 0. \tag{17.8}
\end{aligned}
$$

Measuring marginal and three-way dependence. The total dependence of cell Π_{ijk} can be split into separate contributions of the two-way interactions and the three-way interaction,

$$\Pi_{ijk} = \frac{p_{ij.} - p_{i..}p_{.j.}}{p_{i..}p_{.j.}} + \frac{p_{i.k} - p_{i..}p_{..k}}{p_{i..}p_{..k}} + \frac{p_{.jk} - p_{.j.}p_{..k}}{p_{.j.}p_{..k}} + \frac{p_{ijk} - {}_\alpha p_{ijk}}{p_{i..}p_{.j.}p_{..k}}, \tag{17.9}$$

where $_\alpha p_{ijk}$ is equal to $p_{ij.}p_{..k} + p_{i.k}p_{.j.} + p_{.jk}p_{i..} - 2p_{i..}p_{.j.}p_{..k}$. The terms referring to the two-way margins are equivalent to those defined by expression (17.2). The last term $_\alpha p_{ijk}$ measures the size of the three-way interaction for cell (i, j, k).

Due to the additive splitting of the dependence of individual cells, the inertia Φ^2, which measures the total dependence of the table can be partitioned as

$$
\begin{aligned}
\Phi^2 &= \sum_{ij} p_{i..}p_{.j.}\left(\frac{p_{ij.} - p_{i..}p_{.j.}}{p_{i..}p_{.j.}}\right)^2 + \sum_{ik} p_{i..}p_{..k}\left(\frac{p_{i.k} - p_{i..}p_{..k}}{p_{i..}p_{..k}}\right)^2 \\
&\quad + \sum_{jk} p_{.j.}p_{..k}\left(\frac{p_{.jk} - p_{.j.}p_{..k}}{p_{.j.}p_{..k}}\right)^2 + \sum_{ijk} p_{i..}p_{.j.}p_{..k}\left(\frac{p_{ijk} - {}_\alpha p_{ijk}}{p_{i..}p_{.j.}p_{..k}}\right)^2 \\
&= \Phi^2_{IJ} + \Phi^2_{IK} + \Phi^2_{JK} + \Phi^2_{IJK}. \tag{17.10}
\end{aligned}
$$

The importance of the partitioning (17.10) is that it provides measures of fit for each of the interactions and thus indicates the contributions of these interactions to the total dependence.

Modeling dependence. Given measures for total dependence, marginal dependence, and three-way dependence, a model for these measures has to be found to enable us to construct graphs depicting the dependence. In two-way correspondence analysis the generalized singular value decomposition is used for this purpose. Thus, for the three-way case, a three-way analogue of the GSVD is desired. There are several candidates of which we will consider only the Tucker3 model (see Section 4.5.3, p. 54), which can be seen as a three-mode singular value decomposition.

Modeling total dependence. We will use the Tucker3 model in this chapter only in its summation form:

$$\Pi_{ijk} = \sum_{p=1}^{P} \sum_{q=1}^{Q} \sum_{r=1}^{R} g_{pqr} a_{ip} b_{jq} c_{kr} + e_{ijk}. \tag{17.11}$$

The difference between its properties here and the standard usage is that the components are weighted orthonormal with respect to the marginal proportions rather than unweighted. In particular, the $\{a_p\}$ are pair-wise orthonormal with respect to the weight $(p_{i..})$, the $\{b_q\}$ are pair-wise orthonormal with respect to $(p_{.j.})$, and the $\{c_r\}$ are pair-wise orthonormal with respect to $(p_{..k})$. The g_{pqr} are thus the three-way analogue of the singular values. In three-way correspondence analysis, a weighted least-squares criterion is used: the parameters g_{pqr}, a_{ip}, b_{jq}, and c_{kr} are those that minimize

$$\sum_{i,j,k} p_{i..} p_{.j.} p_{..k} e_{ijk}^2.$$

As in two-way correspondence analysis, the global measure of dependence, the inertia Φ^2 can be split into a part fitted with three-mode singular value decomposition (here the Tucker3 model) and a residual part.

Modeling marginal dependence. One of the attractive features of the additive partitioning of the dependence in Eq. (17.9) is that the single decomposition of the total dependence can be used to model the marginal dependence as well.

The marginal dependence of the rows i and columns j is contained in a matrix $\mathbf{\Pi}_{IJ}$ with elements $\Pi_{ij.} = (p_{ij.} - p_{i..} p_{.j.})/(p_{i..} p_{.j.})$ with similar expressions for the other two matrices with marginal dependencies $\mathbf{\Pi}_{IK}$ and $\mathbf{\Pi}_{JK}$. The elements $\Pi_{ij.}$ are derived from the total dependence via a weighted summation over k:

$$\Pi_{ij.} = \sum_{k} p_{..k} \Pi_{ijk}. \tag{17.12}$$

We have modeled the total dependence Π_{ijk} using the Tucker3 model in Eq. (17.11) and we may use Eq. (17.12) to find the model for the marginal dependence:

$$\Pi_{ij.} = \sum_{p=1}^{P}\sum_{q=1}^{Q}\sum_{r=1}^{R} g_{pqr}a_{ip}b_{jq}c_{.r} + e_{ij.} \tag{17.13}$$

with $c_{.r} = \sum_k p_{..k}c_{kr}$ and $e_{ij.} = \sum_k p_{..k}e_{ijk}$. Inspection of this formula leads to the conclusion that the marginal model is derived from the overall model by averaging the appropriate components (here, $\{c_r\}$).

Modeling partial dependence.   In some applications the mode k is a time mode, and we are interested in investigating that part of the dependence which explicitly depends on time. Thus, the dependence not associated with time (i.e., the dependence due to the $I \times J$ margin), has to be removed from the total dependence. The removal of a margin, say, the $I \times J$ one, to investigate the *partial dependence*, is realized as follows:

$$\Pi_{ijk} - \Pi_{ij.} = \sum_{p,q,r} g_{pqr}a_{ip}b_{jq}(c_{kr} - c_{.r}) + (e_{ijk} - e_{ij.}). \tag{17.14}$$

As Eq. (17.14) shows, the modeling of the partial dependence is achieved by centering the components of one of the modes (here the time mode, $\mathbf{c}_r$).

When we have one criterion and two predictor variables such as in Davis's Happiness data (see Section 17.4) we may want to eliminate the influence of the design variables on the total dependence in a three-way contingency table. This can also be done by eliminating the two-way interaction due to the predictor variables and investigating the remaining partial dependence.

In this situation, the usual approach in loglinear modeling is to transform the problem into a logit one by building a model for the logits using only the interactions of the predictor variables with the response variable. In this case, too, the margins and terms exclusively associated with the design variables are removed from the model. From this perspective, one could say that modeling partial dependence by removing the design variables is one way to introduce a prediction structure into three-way correspondence analysis, but this will not pursued here.

Plotting dependence.   With respect to dependence and its modeling, the three ways of the contingency table behave in an entirely symmetric fashion. This symmetry, however, is difficult to maintain when graphing the dependence, because no viable spatial representations have been worked out to portray all three ways simultaneously and symmetrically in one subspace. A strict parallel with two-way correspondence analysis can therefore not be maintained. To display the dependence or its approximation in three-way correspondence analysis, we can make use of two kinds of biplots: the *nested-mode biplot* (see Section 11.5.4, p. 276) and the *joint biplot* (see Section 11.5.3, p. 273).

Plotting total dependence: Nested-mode biplot. The *nested-mode biplot* aims to portray all three modes in a single biplot. As a biplot has only two types of markers, two modes have to be combined into a single one via interactive coding, i.e. two modes are condensed to one by fully crossing them. For the Peer play data, each pair of indices of the display modes (group and measurement times), (i, k), were interactively combined into a single marker. The remaining reference mode (play qualities) supplies the other set of markers. The choice of reference mode is always data dependent. Given that an ordered (here, time) mode will always be coded interactively, the choice between the remaining two depends on which of the two modes produces the clearest patterns in their changes over time.

The construction of the biplot for the total dependence $\widehat{\Pi}_{ijk}$ follows directly from three-mode singular value decomposition of the total dependence.

$$
\widehat{\Pi}_{ijk} = \sum_{q=1}^{Q} \left[\sum_{p=1}^{P} \sum_{r=1}^{R} g_{pqr} a_{ip} c_{kr} \right] b_{jq} = \sum_{q=1}^{Q} d_{(ik)q} b_{jq}
$$

$$
\widehat{\Pi}_{\ell j} = \sum_{q=1}^{Q} d_{\ell q} b_{jq} . \tag{17.15}
$$

By replacing the (ik) with a new (interactively coded) index ℓ, we see that the coordinates of the row markers are the $d_{\ell q}$ and those of the column markers b_{jq}. Note that the g_{pqr} are absorbed in the coordinates of the row markers ℓ. The nested-mode biplot is therefore a row-metric preserving one with respect to the weights $p_{.j.}$. Thus, the rows are in principal coordinates and the columns in normalized coordinates. The number of two-dimensional biplots does not depend on P or R but only on Q, the number of components of the reference mode, and it is equal to $Q/2$ if Q is even. The choice between three or four components in the reference mode could be guided by whether it is easier to inspect a three-dimensional plot or two independent two-dimensional plots, one with the first two and the other with the last two components.

The nested-mode biplot is especially useful when the number of elements in $I \times K$ is not too large, or when one of the two sets I or K is ordered (e.g., is associated with time, as it is in our examples). Assuming k is an ordered mode, trajectories can be drawn in the biplot by connecting, for each i, the points (i, k) in their given order (see Fig. 17.5).

Plotting marginal dependence. Equation (17.13) (p. 440) shows that the marginal dependence of the rows i and the columns j, where the second mode (columns) represents the reference mode, can be modeled by

$$
\widehat{\Pi}_{ij.} = \sum_{p=1}^{P} \sum_{q=1}^{Q} \sum_{r=1}^{R} g_{pqr} a_{ip} b_{jq} c_{.r} = \sum_{q=1}^{Q} d_{(i.)q} b_{jq} , \tag{17.16}
$$

with $c_{.r} = \sum_k p_{..k} c_{kr}$. This leads to the conclusion that the marginal model can be derived from the overall model by averaging the appropriate components (here: $\{c_r\}$). The importance of this result should not be overlooked. It means that the coordinates of the two-way interactions are directly contained in the plot of the total dependence. All we have to do is compute the marginal coordinates from the global ones and plot the coordinates of the $d_{(i.)}$. Thus, each of the $d_{(i.)}$ is at the centroid of the row points $d_{(i1)} \cdots d_{(iK)}$. To evaluate the interaction one can either do this in the full nested-mode biplot, or make a special graph displaying only the marginal dependence by graphing the points j of the reference mode together with the centroids of i. The interpretation can then proceed as in a standard two-way correspondence analysis. Note, however, that even though one may make a separate plot, the space portrayed is still that of the total dependence, and the separate plot is made purely to avoid the clutter of the total dependence solution. In small examples, separate plots may not be necessary.

The procedures and considerations are the same for the marginal dependence of tubes and columns, (i.e., $\widehat{\Pi}_{.jk}$), but the dependence between rows and tubes, $\widehat{\Pi}_{i.k}$, is different, because the averaging is over the index j belonging to the reference mode, a result of the asymmetry created by the way we chose to plot. In that case,

$$\widehat{\Pi}_{i.k} = \sum_{p=1}^{P}\sum_{q=1}^{Q}\sum_{r=1}^{R} g_{pqr} a_{ip} c_{kr} b_{.q} = \sum_{q=1}^{Q} d_{(ik)q} b_{.q}, \qquad (17.17)$$

with $b_{.q} = \sum_k p_{.j.} b_{jq}$. Thus, in this case the centroid is the vector $(b_{.1} \cdots b_{.Q})$, the vector containing the average coordinate values of the elements of the reference mode. In the plot this will be a single arrow from the origin with the $b_{.j}$ as coordinates. Obviously, a small arrow indicates that the row by tube interaction of the display modes is small and vice versa. To get the size of interaction between a particular row i across all k one has to project this trajectory onto the arrow. Therefore, trajectories perpendicular to the arrow do not contribute to the interaction.

Plotting partial dependence. In Section 17.3.2 (p. 440) we indicated that at times one might want to investigate only the time-dependent part of the interaction or that part of the interaction that is not influenced by the design margin. This can be achieved by subtracting the relevant marginal dependence from the total dependence:

$$\Pi_{ijk} - \Pi_{ij.} = \frac{\Pr[ij]}{\Pr[i]\Pr[j]}\left(\frac{\Pr[ij|k]}{\Pr[ij]} - 1\right). \qquad (17.18)$$

In case the third mode is time, the part on the right-hand side in square brackets can be considered a growth index for the row–column combination (i, j). If there is no dependence on time, then the conditional probability of (i, j) given k, $\Pr[ij|k]$, is equal to $\Pr[ij]$ and the difference in Eq. (17.18) is 0 for all (i, j). In that case, all trajectories will consist only of their centroids. In other words, the variability around the centroids is made up of the changes over time due to all interactions.

Again, there is no need to create special graphs, but the dependence on time can be evaluated in the basic nested-mode biplot by evaluating the variability around the centroid. This is contained in the length and direction of the trajectory with respect to its centroid.

Joint biplots. To construct joint biplots we again need to choose a reference mode. As before, we will use the second mode for this, but another mode could have been chosen as well. The best choice of reference mode is again data dependent. For joint biplots, it is especially important that the components of the reference mode have a clear interpretation, because without it the overall interpretation may be difficult, as the interpretation of these plots is conditional on the component of the reference mode. The joint biplots are based on the same equation as the nested-mode biplot — Eq. (17.15) (see also Kroonenberg, 1983c, p. 164ff.). The construction of joint biplots proceeds as follows (see Section 11.5.3, p. 273):

$$
\widehat{\Pi}_{ijk} = \sum_{q=1}^{Q} \left[\sum_{p=1}^{P} \sum_{r=1}^{R} g_{pqr} a_{ip} c_{kr} \right] b_{jq}
$$

$$
= \sum_{q=1}^{Q} d_{(ik)q} b_{jq}.
$$

In this case we consider each slice j of the array $\widehat{\Pi}$ to be a linear combination of the $\mathbf{D}_q = (d_{(ik)q})$ with as coefficients the b_{jq}; thus, if $q = 2$, $\widehat{\Pi}_j = b_{j1}\mathbf{D}_1 + b_{j2}\mathbf{D}_2$. For further technical details about the construction of joint biplots, see Section 11.5.3 (p. 273).

The marginal effects of the plotted modes with the reference mode, here $(I \times J)$ and $(K \times J)$, can be studied in the same joint biplot via combined usage of one set of markers and the axis determined by the centroid of the other set. Recall that the marginal effect of, for instance, the $(I \times J)$ margin is given by $\Pi_{ij.} = \sum_k p_{..k}\Pi_{ijk}$ (see Eq. (17.7)). Its approximation can be expressed in terms of $d_{(ik)q}$:

$$
\widehat{\Pi}_{ij.} = \sum_{k} p_{..k} \sum_{q=1}^{Q} d_{(ik)q} b_{jq} = \sum_{q=1}^{Q} d_{(i.)q} b_{jq} = \sum_{q=1}^{Q} b_{jq} \sum_{\ell} \tilde{a}_{i\ell} \tilde{c}_{.\ell}, \tag{17.19}
$$

where ℓ indicates the ℓth axis of the joint biplot. Thus, by taking the centroid of the third mode, and combining it with the markers of the first mode, it is possible to assess the value of $d_{(i.)q}$ from the joint biplot of $\mathbf{D}_q$, and to evaluate the $(I \times J)$ interaction. The evaluation of the $(K \times J)$ interaction proceeds analogously.

17.4   EXAMPLE: SOURCES OF HAPPINESS

To get acquainted with the general approach toward three-way correspondence analysis in practice, a relatively small example is presented by which some, but not all, of the features of the technique will be illustrated. This same data set will be used in a later section to illustrate specific practical aspects and problems in three-way correspondence analysis.

17.4.1   Stage 1: Objectives

The object of the analysis of the Happiness data is to gain insight into the relationship between happiness as reported by participants in a survey, and the number of their siblings and the number of years of schooling they had completed. The main interest lies in the three-way interaction between the two predictors and the dependent variable Happinessc. Of the two-way interaction terms, that between the two predictors is of limited interest because it bears no relation to the prediction of the criterion because it is the design interaction. Its relevance very much depends on the sampling frame.

The elements to be considered in a three-way correspondence analysis are the marginal proportions, the two-way interactions (which we will skip in this example), the relative size of the X^2 contribution of each interaction (see Table 17.3), the fit of the three-mode model to each of the interactions (see again Table 17.3), the biplot of the total dependence (see Fig. 17.1), the marginal dependence of each two-way interaction in the space of the total dependence (discussed in Section 17.3.2), and sometimes the partial dependence with respect to the total dependence. In practice it may not be necessary to present all these elements in a research paper, but the researcher should at least have seen them all and thought about their relevance for the substantive problem at hand.

17.4.2   Stage 2: Data description and design

Davis (1977) presented data on the Happiness reported by participants in a survey as a function of Number of siblings (Siblings) and the Years of schooling completed (Schooling). The data were originally published in Clogg (1982, Table 2). They are reproduced here as Table 17.1.

The design of the data is such that there is an ordered response variable Happiness and two ordered predictor variables Years of schooling and Number of siblings. For the moment we will not make a distinction between the two types of variables, and the ordinality of the variables will only enter our analyses at the interpretation stage, not during the analysis itself. Inspecting the marginal proportions (Table 17.2) we see that the variable Years of schooling is relatively evenly distributed except that there are only 6% college-educated persons in the sample. A sizeable number of people (30%) come from families with 6 or more children, and one-third of the respondents are not too happy, half are pretty happy, and for 12% things could not be better.

Table 17.1   Davis's Happiness data

Years of schooling	Number of siblings				
	0-1	2-3	4-5	6-7	8+
Not too happy					
<12	15	34	36	22	61
12	31	60	46	25	26
13-16	35	45	30	13	8
17+	18	14	3	3	4
Pretty happy					
<12	17	53	70	67	79
12	60	96	45	40	31
13-16	63	74	39	24	7
17+	15	15	9	2	1
Very happy					
<12	7	20	23	16	36
12	5	12	11	12	7
13-16	5	10	4	4	3
17+	2	1	2	0	1

Source: Clogg (1982, Table 2)

The Years of schooling by Number of siblings interaction is a result of the design of the study. The major trend there is that the less schooling the respondents have, the more siblings they have. For instance, respondents with less than 12 years of schooling (did not finish high school) tend to come more from large families, and those who have more than 17 years of schooling (university graduates) tend to come from families with four children or less.

Both Clogg (1982) and Beh and Davy (1998) gave analyses of these data exploiting the ordinality and interpreting the parameters, but neither made an attempt to provide a graphical representation of the interactions.

Table 17.2   Davis's Happiness data: Marginal proportions

Years of schooling		Number of siblings		Happiness	
	Prop.		Prop.		Prop.
<12	0.37	0–1	0.18	Not too happy	0.35
12	0.33	2–3	0.29	Pretty happy	0.53
13–16	0.24	4–5	0.21	Very happy	0.12
17+	0.06	6–7	0.15		
		8+	0.17		

Table 17.3 Davis's Happiness data: Partitioning of the inertia

Source	df	Total dependence			Residual dependence		Percentage fit
		X^2	%	X^2/df	X^2	%	%
Schooling×Siblings	12	235	72	20	11	22	96
Happiness×Schooling	6	41	13	7	6	12	87
Happiness×Siblings	8	26	8	3	9	18	67
Three-way interaction	24	26	8	1	23	48	13
Total	50	329	100	7	47	100	86

For the decomposition of the dependence a $2 \times 2 \times 2$-Tucker3 model was used. Percentage fit (last column) indicates to what extent an interaction term was fit by the $2 \times 2 \times 2$ model.

17.4.3 Stage 3: Model and dimensionality selection

Before looking at the interactions, we have to establish their relative sizes, so that we know which interactions we have to focus on. Table 17.3 shows the partitioning of the total $X^2(= 329)$ into the contributions of the interactions, for which we have used Eq. (17.10). The last two columns refer to the extent that the marginal dependence could be fit by a $2\times2\times2$-Tucker3 model. The fit of this model was 86% and the well-fitting to reasonably well-fitting two-way interactions show that this choice of numbers of components was a reasonable one. Details of model selection will be delegated to Section 17.5.3.

The design interaction (Years of schooling×Number of siblings) is the largest ($X^2 = 235$) and is fitted well as the chosen model explains 96% of this interaction. Of the other two interactions, Happiness×Years of schooling is the most important, both if we take the absolute X^2 (41 versus 26) and if we look at X^2/df (7 versus 3). This interaction is also fit best by the model (87% versus 67%). The X^2 of the three-way interaction is equal to the Happiness×Siblings interaction but much smaller in X^2 per df (1), and is hardly fitted by the model (13%), which is not surprising as most error is concentrated in this interaction. The latter values indicate that in any graphics we make the three-way interaction will not play an important role.

17.4.4 Stage 4: Results and their interpretation

To get an insight into the interactions themselves, we either have to present tables of all four interactions or we could, as discussed earlier, look at the nested-mode biplot to get a full overview of the dependence. Such a biplot is presented in Fig. 17.1 with Years of schooling chosen as the reference mode. This mode is represented in the plot via the dotted straight lines running through the location of their categories.

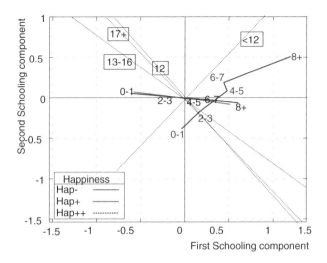

Figure 17.1   Davis's happiness data: Total dependence of Happiness by Years of schooling by Number of siblings. *Years of schooling*: dotted straight lines, boxed values; *Happiness*: Solid thick lines: The two nearly coinciding solid trajectories are Not very happy and Pretty happy, the other solid trajectory is Very happy. *Number of siblings*: nested in happiness

Trajectories have been made for the Happiness categories such that Number of siblings is nested in Happiness.

Before discussing the patterns, it may be observed that the analysis suggests that the original table could be simplified by taking several categories together: in particular, this applies to the lowest two happiness categories and the medium-sized family categories (4–5 and 6–7) as they occupy the same places in the plot. This would clearly simplify visual inspection of the table itself.

The analysis of Fig. 17.1 should start with the reference mode, in this case Years of schooling. The axis through the group who did not finish high school (< 12) and the axes for the other groups are virtually perpendicular, so that radically different patterns can be observed for them. There is an almost linear relationship for the points representing 12 years of schooling or over, and their common axis.

The patterns of dependence for Not very happy and Pretty happy are identical with an almost equal spacing of the number of siblings on their joint axis. Combining the two predictors for these states of the dependent variable Happiness, the conclusion is that the more years of education and the fewer siblings, the more participants stated that they were rather happy in contrast with those who have many siblings. This conclusion is (almost) independent of Years of schooling as is evident from the fact that the Happiness lines nearly bisect the angle between the two Schooling axes.

The situation is radically different with respect to the Very happy category. Its biplot axis is more or less perpendicular to the more than 12 year schooling axes, so that for the very happy persons, there is no relationship between Number of siblings and these schooling categories. On the other hand, of the minimally schooled participants those with more siblings are more inclined to say that they are very happy.

Further discussion of the technical details of this example will follow in the next sections, when we discuss several practical aspects of three-way correspondence analysis.

17.5 THREE-WAY CORRESPONDENCE ANALYSIS: PRACTICE

17.5.1 Stage 1: Objectives

The object of three-way correspondence analysis is to gain insight into the interactions between the three variables that make up the table. Because such interactions, especially the three-way interaction, are inherently complex, one cannot expect that their analysis will yield a simple answer. Simple answers can only be had in the absence of the three-way interaction and/or several two-way interactions. Establishing the relative importance of these interactions and their portrayal in low-dimensional space within the single framework can be seen as the major purposes of a three-way correspondence analysis.

17.5.2 Stage 2: Data description and design

Types of data and their requirements.   So far, correspondence analysis for more than two modes has been developed only for three-way contingency tables, and all applications have been for that type of data. The contents of the cells are generally frequencies or proportions. It is implied that the observations come from a (product-)multinomial or Poisson process. Most of the interpretations mentioned in the examples lean on the fact that we are dealing with proportions, and are often given in terms of (estimated) conditional probabilities.

Ordinary correspondence analysis has also been applied to arbitrary two-way tables provided the cells contained nonnegative numbers. Whether it is advisable to do so in three-way correspondence analysis depends on the assumptions one is willing to make about the data. When other types of data are used, it has to be assessed *a priori* whether the basic concepts of correspondence analysis can be given an appropriate interpretation given the contents of the data set at hand. If the categories of the modes are seen as variables, Lebart, Morineau, and Warwick (1984) suggest that correspondence analysis is only appropriate if these variables are "homogenous", so that it makes sense to calculate and interpret a statistical distance between them. It might help to realize that the correspondence analysis graph is essentially constructed from profiles (i.e., vectors $p_{ij}/p_{i.}$ in two-mode analysis), and that two points with identical profiles will be placed on top of each other in the plot (known as *distributional equivalence*). If the concept of a profile is meaningful for a particular table with nonnegative numbers, correspondence analysis may be fruitful. However, it seems that the only type of profiles that can be meaningfully defined in three-way tables are slice profiles, for example, $p_{ijk}/p_{i..}$ (see Kroonenberg, 1989), a fact that should be taken into consideration. So far very little work has been done in this area.

Some examples of applications of two-way correspondence analysis to nonfrequency data are included in Greenacre (1984). Nishisato (1980, 1994) discuss two-way correspondence analysis (or dual scaling) of rank-order data, sorting data, paired-comparison data, and successive categories data. As yet there are no examples of analyses with these types of data in the three-mode context.

From multiway data to three-way tables.   If instead of a three-way table one has a multiway table with a moderate number of categories in each variable, one may consider combining two variables, say, U with I categories and V with K categories, into a new variable with $I \times K$ categories. Because the two modes are combined into a single one (or interactively coded), the initial two-way interaction embodied in their two-way margin becomes the new one-way margin. As correspondence analysis is concerned with the dependence after removing the effects of the one-way margins, the effect of interactively coding U and V into a single mode is that their interaction is removed from the analysis. Selecting which two or more variables should be combined is therefore of considerable concern. Typical candidates for combination are design variables, because combining them can lead to removing design effects from the analysis (for an application, see Section 17.6.2).

Sample size.   Using three-way correspondence analysis for very small three-way contingency tables is like using a cannon to shoot at a mosquito. When the number of categories of the variables increases, it becomes more and more difficult to get a proper overview of the patterns in the table without appropriate tools. To get stable results, one should have a reasonable to a good number of observations per cell. Technically, there is no strict reason to have large numbers of observations per cell, because the correspondence analysis operates on proportions and on profiles of proportions, and these can always be computed. Furthermore, as no formal testing is done, there is no problem either with respect to correctness of asymptotic approximations to test statistics for small numbers of observations. The major concern with respect to sample size is replicability or generalizability and stability. If only very few observations are present in a cell, the proportions are ill-determined, and in a new sample they may be considerably different. To investigate the stability one might use bootstrap procedures, but no examples of this are yet in existence. In several cases, the number of observations per cell may not matter if the data set is one of a kind and can be considered a population. The data on censuses of the work force in Languedoc-Roussillon analyzed in Carlier and Kroonenberg (1998) are an example.

17.5.3   Stage 3: Model and dimensionality selection

Partitioning of dependence.   To evaluate the size of the dependence, there are two aspects that need to be considered. First, we have to look at the dependence without

Table 17.4   Davis's Happiness data: Partitioning dependence

Source	df	Total dependence			Fitted dependence		Percentage fitted
		X^2	%	X^2/df	X^2	%	%
Schooling	3	0	0	0	0	0	
Happiness	2	0	0	0	0	0	
Siblings	4	0	0	0	0	0	
Schooling×Siblings	12	235	72	20	225	80	96
Happiness×Schooling	6	41	13	7	36	13	88
Happiness×Siblings	8	26	8	3	17	6	65
Three-way interaction	24	26	8	1	4	1	15
Total	50	329	100	7	281	100	86

attempting to model it, because we need to know the relative contributions of the interactions to the dependence (see Eq. (17.10)). The results of this equation for Davis's Happiness data are contained in the first section of Table 17.4. Due to the removal of the main effects, the model of independence, we have zeroes in the first three rows (see also Eq. (17.8)). The next three lines indicate the contributions to the dependence by the two-way interactions, followed by the contribution of the three-way interaction. The partitioning can be evaluated in absolute terms (i.e., which interactions contribute most), but also via the contribution per degree-of-freedom. The latter should be approached with some caution, because large interaction tables generally contain both structure and random errors. Often the structure takes only a relatively small number of degrees of freedom, leaving the rest for the errors. Therefore, a low value of X^2/df for large interactions might mistakenly be interpreted as an interaction made up solely of random error. Of course, when X^2 is small and explains a small percentage of the dependence, the interaction can safely be ignored. In the Davis data, the three-way interaction has the same size as the Happiness×Number of siblings interaction, but it has a much lower X^2/df. A large part of the three-way interaction will be error, and this makes it difficult to assess the size of the structural part is in terms of X^2/df.

Numbers of components.   As explained in Section 17.3.2, the total dependence is modeled using the Tucker3 model in which for each mode a number of components have to be selected. In Table 17.5 an overview is given of all possible Tucker3 analyses for these data, plus some additional information that might be helpful in deciding which is the most appropriate one. Note that we do not think that a *correct* model necessarily exists. With a 2×2×2 model we are able to fit 86% of the dependence,

Table 17.5   Davis's Happiness data: Summary of analyses with different numbers of components

	Model size	Sum of components		Proportional SS(Fit)	Difference in Proportional fit	*df*
1	$1 \times 1 \times 1$	3	*	**0.69**	0.69	50
2	$2 \times 2 \times 1$	5		0.71		44
3	$2 \times 1 \times 2$	5		0.74		46
4	$1 \times 2 \times 2$	5	*	**0.80**	0.11	45
5	$2 \times 2 \times 2$	6	*	**0.86**	0.06	40

Order of the modes is Years of schooling × Happiness × Number of siblings; * = best solution for a given value of the sum of number of components.

which is quite satisfactory (Carlier, 1995). With larger tables, such high percentages of explained variability are of course rare.

Given that a model has sufficient components to make a detailed inspection of the total dependence and that the proportional fit is high, it is the preferred model. If one is less interested in detail, the $1 \times 2 \times 2$ model might do as well.

Partitioning of dependence.   Given that the total dependence of the $2 \times 2 \times 2$ model has a fit of 86%, we would like to know how well each of the interactions is fitted by this model. This information is contained in the last columns of Table 17.4. From the last column we see that the chosen model manages to fit almost all of the design interaction (96%), which is in a way good because that parallels the situation in loglinear modeling where in logistic models one has to include this margin in the prediction of the dependent variable. Of the two predictive (two-way interaction) margins, Years of schooling is fitted very well (88%) and Number of siblings also fits quite good (65%). On the other hand, only 15% of the three-way interaction is fitted by the three-mode model. In terms of comparative contribution to the total, we see that three-way interaction only contributes 1%. The strong linearity in the plot already suggested as much (see Gabriel, 1981, for detailed statements about the interpretation of specific patterns in biplots).

Summary.   The evaluation of a model for three-way correspondence analysis has to take into account the relative sizes of the interactions, how many components are necessary to get an adequate fit to the dependence, and how well the chosen model fits each of the interactions.

17.5.4   Stage 4: Results and their interpretation

The interpretation of a three-way correspondence analysis centers around the examination of the size of the various interactions, the fit of the model to the total dependence, and the various interactions, as discussed in the previous section. After we evaluated all this, decided that the solution is the one that we want, and that we know which interactions deserve our attention, the next task is to interpret the solution itself. For three-way correspondence analysis, the biplot (also called *asymmetric map*) is the motor for displaying the results, even though some might argue that symmetric (or barycentric) plots should be used (see Greenacre, 1993, Chapters 9 and 13, for an enlightening discussion of the differences between them). Two kinds of biplots are particularly important: the nested-mode biplot and the joint biplot. At present, it seems that the nested-mode biplot is more versatile and complete, but there are also situations when the joint biplot will be more effective. In this section, the former will be presented in more detail than the latter.

Total dependence: Nested-mode biplot.   As worked out in detail in Section 17.3.2 the nested-mode biplot is a plot which displays the coordinates of a matrix whose rows consist of the interactively coded levels of two of the modes and whose columns consist of the levels of the remaining reference mode (See also Section 11.5.4, p. 276). In the case of three-way correspondence analysis, the nested-mode biplot is the basis for the interpretation because in it the total dependence can be displayed as well as all derived dependencies, such as the marginal and partial dependence.

Marginal dependence.   One of the key features of three-way correspondence analysis is that the coordinates for the marginal dependence may be derived from those of the total dependence and that, in principle, all two-way interactions can be displayed in the same plot as the total dependence (see Section 17.3.2). However, this may lead to an extremely messy plot, and therefore one could choose to make separate plots for the marginal dependence. In that case, one should try to keep the scale of these plots the same as the master plot.

In Fig. 17.2, the Happiness × Years of schooling margin is portrayed with Years of schooling as reference mode in normalized coordinates and Happiness in principal coordinates. For the interpretation it is often convenient to draw the axes through the reference mode points so that the relative positions of the rows can be evaluated on these axes. In the present plots, this has been done and the location of these points is given by the labels. A powerful addition is to calibrate or grade the axes (see Carlier, 1995, for some examples), but this has not been done in the figures in this book.

In Table 17.6 the chi-terms of the Happiness × Years of schooling are given to help understanding Fig. 17.2. The plot is only an indirect approximation of the marginal dependence as it is based on the fitting of the total dependence using a three-mode model, rather than on a two-way correspondence analysis applied directly to the margin. Notwithstanding, it can clearly be seen that the highest value in the table corresponds with largest inner product, that is, the combination of Least schooling and Very happy.

The Happiness × Years of schooling interaction portrayed in Fig. 17.2 shows that this interaction primarily consists of the comparatively higher incidence of reported happiness of the group with less than 12 years of education, combined with a reversed trend for the reported very happiness of the other groups.

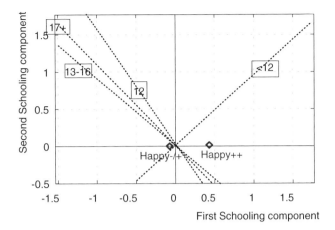

Figure 17.2   Davis's Happiness data: Marginal dependence – Happiness×Years of schooling.

The Number-of-siblings×Years-of-schooling interaction (Fig. 17.3) shows that the individuals with the least amount of education (< 12) tend to come more frequently from large families while the trend is reversed for the other educational groups. This trend looks nearly linear.

The Number of siblings×Happiness interaction takes a different form than the other two interactions because it involves summing over the reference mode, while both other variables are the rows of the plotting coordinates. As explained in Section 17.3.2, this summing results in the average row coordinates of the components; in this case these values −0.003 and 0.978 for the first and second component, respectively. In the nested-mode biplot, this is portrayed by an arrow from the origin to this point (-0.003,0.978), which thus coincides with the positive side of the second axis.

Table 17.6   Davis's Happiness data: Chi-terms for Happiness×Years of schooling margin

Years of schooling	Marginal proportion	Happiness		
		Not Very: −	Pretty: +	Very: ++
< 12	0.367	−0.05	−0.02	**0.11**
12	0.334	0.02	0.00	−0.05
13–16	0.240	0.01	0.03	−0.07
17+	0.059	0.05	−0.02	−0.04
Marginal proportions	1.000	0.35	0.53	0.12

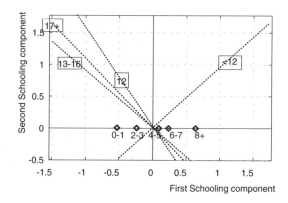

Figure 17.3   Davis's Happiness data: Marginal dependence – Number of siblings × Year of schooling.

A general problem in correspondence analysis is that the mode in principal coordinates has often a much smaller spread than the mode in normalized coordinates. One way to circumvent this is to multiply all coordinates with a single appropriate number. This will not affect the relative distances and properties, but it might give an inflated idea of the strength of the interaction.

Joint biplot.   In the case of constructing joint biplots for the Happiness data, we should again choose Years of schooling as the reference mode. Unlike in the nested-mode biplot, the reference mode cannot be displayed in the dependence plot itself, so that one has to interpret these components first to understand the meaning of components on which the conditioning takes place. In fact, there seems generally little point in making joint biplots if it is impossible to put a reasonable interpretation to the components of the reference mode. An example of the use of a joint biplot for a three-way correspondence analysis can be found in Carlier and Kroonenberg (1996).

17.6   EXAMPLE: PLAYING WITH PEERS

17.6.1   Stage 1: Objectives

Kramer and Gottman (1992) studied to what extent the quality of social interaction with peers influences the attitude of children toward their newly born siblings. To this end, they collected data on the quality of play of children with their best friends. In our analyses we collapsed over the response variable "attitude toward sibling" in order to concentrate on finding patterns in quality of play between children as a function of sex, age, and time of measurement. The reference point for the measurements was taken as the moment of the sibling's birth, and the major question was whether the quality of play of the elder child was different before and after the sibling's birth.

17.6.2   Stage 2: Data description and design

The data for the example are a condensation of the data contained in Anderson (1996, Table 3). Here, the original seven types of three-year-old to five-year-old first-born children were reduced to four types of children, because the original groups contained very few children. The groups now form a fully crossed design of girls versus boys, and younger versus older children. These two variables were interactively coded, as they both were design variables and we wanted to reduce the four-way table to a three-way one. Moreover, this removed the design interaction from the data (see Section 17.5.2). The cells of the resulting three-way table contained the information on how often each of eleven play qualities occurred during play sessions of a child with his or her best friend. The children were observed at five occasions (3 and 1 months before a sibling was born to their families and 1, 3, and 5 months after its birth; in the sequel these time points will be indicated by $-3, -1, +1, +3, +5$, respectively). The eleven play qualities were: sustained communication, coordinated or successful gossip, coordinated or positive play, excitement, amity, shared or successful fantasy, unsustained communication, uncoordinated or poor play, negative emotion, conflict, and prohibitions. Thus, the data make up a 4 (Groups) by 11 (Play qualities) by 5 (Time points) contingency table. The full background to the study can be found in Kramer and Gottman (1992).

17.6.3   Stage 3: Model and dimensionality selection

Table 17.7 shows the partitioning of the total inertia of the Peer Play data. The Play quality×Group interaction is clearly very large, indicating that different types of children play in different ways with their favorite friends. The Play quality×Time point interaction is about half in absolute size and about two times as small in the X^2/df ratio. This indicates that the quality of play changes over time for all groups taken together. The Group×Time point interaction is a design one and not of substantive interest. Finally, the three-way interaction is huge, but in terms of degrees of freedom about the same size as the Play-quality×Time point interaction. However, it contains both systematic and random variability, so that the X^2/df ratio underestimates the systematic part of the interaction.

The right-hand side of Table 17.7 shows how well a 2×2×2 solution for the Tucker3 model fits the dependence. The model is able to fit about half the variability (46%) in the three-way table. (The fit of the 3×4×3 model is 68%, but it requires a large number of additional parameters.). Note that next to the good fit of the Play-quality×Group interaction (81% is fitted by the model), the three-way interaction has also a comparatively good fit (41%), emphasizing that there is considerable systematic variation in the three-way interaction, and that the differences between the groups are complex.

Table 17.7   Kramer's Peer play data: Partitioning of the total inertia

Source	df	X^2_{Total}	% of	X^2_{error}	% of	$X^2_{\text{Fit}}/X^2_{\text{Total}}$
Constant	1	0	0	0	0	
Main effects						
Groups (G)	3	0	0	0	0	
Play quality (P)	10	0	0	3	2	
Time points(T)	4	0	0	0	0	
Two-way interactions						
Play quality × Groups	30	93	33	18	12	81
Play quality × Time points	40	41	14	35	23	15
Time points × Groups	12	20	7	19	13	3
Three-way interaction	120	132	46	78	51	41
Total	202	286	100	153	100	46

The one-dimensional margins or main effects are zero by definition.

17.6.4   Stage 4: Results and their interpretation

The results of the analysis of dependence can be compressed into two graphs: the space of the reference mode Play qualities (Fig. 17.4), and the nested-mode biplot showing the total dependence (Fig. 17.5).

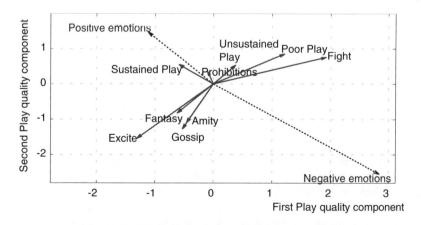

Figure 17.4   Peer play data: Reference mode — Play qualities.

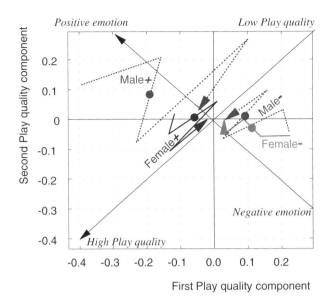

Figure 17.5   Peer play data: Nested-mode biplot - Groups, Time points, and Play qualities. Trajectories run from -3, -1, $+1$, $+3$, $+5$ months with respect to the birth of the second child, indicated by a dot in the trajectory. Play quality axes are based on Fig. 17.4. The older children are located in the left-upper corner, and younger children in the right-bottom corner.

By drawing axes at an angle of 45 degrees in Fig. 17.4, it can be seen that play qualities fall into two groups of categories: emotional categories and actual play categories. The nested-mode biplot (Fig. 17.5) portrays the total dependence in the table using all three aspects of the data, Groups, Time points, and Play qualities. The latter variable is represented by the two axes mentioned above, rather than the 11 Play qualities themselves, in order not to clutter the picture too much. All four age–sex groups start at the outside of the plot but end up in the middle. To interpret this, it should be recalled that a category far away from the origin means that the relative frequency deviates considerably from the model of independence, and a category near the origin more or less follows the model of independence. Thus, while the qualities of play were initially very different, at the end of the period they all end up with a similar distribution across the play categories, specifically that of the marginal proportions. The major direction of change over time is in the emotion categories, that is, the groups differed considerably in the emotion categories of the Play quality variable. In particular, the older groups showed initially more emotion, while the younger ones showed comparatively less emotion at first. Moreover, the older boys and girls differed more in their play style than did the younger ones. Five months after the birth of the second child, the levels of emotion seem to have converged for all children. Secondly, there are clear and constantly varying differences over time in the quality of play between the groups. Even though the different patterns can be clearly seen in Fig. 17.5, their interpretation remains unclear, and ideally one would like to repeat the study to understand whether the fluctuations on the Play quality variables are systematic or random (see Kramer & Gottman, 1992).

17.7 CONCLUSIONS

In this chapter, it was describe how three-way contingency tables can be analyzed with three-way correspondence analysis. This technique is especially useful if such contingency tables have many categories and the interactions are far from trivial. Particularly useful is the possibility of partitioning the departure from independence into independent contributions for each of the interactions. Moreover, the interactions can be simultaneously modeled with a single three-mode model. In addition, the dependence can be portrayed in various types of biplots which not only contain the total dependence but also the marginal and partial dependence, so that the various interactions can be assessed with respect to one another.

CHAPTER 18

THREE-WAY BINARY DATA

18.1 INTRODUCTION[1]

In this chapter a brief overview is presented of a group of hierarchical classes (HICLAS models for three-way binary data developed under the direction of De Boeck and Van Mechelen at the University of Leuven. Whether they will be hierarchical in any particular application is an empirical matter, but with a limited number of components this will be nearly always the case. There is much more to these models than can be discussed in this brief chapter; the major references to the multiway variants of the fully crossed models, which are the focus of this chapter are Leenen, Van Mechelen, De Boeck, and Rosenberg (1999), Ceulemans et al. (2003), and Ceulemans and Van Mechelen (2004, 2005) .

[1] This chapter was written with the assistance of Eva Ceulemans and Iven van Mechelen. The examples and portions of the text have been taken and adapted from the original papers which appeared in *Psychometrika*. Reproduced and adapted with kind permission from the authors and the Psychometric Society.

The models are radically different from the other models discussed in this book in that they are based on Boolean algebra. In *Boolean algebra*, the following operations on 0-1 data are defined: the union ($\vee$) and intersection ($\wedge$) of two elements and the negation of a single one ($\neg$). Thus, Boolean algebra defines how zeroes and ones are combined. In the context of multiway, in particular, three-way, binary data, the models are specified analogously to the Tucker and Parafac models, but for the combination of components and core array Boolean algebra is used. Furthermore, the entries of the component matrices and core arrays in such models are also binary. Rules are defined for how the values in the components and core array combine to create structural images or model arrays of the original data. It is especially the binary structure in the components which makes the hierarchical structure possible.

18.2   CHAPTER PREVIEW

First we will present a synthetic example of the results of a Tucker2 hierarchical classes analysis in order to explain the basic principles in a relatively simple fashion. Subsequently, the basic theory will be presented, as well as a number of variants of the hierarchical classes models. and finally we will present an example with real data for the Tucker3–HICLAS model. All illustrations, examples, and data have been directly taken from the original papers, and no additional analyses were carried out by the author.

18.3   A GRAPHICAL INTRODUCTION

To introduce the basic ideas behind the hierarchical classes models, we assume we have collected data from four subjects (Eva, Katrijn, Iven, Paul) whose present suitability for four jobs (university administrator, full professor, assistant professor, postdoc) has been assessed by seven raters (A–G). The results of the analysis are depicted in Fig. 18.1. The top part of the figure shows the subjects hierarchy, the bottom part shows an upside-down representation of the job hierarchy. The associations between the subjects and the jobs proceed via the raters and are indicated by the line segments in the middle part of Fig. 18.1. The associations are expressed by the raters who are shown in the hexagons. Raters with the same views are located in the same hexagon. Each of the thick-lined boxes represents one of the component-specific classes. For the subject mode the left-hand component can be interpreted as "Suitable to working at university level" and the right-hand component as "Suitable to work at departmental level". In other words, these characteristics indicate the qualifications of the subjects. In reality such an interpretation is often validated via external information. For the job mode, the grouping of the jobs is supposedly generated by the left-hand job component "Administrative duties" and the right-hand job component "Research and teaching" (b_2).

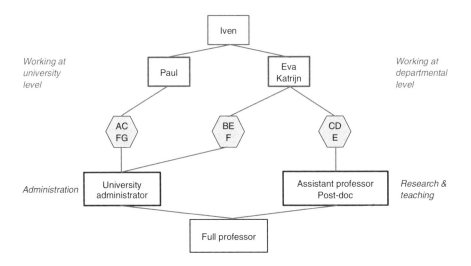

Figure 18.1   Overall graphical representation of the Tucker2–HICLAS model for the synthetic university staff data.

The graph can be interpreted as follows. A subject i is suitable for job j according to rater k if and only if there exists a path from subject i to job j that passes through at least one hexagon containing the rater k. Thus, according to raters B and F, Eva and Katrijn are suitable for administration, whereas E deems them suitable for all jobs. Raters A and G do not think them suitable for any of the jobs, because there are no paths for the raters connecting Eva and Katrijn with any of the jobs. C and D consider Eva and Katrijn suitable for research and teaching but not for administration. Note, that the opinions of E are different, because E appears also in the middle hexagon. On the other hand, all raters consider Iven suitable for any of the jobs, but none of the raters consider Paul suitable for research and teaching. Rater B considers that being fit for working at the departmental level is both necessary and sufficient for an administrative job, because only persons working at the departmental level have a path to Administration. Therefore, in his eyes Paul does not qualify on this ground. On the other hand, for rater A, Paul and Iven are the ones who qualify but Eva and Katrijn do not, due to their lack of experience at the university level.

There are a few general things to note about the figure. First of all, there are *hierarchical relationships* in both modes: Iven is hierarchically above the other subjects, and "full professor" is hierarchically above the other university jobs. This indicates that any subject suitable for any job is suitable for full professor, so long as there is an association between that subject and any of the lower level of the jobs. This might seem a bit counterintuitive, but that is what these data tell. It is important to realize

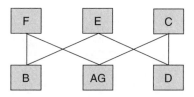

Figure 18.2 Rater structure of the Tucker2–HICLAS model for the synthetic university staff data.

that the top of the trees represent the person or job with the most general characteristics, while lower levels indicate a certain amount of specialization. Apart from hierarchical relationships, there are also *equivalence relations*. For instance, some subjects belong to the same class (as indicated by the boxes, here Eva and Katrijn). These subjects are equivalent because they are deemed suitable for the same jobs by the same raters. Representations like the one in Fig. 18.1 do not make all relationships visible among the raters, but this may be done in a separate graph (Fig. 18.2).

18.4 FORMAL DESCRIPTION OF THE TUCKER–HICLAS MODELS

All models in this book can be written as $\mathcal{X} = \hat{\mathcal{X}} + \mathcal{E}$ or in the notation of the proposers of the HICLAS models: $\mathcal{X} = \mathcal{M} + \mathcal{E}$, where the $I \times J \times K$ array $\mathcal{M}$ is the *structural image* of the data or *model array*.

18.4.1 The Tucker2-HICLAS model

Parallel to the Tucker2 model for continuous data (Section 4.5.2, p. 53), the Tucker2–HICLAS model for binary data can be written as

$$\hat{x}_{ijk} = \bigoplus_{p=1}^{P} \bigoplus_{q=1}^{Q} a_{ip} b_{jq} g_{pqk}, \qquad (18.1)$$

where $\oplus$ denotes the Boolean sum. The meaning of the equation can be formulated as an association rule between the three modes of the data. In particular, an element $\hat{x}_{ijk}$ of the model matrix $\hat{\mathcal{X}}$ has a 1 if it has a 1 at all the appropriate places in the two component matrices and the binary extended core array, that is,

$\hat{x}_{ijk} = 1$ if and only if   there exist a p, q such that

$$a_{ip} = 1 \text{ and } b_{jq} = 1 \text{ and } g_{pqk} = 1 \quad \text{or formally}$$

$\hat{x}_{ijk} = 1$ $\Leftrightarrow$ $\exists p, q : a_{ip} = 1 \wedge b_{jq} = 1 \wedge g_{pqk} = 1.$ (18.2)

Eq. (18.2) means that a rater k only finds a subject i suitable for a job j if and only if the subject has at least one of the qualities p ($a_{ip} = 1$) (fit for working at

Table 18.1   Hypothetical binary suitability data: Subject by university position by rater

Raters	Paul				Iven				Eva				Katrijn			
	full professor	univ. administrator	assistant professor	post-doc	full professor	univ. administrator	assistant professor	post-doc	full professor	univ. administrator	assistant professor	post-doc	full professor	univ. administrator	assistant professor	post-doc
A	1	1	0	0	1	1	0	0	0	0	0	0	0	0	0	0
G	1	1	0	0	1	1	0	0	0	0	0	0	0	0	0	0
B	0	0	0	0	1	1	0	0	1	1	0	0	1	1	0	0
F	1	1	0	0	1	1	0	0	1	1	0	0	1	1	0	0
E	0	0	0	0	1	1	1	1	1	1	1	1	1	1	1	1
C	1	1	0	0	1	1	1	1	1	0	1	1	1	0	1	1
D	0	0	0	0	1	0	1	1	1	0	1	1	1	0	1	1

Source: Ceulemans and Van Mechelen (2004, Table 1). Reproduced and adapted with kind permission of the Psychometric Society

the university level or departmental level) for one of the job specifications q (either administrative or research and teaching qualities) in the eye of the kth rater, g_{pqk}. Note that the subject only needs one and not all qualifications, which makes Eq. (18.2) a *disjunctive association rule*. From the core array one may derive the implicit rules (as we did in the previous section) for job suitability, as $\mathcal{G}$ indicates for each rater k separately which qualifications p are sufficient to make a subject i suitable for which job type q.

In Table 18.1 the complete data set which was behind Figs. 18.1 and 18.2 is given. Our first task is to decompose the data into its constituent parts: the two binary component matrices and the binary extended core array according to Eq. (18.1). The result of this is given in Table 18.2.

Figure 18.1 clearly shows the hierarchical and equivalence relationships. These relationships are based on the component matrices in Table 18.2. For instance, that Iven is hierarchically related to all the other subjects follows from the ones in each of the components. Said differently each of the others' profiles is contained in his profile. The patterns of Eva and Katrijn are the same in the subject component matrix, indicating that they can be considered equivalent. Similarly, the profile of Full professor subsumes the other two patterns and thus it is hierarchically higher than all other jobs. Hierarchical and equivalence relations between the raters as reflected by the relations between the rows of the core array in Table 18.2 are graphically represented by the links in Fig. 18.2.

Table 18.2   Hypothetical binary suitability data: Component matrices and core array for the $2 \times 2 \times 2$-Tucker2–HICLAS model

	Component matrices					Core array				
	University level			Job activities			Admini- stration		Research Teaching	
Subjects	Univ.	Dept.	Position	Admin.	R&T	Raters	Univ.	Dept.	Univ.	Dept.
Paul	1	0	Full professor	1	1	A	1	0	0	0
Iven	1	1	Univ. administrator	1	0	G	1	0	0	0
Eva	0	1	Assistant professor	0	1	B	0	1	0	0
Katrijn	0	1	Post-doc	0	1	F	1	1	0	0
						E	0	1	0	1
						C	1	0	0	1
						D	0	0	0	1

Source: Ceulemans and Van Mechelen (2004, Table 2). Reproduced and adapted with kind permission of the Psychometric Society.

18.4.2   The Tucker3–HICLAS model

Parallel to the Tucker3 model for continuous data (Section 4.5.3, p. 54), the Tucker3–HICLAS model for binary data can be written as

$$\hat{x}_{ijk} = \bigoplus_{p=1}^{P} \bigoplus_{q=1}^{Q} \bigoplus_{r=1}^{R} a_{ip} b_{jq} c_{kr} g_{pqr}. \tag{18.3}$$

The meaning of the equation is that an element $\hat{x}_{ijk}$ of the model matrix $\hat{\mathcal{X}}$ has a one if it has a one at all the appropriate places in the three component matrices and the binary core array, that is,

$$\hat{x}_{ijk} = 1 \Leftrightarrow \exists p, q, r : a_{ip} = 1 \wedge b_{jq} = 1 \wedge c_{kr} = 1 \wedge g_{pqr} = 1. \tag{18.4}$$

Ceulemans and Van Mechelen (2004, p. 383) indicate to what extent the two Tucker models differ, that is, when the two models are equal and to what extent they are different under which circumstances. In particular, they show that "each (P, Q) Tucker2–HICLAS model is a (P, Q, R) Tucker3–HICLAS model with $R \leq \min(K, PQ)$", and "[c]onversely, each (P, Q, R) Tucker3–HICLAS model can be converted into a (P, Q) Tucker2–HICLAS model." Note that Ceulemans and Van Mechelen (2004) also point out that a similar relationship exists between the Tucker3 and Tucker2 models for continuous data.

18.5 ADDITIONAL ISSUES

18.5.1 Disjunctive models versus conjunctive models

In the models discussed above a subject needed at least one qualification to be considered suitable, but in real life one often needs to fulfill all qualifications to be considered suitable. If that is the case, the use of a *conjunctive association rule* is more appropriate. We will not go into details here but refer to the original papers mentioned at the beginning of this chapter.

18.5.2 The Parafac–HICLAS model

One of the first HICLAS models that was developed was the Parafac–HICLAS model (Leenen et al., 1999), which was introduced under the name INDCLAS. The acronym refers to individual differences in classifications, but not to the kind of data for which the apparent namesake INDSCAL and INDCLUS models (see Section 2.4.5 (p. 23) and Section 2.4.6, p. 23) were designed, that is, sets of dissimilarity or similarity matrices. Vansteelandt and Van Mechelen (1998) published a detailed example of this model, which example was reanalyzed by the Tucker3–HICLAS model in Ceulemans et al. (2003). Ceulemans et al. (2003) gave a comparison of the INDCLAS model and the Tucker3–HICLAS model for the same data. The authors concluded that "Tucker3–HICLAS can result in more parsimonious hierarchical classifications for one or more of the modes, since part of the complexity can be moved to the linking structure. Moreover, the Tucker3–HICLAS relaxations of the INDCLAS restrictions also make sense from a substantive view" (Ceulemans et al., 2003, p. 432).

18.5.3 Technical details

In the synthetic example the decomposition was exact. This will, however, not generally be the case in reality, so that a definition of a loss function is needed as well as an algorithm to find the minimum. The loss function is simply the sum of the squared differences between the binary data x_{ijk} and the binary model $\hat{x}_{ijk}$. Details of the algorithms and simulation studies of their behavior can be found in the original papers. Ceulemans et al. (2003) report a study of the uniqueness of the solutions for multiway hierarchical classes models, in which they come to the conclusion that the Tucker hierarchical models suffer less from the lack of identifiability than their continuous counterparts, while the uniqueness conditions of Parafac hierarchical models are more restrictive than those for their continuous counterparts.

18.6 EXAMPLE: HOSTILE BEHAVIOR IN FRUSTRATING SITUATIONS

In this section we will give a brief account of an analysis of a data set first analyzed by Vansteelandt and Van Mechelen (1998) and later by Ceulemans et al. (2003). The description of the hostility data and their analyses have been taken from the latter publication.

18.6.1   Stage 1: Objectives

The aim of the study was to examine the extent of individual differences in the self-reported hostile behavior of subjects when they were placed in various frustrating situations. External information was used to validate the results of the analyses (Stage 5), but that part of the study will not be reported explicitly here.

18.6.2   Stage 2: Data description and design

In the study 54 subjects were asked to indicate with a yes or no whether they displayed 15 hostile behaviors in 23 frustrating situations, which thus created a $54 \times 15 \times 23$ binary data array. We will not provide all the detailed descriptions of the behaviors and situations but use only their abbreviations, especially in the resulting diagram, Fig. 18.3.

18.6.3   Stage 3: Model and dimensionality selection

Ceulemans et al. (2003) investigated a series of Tucker3–HICLAS models with dimensions $1 \times 1 \times 1$ through to $6 \times 6 \times 6$. Even though more complicated models seemed to be indicated by the selection rules in the paper, a $2 \times 3 \times 2$ model was selected for presentation for reasons of parsimony and interpretability. The equally well-fitting Parafac–HICLAS model studied by Vansteelandt and Van Mechelen (1998) had 3 components for all the modes.

18.6.4   Stage 4: Results and their interpretation

The results of the $2 \times 3 \times 2$ Tucker3–HICLAS analysis can be graphically represented, as was done by Ceulemans et al. (2003).

The lower class of the frustrating situations can be characterized as being only mildly frustrating, while the other class consists of the more strongly frustrating situations of longer duration, more severe consequences, more external in origin, and more ego-threatening. Note again that the top of the hierarchy constitutes the more general class, while the classes lower in the hierarchy represent more specific aspects. Three of the four classes of behaviors (except the Grimace class) are arranged in a hierarchy of physiological reactions to frustration, with the lowest class having the behaviors that are the most physiological in nature. Having that behavior implies having all the other ones in the hierarchy. The class consisting of Turn away, Lose patience, etc. at the top of the hierarchy is the most general one. It might be considered the most common or elementary reaction to a frustrating behavior.

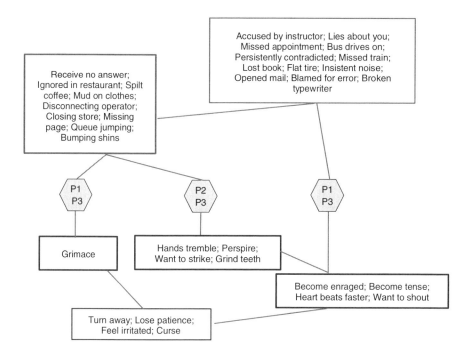

Figure 18.3   Overall graphical representation of the Tucker3–HICLAS model for the hostility data. Top: Hierarchical situations tree, Bottom: Inverted hierarchical response tree. Source: Ceulemans et al. (2003), Fig. 8. Redrawn with kind permission from the authors and the Psychometric Society.

The person hierarchy can also be portrayed in a separate figure, which is not included here. It shows that there are three groups, labeled P1, P2, P3 in Fig. 18.3, and P3 is hierarchical with respect to the other two.

18.7   CONCLUSION

The models for binary data discussed in this chapter have a great potential for providing insight into various psychological phenomena. One of the reasons is that due to their binary character the data collection is relatively straightforward. Furthermore, the techniques are specifically geared toward this kind of data and the way they represent the results is very intuitive. The usefulness of the multiway HICLAS approach in substantive research has been demonstrated by successful applications in personality and emotion psychology: Vansteelandt and Van Mechelen (1998, 2006), and Kuppens, Van Mechelen, Smits, De Boeck, and Ceulemans (2007).

CHAPTER 19

FROM THREE-WAY DATA TO FOUR-WAY DATA AND BEYOND

19.1 INTRODUCTION

The larger part of this book has been focussed on three-way data, and one may wonder why higher-way data have not been more prominent. The basic reason for this is that four-way and higher-way data are much rarer, and consequently much less attention has been paid in the literature to their analysis. To gain an understanding of the nature of multiway data, it is useful to consider how higher-way data arise and what relationships they describe. Cattell (1946) initially only considered three-way research designs and data, which he described with his three-dimensional *Covariation Chart*, which pertained to three-way profile data, that is, variables by persons by occasions (see Chapter 13). Later, he proposed probably the most comprehensive system for describing relations (Cattell, 1966), in particular he claimed that all (psychological) research designs can be described by what he called the ten-dimensional *Basic data relation matrix*, BDRM. He also contended that "although the ten-dimensional treatment may often be more than is needed, (1) it should be kept in mind as representing the true dimensionality, and (2) there is nothing sacred about three dimensions [of the

Covariation Chart], and if abridgment is desired, then four, five, or six may often be more apt simplifications for particular purposes." (p. 69).

The majority of two-way data consist of scores on variables or attributes provided by the data generators (or organisms as Cattell (1966) calls them), such as subjects, objects, or plant varieties. Some data sets are more complex than two-way data, in that they have been collected at various times or under various conditions. When one of these facets is used the data become three-way, and four-way when both of them are present. Another extension occurs when the variables have a *design* in the sense of a stimulus–response design. In such a case, subjects have to judge variables in various situations (see Chapter 14). By employing such designs on the variables, we can construct five-way data of subjects by variables by situations by conditions by time. To extend this even further, one might also think of constructing a design on the conditions so as to make it a six-way data set set. Cattell (1966, p. 77) takes a somewhat different approach in that each of the five basic dimensions or entities (organism, stimulus, response, background, and observer) can occur in two forms: one showing a relatively permanent pattern and another showing transient states. As Cronbach (1984) remarked: "In this grandest of his presentation on the subject, Cattell's conceptualization sported more embellishments than Ludwig's castle at Ehrenbreitstein[1]. As he ruefully foresaw, the complications 'priced the system out of business'. Users of Cattell's scheme have stayed close to the 'nursery' version." (pp. 225–226), that is, the Covariation Chart of subjects by variables by occasions. Even though this is still largely true, in psychology there is clearly a place for four-way data of subjects by stimuli by responses (or scales) and conditions or time points. Outside psychology, multiway data arise naturally in chemistry, for instance, in spectroscopy and chromatography, and multiway analyses of EEG and fMRI data are becoming more and more common (Bro, 2006) (see also Section 19.3.1, p. 472).

Collecting fully crossed three-way data is already a considerable task, but organizing multiway data is even harder and more laborious, especially in the social and behavioral sciences. Analyzing higher-way data in itself is eminently feasible, but interpretation obviously gets rather complicated. Some authors already consider three-mode analysis a lost cause because of the complexity of interpreting the core array, and extending such core arrays to even higher orders is clearly not going to be very popular in those circles. Initially computer programs were only designed for three-way analysis, but extending programs to four and more ways was more of a combinatorial problem than a technical-mathematical problem, and several programs are now available (see Section 19.7.2). However, designing graphical aids for multiway extensions to enhance interpretation is a major undertaking. Dynamic graphics, which require presentation programs rather than paper, will be very useful to fathom the results from multiway analyses.

---

[1] I suspect Cronbach actually meant Neuschwanstein in Schwangau, Bavaria, built by Ludwig II von Bayern, rather than the more military castle at Koblenz.

19.2   CHAPTER PREVIEW

In this chapter, first a number of examples of higher-way data will be presented. This will be followed by a brief overview of the simpler higher-way techniques. The two illustrations in this chapter should give a flavor of multiway analysis for "multi" greater than 3. The first deals with the data of a multiple personality (Osgood & Luria, 1954) of which three-way analyses were presented in Kroonenberg (1983c, 1985a) and in Section14.6 (p. 361). The second example deals with environmental data of air pollution due to airborne chemical compounds (Stanimirova & Simeonov, 2005).

19.3   EXAMPLES OF MULTIWAY DATA

19.3.1   Profile data and rating scale data

Multiple personality.   The classic rating scale data collected by Osgood and Luria (1954) on Eve White, Eve Black, and Jane, a woman with a multiple personality, which were analyzed as three-way data in Section 14.6 (p. 361), are in essence four-way: Personalities (3) by Scales (15) by Concepts (10) by Repeated measurement (2). Earlier, Kroonenberg (1983c, 1985a) analyzed these as three-way data by interactively coding the personalities and measurement times. By separating out the personalities and measurement times, one may establish whether the responses across measurements were sufficiently different and if so, which aspects of the data changed most. This question will be tackled in Section 19.5 with a Tucker4 model.

Social relations.   An early four-way example of rating scale data was presented by Iacobucci (1989), who analyzed the sociometric data consisting of judgments of a group of monks about each other. The original data appeared in Sampson's (1968) thesis. Iacobucci cites various earlier analyses, and part of the data are publicly available[2]. As analyzed by Iacobucci, the data consist of 18 actors (judges) by 18 partners (judged) by 3 time points and 8 types of judgments. These data can be characterized as four-way rating scale data. An example of four-way factorial data (see Section 3.6.3, p. 35) on cross-cultural differences in respect for others can be found in Kroonenberg and Kashima (1997), who analyzed them as three-way data with a Tucker3 model.

Plant breeding.   In agriculture, especially plant breeding, four-way profile data are a fairly common occurrence even though they are seldom, if ever, analyzed that way. In multienvironment trials, several varieties of a crop are planted over a series of years in a number of environments and measured on a series of attributes. The aim is to establish both general and specific adaptation. With respect to the latter, plant

---

[2]*http://www.casos.cs.cmu.edu/computational_tools/datasets/sets/sampson/index.html.*   Accessed May 2007.

breeders try to find out which varieties do particularly well in which environments on which attributes and to what extent this is influenced by yearly fluctuation.

Spectroscopy.   In his review paper on multiway analysis in analytical chemistry Bro (2006, p. 279) states that "[f]luorescence spectroscopy is by far the most abundant type of data used for multiway analysis [..]. This is undoubtedly due to the close relationship between the unique PARAFAC model and the fundamental structure of common fluorescence spectroscopic data." It is therefore no wonder that in analytical chemistry, several examples of four-way data can be found. In the review paper at least four instances of such data are mentioned. One type of such four-way data may be characterized by excitation–emission matrices for different samples taken at several time points. A study by Arancibia et al. (2006, p. 77) using kinetic measurement of evolution–emission fluorescence with time for mixtures of two anticancer drugs may serve as an example. A much earlier spectroscopic example was presented by Durell et al. (1990) who used a four-way Parafac model to analyze four-way data representing the amount of absorption by a protein for a number of plant species at several wavelengths given for several oxidation states and environmental pH concentrations. The Durell et al. data are *multiway factorial data* in the sense that there is one dependent variable — amount of absorption. The ways of the data array constitute the design of the study.

Monitoring air quality.   Stanimirova and Simeonov (2005) present an analysis of monitoring air quality. They analyzed a $5 \times 16 \times 4 \times 2$ four-way profile data consisting of the concentrations of 16 chemical compounds for each of 5 particle sizes, which were measured in each of the 4 seasons at 2 locations in Kärnten (Carynthia), Austria. Using the core consistency diagnostic (see Section 8.6.3, p. 185), the authors decided against using the only obvious Parafac model, that is, $2 \times 2 \times 2 \times 2$ model, and reported a $2 \times 4 \times 2 \times 2$ Tucker4 solution, which after preprocessing fitted 85% of the data. For a reanalysis of these data see Section 19.6.

EEG and fMRI.   Other types of common multiway data are scans of three-dimensional objects, for instance using functional magnetic resonance (fMRI) or electroencephalographic (EEG) data (Leibovici, 2000; Martínez-Montes, Valdés-Sosa, Miwakeichi, Goldman, & Cohen, 2004). A fascinating example of a six-way data set is presented by Estienne, Matthijs, Massart, Ricoux, and Leibovici (2001). The research group collected EEG data from 12 male subjects and measured their brain activities in response to various doses of an antidepressant drug. The data formed a six-way array of averaged energy measurements on 7 (standard) frequency bands measured by 12 subjects on 28 leads using 4 different doses at 12 moments in time under 2 conditions. Thus, there were 32256 measurements on 7 variables. After ascertaining that Parafac models were too restricted, the parameters of an uncentered Tucker6 model with $3 \times 3 \times 3 \times 2 \times 2 \times 1$ components with an explained variability of 95% were estimated using Andersson and Bro's MATLAB N-way Toolbox[3] (Andersson & Bro, 2000). Detailed interpretations of the components of each of the ways were provided and the high-dimensional core array could also be readily interpreted. These data are again ANOVA-like in the sense that there is one dependent variable energy level. Subjects in this case are both the organisms and a fixed factor.

Another intriguing example is contained in Andersen and Rayens (2004). In their abstract they state that in comprehensive fMRI studies of brain function data structures often contain higher-order ways such as trial, task condition, subject, and group in addition to the intrinsic dimensions of time and space. Multiway analysis of fMRI data from multiple runs of a bilateral finger-tapping paradigm was performed using the Parafac model. A trilinear Parafac model was fitted to a data cube with as dimensions voxels by time by run. Similarly, a quadrilinear Parafac model was fitted to a higher-way structure with as dimensions voxels by time by trial by run. The spatial and temporal response components were extracted and validated by comparison to results from traditional SVD/PCA analyses based on scenarios of matricizing into lower-order bilinear structures. Interesting from a technical point of view, is that in their four-way Parafac model they extracted three components while the smallest dimension of the data set was only two. The theoretical basis for this is contained in a paper by Kruskal (1976).

19.3.2 Other types of higher-way data

Interactions from fixed-effect higher-way factorial experimental designs constitute a different type of higher-way data. However, psychological experiments, for instance, rarely have a large enough number of levels to make such an analysis necessary. In other fields such large higher-way interactions occur from time to time, and they can be analyzed with multiway methods to show the patterns in these interactions. Most applications so far come from the research group at KVL, University of Copenhagen, under the direction of Rasmus Bro, who use the GEMANOVA model for the decomposition of the interactions; see Bro (1998c, p. 192-196; 248-253), or Smilde et al. (2004, p. 340ff.). Bro (1998c, Section 7.6) examined the five-way interaction term of a multiway factorial design in a study of enzymatic browning of vegetables. In addition, Bro and Jakobsen (2002a) analyzed a five-way factorial data set to study factors influencing color changes in fresh beef during storage, and Nannerup et al. (2004) examined a similar six-factorial data set for examining factors affecting color in packaged sliced ham.

In theory such higher-order interactions also occur in data sets with several categorical variables, but again, to make a higher-way analysis worthwhile the categorical variables must have more than a few categories, which is not always the case. The three-way case for categorical data is examined in depth in Chapter 17.

---

[3] *http://www.models.life.ku.dk/source/nwaytoolbox/index.asp.* Accessed May 2007.

19.4 MULTIWAY TECHNIQUES: THEORY

In this section we briefly indicate what the Parafac and Tucker models for four-way data look like. The topic of fitting these models will not be discussed here, as the algorithms in essence parallel to those for the three-mode models (see Chapter 5). Full expositions can be found in Smilde et al. (2004).

19.4.1 Four-mode Parafac

Four-way profile data can be collected in an $I \times J \times K \times L$ four-way data array $\mathcal{X}$ = (x_{ijkl}) of I subjects by J variables by K conditions by L occasions. The variables do not necessarily have the same scales, so that some type of normalization will generally be required. The basic Parafac model for a four-way array $\mathcal{X}$ with elements x_{ijkl} has the form

$$x_{ijkl} = \sum_{s=1}^{S} a_{is}b_{js}c_{ks}d_{ls}g_{ssss} + e_{ijkl}, \qquad (19.1)$$

where the subject scores a_{is}, the variable loadings b_{js}, the condition coefficients c_{ks}, and the occasion coefficients d_{ls} are the elements of the unit-length component matrices $\mathbf{A}, \mathbf{B}, \mathbf{C}, \mathbf{D}$, respectively. The weight of the sth component, g_{ssss}, indicates its importance. The g_{ssss} can be collected in a superdiagonal core hypercube $\mathcal{G}$, that is, $g_{pqrs} = 0$ if $p \neq q \neq r \neq t$. Note that because Parafac components are generally correlated, the overall fit of the model to the data is not equal to the sum of the g_{ssss}^2. However, if the components are approximately uncorrelated across the levels of any mode, then the variance components are approximately additive. When components are exactly orthogonal or have zero correlations in any mode, the variances will be strictly additive. Many technical issues are addressed in Smilde et al. (2004). As in the three-way Parafac model, all modes take identical roles so that there is no *a priori* designation of variable coefficients as loadings in the sense of variable–component correlations and subject coefficients as normalized scores, as is common in two-mode analysis.

As in the three-way case, the four-way Parafac model has, given the number of components, a unique solution apart from trivial rescalings and reordering of the components. For the uniqueness, there has to be sufficient system variation in each of the modes, and at least one of the modes has to have nonproportional components; for further discussions of uniqueness, see Sections 4.6.1 (p. 61), and 5.6.4 (p. 88).

The uniqueness makes the model very attractive if it is known or hypothesized that such a model is appropriate for the data. A consequence of the parallel proportionality of the components is that the correlations between the components within a mode are constant across all levels of the other modes (see Section 13.5.2, p. 323). Clearly, this restriction is far more demanding in the four-way and higher-way cases than in the three-way case. So far it seems that the multiway Parafac model is most useful for those situations in which already explicit models with a Parafac structure exist.

An interesting example of chromatographic data with retention time shifts is presented in Bro et al. (1999b), who show that a satisfactory four-way Parafac model could not be found, but a good-fitting three-way Parafac model exists after interactively coding two of the modes. Moreover, a four-way Parafac2 model could be fitted as well; see Section 4.6.2 (p. 62) for a discussion of three-way Parafac2, and Kiers et al. (1999) and Bro et al. (1999b) for details on the four-way Parafac2 model.

19.4.2 Tucker4 models

If we look at published applications of multiway component analyses, the Tucker$n$ model seems to be a most versatile and effective model. In the full Tucker4 model the numbers of components may be different in all modes, which makes it suitable for many multiway data sets. The Tucker4 model applied to a four-way array $\mathcal{X}$ with elements x_{ijkl} has the form

$$x_{ijkl} = \sum_{p=1}^{P}\sum_{q=1}^{Q}\sum_{r=1}^{R}\sum_{t=1}^{T} a_{ip}b_{jq}c_{kr}d_{lt}g_{pqrt} + e_{ijkl}, \qquad (19.2)$$

where the scores a_{ip}, the loadings b_{jq}, the condition coefficients c_{kr}, and the occasion coefficients d_{lt} are the elements of the component matrices $\mathbf{A}, \mathbf{B}, \mathbf{C}, \mathbf{D}$, respectively. The elements of g_{pqrt} of the four-way core array provide the linkages between the components of the four modes. In particular, g_{pqrt} indicates the importance or weight of the combination of the pth component of the first mode, qth component of the second mode, the rth component of the third mode, and the tth component of the fourth mode. The e_{ijkl} are the elements of the four-way with errors of approximation.

The Tucker4 model can also be written in matrix form using the vec operator, which strings out a multiway array column-wise. The same applies for the core array $\mathcal{G}$ and error array $\mathcal{E}$.

$$vec(\mathcal{X}) = (\mathbf{A} \otimes \mathbf{B} \otimes \mathbf{C} \otimes \mathbf{D})vec(\mathcal{G}) + vec(\mathcal{E}). \qquad (19.3)$$

The symmetry with respect to the modes (see Eq. (19.2)) makes all modes take identical roles so that there is no *a priori* designation of loadings as variable coefficients and scores as subject coefficients, as is common in two-mode analysis. However, for the Tucker4 model, we also often use these terms for ease of presentation without assuming the specific interpretations attached to them in two-mode analysis.

What makes the Tucker$n$ model effective is that it is a decomposition model and that it can cope with varying numbers of levels in the n ways. Thus, it is not problematic when one of the ways consists only of two or three levels while the other ways have many more.

19.4.3   Other Tucker models and hybrid models

Besides analyzing data from fully crossed designs, which are the mainstay of this book, several other multiway models have been developed such as multiway multiblock models, an extension of multiset canonical correlation analysis (Smilde et al., 2000), and several multiway regression analysis procedures, such as N-way PLS (Bro, 1996c). So far these procedures have only been applied within a chemical context. These techniques are discussed in Smilde et al. (2004).

19.5   EXAMPLE: DIFFERENCES WITHIN A MULTIPLE PERSONALITY

19.5.1   Stage 1: Objectives of the Tucker4 analysis

The aim of the analyses in this section is to present a Tucker4 analysis of probably the most famous case of a multiple personality, Eve White, Eve Black, and Jane (Osgood & Luria, 1954). In particular, the intention is to establish whether there is a substantial change in the structure of the concepts scored by the different personalities from one measurement occasion to the next three months later. From previous analyses we already know the large difference between Eve White and Jane vis-à-vis Eve Black, but we are now especially interested in a possible difference between Eve White and Jane. To this end we will first search for an adequate model, and then evaluate this model with respect to the questions posed.

19.5.2   Stage 2: Multiple personality data: Description

In each of the three personalities, the woman in question, produced scores on 10 semantic differential scales (see Table 14.2, p. 366, for the scales used) about concepts related to her present life and her mental state (see Table 14.8, p. 365, for the concepts used). These measurements were obtained twice with three months apart. Three-mode analyses of these data can be found in Section 14.6 (p. 361). The analysis presented here is the first four-mode analysis of these data.

19.5.3   Stage 3: Model and dimensionality selection

As in the three-mode case, we will base our dimensionality selection both on the deviance plot (see Section 8.5.2, p. 181) including the st-criterion (see Section 8.5.3, p. 182) and the multiway scree plot (see Section 8.5.1, p. 179). Even though these plots and procedures were developed for the three-way case there is no a priori reason why they could not be used for the multiway selection.

Figures 19.1 and 19.2 show that the same models lie on the convex hulls in the figures, and those are the ones listed in Table 19.1. This table provides the information necessary to evaluate the models and calculate the Ceulemans and Kiers (2006) st-coefficient (see Eq. 8.2, p. 182). From the table we see that the $2 \times 1 \times 1 \times 2$ model

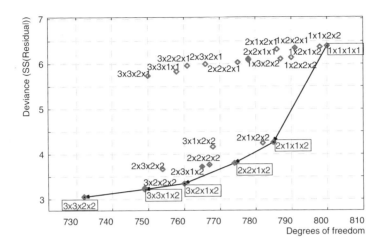

Figure 19.1   Multiple personality four-way data: Deviance plot.

(SS(Fit) = 57%) has the highest st-coefficient, followed by the $3 \times 2 \times 1 \times 2$ model (SS(Fit) = 67%).In practical applications the choice of the preferred model should also be based on substantive considerations. In the present case, the choice between the two models should be based on the detail with which one intends to describe the

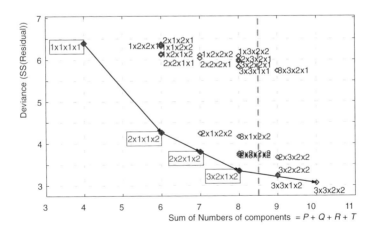

Figure 19.2   Multiple personality four-way data: Multiway scree plot. The models on the convex hull at the far right fail to pass the Timmerman–Kiers minimum DifFit criterion indicated by the vertical dashed line (see Section 8.5.1, p. 179), and have therefore not been taken into consideration.

Table 19.1   Multiple personality data: Dimensionality selection information

Model complexity	Sum comps.	SS(Fit)	PSS(Fit)	Diff. PSS(Fit)	df	Diff. df	st- coeff.
1×1×1×1	4	3.61	0.36		803		-.-
2×1×1×2	**6**	**5.73**	**0.57**	**0.21**	**788**	**15**	**3.38**
2×2×1×2	7	6.21	0.62	0.05	777	11	1.31
3×2×1×2	**8**	**6.67**	**0.67**	**0.05**	**763**	**14**	**2.67**
3×3×1×2	9	6.79	0.68	0.01	752	11	1.33
3×3×2×2	10	6.95	0.69	0.01	735	17	-.-

The order of the modes is: Concepts, Scales, Measurement Times, Personalities.
Sum Comps. $= P + Q + R + T$; PSS(Fit) = SS(Fit)/SS(Total); df = Degrees of freedom;
Diff. PSS(Fit) = PSS(Fit$_{i-1}$)−PSS(Fit$_i$); Diff. $df = df_{i-1} - df_i$.

concepts and scales, and on the consideration whether an additional 10% fit is worth the additional complexity of the interpretation.

Looking at the explained variability per component (Table 19.2), we see that the additional component for the scales of the more complex solution nearly accounts for the full extra 10%, but the extra concept component only half of that (5%). In other words, the earlier concept components are now able to explain more of the original data (57% versus 62%, respectively).

In data sets such as these it is questionable to insist that one particular model is the one and only true model. It makes much more sense to see them as descriptions of the data at hand with more and less detail. The crucial question is where information stops and random error begins. In inferential statistics this is solved by tests, but in the descriptive analyses we are concerned with here, one can only embark on a statistical stability analysis via a bootstrap analysis (see Sections 8.8.1 and 9.8.2).

Table 19.2   Multiple personality data: Fit per components for each of the modes

Model complexity	Concept components			Scale components		Measurement component	Personality components	
	1	2	3	1	2	1	1	2
2×1×1×2	0.35	0.22	0.00	0.57	0.00	0.57	0.35	0.22
3×2×1×2	0.38	0.24	0.05	0.57	0.09	0.67	0.37	0.29

Table 19.3   Personality components for the $2 \times 1 \times 1 \times 2$- and the $3 \times 2 \times 1 \times 2$-Tucker4 models

	Models			
	$2 \times 1 \times 1 \times 2$		$3 \times 2 \times 1 \times 2$	
Personality	1	2	1	2
Eve White	0.33	0.70	0.44	0.62
Jane	0.38	0.52	0.44	0.48
Eve Black	-0.87	0.49	-0.78	0.62
Component fit	35%	22%	37%	29%

19.5.4   Stage 4: Results and their interpretation

As the interpretation of the relationship between scales and concepts is not our primary interest at present (see, however, Section 14.6, p. 361), but we are instead concerned with the differences between personalities and measurement times, we will only inspect the components for those two modes and the core array.

Measurement times.   For both models there is only one single component with the normalized coordinates for the first measurement time equal to 0.67 and 0.69 for the simple and more complex model, respectively. For the second measurement time, these coordinates are 0.74 and 0.73, respectively. In other words, the coefficients and their ratios are practically invariant for the simple and complex models, so that we can safely say that the two measurements of the multiple personality can be seen as replications.

Personalities.   The next question was whether it was possible to separate Eve White and Jane. The difference between either of them and Eve Black is striking, but can we make a distinction between the other two? This question is the more interesting because a footnote in an article by Osgood, Luria, and Smith (1976) says that "Jane was actually a role being played by Eve Black in her attempt to "win" the competition with Eve White in therapy" (p. 256). If this is indeed true, then it is of interest to see to what extent Eve Black was aware of Eve White and all her opinions.

In both models, we have two components for the personalities and their normalized coordinates are presented in Table 19.3, and Eve White and Jane's coefficients on the first personality component are practically equal. Moreover, they are very different from those of Eve Black. On the second personality component, Jane has a somewhat lower weight than Eve White, independent of the model chosen. Thus, Jane has the same structure in her judgments about the relationships between concepts and scales as Eve White but they are somewhat less outspoken (her coefficients are smaller). The conclusion must be that if Eve Black did indeed created Jane, she must have known Eve White very well, but she was slightly off regarding the intensity of Eve White's opinion. Should one want to pursue this issue further, than a detailed analysis of the residuals seems in order.

19.6 EXAMPLE: AUSTRIAN AEROSOL PARTICLES

In this example, a four-way data set on air quality is analyzed by a Tucker4 model and its results are compared to the results from a Tucker3 model for the same data. The description of the data[4] is taken from Stanimirova and Simeonov (2005). The analysis presented here is a simplified version of the one given in that paper, and most of the interpretation is directly taken from the original. The differences between the presentation here and that in the original paper are primarily that a simpler model is presented, and the presentation focusses on methodology rather than the content. This study can be seen as a part of the area of receptor modeling in which not only the content of aerosol samples are analyzed, but also the sources of pollution are identified (see Hopke, 2003). Receptor modeling has greatly benefitted from the use of multiway models as is evident from the work by Hopke and colleagues[5].

19.6.1 Stage 1: Objectives

The aim of Stanimirova and Simeonov's (2005) analysis of the air quality data from Kärnten (Carynthia), Austria was to shed light on the nature and origin of the air pollution by use of four-way methods. More in particular, they wanted to characterize air quality on the basis of particle size, seasonality, and chemical composition. To this end the airborne concentrations of the chemicals were measured between March 1999 and February 2000 in Unterloibach and Arnoldstein, both close to industrial centers. By relating the four aspects of the data, it was hoped that an insight could be gained into the nature of the air pollution in the Carynthian province.

19.6.2 Stage 2: Data description

The data form a four-way array: five groups of particle size fractions (0.04–0.1; 0.1–0.4; 0.4–1.6; 1.6–6.4; 6.4–25); 17 chemical components (Na^+, NH_4^+, K^+, Ca^{2+}, Mg^{2+}, Cl^-, NO_3^-, SO_4^{2-}, C, Cd, Cu, Fe, Mn, Pb, V, Zn, and dust); four seasons (Spring 1999, Summer 1999, Autumn, 1999, Winter 2000); and two sample locations (Unterloibach, Arnoldstein). Procedures with respect to the data collection and preparation are described in Stanimirova and Simeonov (2005). We are dealing here with

---

[4]Dr. I. Stanimirova kindly supplied the data and Prof. H. Puxbaum generously allowed their reanalysis.
[5]For an introduction and relevant publications see *http://people.clarkson.edu/ hopkepk/project1.html*. Accessed September 2007.

a 5 (Fractions)×17 (Chemical compounds)×2 (Locations) ×4 (Seasons) four-way data set. This order of modes will be used throughout this section.

For stability reasons, each data point was the average of measurements taken on four or six, mostly consecutive, days. Only valid data were averaged, resulting in only 26 missing data points, which were estimated during the alternating least-squares procedure (see Section 7.3.2, p. 150). There were two genuinely missing (i.e. unmeasured) data points, the other 24 missing values were due to too low concentrations for valid reliable observations. To start the analysis, these initial values for the missing data were set at 0.001, which was preferable over substituting the mean for that chemical. In fact, using the means as starting values led to nonconvergent solutions (see Section 7.7.3, p. 164, for a detailed example of the effect of inappropriate starting values).

Before the analysis proper, the data were normalized per chemical compound as their concentrations were measured in different units. Thus, the overall sum of squares of the data used in the analysis was equal to 17. No centering was carried out as all concentrations were considered to be ratio-scaled quantities with an absolute zero point.

19.6.3   Stage 3: Model and dimensionality selection

In the original paper it was established that a two-component Parafac model did not provide an adequate model. This judgment was based on the core consistency plot; for such plots see Bro (1998c), Bro and Kiers (2003b), and also Section 8.6.3 (p. 185). As Parafac models always have the same number of components for all modes, more than two components is not always a practical possibility. See, however, the paper by Kruskal (1976), and the example by Andersen and Rayens (2004) mentioned earlier, for the theory and an application of extracting more components than levels for a mode. In this case, it seemed reasonable to consider only Tucker4 models.

The deviance plot (Fig. 19.3) shows that reasonable models for the data are the $2\times2\times2\times2$, the $2\times3\times2\times3$, and possibly the $2\times4\times2\times3$ model, even though the latter has a rather large number of components to interpret. The st-criterion (see Ceulemans and Kiers (2006), and Section 8.5.3, p. 182) which selects the model for which the angle in the convex hull is sharpest, favors the $2\times2\times2\times2$ model. Note that Stanimirova and Simeonov (2005) preferred the $2\times4\times2\times2$ model with 85% fit, which lies very close to the convex hull. Here we will present the simpler $2\times2\times2\times2$ model with 80% fit, but it should be borne in mind that there are other considerations regarding model selection than just the deviance-df ratio.

Another tool to assist in choosing between the models presented here and that in the original paper is to look at the explained variability per component of the two models (Table 19.4). From this table we see that the Chemical compound components 3 and 4 each explain less than any of the other components, but again, substantive considerations together with knowledge outside the present data should be used to

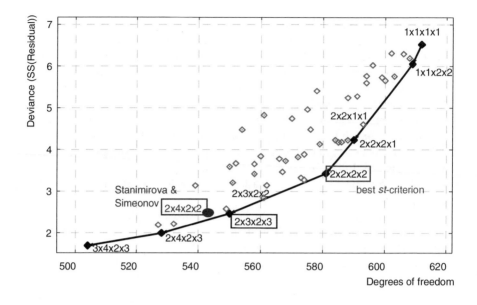

Figure 19.3   Aerosol particles: Deviance plot with convex hull.

select an appropriate model. For instance, the $2\times2\times2\times2$ model might just be too simple and miss subtleties that are important to environmental agencies.

19.6.4   Stage 4: Interpretation of the results

Graphs of each of the four two-dimensional component spaces are displayed to interpret the results from the $2\times2\times2\times2$ model. All components are in principal coordinates so that the inner products of the elements of a mode can be assessed with respect

Table 19.4   Aerosol particles: Proportional fit of the components of each mode for the $2\times2\times2\times2$ solution and the $2\times4\times2\times2$ solution

| | $2\times2\times2\times2$ | | $2\times4\times2\times2$ | | | |
Mode	1	2	1	2	3	4
1 Fractions	0.673	0.126	0.766	0.088		
2 Chemicals	0.651	0.148	0.650	0.142	0.039	0.024
3 Samples	0.743	0.056	0.705	0.149		
4 Seasons	0.746	0.053	0.772	0.082		
Total Fit	0.799		0.854			

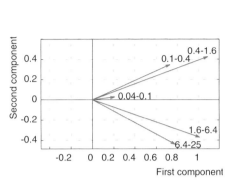

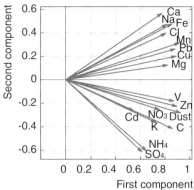

(a) Particle size fractions - 67% versus 13% explained variability

(b) Chemical compounds - 65% versus 15% explained variability

Figure 19.4   Aerosol particles: Component spaces for Fractions and Chemical compounds; both are in principal coordinates.

to each other in a correct way (see Kiers, 2000a, and Section 11.4, p. 260). Such representations are not necessary for all modes, but is generally strongly advised for the variables, because the angles between the variables are related to their correlations (see Appendix B, p. 500).

The four component spaces.   First we will discuss the four component spaces one by one, basing ourselves on the plots of the component spaces in principal coordinates (Figs. 19.4 and 19.5).

Fractions.   The concentration of the finest particles, 0.01–0.4, is so small (its sum of squares is 10–16 times smaller than that of the other particle sizes) that it does not really play a role in the analysis. This is also evident from its location close to the origin of the plot (Fig. 19.4(a)). Being close to the origin means that its coefficients $a_{11} \approx a_{12} \approx 0$, so that each term $a_{1p}b_{jq}c_{kr}d_{lt} \approx 0$ and thus does not contribute to the structural image. Moreover, only 41% of the total sum of squares of the smallest fraction is fitted by the model, in contrast with 80% for the data as a whole.

The high coefficients for all fractions on the first Fraction component (67%) indicate that the fractions have much variability in common. There are marginally higher values for the middle-size particles compared to the lighter and heavier ones. The second Fraction component (13%) marks the difference between fine particles at the positive pole and the coarse particles at the negative pole of the component.

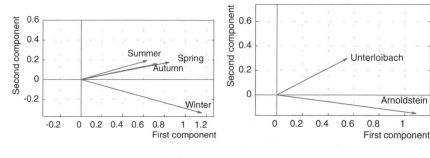

(a) Seasons - 75% versus 5% explained variability

(b) Locations - 74% versus 6% explained variability

Figure 19.5   Aerosol particles: Component spaces for Seasons and Locations, both are in principal coordinates.

Chemical compounds.   The first Chemicals component (65%) in Fig. 19.4(b) shows that there is a high correlation between the concentrations of the chemical compounds with more or less equal values on the first component, but a clear differentiation in the second component (15%), with Ca, Na, Fe (referred to as crustal content by Stanimirova & Simeonov, 2005) at the positive extreme and NH_4 and SO_4 (secondary emission) at the other, negative extreme.

Seasons.   Overall the concentrations are highest in winter and lowest in summer (first Seasons component — 75%), but the differences are not very large, and the winter concentrations (negative on the second Seasons component — 5%) are markedly different from those in the other three seasons (positive on the second Seasons component).

Locations.   The concentrations of particles shows both similarities and differences in both localities, with higher values shown on the first Locations component by Arnoldstein. The second Locations component shows the contrast between the two sampling sites. To facilitate the discussion of the differences between the sites, the normalized location space has been orthonormally rotated in such a way that each rotated component aligns perfectly with a location (Arnoldstein 60% and Unterloibach 20%). The core array shown in the next section has been accordingly adjusted via a counterrotation.

Detailed analysis using the core array.   The core array (Table 19.5) provides the basis for a detailed interpretation of the differences and similarities of the two locations with respect to differences in particle size, the chemicals found, and the seasonal differences.

Table 19.5 Aerosol particles: Core array

All sizes — Fractions Component 1

Location	Unstandardized Core Slice		Proportion Explained Variation	
	All	+: SO_2, NH_4 −: Ca, Fe, Na, Cl	All	+: SO_2, NH_4 −: Ca, Fe, Na, Cl
All seasons — Season component 1				
Arnoldstein	**2.90**	−0.25	0.493	0.004
Unterloibach	**1.44**	*0.45*	0.122	0.012
Spring, summer, autumn(+) versus winter(−) — Season component 2				
Arnoldstein	−0.28	*−0.49*	0.005	0.014
Unterloibach	*0.53*	0.35	0.017	0.007

Fine (+) versus coarse (-) — Fractions component 2

Location	Unstandardized Core Slice		Proportion Explained Variation	
	All	+: SO_2, NH_4 −: Ca, Fe, Na, Cl	All	+: SO_2, NH_4 −: Ca, Fe, Na, Cl
All seasons — Season component 1				
Arnoldstein	−0.22	**1.18**	0.003	0.082
Unterloibach	*0.38*	**0.61**	0.009	0.022
Spring, summer, autumn(+) versus winter(−) — Season component 2				
Arnoldstein	0.13	−0.17	0.001	0.002
Unterloibach	0.19	0.29	0.002	0.005

Larger values are shown in **bold**. Values that show a contrast between the two samples are in *italics*; (+)[(−)] indicates the levels have positive [negative] values on the component. The indication "All" for the first Chemicals component should be understood to mean that all chemicals have positive values on this component, but that the values for Cd, K, and for the secondary emissions (SO_4 NO_3, and NH_4) are somewhat lower than those for the other chemicals.

The interpretation will proceed along two lines. The first describes what the two locations have in common, and the second in which aspects the concentrations differ at the two locations. Thus, we will look at both the high values in the core array and at contrasting pairs between localities. When considering contrasts occasionally somewhat smaller values in the core array are taken into account, rather than only the higher values.

Pollution common to Arnoldstein and Unterloibach. There are primarily two aspects that are common to both locations. The first aspect follows from the $g_{1111} = 2.90$ and $g_{2111} = 1.44$ core elements. Both locations have high concentrations of all chemicals in all seasons for all particle sizes, but the values are about twice as high for Arnoldstein as for Unterloibach. This is also evident from the higher total sum of squares of Arnoldstein (SS(Total = 12.5, or 74%, of which 82% is fitted by the model) compared to Unterloibach (SS(Total) = 4.5 or 26%, of which 74% is fitted by the model). Furthermore, from the core elements $g_{1212} = 1.18$ and $g_{2212} = 0.61$ we may deduce that there is more fine than coarse emission for secondary emissions such as SO_2, NH_4 compared to for the crustal content (i.e., Ca, Fe, Na, Cl) in all seasons in both locations, but again more for Arnoldstein than for Unterloibach.

Differences between Arnoldstein and Unterloibach. There are three aspects on which the two locations differ if we demand that at least one of the core elements exceeds 1% explained variability. In all cases this involves a correction of the general patterns described in the previous paragraph, in the sense that the correction leads to higher values for Unterloibach and lower values for Arnoldstein than is described by the general pattern. In other words, in these particular cases the differences are smaller than suggested by the general patterns.

- $g_{1211} = -0.25$ and $g_{2211} = 0.45$: Even though the overall level of pollution is higher in Arnoldstein than in Unterloibach in all seasons for all particle sizes, the differences are less marked for the secondary emissions than for the crustal content of the emissions.

- $g_{1221} = -0.49$ and $g_{2221} = 0.35$: Moreover, this difference is even further reduced in spring, summer, and autumn compared to the winter, where the difference is emphasized.

- $g_{1121} = -0.28$ and $g_{2121} = 0.53$: A similar reduction of the difference can be observed for all particles in spring, summer, and autumn compared to winter, where again the contrast is emphasized.

There is a minor difference between the locations with respect to the size of all particles independent of seasons, as for all particles there seem to be comparatively more fine particles in Unterloibach than in Arnoldstein (core elements: $g_{1122} = -0.22$ and $g_{2122} = 0.38$). Thus, the seasonal effects are not the dominant ones. However, in winter values tend to be raised somewhat for Unterloibach and lowered for Arnoldstein. All chemicals are affected but the secondary emissions more than the crustal content, and particle size only plays a limited role here.

19.6.5   Stage 5: Validation of the results

For the present data independent substantive validation is difficult. The obvious way to go about this is to talk to the experts, confront them with the results, and take it from there. Numerical validation could take place by investigating the numerical stability of the solution. In particular, one might want to use the replicates, which do not feature in the present data, and create a number of data sets by sampling with replacement from these replicates. However, there are only four of them so that the number of different data sets of the same size is limited. One possible internal check is to evaluate the estimated missing values against the detection limits. A cursory check showed that 5 of the 26 missing values had estimated values of more than 20 times the detection limit, and this is probably reasonable. No value was way out of line.

Another interesting question is whether a Tucker4 analysis is really necessary or whether a Tucker3 analysis with location and seasons as a single mode would fit equally well. A quick and dirty analysis revealed that with a $2 \times 3 \times 3$ model again 85.4% can be explained with 556 df, while the four-mode model with the same amount explained variability needed has only 541 df. The model presented here had a fit of 79.9% with 581 df and the $2 \times 2 \times 2$ three-mode model with the same degrees of freedom has a fit of 80.2%. In other words, in terms of fit versus degrees of freedom, there is not much to choose, and a three-mode model might have served as well, be it that there are fewer parameters in the four-way analysis.

19.7 FURTHER READING AND COMPUTER PROGRAMS

19.7.1 Further reading

The primary integrated source for the technical side of multiway data analysis is Smilde et al. (2004). It contains extensive descriptions of several more complex models such as more-sets multiway data, and multiway multiblock models. However, the book does not contain actual analyses of higher-way data. Bro's (1998) publicly available thesis on multiway methods[6] contains an example of a five-way Parafac analysis as an alternative to the ANOVA (see also Heimdal, Bro, Larsen, & Poll, 1997).

There are a number of other papers dealing specifically with multiway techniques, where "multi" is greater than 3. Carroll and Chang (1970) presented a program to solve the seven-way Parafac model (under the name CANDECOMP). Lastovicka (1981) published a four-way Tucker model using Tucker's non-least-squares approach. Later, Kapteyn et al. (1986) solved the higher-way model with least-squares techniques. Leibovici and Sabatier (1998) presented a truly multiway paper describing the package PTAk in R.

Several authors have paid specific attention to the notation for multiway analysis, such as Kapteyn et al. (1986), Wansbeek and Verhees (1990), with Alsberg (1997) proposing Feynman-like diagrams and Harshman (2001) putting forward an index-

---

[6]*http://www.models.life.ku.dk/users/rasmus/brothesis.pdf.* Accessed May 2007.

based proposal. Kiers's (2000b) general paper discusses notation for higher-way arrays, but primarily with a view toward standardization of notation in the field.

19.7.2 Computer programs

The analyses in this chapter were carried out using an experimental extension to the TUCKALS3 program contained in 3WayPack, as developed by the author. Other publicly available programs or building blocks with facilities for higher-way analyses can be found in the MATLAB N-way Toolbox (Andersson & Bro, 2000)[7]. An important extension of the standard MATLAB functions to perform multilinear analyses is contained in Bader and Kolda's (2006) Tensor Toolbox[8]. An R add-on PTAk has been developed by Leibovici (Leibovici & Elmaache, 1997; Leibovici, 2007)[9], and Paatero (1999) has developed a sophisticated and versatile program for multiway analysis called the Multilinear Engine[10].

19.8   CONCLUSIONS

In this chapter we have touched on the most important multiway models and shown how an analysis may proceed. From the literature it has become clear that the theory is far more advanced than their applications in practice. Multiway counts are really more like "one, two, multi" than like "one, two, three, four, multi". There is no doubt that for specific questions in large data sets with many levels in most if not all modes, there will be an advantage in using higher-way models, but high-quality data of that kind seem to be rare, but in areas like analytical chemistry there is an increase of higher-way data due to an increase in sophisticated measuring instruments.

---

[7] *http://www.models.life.ku.dk/source/.* Accessed May 2007.; from the site a menu-driven interface Cu-Batch can also be downloaded and other multiway related software, nearly all in MATLAB code.
[8] *http://csmr.ca.sandia.gov/~tgkolda/TensorToolbox.* Accessed May 2007.
[9] *http://cran.r-project.org/.* Accessed May 2007.
[10] *ftp://rock.helsinki.fi/pub/misc/pmf/me2.* Accessed October 2007.

APPENDIX A

STANDARD NOTATION FOR

MULTIWAY ANALYSIS

Standard notation for multiway analysis

The notation is based on Kiers (2000b) and Smilde et al. (2004). However, arrays are indicated by bold calligraphic letters, $(\mathcal{X})$, rather than bold underlined letters $\underline{\mathbf{X}}$.

$i, j, k, l, \cdots$	running indices for modes 1, 2, 3, 4, ...
$I, J, K, L, \cdots$	sizes of modes 1, 2, 3, 4, ...
$p, q, r, t, \cdots$	running indices for components of modes 1, 2, 3, 4, ...
$P, Q, R, T, \cdots$	number of components for modes 1, 2, 3, 4, ...
s	running index for components for the Parafac model and PCA
S	number of components for Parafac model and two-mode PCA

$\mathbf{x} = (x_i)$	(column) vector (size: e.g., $I \times 1$)
$\mathbf{1} = 1$	a column vector of ones
$\mathbf{X} = (x_{ij})$	(data) matrix (size: e.g., $I \times J$)
$\mathbf{G} = (g_{ss})$	two-way core array with (rotated) singular values
$\mathcal{X} = (x_{ijk}) \, or \, (x_{ijkl})$	three-way arrays (size: $I \times J \times K$); multiway arrays (size: $I \times J \times K \times L$)
$\mathcal{G} = (g_{pqr}) \, or \, (g_{pqrt})$	core array in the Tucker3 model (size: $P \times Q \times R$); core array in the Tucker 4 model (size: $P \times Q \times R \times T$)
$\mathcal{H} = (h_{pqk})$	extended core array in the Tucker2 model (size: $P \times Q \times K$)
$\tilde{\mathcal{H}} = (h_{ppk})$	diagonal extended core array in the Tucker2 model
$\mathcal{D} = (d_{sss})$	three-way or multiway superdiagonal array (size: $S \times S \times S$)
$\mathcal{I} = (i_{sss})$	identity array, with 1s on the superdiagonal
$\mathcal{E} = (e_{ijk}) \, or \, (e_{ijkl})$	three-way or multiway array of residuals
$\mathbf{X}_1, \ldots, \mathbf{X}_k, \ldots$	matricized (frontal-slice)version of $\mathcal{X}$
$\mathbf{G}_1, \ldots, \mathbf{G}_r, \ldots$	$P \times QR$ matricized (frontal-slice) version $\mathbf{G}_a$ of $\mathcal{G}$
$\mathbf{G}_1, \ldots, \mathbf{G}_q, \ldots$	$P \times QR$ matricized (lateral-slice) version $\mathbf{G}_b$ of $\mathcal{G}$
$\mathbf{G}_1, \ldots, \mathbf{G}_p, \ldots$	$R \times QR$ matricized (horizontal-slice) version $\mathbf{G}_c$ of $\mathcal{G}$
$\mathbf{H}_1, \ldots, \mathbf{H}_k, \ldots$	matricized version of extended core array $\mathcal{H}$
$\tilde{\mathbf{H}}_1, \ldots, \tilde{\mathbf{H}}_k, \ldots$	matricized version of diagonal extended core array $\tilde{\mathcal{H}}$
$Vec(\mathbf{X})$	vectorized version of a matrix $\mathbf{X}$ (size: e.g., $IJ \times 1$)
$\mathbf{A}, \mathbf{B}, \mathbf{C}, \mathbf{D}, \ldots$	component matrices for modes 1, 2, 3, 4, ...
$\mathbf{F} = (f_{is})$	component matrix for two-way containing the variable–component correlations

$\otimes$	Kronecker product
SS_{tot}, SS(Tot)	total sum of squares in a data set
SS_{fit}, SS(Fit)	fitted sum of squares
SS_{res}, SS(Res)	residual sum of squares or deviance
SS_{resf}, SS(Res)$_f$	residual sum of squares of level f with $f = i, j, k, \ldots$

APPENDIX B

BIPLOTS AND THEIR

INTERPRETATION

Applied Multiway Data Analysis. By Pieter M. Kroonenberg
Copyright © 2007 John Wiley & Sons, Inc.

BIPLOTS AND THEIR INTERPRETATION

B.1   INTRODUCTION[1]

Biplots (see e.g., Gabriel, 1971) allow for the analysis of the two-way interaction in a table of I objects by J variables such that systematic patterns between rows, between columns, and between rows and columns can readily be assessed and evaluated. The prefix *bi* refers to the simultaneous display of both rows and columns of the table, not to a two-dimensionality of the plot. The number of dimensions is at most $\min(I, J)$. Arbitrarily, we will assume that there are more objects than variables, so that I is greater than J, and thus at most J dimensions are possible. As displays of more than two dimensions are generally difficult to make and even more difficult to interpret, most biplots show only the two dimensions which account for the maximum amount of variation in the table (see Section 11.4.2, p. 262). By using the *singular value decomposition* (SVD), it is possible find such a "best" representation in low-dimensional space. The technique provides the coordinates on dimensions (or directions in space); in the mathematical literature these dimensions are called *singular vectors*. The dimensions are arranged in such a way that they are *orthogonal* (i.e., at right angles), and successively represent as much of the variation as possible. Moreover, the technique provides us with measures (*singular values*), which, if squared, indicate the amount of variability accounted for by each dimension. To display the main variability in the table in a two-dimensional graph, we should use the first two dimensions.

B.2   SINGULAR VALUE DECOMPOSITION

B.2.1   Basic theory

Suppose that we have a two-way data matrix $\mathbf{X}$ with information of I objects on J variables, and that there are more objects than variables, so that $\min(I, J) = J$. The singular value decomposition SVD of the matrix $\mathbf{X}$ is defined as

$$\mathbf{X} = \mathbf{U}\Lambda\mathbf{V}', \tag{B.1}$$

which may be written in summation notation as

$$x_{ij} = \sum_{s=1}^{\tilde{S}} \lambda_s u_{is} v_{js}, \tag{B.2}$$

where $\tilde{S}$ is in most cases equal to J; that is, we generally need J terms to perfectly reproduce the original matrix $\mathbf{X}$. The scalars λ_s are the singular values arranged in decreasing order of magnitude, $\mathbf{u}_s$ is a set of object vectors (the left singular vectors),

---

[1]This appendix is based on Kroonenberg (1995c).

and $\mathbf{v}_s$ is a set of variable vectors (the right singular vectors). In both sets the vectors are *orthonormal*, that is, they are pair-wise at right angles and have lengths equal to 1. $\mathbf{U}$ and $\mathbf{V}$ are matrices which have the vectors $\mathbf{u}_s$ and $\mathbf{v}_s$ as their columns, respectively. If the entries in the table are the interactions from a univariate two-way analysis of variance on the original table, then both $\mathbf{u}_s$ and the $\mathbf{v}_s$ are *centered*; that is, each column of $\mathbf{U}$ and $\mathbf{V}$ has a zero mean, because the original table of interaction effects is centered. Moreover, in this case $\tilde{S}$ is at most $J - 1$, because centering reduces the number of independent dimensions by 1.

B.2.2 Low-dimensional approximation

To find a low-dimensional approximation of $\mathbf{X}$ we have to minimize the distance between the original matrix and an approximating matrix, $\hat{\mathbf{X}}$. This (Euclidean) distance between two matrices, $\mathbf{X} = x_{ij})$ and $\hat{\mathbf{X}} = (\hat{x}_{ij})$, is defined as

$$d(\mathbf{X}, \hat{\mathbf{X}}) = \sqrt{\sum_{i=1}^{I} \sum_{j=1}^{J} (x_{ij} - \hat{x}_{ij})^2}, \tag{B.3}$$

and the Eckart and Young (1936) theorem shows that the best S-dimensional least-squares approximation of the matrix $\mathbf{X}$ can be obtained from the SVD of $\mathbf{X}$ by summing only the first S terms of Eq. (B.2) ($S \leq \tilde{S}$). Such a S is also referred to as the best rank-S approximation of the matrix $\mathbf{X}$.

The first S $\mathbf{u}_s$ and $\mathbf{v}_s$, with S usually two or three, are used as the coordinates for graphical representations of the data. They can be combined with the singular values λ_s in different ways, of which the following two versions are the most common:

$$\hat{x}_{ij} = \sum_{s=1}^{S} u_{is}(\lambda_s v_{js}) = \sum_{s=1}^{S} y_{is} z_{ij}, \tag{B.4}$$

$$\hat{x}_{ij} = \sum_{s=1}^{S} (u_{is}\lambda_s^{1/2})(v_{js}\lambda_s^{1/2}) = \sum_{s=1}^{S} y_{is}^* z_{js}^*, \tag{B.5}$$

where the $\mathbf{y}$ and $\mathbf{z}$ are the object and variable coordinates of the first *principal coordinate scaled* version, and $\mathbf{y}^*$ and the $\mathbf{z}^*$ those of the second *symmetrically scaled* version, respectively (see Section B.3.3). In this book the matrix $\hat{\mathbf{X}}$ is called the *structural image* of the data $\mathbf{X}$.

B.2.3 Quality of approximation

To evaluate the quality of the S-dimensional approximation we have to know how much of the original variability of $\mathbf{X}$ is contained in the structural image $\hat{\mathbf{X}}$. The total variability in a matrix, here defined as the uncorrected sum of squares, is equal to the sum of squared entries in the table,

$$\text{Total variability} = \text{SS}_x = ||X||^2 = \sum_{i=1}^{I} \sum_{j=1}^{J} x_{ij}^2, \qquad \text{(B.6)}$$

where ||X|| is called the *norm* of **X**. Because of the least-squares properties of the singular value decomposition, the norm can be split into an explained and a residual part,

$$||\mathbf{X}||^2 = ||\hat{\mathbf{X}}||^2 + ||\mathbf{X} - \hat{\mathbf{X}}||^2. \qquad \text{(B.7)}$$

Furthermore, one can use the orthonormality of **U** and **V** to show that this equation may be expressed in terms of the squared singular values,

$$\sum_{s=1}^{S} \lambda_s^2 = \sum_{s=1}^{S} \lambda_s^2 + \sum_{s=S+1}^{\tilde{S}} \lambda_s^2. \qquad \text{(B.8)}$$

Equation (B.8) shows that the sum of the first two squared singular values divided by the total sum of the squared singular values will give the proportion of the variability accounted for by the first two singular vectors. Large proportions of explained variability will indicate that the plot based on these two singular vectors will give a good representation of the structure in the table. If only a moderate or low proportion of the variability is accounted for the main structure of the table will still be represented in the graph, but some parts of the structure may reside in higher dimensions. If the data are centered per variable, objects located near the origin might either have all their values close to the variable means, or their variability is located in another dimension. Similarly, variables close to the origin may have little variability or may not fit well in two dimensions.

B.3   BIPLOTS

B.3.1   Standard biplots

A standard biplot is the display of an object by variable (interaction) table, **X** decomposed into a product **YZ′** of an $I \times S$ matrix $\mathbf{Y} = (y_{is})$ and a $J \times S$ matrix $\mathbf{Z} = (z_{js})$. Using a two-dimensional decomposition for the structural image $\hat{\mathbf{X}}$, each element $\hat{x}_{ij}$ of this matrix can be written as

$$\hat{x}_{ij} = y_{i1} z_{j1} + y_{i2} z_{j2}, \qquad \text{(B.9)}$$

which is the *inner (or scalar) product* of the row vectors (y_{i1}, y_{i2}) and (z_{j1}, z_{j2}). A biplot is obtained by representing each row as a point Y_i with coordinates (y_{i1}, y_{i2}), and each column as point Z_j with coordinates (z_{j1}, z_{j2}) in a two-dimensional graph (with origin O). These points are generally referred to as *row markers* and *column*

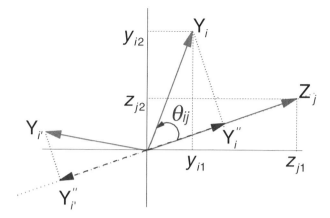

Figure B.1   Representation of two object markers and one variable marker in a biplot.

markers, respectively. Sometimes the word "markers" is also used for the coordinate vectors themselves. Because it is not easy to evaluate markers in a three-dimensional space, the most commonly used biplots are two-dimensional. With the current state of graphics software, it is likely that three-dimensional biplots will become more common. A straight line through the origin O and a point, say, Z_j, is often called a *biplot axis* and is written as OZ_j, not to be confused with a coordinate axis.

If we write Y_i'' for the orthogonal projection of Y_i on the biplot axis OZ_j, θ_{ij} for the angle between the vectors OY_i and OZ_j, and write $|OZ|^2$ for the length of a vector OZ, we have the geometric equivalent of Eq. (B.9) as displayed in Fig. B.1:

$$\hat{x}_{ij} = |OZ_j||OY_i|\cos(\theta_{ij}) = |OZ_j||OY_i''|. \tag{B.10}$$

Equation (B.10) shows that $\hat{x}_{ij}$ is proportional to the length of OY_i'', $|OY_i''|$. This relationship is of course true for any other object i' as well. The relationships or interactions of two objects with the same variable can be assessed simply by comparing the lengths of their projections onto that variable. Furthermore, the relationship or interaction between an object vector OY_i and a variable vector OZ_j is positive if their angle is acute, and negative in the case of an obtuse angle. When the projection of a marker Y_g onto the variable vector OZ_j coincides with the origin, $\hat{x}_{ij}$ is equal to 0 and the object has approximately a mean value for that variable, if the data were variable centered. A positive value for $\hat{x}_{ij}$ indicates that object i has a high score on variable j relative to the average score in that variable, and a negative value indicates object i has a relatively low score on variable j.

In graphs, the object markers Y_i are generally represented by points, and the variable markers Z_j by vectors, so that the two types of markers can be clearly distinguished. This arrangement is preferred because objects are compared with respect to a variable rather than the reverse.

B.3.2   Calibrated biplots

Because the inner products between the coordinates of the object markers Y_i and those of a column marker Z_j vary linearly along the biplot axis OZ_j, it is possible to mark (or calibrate) the biplot axis OZ_j linearly in such a way that the $\hat{x}_{ij}$ can be read directly from the graph (Gabriel & Odoroff, 1990; Greenacre, 1993). Note that the approximate value $\hat{x}_{ij}$ does not depend on the position of Y_i, but only on the orthogonal projection Y_i'' onto the axis OZ_j. When a data matrix is centered, as is the case with data centered per variable, the approximating matrix is centered as well, and a value of $\hat{x}_{ij}$ equal to zero means that on the ith uncentered variable object i has a value approximately equal to the mean of the jth variable. One option is to mark biplot axes with (approximations of) the centered variables. However, sometimes it is also informative to replace the centered values by their "real" values by adding the observed means. After this *decentering*, the origin indicates the true mean values for the variables, rather than zero for all of them. (see Carlier, 1995, for some three-mode examples).

B.3.3   Two different versions of the biplot

In Section B.2.2 the two most common decompositions of $\mathbf{X}$ were presented, both based on the SVD. These two decompositions lead to different biplots with different properties. Equations (B.4) and (B.5) show that the values of the inner products between object and variable markers are independent of the version used, so that in this respect the two versions are equivalent. However, when looking at the relationships within each set of markers, the two decompositions lead to different interpretations.

In the case of *principal coordinate scaling* (Eq. (B.4)), the objects are in so-called *standard coordinates*, that is, they have zero means and unit lengths, and the variables are in *principal coordinates*, that is, they have unrestricted means and lengths equal to the associated singular values. If in the data matrix $\mathbf{X}$ the variables are standardized, the coordinates of the variables may be interpreted as correlations between the variables and the coordinate axes.

With the *symmetric scaling* (Eq. (B.5)), the correlation interpretation cannot be used, because both the object components and those of the variables have lengths equal to the square root of the singular values. Therefore, this version should primarily be used when the relations between the objects and the variables are the central focus in the analysis, and not the relations among objects and/or among variables, or when the row and column variables play a comparable role in the analysis. The advantage of the representation is that lengths of the variable and the object vectors in the biplot are approximately equal. With principal component scaling it can easily happen that the objects are concentrated around the origin of the plots, while the variables are located on the rim, and vice versa.

B.3.4   Interpretational rules

An important point in constructing the actual graphs for biplots is that *the vertical and horizontal coordinate axes have the same physical scale.* This will ensure that when objects are projected on a variable vector, they will end up in the correct place. Failing to adhere to this scaling will make it impossible to evaluate inner products from the graph. The ratio of the units of the vertical axis and those on the horizontal axis is also referred to as the *aspect ratio*, and should be equal to 1.

The most basic property of a biplot is that the inner product of a row (object) vector and a column (variable) vector in the plot is the best approximation to the corresponding value in the table. If there is a perfect fit in, say, two dimensions, then the inner products are identical to the values in the table. The majority of the interpretational rules given below follow from this basic property. Additional interpretations become available if special treatments such as centering and standardization have been applied to the rows and/or columns, or principal coordinate scaling and symmetric scaling to the coordinate axes. Below we will only present those interpretational rules that we think are relevant for object by variable tables, in particular, we will not consider the situation when the original table is analyzed without centering.

- *General interpretational rules (irrespective of scaling coordinate axes)*

  - *Vectors and points*

    * objects are preferably *displayed* as points and variables as vectors or arrows;
    * if the angle between two object vectors is small, they have similar response patterns over variables;
    * if the angle between two variable vectors is small, they are strongly associated.

  - *Centered per variable*

    * the biplot displays the table of object main effect plus the two-way interaction;
    * object scores are in deviation from their average for each of the variables;
    * the origin represents the average value for each variable, that is, it represents the object that has an average value in each variable; this average object has a value of 0 in the centered data matrix;
    * an object at a large distance from the origin has a large object-plus-interaction effect;
    * the larger the projection of an object on a variable vector, the more this object deviates from the average in the variable.

– *Centered per variable and per object*

* the biplot displays the two-way interaction table; there are at most $\min(I, J)$ dimensions or coordinate axes;
* both object scores and variable coefficients are in deviation from their averages;
* the origin represents the average value both for each variable and for each object across all variables;
* an object (variable) at a large distance from the origin has a large interaction effect with at least one variable (object);
* the larger the projection of an object on a variable vector, the more this object deviates from the average in the variable, and vice versa.

• *Principal coordinate scaling:* **U** *and* **VΛ** *(Principal component biplot)*

– *Centered per variable*

* the cosine of the angle between any two variables approximates their correlation with equality if the fit is perfect;
* the lengths of the variable vectors are approximately proportional to the standard deviations of the variables with exact proportionality if the fit is perfect;
* the inner product between two variables approximates their covariance with equality if the fit is perfect;
* the Euclidean distance between two objects does not approximate the distances between their rows in the original matrix but their standardized distance, which is the square root of the Mahalanobis distance (for further details, see Gabriel, 1971, p. 460ff.);
* variables can have much longer vectors than objects, making visual inspection awkward; a partial remedy is to multiply all variable coordinates by an *arbitrary* constant, which will make the relative lengths of the variable and object vectors comparable. Note, however, that there is no obligation to use such a constant, and that it is an ad-hoc measure.

– *Standardized per variable*

* the lengths of the variable vectors indicate how well the variables are represented by the graph — with a perfect fit if all vectors have equal lengths;
* the inner product between two variables (and the cosine of the angle between them) approximates their correlation with equality if the fit is perfect.

- *Symmetric scaling*: $\mathbf{U\Lambda}^{1/2}$ *and* $\mathbf{V\Lambda}^{1/2}$

  - *General*

    * if the angle between two variable vectors is small, they are highly correlated, but their correlation cannot be deduced from the graph; similarly, the association between the objects cannot be properly read from the graph;

    * due to the symmetric scaling of variables and objects, both are located in the same part of the space and inner products are easily assessed.

B.4 RELATIONSHIP WITH PCA

In principal component analysis the linear combination $\mathbf{c} = \mathbf{Xb}$ is required which accounts for the largest amount of variation in a set of variables $\mathbf{X}$. The standard solution to this problem is to construct the sums-of-squares-and-cross-products matrix (or after centering and scaling, the correlation matrix) $\mathbf{X'X}$, and decomposing it (via the eigenvectors and eigenvalues) into $\mathbf{V\Lambda}^2\mathbf{V'}$; furthermore, $\mathbf{XX'}$ can be decomposed into $\mathbf{U\Lambda}^2\mathbf{U'}$. It can be shown that $\mathbf{U}$, $\mathbf{V}$, and $\mathbf{\Lambda}$ are the same as the matrices defined in Eq. (B.1). Moreover, $\mathbf{c}$ is equal to the first column of $\mathbf{U}$, and $\mathbf{b}$ is equal to λ_1 times the first column of $\mathbf{V}$. In other words, principal component analysis corresponds to the factorization of Eq. (B.9). The parameters for a principal component analysis can thus be directly derived from the singular value decomposition (see also Fig. 4.4, p. 49). However, in PCA it is general practice that $\mathbf{X'X}$ is a correlation matrix, whereas this assumption is not made for the singular value decomposition. What this shows is that PCA is a procedure with two steps: a centering and scaling followed by a (singular value) decomposition. The separation of these two steps is generally not emphasized in object-by-variable analyses but it becomes essential when analyzing three-way data of objects by variables by occasions (see Section 6.1, p. 109).

B.5 BASIC VECTOR GEOMETRY RELEVANT TO BIPLOTS

The interpretation of biplots depends heavily on properties of vectors in the plane or three-dimensional space. This section describes the basic properties of vectors, inner products, and projections.

Vector. Vectors and their properties

  Symbol. $\mathbf{x}$ or $\vec{x}$.

  Definition. A *vector* is a directed line segment; it has a *length* and a *direction*. In biplots vectors start at the origin, the point (0,0) in a two-dimensional biplot. The coordinates of $\vec{x}$ in the two-dimensional case are (x_1, x_2), where x_1 is the value on the horizontal coordinate axis and x_2 the value on the vertical coordinate axis. Therefore, a vector $\vec{x}$ runs from (0,0) to (x_1, x_2).

Length. The length of a vector, $|\vec{x}|$, is found via the Pythagorean theorem $(a^2 = b^2 + c^2)$: $|\vec{x}| = \sqrt{(x_1^2 + x_2^2)} = \sqrt{(\Sigma_i x_i^2)}$.

Angles. Angles and their properties

Angle. The angle between two vectors can be directly inferred from a graph; its angle between $\vec{x}$ and $\vec{y}$ is θ_{xy}. The angle can be computed via the inner product or dot product.

Inner product/Dot product/Scalar product. These terms are equivalent. The term dot product is used for the product between two vectors when using vector geometry and is written as $\vec{x} \bullet \vec{y}$. When using the term inner product it is mostly written as $\vec{x}'\vec{y}$.

The product is defined as $\vec{x} \bullet \vec{y} = x_1 y_1 + x_2 y_2$ using the coordinates of the vectors or, in more geometric terms, as: $\vec{x} \bullet \vec{y} = |\vec{x}||\vec{y}| \cos \theta_{xy}$, which is the length of $\vec{x}$ times the length of $\vec{y}$ times the cosine of the angle between them.

Calculation. $\cos \theta_{xy} = (\vec{x}\vec{y})/|\vec{x}||\vec{y}|$; convert the cosine to an angle via the "inverse cosine" button on your pocket/desktop calculator, look it up in a table, or use a computer program.

Special angles. $\theta_{xy} = 0° \rightarrow \cos \theta_{xy} = 1$: $\vec{x}$ and $\vec{y}$ are *collinear*, that is, they lie on the same line in the same direction; $\vec{y} = b\vec{x}$ with $b > 0$; $\vec{x}$ is collinear with itself $\theta_{xx} = 0$;

$\theta_{xy} = 180° \rightarrow \cos \theta_{xy} = -1$: $\vec{x}$ and $\vec{y}$ are *collinear*, that is, they lie on the same line but in opposite directions; $\vec{y} = b\vec{x}$ with $b < 0$;

$\theta_{xy} = 90° \rightarrow \cos \theta_{xy} = 0$: $\vec{x}$ and $\vec{y}$ are *orthogonal* (perpendicular); $\vec{x}\vec{y} = 0$.

Projection. Projections and their properties

The projection $\vec{y}'$ of $\vec{y}$ on $\vec{x}$ is a vector collinear with $\vec{x}$, which can be found by dropping a perpendicular line from $\vec{y}$ onto $\vec{x}$. Thus, $\vec{y}' = d\vec{x}$. The length of $\vec{y}'$ is $|\vec{y}| \cos \theta_{xy}$, and $d = (\vec{x} \bullet \vec{y})/|\vec{x}|^2$.

Equality between cosines and correlations. If the variables are centered, the cosine of θ_{xy}, the angle between two variables $\vec{x}$ and $\vec{y}$, is equal to their correlation r_{xy}:

$$r_{xy} = \frac{\sum (x_i - \overline{x})(y_i - \overline{y})}{\sqrt{\sum (x_i - \overline{x})^2} \sqrt{\sum (y_i - \overline{y})^2}} = \frac{\sum x_i y_i}{\sqrt{\sum x_i^2} \sqrt{\sum y_i^2}} = \frac{\vec{x} \bullet \vec{y}}{|\vec{x}||\vec{y}|} = \cos \theta_{xy}$$

where we have used the fact that the means are 0.

REFERENCES

Achenbach, T. M., & Edelbrock, C. (1983). *Manual for the Child Behavior Check List and revised child behavior profile* (Technical report). Burlington, VT: Department of Psychiatry, University of Vermont[1].

Adams, E., Walczak, B., Vervaet, C., Risha, P. G., & Massart, D. L. (2002). Principal component analysis of dissolution data with missing elements. *International Journal of Pharmaceutics, 234*, 169–178.

Agresti, A. (1990). *Categorical data analysis*. New York: Wiley.

Ainsworth, M. D. S., Blehar, M. C., Waters, E., & Wall, S. (1978). *Patterns of attachment. A psychological study of the Strange Situation*. Hillsdale, NJ: Erlbaum.

Alsberg, B. K. (1997). A diagram notation for N-mode array equations. *Journal of Chemometrics, 11*, 251–266.

Alsberg, B. K., & Kvalheim, O. M. (1993). Compression of *n*th-order data arrays by B-splines. Part 1: Theory. *Journal of Chemometrics, 7*, 61–73.

Alsberg, B. K., & Kvalheim, O. M. (1994). Compression of three-mode data arrays by B-splines prior to three-mode principal component analysis. *Chemometrics and Intelligent Laboratory Systems, 23*, 29–38.

Andersen, A. H., & Rayens, W. S. (2004). Structure-seeking multilinear methods for the analysis of fMRI data. *NeuroImage, 22*, 728–739.

---

[1]Assignment of a,b,c... within a year is exclusively done on the basis of the first author.

Applied Multiway Data Analysis. By Pieter M. Kroonenberg
Copyright © 2007 John Wiley & Sons, Inc.

Anderson, C. J. (1996). The analysis of three-way contingency tables by three-mode association models. *Psychometrika, 61,* 465–483.

Andersson, C. A., & Bro, R. (2000). The *N*-way Toolbox for MATLAB. *Chemometrics and Intelligent Laboratory Systems, 52,* 1–4.

Andersson, C. A., & Henrion, R. (1999). A general algorithm for obtaining simple structure of core arrays in *N*-way pca with application to fluorometric data. *Computational Statistics and Data Analysis, 31,* 255–278.

Andrews, D. T., & Wentzell, P. D. (1997). Applications of maximum likelihood principal component analysis: Incomplete data sets and calibration transfer. *Analytica Chimica Acta, 350,* 341–352.

Appellof, C. J., & Davidson, E. R. (1981). Strategies for analyzing data from video fluorometric monitoring of liquid chromatographic effluents. *Analytical Chemistry, 53,* 2053–2056.

Arabie, P., Carroll, J. D., & DeSarbo, W. S. (1987). *Three-way scaling and clustering.* Beverly Hills: Sage.

Arabie, P., Carroll, J. D., & DeSarbo, W. S. (1990). *San-gen deta no bunseki: Tajigen shakudo koseiho to kurasuta bunseikiho* [Three-way scaling and clustering.]. Tokyo: Kyoritsu Shuppan. (Translated by A. Okada and T. Imaizumi.)

Arancibia, J. A., Olivieri, A. C., Gil, D. B., Mansilla, A. E., Durán-Merás, I., & de la Peña, A. M. (2006). Trilinear least squares and unfolded-PLS coupled to residual trilinearization: New chemometric tools for the analysis of four-way instrumental data. *Chemometrics and Intelligent Laboratory Systems, 80,* 77–86.

Bader, B. W., & Kolda, T. G. (2006). MATLAB tensor classes for fast algorithm prototyping. *ACM Transactions on Mathematical Software, 32,* 635–653.

Bahram, M., Bro, R., Stedmon, C., & Afkhami, A. (2007). Handling of Rayleigh and Raman scatter for PARAFAC modeling of fluorescence data using interpolation. *Journal of Chemometrics, 20,* 99–105.

Bartussek, D. (1973). Zur Interpretation der Kernmatrix in der dreimodalen Faktorenanalyse von R. L. Tucker [On the interpretation of the core matrix in the three-mode factor analysis of R. L. Tucker]. *Psychologische Beiträge, 15,* 169–184.

Basford, K. E., Greenway, D. R., McLachlan, G. J., & Peel, D. (1997). Standard errors of fitted component means of normal mixtures. *Computational Statistics, 12,* 1–17.

Basford, K. E., & Kroonenberg, P. M. (1989). An investigation of multi-attribute genotype response across environments using three-mode principal component analysis. *Euphytica, 44,* 109–123.

Basford, K. E., Kroonenberg, P. M., & Cooper, M. (1996). Three-mode analytical methods for crop improvement programs. In M. Cooper & G. L. Hammer (Eds.), *Plant adaptation and crop improvement* (pp. 291–305). Wallingford, UK: CAB International.

Basford, K. E., Kroonenberg, P. M., & DeLacy, I. H. (1991). Three-way methods for multiattribute genotype by environment data: An illustrated partial survey. *Field Crops Research, 27,* 131–157.

Basford, K. E., & McLachlan, G. J. (1985a). Estimation of allocation rates in a cluster analysis context. *Journal of the American Statistical Association, 80,* 286–293.

Basford, K. E., & McLachlan, G. J. (1985b). The mixture method of clustering applied to three-way data. *Journal of Classification, 2,* 109–125.

Basford, K. E., & Tukey, J. W. (1999). *Graphical approaches to multiresponse data: Illustrated with a plant breeding trial.* London: Chapman & Hall.

Beh, E. J., & Davy, P. J. (1998). Partitioning Pearson's chi-squared statistic for a completely ordered three-way contingency table. *Australian and New Zealand Journal of Statistics*, *40*, 465–477.

Bentler, P. M. (1978a). Assessment of developmental factor change at the individual and group level. In J. R. Nesselroade & H. W. Reese (Eds.), *Life-span developmental psychology. Methodological issues* (pp. 145–174). New York: Academic Press.

Bentler, P. M., & Lee, S.-Y. (1978b). Statistical aspects of a three-mode factor analysis model. *Psychometrika*, *43*, 343–352.

Bentler, P. M., & Lee, S.-Y. (1979). A statistical development of three-mode factor analysis. *British Journal of Mathematical and Statistical Psychology*, *32*, 87–104.

Bentler, P. M., Poon, W.-Y., & Lee, S.-Y. (1988). Generalized multimode latent variable models: Implementation by standard programs. *Computational Statistics and Data Analysis*, *6*, 107–118.

Bijleveld, C. C. J. H., Mooijaart, A., Van der Kamp, L. J. T., & Van der Kloot, W. A. (1998). Structural equation models for longitudinal data. In C. C. J. H. Bijleveld & L. J. T. Van der Kamp (Eds.), *Longitudinal data analysis: Designs, models, methods* (pp. 207–268). London: Sage.

Bloxom, B. (1968). A note on invariance in three-mode factor analysis. *Psychometrika*, *33*, 347–350.

Bloxom, B. (1984). Tucker's three-mode factor analysis model. In H. G. Law, C. W. Snyder Jr., J. A. Hattie, & R. P. McDonald (Eds.), *Research methods for multimode data analysis* (pp. 104–121). New York: Praeger.

Boik, R. J. (1990). A likelihood ratio test for three-mode singular values: Upper percentiles and an application to three-way ANOVA. *Computational Statistics and Data Analysis*, *10*, 1–9.

Bollen, K. A. (1989). *Structural equations with latent variables*. New York: Wiley.

Boomsma, A., & Hoogland, J. J. (2001). The robustness of LISREL modeling revisited. In R. Cudeck, S. du Toit, & D. Sörbom (Eds.), *Structural equation models: Present and future. A Festschrift in honor of Karl Jöreskog* (pp. 139–168). Chicago, IL: Scientific Software International.

Borg, I., & Groenen, P. (2005). *Modern multidimensional scaling: Theory and applications* (2nd ed.). New York: Springer. (Chapters 21 and 22; First edition 1997)

Bro, R. (1996a). Multiway calibration. Multilinear PLS. *Journal of Chemometrics*, *10*, 47–61.

Bro, R. (1996c). Multiway calibration. Multilinear PLS. *Journal of Chemometrics*, *10*, 47–61.

Bro, R. (1997). PARAFAC. Tutorial and applications. *Chemometrics and Intelligent Laboratory Systems*, *38*, 149–171.

Bro, R. (1998c). *Multi-way analysis in the food industry. Models, algorithms, and applications.* Unpublished doctoral dissertation, University of Amsterdam, Amsterdam, The Netherlands. (*http://www.models.kvl.dk/users/rasmus/brothesis.pdf.* Accessed May 2007.)

Bro, R. (1999a). Exploratory study of sugar production using fluorescence spectroscopy and multi-way analysis. *Chemometrics and Intelligent Laboratory Systems*, *46*, 133–147.

Bro, R. (2006). Review on multiway analysis in chemistry—2000—2005. *Critical Reviews in Analytical Chemistry*, *36*, 279–293.

Bro, R., & Andersson, C. A. (1998a). Improving the speed of multiway algorithms. Part II: Compression. *Chemometrics and Intelligent Laboratory Systems*, *42*, 105–113.

Bro, R., Andersson, C. A., & Kiers, H. A. L. (1999b). PARAFAC2 - Part II. Modeling chromatic data with retention time shifts. *Journal of Chemometrics*, *13*, 295–309.

Bro, R., Harshman, R. A., & Sidiropoulos, N. (2007). *Modeling multi-way data with linearly dependent loadings* (KVL Technical report, 2005-176, Version 2). Copenhagen: Royal Veterinary & Agricultural University (*www.models.life.ku.dk/users/rasmus/paralind_KVLTechnicalReport2005-176.pdf.* Accessed May 2007.)

Bro, R., & Heimdal, H. (1996b). Enzymatic browning of vegetables. Calibration and analysis of variance by multiway methods. *Chemometrics and Intelligent Laboratory Systems*, *34*, 85–102.

Bro, R., & Jakobsen, M. (2002a). Exploring complex interactions in designed data using GEMANOVA. Color changes in fresh beef during storage. *Journal of Chemometrics*, *16*, 294–304.

Bro, R., & Kiers, H. A. L. (2003b). A new efficient method for determining the number of components in PARAFAC models. *Journal of Chemometrics*, *17*, 273–286.

Bro, R., & Sidiropoulos, N. D. (1998b). Least squares algorithms under unimodality and non-negativity constraints. *Journal of Chemometrics*, *12*, 223–247.

Bro, R., Sidiropoulous, N. D., & Smilde, A. K. (2002b). Maximum likelihood fitting using ordinary least squares algorithms. *Journal of Chemometrics*, *16*, 387–400.

Bro, R., & Smilde, A. K. (2003a). Centring and scaling in component analysis. *Journal of Chemometrics*, *17*, 16–33.

Bro, R., Smilde, A. K., & De Jong, S. (2001). On the difference between low-rank and subspace approximation: improved model for multi-linear PLS Regression. *Chemometrics and Intelligent Laboratory Systems*, *58*, 3–13.

Brouwer, P., & Kroonenberg, P. M. (1991). Some notes on the diagonalization of extended three-mode core matrices. *Journal of Classification*, *8*, 93–98.

Browne, M. W. (1982). Covariance structures. In D. M. Hawkins (Ed.), *Topics in applied multivariate analysis* (pp. 72–141). Cambridge, UK: Cambridge University Press.

Browne, M. W. (1984). The decomposition of multitrait-multimethod matrices. *British Journal of Mathematical and Statistical Psychology*, *37*, 1–21.

Browne, M. W. (1993). Structured latent curve models. In C. M. Cuadras & C. R. Rao (Eds.), *Multivariate analysis: Future directions 2* (pp. 171–198). Amsterdam: North-Holland.

Browne, M. W. (2001). An overview of analytic rotation in exploratory factor analysis. *Multivariate Behavioral Research*, *36*, 111–150.

Browne, M. W., & Toit, S. H. C. D. (1991). Models for learning data. In L. M. Collins & J. L. Horn (Eds.), *Best methods for the analysis of change* (pp. 47–68). Washington DC: American Psychological Association.

Carlier, A. (1986). Factor analysis of evolution and cluster methods on trajectories. In F. de Antoni, N. Lauro, & A. Rizzi (Eds.), *Compstat 1986* (pp. 140–145). Heidelberg: Physica Verlag.

Carlier, A. (1995). *Examples of 3-way correspondence analysis.* (Technical report). Toulouse, France: Laboratoire of Statistique et Probabilité, Université Paul-Sabatier.

Carlier, A., & Kroonenberg, P. M. (1996). Decompositions and biplots in three-way correspondence analysis. *Psychometrika*, *61*, 355–373.

Carlier, A., & Kroonenberg, P. M. (1998). The case of the French cantons: An application of three-way correspondence analysis. In J. Blasius & M. Greenacre (Eds.), *Visualization of categorical data* (pp. 253–275). New York: Academic Press.

Carroll, J. D. (1972a). Individual differences and multidimensional scaling. In R. N. Shepard, A. K. Romney, & S. B. Nerlove (Eds.), *Multidimensional scaling: Theory and applications in the behavioral sciences, 1: Theory* (pp. 105–155). New York: Seminar Press.

Carroll, J. D., & Arabie, P. (1983). INDCLUS: An individual differences generalization of the ADCLUS model and the MAPCLUS algorithm. *Psychometrika, 48*, 157–169.

Carroll, J. D., & Chang, J. J. (1970). Analysis of individual differences in multidimensional scaling via an N-way generalization of "Eckart-Young" decomposition. *Psychometrika, 35*, 283–319.

Carroll, J. D., & Chang, J. J. (1972b, March). *IDIOSCAL: A generalization of INDSCAL allowing IDIOsyncratic reference system as well as an analytic approximation to INDSCAL.* Paper presented at the Spring Meeting of the Classification Society of North America, Princeton NJ.

Carroll, J. D., & Chaturvedi, A. (1995). A general approach to clustering and multidimensional scaling of two-way, three-way, or higher-way data. In R. D. Luce, M. D'Zmura, D. D. Hoffman, G. J. Iverson, & A. K. Romney (Eds.), *Geometric representations of perceptual phenomena: Papers in honor of Tarow Indow on his 70th birthday* (pp. 295–318). Mahwah, NJ: Erlbaum.

Carroll, J. D., Clark, L. A., & DeSarbo, W. S. (1984a). The representation of three-way proximity data by single and multiple tree structure models. *Journal of Classification, 1*, 25–74.

Carroll, J. D., De Soete, G., & Kamensky, V. (1992). A modified CANDECOMP algorithm for fitting the latent class model: Implementation and evaluation. *Applied Stochastic Models and Data Analysis, 8*, 303–309.

Carroll, J. D., De Soete, G., & Pruzansky, S. (1989). Fitting of the latent class model via iteratively reweighted least squares CANDECOMP with nonnegativity constraints. In R. Coppi & S. Bolasco (Eds.), *Multiway data analysis* (pp. 463–472). Amsterdam: Elsevier.

Carroll, J. D., & Pruzansky, S. (1984b). The CANDECOMP-CANDELINC family of models and methods for multidimensional data analysis. In H. G. Law, C. W. Snyder Jr., J. A. Hattie, & R. P. McDonald (Eds.), *Research methods for multimode data analysis* (pp. 372–402). New York: Praeger.

Carroll, J. D., Pruzansky, S., & Kruskal, J. B. (1980). CANDELINC: A general approach to multidimensional analysis of many-way arrays with linear constraints on parameters. *Psychometrika, 45*, 3–24.

Carroll, J. D., & Wish, M. (1974). Models and methods for three-way multidimensional scaling. In D. H. Krantz, R. C. Atkinson, R. D. Luce, & P. Suppes (Eds.), *Contemporary developments in mathematical psychology. Vol. 2* (pp. 57–105). San Francisco: W. H. Freeman.

Cattell, R. B. (1944). "Parallel proportional profiles" and other principles for determining the choice of factors by rotation. *Psychometrika, 9*, 267–283.

Cattell, R. B. (1946). *The description and measurement of personality.* New York: World Book.

Cattell, R. B. (1966). The data box: Its ordering of total resources in terms of possible relational systems. In R. B. Cattell (Ed.), *Handbook of multivariate experimental psychology* (pp. 67–128). Chicago: Rand McNally.

Cattell, R. B., & Cattell, A. K. S. (1955). Factor rotation for proportional profiles: Analytical solution and an example. *The British Journal of Statistical Psychology, 8*, 83–92.

Ceulemans, E., & Kiers, H. A. L. (2006). Selecting among three-mode principal component models of different types and complexities: A numerical convex hull based method. *British Journal of Mathematical and Statistical Psychology, 59*, 133–150.

Ceulemans, E., & Van Mechelen, I. (2004). Tucker2 hierarchical classes analysis. *Psychometrika, 69*, 375–399.

Ceulemans, E., & Van Mechelen, I. (2005). Hierarchical classes models for three-way three-mode binary data: Interrelations and model selection. *Psychometrika, 70*, 461–480.

Ceulemans, E., Van Mechelen, I., & Leenen, I. (2003). Tucker3 hierarchical classes analysis. *Psychometrika, 68*, 413–433.

Chapman, S. C., Crossa, J., Basford, K. E., & Kroonenberg, P. M. (1997). Genotype by environment effects and selection for drought tolerance in tropical maize. II. Three-mode pattern analysis. *Euphytica, 95*, 11–20.

Clogg, C. C. (1982). Some models for the analysis of association in multiway cross-classifications having ordered categories. *Journal American Statistical Association, 77*, 803–815.

Cohen, H. S. (1974). *Three-mode rotation to approximate INDSCAL structure (TRAIS)*. (Technical Memorandum No. 74-3433-14). Murray Hill, NJ: Bell Laboratories.

Cohen, J. (1960). A coefficient of agreement for nominal scales. *Educational and Psychological Measurement, 20*, 37–46.

Commandeur, J. J. F., Kroonenberg, P. M., & Dunn III, W. J. (2004). A dedicated generalized Procrustes algorithm for consensus molecular alignment. *Journal of Chemometrics, 18*, 37–42.

Comon, P. (2001). Tensor decompositions. State of the art and applications. In J. G. McWhirter & I. K. Proudler (Eds.), *Mathematics in signal processing V* (pp. 1–24). Oxford, UK: Clarendon.

Coombs, C. H. (1964). *A theory of data*. New York: Wiley.

Cooper, M., & DeLacy, I. H. (1994). Relationships among analytical methods used to study genotypic variation and genotype-by-environment interaction in plant breeding multi-environment experiments. *Theoretical and Applied Genetics, 88*, 561–572.

Cooper, M., DeLacy, I. H., & Basford, K. E. (1996). Relationships among analytical methods used to analyse genotypic adaptation in multi-environment trials. In M. Cooper & G. L. Hammer (Eds.), *Plant adaptation and crop improvement* (pp. 193–224). Oxford: CAB International.

Coppi, R., & Bolasco, S. (Eds.). (1989). *Multiway data analysis*. Amsterdam: Elsevier.

Coppi, R., & D'Urso, P. (2003). Three-way fuzzy clustering models for LR fuzzy time trajectories. *Computational Statistics & Data Analysis, 43*, 149–177.

Cormen, T. H., Leierson, C. E., & Rivest, R. L. (1989). *Introduction to algorithms*. Cambridge, MA: The MIT Press.

Cox, T. F., & Cox, M. A. (2001). *Multidimensional scaling* (2nd ed.). Boca Raton, FL: Chapman & Hall/CRC.

Cronbach, L. J. (1984). A research worker's treasure chest. *Multivariate Behavioral Research, 19*, 223–240.

Croux, C., Filzmoser, P., Pison, G., & Rousseeuw, P. J. (2003). Fitting multiplicative models by robust alternating regressions. *Statistics and Computing, 13*, 23–36.

Croux, C., & Haesbroeck, G. (2000). Principal component analysis based on robust estimators of the covariance or correlation matrix: Influence functions and efficiencies. *Biometrika, 87*, 603–618.

Croux, C., & Ruiz-Gazen, A. (1996). A fast algorithm for robust principal components based on projection pursuit. In A. Prat (Ed.), *COMPSTAT 1996, proceedings in computational statistics* (pp. 211–217). Heidelberg, Germany: Physica.

Dai, K. (1982). Application of the three-mode factor analysis to industrial design of chair styles. *Japanese Psychological Review, 25*, 91–103 (in Japanese with English abstract).

D'Ambra, L. (1985). Alcuna estensioni dell'analisi in componenti principali per lo studio di sistemi evolutivi. Uno studio sul commercio internationale dell'elettronica. [Some extensions of principal component analysis for the study of longitudinal systems. A study of the international electronics trade.]. *Ricerche Economiche, 34*, 233–260.

D'Ambra, L., & Lombardo, R. (1993, July). Normalized non-symmetrical correspondence analysis for three-way data-sets. In *Bulletin of the ISI. Contributed papers 49th session, Vol. 1* (pp. 301–302). Rome, Italy: International Statistical Society.

D'Aubigny, G. (2004, May). *Une méthode d'imputation multiple en acp.* Paper presented at the XXXVIème Journées de Statistique. Montpellier, France.

Davis, J. A. (1977). *Codebook for the 1977 general social survey.* Chicago, IL: National Opinion Reaerach Center.

De la Torre, F., & Black, M. J. (2003). A framework for robust subspace learning. *International Journal of Computer Vision, 54*, 117–142.

De la Vega, A. J., Hall, A. J., & Kroonenberg, P. M. (2002). Investigating the physiological bases of predictable and unpredictable genotype by environment interactions using three mode pattern analysis. *Field Crops Research, 78*, 165–183.

De Lathauwer, L. (2006). A link between the canonical decomposition in multilinear algebra and simultaneous matrix diagonalization. *SIAM Journal on Matrix Analysis and Applications, 28*, 642–666.

De Lathauwer, L., De Moor, B., & Vandewalle, J. (2000). A multilinear singular value decomposition. *Siam Journal on Matrix Analysis and Applications, 21*, 1253–1278.

De Ligny, C. L., Nieuwdorp, G. H. E., Brederode, W. K., Hammers, W. E., & Van Houwelingen, J. C. (1980). An application of factor analysis with missing values. *Technometrics, 23*, 91–95.

De Ligny, C. L., Spanjer, M. C., Van Houwelingen, J. C., & Weesie, H. M. (1984). Three-mode factor analysis of data on retention in normal-phase high-performance liquid chromatography. *Journal of Chromatography, 301*, 311–323.

De Soete, G., & Carroll, J. D. (1989). Ultrametric tree representations of three-way three-mode data. In R. Coppi & S. Bolasco (Eds.), *Multiway data analysis* (pp. 415–426). Amsterdam: Elsevier.

DeLacy, I. H., & Cooper, M. (1990). Pattern analysis for the analysis regional variety trials. In M. S. Kang (Ed.), *Genotype-by-environment interaction and plant breeding* (pp. 301–334). Baton Rouge, LA: Louisiana State University.

Dempster, A. P., Laird, N. M., & Rubin, D. B. (1977). Maximum likelihood from incomplete data via the EM algorithm (with discussion). *Journal of the Royal Statistical Society, Series B, 39*, 1–38.

Denis, J. B., & Gower, J. C. (1994). Biadditive models. *Biometrics, 50*, 310–311.

DeSarbo, W. S., & Carroll, J. D. (1985). Three-way metric unfolding via alternating weighted least squares. *Psychometrika, 50*, 275–300.

DeSarbo, W. S., Carroll, J. D., Lehmann, D. R., & O'Shaughnessy, J. (1982). Three-way multivariate conjoint analysis. *Marketing Science, 1*, 323–350.

Dias, C. T. S., & Krzanowski, W. J. (2003). Model selection and cross validation in additive main effect and multiplicative interaction models. *Crop Science, 43*, 865–873.

Dillon, W. R., Frederick, D. G., & Tangpanichdee, V. (1985). Decision issues in building perceptual product spaces with multi-attribute rating data. *Journal of Consumer Research, 12*, 47–63.

Dorans, N. J. (2004). *A conversation with Ledyard R Tucker* (ETS Policy and Research Report, Tucker). Princeton, NJ: ETS.

Dueñas Cueva, J. M., Rossi, A. V., & Poppi, R. J. (2001). Modeling kinetic spectrophotometric data of aminophenol isomers by PARAFAC2. *Chemometrics and Intelligent Laboratory Systems, 55*, 125–132.

Dunn, T. R., & Harshman, R. A. (1982). A multidimensional scaling model for the size-weight illusion. *Psychometrika, 47*, 25–45.

Durell, S. R., Lee, C.-H., Ross, R. T., & Gross, E. L. (1990). Factor analysis of the near-ultraviolet absorption spectrum of plastocyanin using bilinear, trilinear, and quadrilinear models. *Archives of Biochemistry and Biophysics, 278*, 148–160.

Eastment, H. T., & Krzanowski, W. J. (1982). Cross-validatory choice of the number of components from a principal component analysis. *Technometrics, 24*, 73–77.

Eckart, C., & Young, G. (1936). The approximation of one matrix by another of lower rank. *Psychometrika, 1*, 211–218.

Efron, B. (1979). Bootstrap methods: Another look at the jackknife. *Annals of Statistics, 7*, 1–26.

Engelen, S., Frosh Möller, S., & Hubert, M. (2007a). Automatically identifying scatter in fluorescence data using robust techniques. *Chemometrics and Intelligent Laboratory Systems, 86*, 35–51.

Engelen, S., & Hubert, M. (2007b). *Detecting outlying sample in a PARAFAC model* (Technical report, TR-06-07). Leuven, Belgium: Department of Mathematics, Universiteit Leuven. (*http://wis.kuleuven.be/stat/robust/Papers/TR0607.pdf*. Accessed May 2007.)

Engelen, S., Hubert, M., & Vanden Branden, K. (2005). A comparison of three procedures for robust PCA in high dimensions. *Austrian Journal of Statistics, 34*, 117–126.

Estienne, F., Matthijs, N., Massart, D. L., Ricoux, P., & Leibovici, D. (2001). Multi-way modelling of high-dimensionality electroencephalographic data. *Chemometrics and Intelligent Laboratory Systems, 58*, 59–72.

Everitt, B. S. (1980). *Cluster analysis* (2nd ed.). London: Heinemann.

Faber, N. M., Bro, R., & Hopke, P. K. (2003). Recent developments in CANDE-COMP/PARAFAC algorithms: A critical review. *Chemometrics and Intelligent Laboratory Systems, 65*, 119–137.

Fisher, R. A., & Mackenzie, W. A. (1923). Studies in crop variation II. the manurial response to different potato varieties. *Journal of Agricultural Science, 13*, 311–320.

Fitzmaurice, G., Laird, N., & Ware, J. (2004). *Applied longitudinal analysis*. Hoboken, NJ: Wiley.

Fleiss, J. L. (1981). *Statistical methods for rates and proportions*. New York: Wiley.

Franc, A. (1989). Multiway arrays: Some algebraic remarks. In R. Coppi & S. Bolasco (Eds.), *Multiway data analysis* (pp. 19–29). Amsterdam: Elsevier.

Franc, A. (1992). *Étude algebrique des multitableaux: Apports de l'algèbre tensorielle* [An algebraic study of multi-way tables: Contributions of tensor algebra]. Unpublished doctoral dissertation, Université de Montpellier II, Montpellier, France.

Gabriel, K. R. (1971). The biplot graphic display with application to principal component analysis. *Biometrika, 58,* 453–467.

Gabriel, K. R. (1981). Biplot display of multivariate matrices for inspection of data and diagnosis. In V. Barnett (Ed.), *Interpreting multivariate data* (pp. 147–173). Chichester, UK: Wiley.

Gabriel, K. R., & Odoroff, C. L. (1990). Biplots in biomedical research. *Statistics in Medicine, 9,* 469–485.

Gabriel, K. R., & Zamir, S. (1979). Lower rank approximation of matrices by least squares with any choice of weights. *Technometrics, 21,* 489–498. (Correction, *22,* 136)

Gallagher, N. B., Wise, B. M., & Stewart, C. W. (1996). Application of multi-way principal components analysis to nuclear waste storage tank monitoring. *Computers & Chemical Engineering, 20, Supplement 1,* S739–S744.

Gauch Jr., H. G. (1992). *Statistical analysis of regional yield trials: AMMI analysis of factorial designs.* Amsterdam: Elsevier.

Geladi, P., Manley, M., & Lestander, T. (2003). Scatter plotting in multivariate data analysis. *Journal of Chemometrics, 17,* 503–511.

Gemperline, P. J., Miller, K. H., West, T. L., Weinstein, J. E., Hamilton, J. C., & Bray, J. T. (1992). Principal component analysis, trace elements, and blue crab shell disease. *Analytical Chemistry, 64,* A523–A531.

Gifi, A. (1990). *Nonlinear multivariate analysis.* Chichester, UK: Wiley.

Gilbert, D. A., Sutherland, M., & Kroonenberg, P. M. (2000). Exploring subject-related interactions in repeated measures data using three-mode principal components analysis. *Nursing Research, 49,* 57–61.

Gilbert, N. (1963). Non-additive combining of abilities. *Genetic Research Cambrige, 4,* 213–219.

Gnanadesikan, R. (1977). *Methods of statistical data analysis of multivariate obvservations.* New York: Wiley. (1997, 2nd ed.)

Gnanadesikan, R., & Kettenring, J. R. (1972). Robust estimates, residuals, and outliers with multiresponse data. *Biometrics, 28,* 81–124.

Goicoechea, H. C., Yu, S., Olivieri, A. C., & Campiglia, A. D. (2005). Four-way data coupled to parallel factor model applied to environmental analysis: Determination of 2,3,7,8-tetrachloro-dibenzo-para-dioxin in highly contaminated waters by solid-liquid extraction laser-excited time-resolved Shpol'skii spectroscopy. *Analytic Chemistry, 77,* 2608–2616.

Gollob, H. F. (1968a). A statistical model which combines features of factor analysis and analysis of variance. *Psychometrika, 33,* 73–115.

Gollob, H. F. (1968b). Confounding sources of variation in factor-analytic techniques. *Psychological Bulletin, 70,* 330–344.

Gollob, H. F. (1968c). Rejoinder to Tucker's "Comments on confounding sources of variation in factor-analytic techniques". *Psychological Bulletin, 70,* 355–360.

Gordon, A. D. (1981). *Classification.* London: Chapman & Hall.

Gower, J. C. (1977). The analysis of three-way grids. In P. Slater (Ed.), *Dimensions of intrapersonal space* (pp. 163–173). New York: Wiley.

Gower, J. C. (1984). Multidimensional scaling displays. In H. G. Law, C. W. Snyder Jr., J. A. Hattie, & R. P. McDonald (Eds.), *Research methods for multimode data analysis* (pp. 592–601). New York: Praeger.

Gower, J. C. (2006). Statistica data analytica est et aliter. *Statistica Neerlandica, 60,* 124–134.

Gower, J. C., & Dijksterhuis, G. B. (2004). *Procrustes problems*. New York: Oxford University Press.

Greenacre, M. J. (1984). *Theory and applications of correspondence analysis*. London: Academic Press.

Greenacre, M. J. (1993). Biplots in correspondence analysis. *Journal of Applied Statistics, 20*, 251–269.

Grung, B., & Manne, R. (1998). Missing values in principal component analysis. *Chemometrics and Intelligent Laboratory Systems, 42*, 125–139.

Guttman, L. (1954). A new approach to factor analysis: The radex. In P. F. Lazarsfeld (Ed.), *Mathematical thinking in the social sciences* (pp. 258–348). New York: Free Press.

Hair, J. F., Anderson, R. E., Tatham, R. L., & Black, W. H. (2006). *Multivariate data analysis* (6th ed.). Upper Saddle River, NJ: Prentice-Hall.

Hand, D. J., & Taylor, C. C. (1987). *Multivariate analysis of variance and repeated measures. A practical approach for behavioural scientists*. London: Chapman & Hall.

Hanson, R. J., & Lawson, C. L. (1974). *Solving least squares problems*. Englewood Cliff, NJ: Prentice-Hall.

Harman, H. H. (1976). *Modern factor analysis* (3rd ed.). Chicago, IL: University of Chicago Press.

Harris, C. W., & Kaiser, H. F. (1964). Oblique factor analytic solutions by orthogonal transformations. *Psychometrika, 29*, 347–362.

Harshman, R. A. (1970). Foundations of the PARAFAC procedure: Models and conditions for an "explanatory" multi-modal factor analysis. *UCLA Working Papers in Phonetics, 16*, 1–84.

Harshman, R. A. (1972a). Determination and proof of minimum uniqueness conditions for PARAFAC1. *UCLA Working Papers in Phonetics, 22*, 111–117.

Harshman, R. A. (1972b). PARAFAC2: Mathematical and technical notes. *UCLA Working Papers in Phonetics, 22*, 30–44.

Harshman, R. A. (1984a). "How can I know if it's 'real'?" A catalog of diagnostics for use with three-mode factor analysis and multidimensional scaling. In H. G. Law, C. W. Snyder Jr., J. A. Hattie, & R. P. McDonald (Eds.), *Research methods for multimode data analysis* (pp. 566–591). New York: Praeger.

Harshman, R. A. (1994a). Substituting statistical for physical decomposition: Are there applications for parallel factor analysis (PARAFAC) in non-destructive evaluation? In P. V. Malague (Ed.), *Advances in signal processing for nondestructive evaluation of materials* (pp. 469–483). Dordrecht: Kluwer.

Harshman, R. A. (2001). An index to formalism that generalizes the capabilities of matrix notation and algebra to n-way arrays. *Journal of Chemometrics, 15*, 689–714.

Harshman, R. A. (2004, July). *The problem and nature of degenerate solutions or decompositions of 3-way arrays*. Paper presented at Workshop on Tensor decompositions, American Institute of Mathematics, Palo Alto, CA . (*http://csmr.ca.sandia.gov/~tgkolda/tdw2004/Harshman - Talk.pdf*. (Accessed May 2007.))

Harshman, R. A. (2005a, August). *The Parafac model and its variants*. Paper presented at Workshop on Tensor decompositions and applications, CIRM, Luminy, France. (*http://publish.uwo.ca/ harshman/pftut05.pdf*. Accessed May 2007.)

Harshman, R. A. (2005b, August). *Multilinear generalization of the General Linear Model*. Paper presented at Workshop on Tensor decompositions and applications, CIRM, Luminy,

France.

Harshman, R. A., & DeSarbo, W. S. (1984b). An application of PARAFAC to a small sample problem, demonstrating preprocessing, orthogonality constraints, and split-half diagnostic techniques. In H. G. Law, C. W. Snyder Jr., J. A. Hattie, & R. P. McDonald (Eds.), *Research methods for multimode data analysis* (pp. 602–642). New York: Praeger.

Harshman, R. A., Hong, S., & Lundy, M. E. (2003). Shifted factor analysis - Part I: Models and properties. *Journal of Chemometrics, 17*, 363–378.

Harshman, R. A., Ladefoged, P., & Goldstein, L. (1977). Factor analysis of tongue shapes. *Journal of the Acoustical Society of America, 62*, 693–707.

Harshman, R. A., & Lundy, M. E. (1984c). The PARAFAC model for three-way factor analysis and multidimensional scaling. In H. G. Law, C. W. Snyder Jr., J. A. Hattie, & R. P. McDonald (Eds.), *Research methods for multimode data analysis* (pp. 122–215). New York: Praeger.

Harshman, R. A., & Lundy, M. E. (1984d). Data preprocessing and the extended PARAFAC model. In H. G. Law, C. W. Snyder Jr., J. A. Hattie, & R. P. McDonald (Eds.), *Research methods for multimode data analysis* (pp. 216–284). New York: Praeger.

Harshman, R. A., & Lundy, M. E. (1994b). PARAFAC: Parallel factor analysis. *Computational Statistics and Data Analysis, 18*, 39–72.

Harshman, R. A., & Lundy, M. E. (1996). Uniqueness proof for a family of models sharing features of Tucker's three-mode factor analysis and PARAFAC-CANDECOMP. *Psychometrika, 61*, 133–154.

Harshman, R. A., Lundy, M. E., & Kruskal, J. B. (1985, July). *Comparison of trilinear and quadrilinear methods: Strengths, weaknesses, and degeneracies.* Paper presented at the annual Meeting of the North American Classification Society. St. John's Newfoundland, Canada.

Hastie, T. J., & Tobshirani, R. J. (1990). *Generalized additive models.* London: Chapman & Hall.

Heimdal, H., Bro, R., Larsen, L. M., & Poll, L. (1997). Prediction of polyphenol oxidase activity in model solutions containing various combinations of chlorogenic acid, $(-)$-epicatechin, O_2, CO_2, temperature, and pH by multiway data analysis. *Journal of Agricultural and Food Chemistry, 45*, 2399–2406.

Heinen, T. (1996). *Latent class and discrete latent trait models : Similarities and differences.* Thousand Oaks, CA: Sage.

Heise, D. R. (1969). Some methodological issues in semantic differential research. *Psychological Bulletin, 72*, 406–422.

Heiser, W. J., & Kroonenberg, P. M. (1997). *Dimensionwise fitting in Parafac/Candecomp with missing data and constrained parameters* (Leiden Psychological Reports, PRM 97-01). Leiden, The Netherlands: Department of Psychology, Leiden University.

Hendrickson, A. E., & White, P. O. (1964). Promax: A quick method for rotation to orthogonal oblique structure. *British Journal of Statistical Psychology, 17*, 65–70.

Henrion, R., & Andersson, C. A. (1999). A new criterion for simple-structure transformations of core arrays in N-way principal components analysis. *Chemometrics and Intelligent Laboratory Systems, 47*, 189–204.

Herrmann, W. M., Röhmel, J., Streitberg, B., & Willmann, J. (1983). Example for applying the COMSTAT multimodal factor analysis algorithm to EEG data to describe variance sources. *Neuropsychobiology, 10*, 164–172.

Hitchcock, F. L. (1927a). Multiple invariants and generalized rank of a p-way matrix or tensor. *Journal of Mathematics and Physics*, 7, 39–79.

Hitchcock, F. L. (1927b). The expression of a tensor or a polyadic as a sum of products. *Journal of Mathematics and Physics*, 6, 164–189.

Ho, P., Silva, M. C. M., & Hogg, T. A. (2001). Multiple imputation and maximum likelihood principal component analysis of incomplete multivariate data from a study of the ageing of port. *Chemometrics and Intelligent Laboratory Systems*, 55, 1–11.

Hohn, M. E. (1979). Principal components analysis of three-way tables. *Journal of the International Association of Mathematical Geology*, 11, 611–626.

Holbrook, M. B. (1997). Three-dimensional stereographic visual displays in marketing and consumer research. *Academy of Marketing Science*, 11, 1–35. (*http://oxygen.vancouver.wsu.edu/amsrev/theory/holbrook11-97.html*. Accessed May 2007.)

Hopke, P. K. (2003). Recent developments in receptor modeling. *Journal of Chemometrics*, 17, 255–265.

Hopke, P. K., Paatero, P., Jia, H., Ross, R. T., & Harshman, R. A. (1998). Three-way (PARAFAC) factor analysis: examination and comparison of alternative computational methods as applied to ill-conditioned data. *Chemometrics and Intelligent Laboratory Systems*, 43, 25–42.

Horan, C. B. (1969). Multidimensional scaling: Combining observations when individuals have different perceptual structures. *Psychometrika*, 34, 139–165.

Hu, L. T., & Bentler, P. M. (1999). Cutoff criteria for fit indices in covariance structure analysis: Conventional versus new alternatives. *Structural Equation Modeling*, 6, 1–55.

Huber, P. J. (1987). Experiences with three-dimensional scatterplots. *Journal of the American Statistical Society*, 82, 448–453.

Huber, P. J. (1996). *Robust statistical procedures*. Philadelphia, PA: SIAM.

Hubert, M., Rousseeuw, P. J., & Vanden Branden, K. (2005). ROBPCA: A new approach to robust principal component analysis. *Technometrics*, 47, 64–79.

Hubert, M., Rousseeuw, P. J., & Verboven, S. (2002). A fast method for robust principal components with applications to chemometrics. *Chemometrics and Intelligent Laboratory Systems*, 60, 101–111.

Hunt, L. A., & Basford, K. E. (1999). Fitting a mixture model to three-mode three-way data with categorical and continuous variables. *Journal of Classification*, 16, 283–296.

Hunt, L. A., & Basford, K. E. (2001). Fitting a mixture model to three-mode three-way data with missing information. *Journal of Classification*, 18, 209–226.

Iacobucci, D. (1989). Modeling multivariate sequential dyadic interactions. *Social Networks*, 11, 315–362.

Janssen, J., Marcotorchino, F., & Proth, J. M. (Eds.). (1987). *Data analysis. The ins and outs of solving real problems*. New York: Plenum.

Jennrich, R. I. (1970). Orthogonal rotation algorithms. *Psychometrika*, 35, 229–235.

Jordan, C. (1874). Mémoire sur les formes bilinéaires. *Journal de Mathématiques Pures et Appliquées*, 19, 35–54.

Jöreskog, K. G. (1970). Estimation and testing of simplex models. *British Journal of Mathematical and Statistical Psychology*, 23, 121–145.

Jöreskog, K. G. (1971). Simultaneous factor analysis in several populations. *Psychometrika*, 36, 409–426.

Kaiser, H. F. (1958). The varimax criterion for analytic rotation in factor analysis. *Psychometrika, 23*, 187–200.

Kapteyn, A., Neudecker, H., & Wansbeek, T. (1986). An approach to *n*-mode components analysis. *Psychometrika, 51*, 269–275.

Kendall, M. G. (1957). *A course in multivariate analysis*. London, UK: Griffin.

Kettenring, J. R. (1971). Canonical analysis of several sets of variables. *Biometrika, 58*, 433–460.

Kettenring, J. R. (1983a). A case study in data analysis. *Proceedings of Symposia in Applied Mathematics, 28*, 105–139.

Kettenring, J. R. (1983b). Components of interaction in analysis of variance models with no replications. In P. K. Sen (Ed.), *Contributions to Statistics: Essays in Honour of Norman L. Johnson* (pp. 283–297). Amsterdam: North-Holland.

Kiers, H. A. L. (1988). Comparison of "Anglo-Saxon" and "French" three-mode methods. *Statistique et Analyse des Données, 13*, 14–32.

Kiers, H. A. L. (1991a). Hierarchical relations among three-way methods. *Psychometrika, 56*, 449–470.

Kiers, H. A. L. (1992a). TUCKALS core rotations and constrained TUCKALS modelling. *Statistica Applicata, 4*, 659–667.

Kiers, H. A. L. (1993a). An alternating least squares algorithm for PARAFAC2 and three-way DEDICOM. *Computational Statistics and Data Analysis, 16*, 103–118.

Kiers, H. A. L. (1993b). A comparison of techniques for finding components with simple structure. In C. M. Cuadras & C. R. Rao (Eds.), *Multivariate analysis: Future directions Vol. 2* (pp. 67–86). Amsterdam: Elsevier.

Kiers, H. A. L. (1997a). Three-mode orthomax rotation. *Pychometrika, 62*, 579–598.

Kiers, H. A. L. (1997b). Weighted least squares fitting using ordinary least squares algorithms. *Psychometrika, 62*, 251–266.

Kiers, H. A. L. (1998a). Recent developments in three-mode factor analysis: Constrained three-mode factor analysis and core rotations. In C. Hayashi, N. Ohsumi, K. Yajima, Y. Tanaka, H.-H. Bock, & Y. Baba (Eds.), *Data science, classification, and related methods* (pp. 563–574). Tokyo: Springer.

Kiers, H. A. L. (1998b). Three-way SIMPLIMAX for oblique rotation of the three-mode factor analysis core to simple structure. *Computational Statistics and Data Analysis, 28*, 307–324.

Kiers, H. A. L. (1998c). A three-step algorithm for CANDECOMP/PARAFAC analysis of large data sets with multicollinearity. *Journal of Chemometrics, 12*, 155–171.

Kiers, H. A. L. (1998d). Joint orthomax rotation of the core and component matrices resulting from three-mode principal components analysis. *Journal of Classification, 15*, 245–263.

Kiers, H. A. L. (1998e). An overview of three-way analysis and some recent developments. In A. Rizzi, M. Vichi, & H.-H. Bock (Eds.), *Advances in data science and classification* (pp. 593–602). Berlin: Springer.

Kiers, H. A. L. (2000a). Some procedures for displaying results from three-way methods. *Journal of Chemometrics, 14*, 151–170.

Kiers, H. A. L. (2000b). Towards a standardized notation and terminology in multiway analysis. *Journal of Chemometrics, 14*, 105–122.

Kiers, H. A. L. (2004a). Bootstrap confidence intervals for three-way methods. *Journal of Chemometrics, 18*, 22–36.

Kiers, H. A. L. (2004b). Clustering all three modes of three-mode data: Computational possibilities and problems. In J. Antoch (Ed.), *COMPSTAT, Proceedings in computational statistics* (pp. 303–313). Heidelberg, Germany: Springer.

Kiers, H. A. L. (2006). Properties of and algorithms for fitting three-way component models with off-set terms. *Psychometrika, 71*, 231–256.

Kiers, H. A. L., & Der Kinderen, A. (2003). A fast method for choosing the numbers of components in Tucker3 analysis. *British Journal of Mathematical and Statistical Psychology, 56*, 119—125.

Kiers, H. A. L., & Harshman, R. A. (1997d). Relating two proposed methods for speedup of algorithms for fitting two- and three-way principal component and related multilinear models. *Chemometrics and Intelligent Laboratory Systems, 36*, 31–40.

Kiers, H. A. L., & Krijnen, W. P. (1991b). An efficient algorithm for PARAFAC of three-way data with large numbers of observation units. *Psychometrika, 56*, 147–152.

Kiers, H. A. L., Kroonenberg, P. M., & Ten Berge, J. M. F. (1992b). An efficient algorithm for TUCKALS3 on data with large numbers of observation units. *Psychometrika, 57*, 415–422.

Kiers, H. A. L., & Smilde, A. K. (1998f). Constrained three-mode factor analysis as a tool for parameter estimation with second-order instrumental data. *Journal of Chemometrics, 12*, 125–147.

Kiers, H. A. L., & Tcn Berge, J. M. F. (1994a). Hierarchical relations between methods for simultaneous component analysis and a technique for rotation to a simple simultaneous structure. *British Journal of Mathematical and Statistical Psychology, 47*, 109–126.

Kiers, H. A. L., & Ten Berge, J. M. F. (1994b). The Harris-Kaiser independent cluster rotation as a method for rotation to simple component weights. *Psychometrika, 59*, 81–90.

Kiers, H. A. L., Ten Berge, J. M. F., & Bro, R. (1999). PARAFAC2 - Part I. A direct fitting algorithm for the PARAFAC2 model. *Journal of Chemometrics, 13*, 275–294.

Kiers, H. A. L., Ten Berge, J. M. F., & Rocci, R. (1997c). Uniqueness of three-mode factor models with sparse cores: The $3 \times 3 \times 3$ case. *Psychometrika, 62*, 349–374.

Kiers, H. A. L., & Van Mechelen, I. (2001). Three-way component analysis: Principles and illustrative application. *Psychological Methods, 6*, 84–110.

Klapper, D. (1998). *Die Analyse von Wettbewerbsbeziehungen mit Scannerdaten*. Heidelberg, Germany: Physica-Verlag.

Kojima, H. (1975). Inter-battery factor analysis of parents' and children's reports of parental behavior. *Japanese Psychological Bulletin, 17*, 33–48.

Kolda, T. G. (2001). Orthogonal tensor decompositions. *Siam Journal of Matrix Analysis and Applications, 23*, 243–255.

Kramer, L., & Gottman, J. M. (1992). Becoming a sibling: "With a little help from my friends". *Developmental Psychology, 28*, 685–699.

Krijnen, W. P. (1993). *The analysis of three-way arrays by constrained PARAFAC methods*. Leiden: DSWO Press.

Krijnen, W. P., & Kiers, H. A. L. (1995). An efficient algorithm for weighted PCA. *Computational Statistics, 10*, 299–306.

Krijnen, W. P., & Kroonenberg, P. M. (2000). *Degeneracy and Parafac* (Technical report). Groningen, The Netherlands: Department of Psychology, University of Groningen.

Krijnen, W. P., & Ten Berge, J. M. F. (1991). Contrastvrije oplossingen van het CANDECOMP/-PARAFAC-model [Contrast-free solutions of the CANDECOMP/PARAFAC-model]. *Kwantitatieve Methoden, 12*, 87–96.

Krijnen, W. P., & Ten Berge, J. M. F. (1992). A constrained PARAFAC method for positive manifold data. *Applied Psychological Measurement*, *16*, 295–305.

Kroonenberg, P. M. (1983a). Annotated bibliography of three-mode factor analysis. *British Journal of Mathematical and Statistical Psychology*, *36*, 81–113.

Kroonenberg, P. M. (1983c). *Three-mode principal component analysis: Theory and applications*. Leiden: DSWO Press. (Errata, 1989; available from the author)

Kroonenberg, P. M. (1985a). Three-mode principal components analysis of semantic differential data: The case of a triple personality. *Applied Psychological Measurement*, *9*, 83–94.

Kroonenberg, P. M. (1987a). Multivariate and longitudinal data on growing children. Introduction to the proposed solutions. In J. Janssen, F. Marcotorchino, & J. M. Proth (Eds.), *Data analysis. The ins and outs of solving real problems* (pp. 7–11). New York: Plenum.

Kroonenberg, P. M. (1987b). Multivariate and longitudinal data on growing children. Solutions using a three-mode principal component analysis and some comparison results with other approaches. In J. Janssen, F. Marcotorchino, & J. M. Proth (Eds.), *Data analysis. The ins and outs of solving real problems* (pp. 89–112). New York: Plenum.

Kroonenberg, P. M. (1989). Singular value decompositions of interactions in three-way contingency tables. In R. Coppi & S. Bolasco (Eds.), *Multiway data analysis* (pp. 169–184). Amsterdam: Elsevier.

Kroonenberg, P. M. (1994). The TUCKALS line: A suite of programs for three-way data analysis. *Computational Statistics and Data Analysis*, *18*, 73–96.

Kroonenberg, P. M. (1995c). *Introduction to biplots for G×E tables* (Research report No. 51). Brisbane, Australia: Centre for Statistics, University of Queensland.

Kroonenberg, P. M. (2005a). *Datadozen: Analyse en historie* [Data boxes: Analysis and history]. Leiden, The Netherlands: Universiteit Leiden (*https://openaccess.leidenuniv.nl/dspace/bitstream/1887/3494/3*. Accessed May 2007.)

Kroonenberg, P. M. (2005b). Three-mode component and scaling models. In B. S. Everitt & D. Howell (Eds.), *Encyclopedia of statistics in behavioral sciences* (pp. 2032–2044). Chichester, UK: Wiley.

Kroonenberg, P. M. (2005c). Model selection procedures in three-mode component models. In M. Vichi, P. Molinari, S. Mignani, & A. Montanari (Eds.), *New developments in classification and data analysis* (pp. 167–172). Springer: Berlin.

Kroonenberg, P. M. (2006b). Three-mode analysis. In N. Balakrishnan, C. B. Read, & B. Vidakovic (Eds.), *Encyclopedia of statistical sciences. Vol. 13* (2nd ed., pp. 8597–8602). Hoboken, NJ: Wiley.

Kroonenberg, P. M., & Anderson, C. J. (2006). Additive and multiplicative models for three-way contingency tables: Darroch (1974) revisited. In M. J. Greenacre & J. Blasius (Eds.), *Multiple correspondence analysis and related methods* (pp. 455–486). Boca Raton FL, USA: Chapman and Hall.

Kroonenberg, P. M., Basford, K. E., & Ebskamp, A. G. M. (1995a). Three-way cluster and component analysis of maize variety trials. *Euphytica*, *84*, 31–42.

Kroonenberg, P. M., Basford, K. E., & Gemperline, P. J. (2004). Grouping three-mode data with mixture methods: The case of the diseased blue crabs. *Journal of Chemometrics*, *18*, 508–518.

Kroonenberg, P. M., Basford, K. E., & Van Dam, M. (1995b). Classifying infants in the Strange Situation with three-way mixture method clustering. *British Journal of Psychology*, *86*, 397–418.

Kroonenberg, P. M., & De Leeuw, J. (1977). *TUCKALS2: A principal component analysis of three mode data.* (Research Bulletin No. RB 001-'77). Leiden, The Netherlands: Department of Data Theory, Leiden University.

Kroonenberg, P. M., & De Leeuw, J. (1980). Principal component analysis of three-mode data by means of alternating least squares algorithms. *Psychometrika, 45,* 69–97.

Kroonenberg, P. M., Dunn III, W. J., & Commandeur, J. J. F. (2003b). Consensus molecular alignment based on generalized Procrustes analysis. *Journal of Chemical Information and Compututer Science, 43,* 2025–2032.

Kroonenberg, P. M., & Heiser, W. J. (1998). Parallel factor analysis with constraints on the configurations: An overview. In C. Hayashi, N. Ohsumi, K. Yajima, Y. Tanaka, H.-H. Bock, & Y. Baba (Eds.), *Data science, classification, and related methods* (pp. 587–597). Tokyo: Springer.

Kroonenberg, P. M., & Kashima, Y. (1997). Rules in context. A three-mode principal component analysis of Mann et al.'s data on cross-cultural differences in respect for others. *Journal of Cross-Cultural Psychology, 28,* 463–480.

Kroonenberg, P. M., Lammers, C. J., & Stoop, I. (1985b). Three-mode principal component analysis of multivariate longitudinal organizational data. *Sociological Methods and Research, 14,* 99–136.

Kroonenberg, P. M., & Oort, F. J. (2003a). Three-mode analysis of multi-mode covariance matrices. *British Journal of Mathematical and Statistical Psychology, 56,* 305–336.

Kroonenberg, P. M., & Van der Kloot, W. A. (1987d, July). *Three-mode reduced-rank and redundancy analysis.* Paper presented at the 5th Meeting of the Psychometric Society, Twente, The Netherlands.

Kroonenberg, P. M., & Van der Voort, T. H. A. (1987e). Multiplicatieve decompositie van interacties bij oordelen over de werkelijkheidswaarde van televisiefilms [Multiplicative decomposition of interactions for judgments of realism of television films]. *Kwantitatieve Methoden, 8,* 117–144.

Kruskal, J. B. (1976). More factors than subjects, tests and treatments: An indeterminacy theorem for canonical decomposition and individual differences scaling. *Psychometrika, 41,* 281–293.

Kruskal, J. B. (1981). Multilinear models for data analysis. *Behaviormetrika, 10,* 1–20.

Kruskal, J. B. (1988). *Simple structure for three-way data: A new method intermediate between 3-mode factor analysis and PARAFAC-CANDECOMP.* Paper presented at the 53rd Annual Meeting of the Psychometric Society, Los Angeles.

Kruskal, J. B. (1989a). Rank, decomposition, and uniqueness for 3-way and N-way arrays. In R. Coppi & S. Bolasco (Eds.), *Multiway data analysis* (pp. 7–18). Amsterdam: Elsevier.

Kruskal, J. B., Harshman, R. A., & Lundy, M. E. (1989b). How 3-MFA data can cause degenerate PARAFAC solutions, among other relationships. In R. Coppi & S. Bolasco (Eds.), *Multiway data analysis* (pp. 115–122). Amsterdam: Elsevier.

Krzanowski, W. J. (2000). *Principles of multivariate analysis. A user's perspective* (revised ed.). Oxford, UK: Oxford University Press.

Kuppens, P., Van Mechelen, I., Smits, D. J. M., De Boeck, P., & Ceulemans, E. (2007). Individual differences in patterns of appraisal and anger experience. *Cognition & Emotion, 21,* 689–713.

Lammers, C. J. (1974). *Groei en ontwikkeling van de ziekenhuisorganisaties in Nederland.* [Growth and development of hospital organizations in the Netherlands] (Interim Report). Leiden, The Netherlands: Institute of Sociology, Leiden University.

Lastovicka, J. L. (1981). The extension of component analysis to four-mode matrices. *Psychometrika, 46*, 47–57.

Lavit, C. (1988). *Analyse conjointe de tableaux quantitatifs.* [Simultaneous analysis of several quantitative matrices]. Paris: Masson.

Lavit, C., Escoufier, Y., Sabatier, R., & Traissac, P. (1994). The ACT (STATIS method). *Computational Statistics and Data Analysis, 18*, 97–119.

Lavit, C., & Pernin, M.-O. (1987). Multivariate and longitudinal data on growing children: Solution using STATIS. In J. Janssen, F. Marcotorchino, & J. M. Proth (Eds.), *Data analysis. The ins and outs of solving real problems* (pp. 13–29). New York: Plenum.

Law, H. G., Snyder Jr., C. W., Hattie, J. A., & McDonald, R. P. (Eds.). (1984). *Research methods for multimode data analysis.* New York: Praeger.

Lebart, L., Morineau, A., & Warwick, K. M. (1984). *Multivariate descriptive statistical analysis.* New York: Wiley.

Lee, C.-H. (1988). *Multilinear analysis of fluorescence spectra of photosynthetic systems.* Unpublished Master's thesis, The Ohio State University, Columbus OH.

Lee, S.-Y., & Fong, W.-K. (1983). A scale invariant model for three-mode factor analysis. *British Journal of Mathematical and Statistical Psychology, 36*, 217–223.

Leenen, I., Van Mechelen, I., De Boeck, P., & Rosenberg, S. (1999). INDCLAS: A three-way hierarchical classes model. *Psychometrika, 60*, 9–24.

Legendre, P., & Legendre, L. (1998). *Numerical ecology* (2nd English ed.). Amsterdam: Elsevier.

Leibovici, D. (2000). *Multiway multidimensional analysis for pharmaco-EEG studies* (Tech. Rep. No. TR00DL2). University of Oxford, UK: FMRIB Centre . (*http://www.fmrib.ox.ac.uk/analysis/techrep/tr00dl2/tr00dl2.pdf.* Accessed May 2007.)

Leibovici, D. (2007). *Principal Tensor Analysis on k modes:PTA-k add-on R-package* (Tech. Rep. No. version 1.1-14). University of Nottingham, UK: Centre for Geospatial Science. (*http://c3s2i.free.fr; http://cran.r-project.org/doc/packages/PTAk.pdf.* Accessed May 2007.)

Leibovici, D., & Elmaache, H. (1997). A singular value decomposition of an element belonging to a tensor product of kappa separable hilbert spaces. *Comptes Rendus de l'Académie des Sciences Série I - Mathématique, 325*, 779–782.

Leibovici, D., & Sabatier, R. (1998). A singular value decomposition of a k-way array for a principal component analysis of multiway data, PTA-K. *Linear Algebra and its Applications, 269*, 307–329.

Leurgans, S. E., & Ross, R. T. (1992). Multilinear models: Applications in spectroscopy (with discussion). *Statistical Science, 7*, 289–319.

Leurgans, S. E., Ross, R. T., & Abel, R. B. (1993). A decomposition for 3-way arrays. *SIAM Journal on Matrix Analysis and Applications, 14*, 1064–1083.

Levin, J. (1965). Three-mode factor analysis. *Psychological Bulletin, 64*, 442–452.

L'Hermier des Plantes, H. (1976). *Structuration des tableaux à trois indices de la statistique: Théorie et application d'une méthode d'analyse conjointe.* [Statistical structuring of three-way tables: Theory and application of a conjoint analysis method]. Unpublished doctoral dissertation, University of Science and Technology of Languedoc, Montpellier, France.

Li, G., & Chen, Z. (1985). Projection-pursuit approach to robust dispersion matrices and principal components: Primary theory and Monte Carlo. *Journal of the American Statistical Association, 80*, 759–766.

Linting, M., Meulman, J. J., Groenen, P. J. F., & Van der Kooij, A. J. (2007). Stability of nonlinear principal components analysis: An empirical study using the balanced bootstrap. *Psychological Methods, 12*, 359–379.

Liu, X., & Sidiropoulos, N. D. (2000). PARAFAC methods for blind beamforming: Multilinear ALS performance and CRB. In *Proceedings 2000 IEEE international conference on acoustics, speech, and signal processing, ICASSP, Vol. 5* (pp. 3128–3131). Piscataway, NJ: IEEE.

Lohmöller, J.-B. (1978a). *Stabilität und Kontinuität in Längsschnittdaten, analysiert durch T- und trimodale Faktorenanalyse.* [Stability and continuity in longitudinal data, analyzed by T-factor analysis and three-mode factor analysis] (Internal report (Revision 1981)). München, Germany: Department of Education, University of the Federal Armed Forces.

Lohmöller, J.-B. (1978b, June). *How longitudinal factor stability, continuity, differentiation, and integration are portrayed into the core matrix of three-mode factor analysis.* Paper presented at the European Meeting on psychometrics and mathematical psychology, Uppsala, Sweden.

Lohmöller, J.-B. (1979). Die trimodale Faktorenanalyse von Tucker: Skalierungen, Rotationen, andere Modelle [Tucker's three-mode factor analysis: Scaling, rotations, other models]. *Archiv für die Psychologie, 131*, 137–166.

Lohmöller, J.-B. (1989). *Latent variable path modeling with partial least squares.* Heidelberg, Germany: Physica.

Lombardo, R., Carlier, A., & D'Ambra, L. (1996). Nonsymmetric correspondence analysis for three-way contingency tables. *Methodologica, 4*, 59–80.

Lorenzo-Seva, U., & Ten Berge, J. M. F. (2006). Tucker's congruence coefficient as a meaningful index of factor similarity. *Methodology, 2*, 57–64.

Louwerse, D. J., Smilde, A. K., & Kiers, H. A. L. (1999). Cross-validation of multiway component models. *Journal of Chemometrics, 13*, 491–510.

Lundy, M. E., Harshman, R. A., & Kruskal, J. B. (1989). A two-stage procedure incorporating good features of both trilinear and quadrilinear models. In R. Coppi & S. Bolasco (Eds.), *Multiway data analysis* (pp. 123–130). Amsterdam: Elsevier.

MacCallum, R. C. (1976). Transformation of a three-mode multidimensional scaling solution to INDSCAL form. *Psychometrika, 41*, 385–400.

Mandel, J. (1971). A new analysis of variance model for non-additive data. *Technometrics, 13*, 1–18.

Março, P. H., Levi, M. A. B., Scarminio, I. S., Poppi, R. J., & Trevisan, M. G. (2005). Exploratory analysis of simultaneous degradation of anthocyanins in the calyces of flowers of the *Hibiscus sabdariffa* species by PARAFAC model. *Analytical Sciences, 21*, 1523–1527.

Martínez-Montes, E., Valdés-Sosa, P. A., Miwakeichi, F., Goldman, R. I., & Cohen, M. S. (2004). Concurrent EEG/fMRI analysis by multiway partial least squares. *NeuroImage, 22*, 1023–1034.

Mayekawa, S. (1987). Maximum likelihood solution to the PARAFAC model. *Behaviormetrika, 21*, 45–63.

McDonald, R. P. (1967). Nonlinear factor analysis. *Psychometric Monographs, 15*.

McDonald, R. P. (1984). The invariant factors model for multimode data. In H. G. Law, C. W. Snyder Jr., J. A. Hattie, & R. P. McDonald (Eds.), *Research methods for multimode data analysis* (pp. 285–307). New York: Praeger.

McLachlan, G. J., & Basford, K. E. (1988). *Mixture models: Inference and applications to clustering.* New York: Marcel Dekker.

McLaren, C. C. (1996). Methods of data standardization used in pattern analysis and AMMI models for the analysis of international multi-environment variety trials. In M. Cooper & G. L. Hammer (Eds.), *Plant adaptation and crop improvement* (pp. 193–224). Oxford: CAB International.

Meredith, W., & Millsap, J. (1982). Canonical analysis of longitudinal and repeated measures data with stationary weights. *Psychometrika, 47,* 47–67.

Meulman, J. J., Heiser, W. J., & Carroll, J. D. (1986). *PREFMAP-3 user's guide* (Technical report). Leiden, The Netherlands: Department of Data Theory, Leiden University. (*http://www.netlib.org/mds/prefmap3a.htm* (prefmap3b.htm).)

Meulman, J. J., Heiser, W. J., & SPSS Inc. (2000). *Categories 10.0.* Chicago, IL: SPSS.

Meulman, J. J., Van der Kooij, A. J., & Heiser, W. J. (2004). Principal component analysis with nonlinear optimal scaling transformations for ordinal and nominal data. In D. Kaplan (Ed.), *The Sage handbook of quantitative methodology for the social sciences* (pp. 49–70). Thousand Oaks, CA: Sage.

Meyners, M., Kunert, J., & Qannari, E. M. (2000). Comparing generalized Procrustes analysis and STATIS. *Food Quality and Preference, 11,* 77–83.

Mitchell, B. C., & Burdick, D. S. (1994). Slowly converging PARAFAC sequences: Swamps and two-factor degeneracies. *Journal of Chemometrics, 8,* 155–168.

Möcks, J. (1988a). Decomposing event-related potentials: A new topographic components model. *Biological Psychology, 26,* 199–215.

Möcks, J. (1988b). Topographic components model for event-related potentials and some biophysical considerations. *IEEE Transactions on Biomedical Engineering, 35,* 482–484.

Murakami, T. (1979). Inshi henka no kijutsu to jun3so inshi bunseki [Description of factor change and quasi three-mode factor analysis]. *Bulletin of the Faculty of Education, Nagoya University, 26,* 1–16.

Murakami, T. (1981). Saisho jijoho ni yoru jun3so inshi bunseki no algorithm [An algorithm of quasi three-mode factor analysis by the least squares methods]. *Bulletin of the Faculty of Education, Nagoya University, 28,* 39–59.

Murakami, T. (1983). Quasi three-mode principal component analysis - A method for assessing the factor change. *Behaviormetrika, 14,* 27–48.

Murakami, T. (1998a). Tucker2 as a second-order principal component analysis. In C. Hayashi, N. Ohsumi, K. Yajima, Y. Tanaka, H.-H. Bock, & Y. Baba (Eds.), *Data science, classification, and related methods* (pp. 575–586). Tokyo: Springer.

Murakami, T., & Kroonenberg, P. M. (2001). *Individual differences in semantic differential studies and their analysis by three-way models.* (Technical report). Leiden, The Netherlands: Department of Education, Leiden Univeristy.

Murakami, T., & Kroonenberg, P. M. (2003). Individual differences in semantic differential studies and their analysis by three-mode models. *Multivariate Behavioral Research, 38,* 87–96.

Murakami, T., Ten Berge, J. M. F., & Kiers, H. A. L. (1998b). A case of extreme simplicity of the core matrix in three-mode principal components analysis. *Psychometrika, 63,* 255–261.

Murphy, K. R., Ruiz, G. M., Dunsmuir, W. T. M., & Waite, T. D. (2006). Optimized parameters for fluorescence-based verification of ballast water exchange by ships. *Environmental*

Science & Technology, *40*, 2357–2362.

Nakamura, J., & Sinclair, J. (1995). The world of *woman* in the bank of english: Internal criteria for the classification in corpora. *Literary and Linguistic Computing*, *10*, 99-110.

Nannerup, L. D., Jakobsen, M., Van den Berg, F., Jensen, J. S., Møller, J. K. S., & Bertelsen, G. (2004). Optimizing colour quality of modified atmosphere packed sliced meat products by control of critical packaging parameters. *Meat Science*, *68*, 577–585.

NICHD Early Child Care Network. (1994). Child care and child development: The NICHD study of early child care. In S. L. Friedman & H. C. Haywood (Eds.), *Developmental follow-up: Concepts, domains and methods* (pp. 377–396). New York: Academic Press.

Nishisato, S. (1980). *Dual scaling and its applications*. Toronto, Canada: University of Toronto Press.

Nishisato, S. (1994). *Elements of dual scaling: An introduction to practical data analysis*. Hillsdale, NJ: Erlbaum.

Nomikos, P., & MacGregor, J. F. (1994). Monitoring batch processes using multiway principal component analysis. *AIChE Journal*, *40*, 1361–1375.

Noy-Meir, I. (1973). Data transformations in ecological ordination I. Some advantages of non-centring. *Journal of Ecology*, *61*, 329–341.

Noy-Meir, I., Walker, D., & Williams, W. T. (1975). Data transformations in ecological ordination: On the meaning of data standardization. *Journal of Ecology*, *63*, 779–800.

Oldenburger, R. (1934). Composition and rank of n-way matrices and multilinear forms. *The Annals of Mathematics, 2nd Series*, *35*, 622–653.

Oort, F. J. (1999). Stochastic three-mode models for mean and covariance structures. *British Journal of Mathematical and Statistical Psychology*, *52*, 243–272.

Oort, F. J. (2001). Three-mode models for multivariate longitudinal data. *British Journal of Mathematical and Statistical Psychology*, *54*, 49–78.

Osgood, C. E., & Luria, Z. (1954). A blind analysis of a case of multiple personality. *Journal of Abnormal and Social Psychology*, *49*, 579–791.

Osgood, C. E., Luria, Z., & Smith, S. W. (1976). II. A blind analysis of another case of multiple personality using the semantic differential technique. *Journal of Abnormal Psychology*, *85*, 256–286.

Paatero, P. (1999). The multilinear engine - a table-driven, least squares program for solving multilinear problems, including the *n*-way parallel factor analysis model. *Journal of Computational and Graphical Statistics*, *8*, 854–888.

Paatero, P. (2000). Construction and analysis of degenerate PARAFAC models. *Journal of Chemometrics*, *14*, 285–299.

Paatero, P., & Hopke, P. K. (2002). Utilizing wind direction and wind speed as independent variables in multilinear receptor modeling studies. *Chemometrics and Intelligent Laboratory Systems*, *60*, 25–41.

Pravdova, V., Estienne, F., Walczak, B., & Massart, D. L. (2001). A robust version of the Tucker3 model. *Chemometrics and Intelligent Laboratory Systems*, *59*, 75–88.

Ramsay, J. O., & Silverman, B. W. (1997). *Fuctional data analysis*. New York: Springer.

Rao, C. R. (1964). The use and interpretation of principal component analysis in applied research. *Sankyā A*, *26*, 329–358.

Raymond, O., Fiasson, J. L., & Jay, M. (2000). Synthetic taxonomy of Rosa races using ACT-STATIS. *Zeitschrift für Naturforschung*, *55*, 399–409.

Riu, J., & Bro, R. (2003). Jack-knife technique for outlier detection and estimation of standard errors in PARAFAC models. *Chemometrics and Intelligent Laboratory Systems*, *65*,

35-69.

Rocci, R. (1992). Three-mode factor analysis with binary core and orthonormality constraints. *Journal of the Italian Statistical Society, 1,* 413–422.

Rocci, R., & Vichi, M. (2003a, July). *Simultaneous component and cluster analysis: The between and within approaches.* Paper presented at the International Meeting of the Psychometric Society, Chia Laguna, Sardinia, Italy.

Rocci, R., & Vichi, M. (2003b, September). *Three-mode clustering of a three-way data set.* Paper presented at the Meeting of the Classification and Data Analysis Group (CLADAG) of the Italian Statistical Society, Bologna, Italy.

Röder, I. (2000). *Stress in children with asthma. Coping and social support in the school context.* Unpublished doctoral dissertation, Department of Education, Leiden University, Leiden, The Netherlands.

Röhmel, J., Streitberg, B., & Herrmann, W. (1983). The COMSTAT algorithm for multimodal factor analysis: An improvement of Tucker's three-mode factor analysis method. *Neuropsychobiology, 10,* 157–163.

Rousseeuw, P. J. (1984). Least median of squares regression. *Journal of the American Statistical Association, 79,* 871–880.

Rousseeuw, P. J., Debruyne, M., Engelen, S., & Hubert, M. (2006). Robustness and outlier detection in chemometrics. *Critical Reviews in Analytical Chemistry, 36,* 221–242.

Rousseeuw, P. J., & Van Driessen, K. (1999). A fast algorithm for the minimum covariance determinant estimator. *Technometrics, 41,* 212–223.

Rousseeuw, P. J., & Van Zomeren, B. C. (2000). Unmasking multivariate outliers and leverage points. *Journal of the American Statistical Association, 85,* 633–651.

Rowe, H. A. H. (1979). Three-mode factor analysis: Problems of interpretation and possible solutions. *Australian Psychologist, 14,* 222–223.

Sampson, S. F. (1968). *A novitiate in a period of change: An experimental and case study of social relationships.* Unpublished doctoral dissertation, Department of Sociology, Cornell University, Ithaca, NY.

Sands, R., & Young, F. W. (1980). Component models for three-way data: ALSCOMP3, an alternating least squares algorithm with optimal scaling features. *Psychometrika, 45,* 39–67.

Sato, M., & Sato, Y. (1994). On a multicriteria fuzzy clustering method for 3-way data. *International Journal of Uncertainty, Fuzziness and Knowledge-Based Systems, 2,* 127–142.

Schaefer, E. S. (1965). Children's reports of parental behavior: An inventory. *Child Development, 36,* 413–424.

Schafer, J. L. (1999). *NORM: Multiple imputation of incomplete multivariate data under a normal model* (Tech. Rep.). University Park, PA: Department of Statistics, Pennsylvania State University. (available *http://www.stat.psu.edu/~jls/misoftwa.html.* Accessed May 2007.)

Schafer, J. L., & Graham, J. W. (2002). Missing data: Our view of the state of the art. *Psychological Methods, 7,* 147–177.

Schepers, J., & Van Mechelen, I. (2004, March). *Three-mode partitioning: Model and algorithm.* Paper presented at the annual conference of the German Classification Society (GfKl), Dortmund, Germany. (*www.stat.ucl.ac.be/ISpub/tr/2004/TR0480.pdf.*)

Schepers, J., Van Mechelen, I., & Ceulemans, E. (2006). Three-mode partitioning. *Computational Statistics & Data Analysis, 51,* 1623–1642.

Schönemann, P. H. (1972). An algebraic solution for a class of subjective metrics models. *Psychometrika, 37*, 441–451.

Sempé, M. (1987). Multivariate and longitudinal data on growing children. Presentation of the French auxiological survey. In J. Janssen, F. Marcotorchino, & J. M. Proth (Eds.), *Data analysis. The ins and outs of solving real problems* (pp. 3–6). New York: Plenum.

Seyedsadr, M., & Cornelius, P. L. (1992). Shifted multiplicative models for nonadditive two-way tables. *Communications in Statistics - Simulation and Computation, 21*, 807–832.

Sidiropoulos, N. D., & Bro, R. (2000a). On the uniqueness of multilinear decomposition of *N*-way arrays. *Journal of Chemometrics, 14*, 229–239.

Sidiropoulos, N. D., & Bro, R. (2000d). On communication diversity for blind identifiability and the uniqueness of low-rank decomposition of N-way arrays. In *Proceedings. 2000 IEEE international conference on acoustics, speech, and signal processing, ICASSP. vol. 5* (pp. 2449–2452). Piscataway, NJ: IEEE.

Sidiropoulos, N. D., Bro, R., & Giannakis, G. B. (2000b). Parallel factor analysis in sensor array processing. *IEEE Transactions on Signal Processing, 48*, 2377–2388.

Sidiropoulos, N. D., Giannakis, G. B., & Bro, R. (2000c). Blind PARAFAC receivers for DS-CDMA systems. *IEEE Transactions on Signal Processing, 48*, 810–823.

Sidiropoulos, N. D., & Liu, X. (2001). Identifiability results for blind beamforming in incoherent multipath with small delay spread. *IEEE Transactions on Signal Processing, 49*, 228–236.

Skočaj, D., Bischof, H., & Leonardis, A. (2002). A robust PCA algorithm for building representations from panoramic images. In A. Heyden, G. Sparr, M. Nielsen, & P. Johansen (Eds.), *Computer vision - ECCV 2002: 7th European Conference on Computer Vision, Copenhagen, Denmark, May 28-31, 2002. Proceedings, Part IV* (pp. 171–178). Berlin: Springer.

Smilde, A. K., Geladi, P., & Bro, R. (2004). *Multi-way analysis in chemistry*. Chichester, UK: Wiley.

Smilde, A. K., Westerhuis, J. A., & Boqué, R. (2000). Multiway multiblock component and covariates regression models. *Journal of Chemometrics, 14*, 301–331.

Snyder, F. W., & Wiggins, N. (1970). Affective meaning systems: A multivariate approach. *Multivariate Behavioral Research, 5*, 453–468.

Spanjer, M. C. (1984). *Substituent interaction effects and mathematical-statistical description of retention in liquid chromatography*. Unpublished doctoral dissertation, University of Utrecht, Utrecht, The Netherlands.

Spanjer, M. C., De Ligny, C. L., Van Houwelingen, H. C., & Weesie, J. M. (1985). Simultaneous description of the influence of solvent, reaction type, and substituent on equilibrium constants by means of three-mode factor analysis. *Journal of the Chemical Society, Perkin Transactions II, 9*, 1401–1411.

SPSS Inc. (2000). *SPSS 10.0. Base manual*. Chicago, IL: SPSS.

Stanimirova, I., & Simeonov, V. (2005). Modeling of environmental four-way data from air quality control. *Chemometrics and Intelligent Laboratory Systems, 77*, 115–121.

Stanimirova, I., Walczak, B., Massart, D. L., & Simeonov, V. (2004). A comparison between two robust PCA algorithms. *Chemometrics and Intelligent Laboratory Systems, 71*, 83–95.

Stegeman, A. (2006). Degeneracy in Candecomp/Parafac explained for $p \times p \times 2$ arrays of rank $p + 1$ or higher. *Psychometrika, 71*, 483–501.

Takeuchi, H., Kroonenberg, P. M., Taya, H., & Miyano, H. (1986). An analysis of Japanese language on thermal sensation. *Mathematical Linguistics*, *15*, 201–209.

Ten Berge, J. M. F. (1986). Some relationships between descriptive comparisons of components from different studies. *Multivariate Behavioral Research*, *21*, 29–40.

Ten Berge, J. M. F. (1989). Convergence of PARAFAC preprocessing procedures and the Deming-Stephan method of iterative proportional fitting. In R. Coppi & S. Bolasco (Eds.), *Multiway data analysis* (pp. 53–63). Amsterdam: Elsevier.

Ten Berge, J. M. F. (1991). Kruskal's polynomial for $2\times2\times2$ arrays and a generalization to $2\times n\times n$ arrays. *Psychometrika*, *56*, 631–636.

Ten Berge, J. M. F. (2004). Simplicity and typical rank of three-way arrays, with applications to Tucker-3 analysis with simple cores. *Journal of Chemometrics*, *18*, 17–21.

Ten Berge, J. M. F., De Leeuw, J., & Kroonenberg, P. M. (1987). Some additional results on principal components analysis of three-mode data by means of alternating least squares algorithms. *Psychometrika*, *52*, 183–191.

Ten Berge, J. M. F., & Kiers, H. A. L. (1996a). Optimality criteria for principal component analysis and generalizations. *British Journal of Mathematical and Statistical Psychology*, *49*, 335–345.

Ten Berge, J. M. F., & Kiers, H. A. L. (1996b). Some uniqueness results for PARAFAC2. *Psychometrika*, *61*, 123–132.

Ten Berge, J. M. F., & Kiers, H. A. L. (1997). Are all varieties of PCA the same? A reply to Cadima and Joliffe. *British Journal of Mathematical and Statistical Psychology*, *50*, 367–368.

Ten Berge, J. M. F., & Kiers, H. A. L. (1999). Simplicity of core arrays in three-way principal component analysis and the typical rank of $p \times q \times 2$ arrays. *Linear Algebra and Its Applications*, *294*, 169–179.

Thioulouse, J., Chessel, D., Dolédec, S., & Olivier, J. M. (1996). ADE-4: A multivariate analysis and graphical display software. *Statistics and Computing*, *7*, 75–83.

Timmerman, M. E. (2001). *Component analysis of multisubject multivariate longitudinal data.* Unpublished doctoral dissertation, Department of Psychology, University of Groningen, Groningen, The Netherlands. (*http://irs.ub.rug.nl/ppn/217649602.* Accessed May 2007.)

Timmerman, M. E., & Kiers, H. A. L. (2000). Three-mode principal components analysis: Choosing the numbers of components and sensitivity to local optima. *British Journal of Mathematical and Statistical Psychology*, *53*, 1–16.

Timmerman, M. E., & Kiers, H. A. L. (2002). Three-way component analysis with smoothness constraints. *Computational Statistics & Data Analysis*, *40*, 447–470.

Timmerman, M. E., & Kiers, H. A. L. (2003). Four Simultaneous Component Models of multivariate time series from more than one subject to model intraindividual and interindividual differences. *Psychometrika*, *86*, 105–122.

Tipping, M. E., & Bishop, C. M. (1999). Probabilistic principal component analysis. *Journal of the Royal Statistical Society, Series B*, *61*, 611–622.

Tomasi, G., & Bro, R. (2005). PARAFAC and missing values. *Chemometrics and Intelligent Laboratory Systems*, *75*, 163–180.

Tomasi, G., & Bro, R. (2006). A comparison of algorithms for fitting the PARAFAC model. *Computational Statistics & Data Analysis*, *50*, 1700–1734.

Torgerson, W. S. (1958). *Theory and methods of scaling.* New York: Wiley.

Tucker, L. R. (1951). *A method for synthesis of factor analysis studies* (Personell Research Section Report No. 984). Washington, D. C.: Department of the Army.

Tucker, L. R. (1958). An inter-battery method of factor analysis. *Psychometrika, 23*, 111–136.

Tucker, L. R. (1960). Intra-individual and inter-individual multidimensionality. In H. Gulliksen & S. Messick (Eds.), *Psychometric scaling: Theory and applications* (pp. 155–167). New York: Wiley.

Tucker, L. R. (1963). Implications of factor analysis of three-way matrices for measurement of change. In C. W. Harris (Ed.), *Problems in measuring change* (pp. 122–137). Madison WI: University of Wisconsin Press.

Tucker, L. R. (1964). The extension of factor analysis to three-dimensional matrices. In H. Gulliksen & N. Frederiksen (Eds.), *Contributions to mathematical psychology* (pp. 110–127). New York: Holt, Rinehart and Winston.

Tucker, L. R. (1966). Some mathematical notes on three-mode factor analysis. *Psychometrika, 31*, 279–311.

Tucker, L. R. (1968). Comments on "Confounding sources of variation in factor-analytic techniques". *Psychological Bulletin, 70*, 345–354.

Tucker, L. R. (1972). Relations between multidimensional scaling and three-mode factor analysis. *Psychometrika, 37*, 3–27.

Vallejo Arboleda, A., Vicente Villardon, J. L., & Galindo Villardon, M. P. (2007). Canonical STATIS: Biplot analysis of multi-table group structured data based on STATIS-ACT methodology. *Computational Statistics & Data Analysis, 51*, 4193 4205.

Van Buuren, S., Brand, J. P. L., Groothuis-Oudshoorn, C. G. M., & Rubin, D. B. (2006). Fully conditional specification in multivariate imputation. *Journal of Statistical Computation and Simulation, 76*, 1049–1064.

Van Dam, M. (1993). *Secundaire analyse van de Strange Situation* [Secondary analyses of the Strange Situation]. Unpublished doctoral dissertation, Department of Education, Leiden University, Leiden, The Netherlands.

Van de Geer, J. P. (1974). Toepassing van drieweg-analyse voor de analyse van multiple tijdreeksen. [Application of three-mode analysis for the analysis of multiple time series.]. In *Groei en ontwikkeling van de ziekenhuisorganisaties in Nederland.* Leiden, The Netherlands: Institute of Sociology, Leiden University.

Van der Burg, E., & Dijksterhuis, G. (1989). Nonlinear canonical correlation analysis of multiway data. In R. Coppi & S. Bolasco (Eds.), *Multiway data analysis* (pp. 245–255). Amsterdam: Elsevier.

Van der Kloot, W. A., & Kroonenberg, P. M. (1982). Group and individual implicit theories of personality: An application of three-mode principal component analysis. *Multivariate Behavioral Research, 17*, 471–491.

Van der Kloot, W. A., & Kroonenberg, P. M. (1985a). External analysis with three-mode principal component models. *Psychometrika, 50*, 479–494.

Van der Kloot, W. A., Kroonenberg, P. M., & Bakker, D. (1985b). Implicit theories of personality: Further evidence of extreme response style. *Multivariate Behavioral Research, 20*, 369–387.

Van Eeuwijk, F. A., & Kroonenberg, P. M. (1995). The simultaneous analysis of genotype by environment interaction for a number [of] traits using [a] three-way multiplicative model [of]. *Biuletyn Oceny Odmian/Cultivar Testing Bulletin, 26/27*, 83–96.

Van Houwelingen, J. C. (1983). Principal components of large matrices with missing elements. In P. Mandl & M. Husková (Eds.), *Proceedings of the Third Prague symposium on asymptotic statistics, Charles University* (pp. 295–302). Amsterdam: North Holland.

Van IJzendoorn, M. H., & Kroonenberg, P. M. (1990). Cross-cultural consistency of coding the strange situation. *Infant Behavior & Development, 13,* 469–485.

Vansteelandt, K., & Van Mechelen, I. (1998). Individual differences in situation-behaviour profiles: A triple typology model. *Journal of Personality and Social Psychology, 75,* 751–765.

Vansteelandt, K., & Van Mechelen, I. (2006). Individual differences in anger and sadness: In pursuit of active situational features and psychological processes. *Journal of Personality, 74,* 871–909.

Vega-Montoto, L., Gu, H., & Wentzell, P. D. (2005a). Mathematical improvements to maximum likelihood parallel factor analysis: Theory and simulations. *Journal of Chemometrics, 19,* 216–235.

Vega-Montoto, L., & Wentzell, P. D. (2003). Maximum likelihood parallel factor analysis (MLPARAFAC). *Journal of Chemometrics, 17,* 237–253.

Vega-Montoto, L., & Wentzell, P. D. (2005b). Mathematical improvements to maximum likelihood parallel factor analysis: Experimental studies. *Journal of Chemometrics, 19,* 236–252.

Veldscholte, C., Kroonenberg, P. M., & Antonides, G. (1998). Converging perceptions of economic activities between East and West: A three-mode principal component analysis. *Journal of Economic Psychology, 19,* 321–351.

Verbeke, G., & Molenberghs, G. (2000). *Linear mixed models for longitudinal data.* New York: Springer.

Verboven, S., & Hubert, M. (2005). LIBRA: A Matlab library for robust analysis. *Chemometrics and Intelligent Laboratory Systems, 75,* 127–136.

Vermunt, J. K. (2007). A hierarchical mixture model for clustering three-way data sets. *Computational Statistics & Data Analysis, 51,* 5368–5376.

Vermunt, J. K., & Magidson, J. (2005). *Latent GOLD 4.0 User's Guide* (Tech. Rep.). Belmont, MA: Statistical Innovations.

Vichi, M., Rocci, R., & Kiers, H. A. L. (2007). Simultaneous component and clustering models for three-way data: Within and between approaches. *Journal of Classification, 24,* 71–98.

Visser, R. A. (1985). *Analysis of longitudinal data in behavioural and social research: An expository survey.* Leiden, The Netherlands: DSWO.

Vivien, M., & Sabatier, R. (2004). A generalization of STATIS-ACT strategy: DO-ACT for two multiblocks tables. *Computational Statistics & Data Analysis, 46,* 155–171.

Von Hippel, P. T. (2007, February). *Filling holes in your data: Imputation and its discontents* (Tech. Rep.). Columbus, OH: Department of Sociology, Ohio State University. (*http://www.sociology.ohio-state.edu/ptv/Brownbags/MI/MI.ppt*. Accessed May 2007.)

Vorobyov, S. A., Rong, Y., Sidiropoulos, N. D., & Gershman, A. B. (2005). Robust iterative fitting of multilinear models. *IEEE Transactions on Signal Processing, 53,* 2678—2689.

Wansbeek, T., & Verhees, J. (1989). Models for multidimensional matrices in econometrics and psychometrics. In R. Coppi & S. Bolasco (Eds.), *Multiway data analysis* (pp. 543–552). Amsterdam: Elsevier.

Wansbeek, T., & Verhees, J. (1990). The algebra of multimode factor analysis. *Linear Algebra and Its Applications, 127,* 631–639.

Ward, J. H. (1963). Hierarchical grouping to optimize an objective function. *Journal of the American Statistical Association, 58,* 236–244.

Weesie, J., & Van Houwelingen, H. (1983). *GEPCAM users' manual: Generalized principal components analysis with missing values*. (Technical report). Utrecht, The Netherlands: Institute of Mathematical Statistics, University of Utrecht.

Wentzell, P. D., Andrews, D. T., Hamilton, D. C., Faber, K., & Kowalski, B. R. (1997). Maximum likelihood principal component analysis. *Journal of Chemometrics*, *11*, 339–366.

Whittle, P. (1952). On principal components and least square methods of factor analysis. *Skandinavisk Aktuarietidskrift*, *35*, 223–239.

Wiggins, N., & Fishbein, M. (1969). Dimensions of semantic space: A problem of individual differences. In J. R. Snider & C. E. Osgood (Eds.), *The semantic differential technique: A book of readings* (pp. 183–193). Chicago: Aldine Press.

Wolfe, J. H. (1970). Pattern clustering by multivariate mixture analysis. *Multivariate Behavioral Research*, *5*, 329–350.

Young, F. W. (1981). Quantitative analysis of qualitative data. *Psychometrika*, *46*, 357–388.

Young, F. W. (1984). The general Euclidean model. In H. G. Law, C. W. Snyder Jr., J. A. Hattie, & R. P. McDonald (Eds.), *Research methods for multimode data analysis* (pp. 440–469). New York: Praeger.

Young, F. W., & Hamer, R. M. (1987). *Multidimensional scaling: History, theory, and applications*. Hillsdale, NJ: Lawrence Erlbaum. (Reprinted 1994.)

Young, G. (1940). Maximum likelihood estimation and factor analysis. *Psychometrika*, *6*, 49–53.

Zijlstra, B. J. H., & Kiers, H. A. L. (2002). Degenerate solutions obtained from several variants of factor analysis. *Journal of Chemometrics*, *16*, 596–605.

GLOSSARY

This glossary primarily contains concepts directly relevant to multiway matters in the present context. No attempt has been to include all data-analytic, statistical and mathematical terms discussed in this book.[1]

All-components plot. A plot in which the values of all components of a single mode are plotted against their sequence numbers or labels. Especially useful for ordered modes containing time series or spectra. Also called a *one-way plot*. In contrast with the usage in this book, other authors sometimes call this type of plot is a *line plot*, especially when only one component or variable is plotted.

Alternating least squares algorithms. Given that the parameters to be estimated can be split into separate blocks, alternating least squares algorithms minimize a least squares loss function by successively and iteratively minimizing the function for each block conditional on the current values of the other blocks.

Array. Multiway analogue of a vector and a matrix. A vector is a single-subscripted one-way array, a matrix is a double-subscripted two-way array. A multiway array has three or more subscripts.

---

[1] The insightful comments and suggestions by Richard Harshman have been invaluable in constructing this glossary.

Aspect ratio. The aspect ratio of an image is its displayed width divided by its height. In component plots it is important that this ratio is 1, for distances and inner products to be correctly displayed.

Asymmetric scaling. As used in optimal scaling or biplot construction, the coordinates of the row and column markers in biplots are asymmetrically scaled when one of the marker sets is in normalized coordinates whereas the other is in principal coordinates. In principle, but not in practice, other types of asymmetric scalings of the coordinates can occur. See also **Symmetric scaling**.

Average subject's profile. The profile of the subject who has an average score on all variables at all conditions.

Basic form. In its basic form a multiway model has unit length components and the size of the modeled data is contained in a set of special parameters, such as the core array in the Tucker models and the weights g_{sss} in the Parafac model. In addition in their basic forms two-mode PCA and the Tucker models have orthogonal component matrices.

Biadditive model. A model for two-way data that contains both additive terms for the rows and columns and multiplicative terms for their interaction.

Biplot. A biplot of a matrix of I objects by J variables is used to display the systematic patterns between rows, columns, and between rows and columns. The prefix *bi* refers to the simultaneous display of *both* rows and columns of the table, not to the dimensionality of the plots. The *singular value decomposition* is used to derive the coordinates on the dimensions.

Bipolar scale. Any scale in which the middle represents a neutral response (or the absence of the property represented), while the ends represent high levels but opposite aspects of the property. Such scales typically have antonyms as anchors; for instance good versus bad, and when numeric labels represent the strength of the rating they run from negative at one end via zero in the middle to positive at the other end. See also **Monopolar scale**.

Boolean algebra. Boolean algebra is the calculus for 0-1 data used when the numbers are interpreted as "absent" versus "present" rather than amounts. Therefore, while 0+0=0 and 0+1=1, we have 1+1=1. The following operations on 0-1 data are defined: the union and intersection of the two elements and the negation of a single one. This calculus is used, for example, in hierarchical class models.

Bootstrap. A bootstrap analysis is a nonparametric procedure used to estimate what would happen if new samples were taken and the analysis repeated; frequently, it is used to compute standard errors for parameters in a model. Starting from the assumption that the sample distribution is the best available estimate of the population distribution, repeated samples with replacement are taken from the sample distribution. The desired model is calculated for each of these samples, so that variations in their parameter estimates can be used to estimate confidence intervals.

Centering. Subtraction of a constant term, mostly the mean, from raw data. However, sometimes a neutral point on a scale is used. The most common kinds of centering are: (1) *fiber centering*: subtracting the mean resulting from averaging across a single mode (i.e., the mean of a single-subscripted subarray) and (2) *slice centering*: subtracting the mean resulting from simultaneously averaging across two modes (i.e. the means of a double-subscripted subarray); (3) *double centering*: subtracting the two means resulting from two fiber centerings (in any order). Generally fiber centering is preferred.

Column-isometric. A (bi)plot[2] is column-isometric if the columns are in principal coordinates and the rows in normalized or standard coordinates.

Combination-mode (ij)**.** Cartesian product of two (observational) modes i and j; "i outer loop, j inner loop". See also **Interactive coding**.

Component uniqueness. The components in a model cannot be transformed or rotated without adversely affecting the fit of the model. See also **Subspace uniqueness** and **Uniqueness**.

Component interpretation. See **Multivariate point of view**.

Compression. Reduction of a multiway array without necessarily losing (much) information by using a high-dimensional multiway model. Typically used as a preliminary step before applying a Parafac algorithm to a large multiway data set. The use of compression can be justified by the maximum-product rule.

Compromise structure. In STATIS, the component solution on a weighted sum of covariance matrices.

Congruence coefficient. This coefficient is used to express the similarity between two components. It is comparable to a correlation coefficient, except that the coefficients are not centered: that is, it is based on cross products rather (co)variances.

Conjunctive association rule. See **Hierarchical classes model**.

Convex hull. In a two-dimensional scatter plot the convex hull consists of straight line segments that connect the outermost points of a set, such that all points in the plot are within the hull and the hull does not have any inward "dents". More technically, the inner angle between line segments at any point may not exceed 180°. In deviance plots and multiway scree plots the bottom part of the convex hull is used to assess candidate models for further analysis.

Core array. A multiway array that contains the linkage information between components from different modes. The size of a full three-way core array is $P \times Q \times R$, where P, Q, and R are the numbers of components of the three modes, respectively. For multiway models the full core array is $S_1 \times S_2 \times \cdots$, etc. A core array is *superdiagonal* if it is a (hyper)cube and only the elements with the same index, that is, $g_{sss\cdots}$, are nonzero. If all superdiagonal elements are equal to 1, the core

---

[2]In this book the brackets in expressions like (bi)plot and (dis)similarity signify both "biplot and plot", and "dissimilarity and similarity",respectively.

array is a superidentity array. A multiway core array is *slice diagonal* if only the elements with two equal indices, e.g., $g_{ssr...}$, are nonzero; this concept is mostly used for three-way arrays. See also **Extended core array**.

Core consistency. A Parafac model shows core consistency, if the core array calculated using its components has a **Superdiagonal** structure, or nearly does.

Correspondence analysis. A technique to analyze the dependence in contingency tables with nonnegative numbers. The two-way version is standard, three-way versions are rare.

Covariance form. The form of the coefficients in a principal component analysis such that the variable coordinates are correlations between the variables and the components, provided the variables have been standardized. The variable coefficients are in principal coordinates and the subject coefficients in standard coordinates. See also **Distance form**.

Decomposition. An exact re-expression of a matrix or array in terms of a product of simpler terms, or a sum of several products of the same form, as in singular value decomposition or Cholesky decomposition. A decomposition is usually an exact re-expression of a matrix or array in terms that reveal useful mathematical properties or are easier to manipulate mathematically. However, some authors use the term *decomposition* as if it were synonymous with *model*.

Decomposition model. A multiway model that allows a complete decomposition into components of fallible data array (i.e., containing both systematic and random variation). The Tucker models are decomposition models, but the Parafac model is not necessarily so.

Degeneracy. A solution of a Parafac model is called degenerate if two or more components are becoming (nearly) congruent. A solution is *divergent degenerate* if the algorithm to calculate the solution produces one or more of the parameter estimates that increase or decrease without bound, and the fit of the solution can always be improved by increasing the number of iterations and the numerical accuracy. A solution is *bounded degenerate* if it is an "unacceptable" solution with highly congruent components, notwithstanding convergence of the algorithm; often from multiple starting points. Sometimes solutions with temporarily highly congruent components are also called *temporary degenerate*.

Dependence, total. In multiway contingency tables, it is the part of the proportions or counts in a table that cannot be fitted by the model of independence; it is also called global dependence. *Marginal dependence*: That part of the total dependence that is contained in one of the margins of a multiway contingency table. *Partial dependence*: That part of the total dependence that remains after one or more of the marginal dependencies have been subtracted. This type of dependence is used in situations when there is a dependent margin in a table and the influence of the marginal dependence due to the predictors (the design margin) has to be eliminated.

Derived data. Data sets for which the values are not measured directly but are calculated from raw data. For instance, correlations are derived from profile data so that the parameters of one mode, here subjects, no longer feature in the model. Frequencies in contingency tables can also be conceived of as derived data. The same is true for certain types of similarity measures.

Deviance plot. A plot with the deviance or residual sum of squares plotted against the degrees of freedom. Models that are candidates for interpretation lie on or near the convex hull. Detailed inspection of the models on the convex hull can aid in model and dimensionality selection. See also **Multiway scree plot**.

Diagonality core array. See **Core array**.

DifFit criterion. A criterion to assist in choosing an appropriate Tucker model in conjunction with a multiway scree plot.

Direct fitting. Fitting a model to the raw data, which may have been centered or size-standardized in various ways. See also **Indirect fitting**.

Direction cosine. The angle between a vector and a coordinate axis.

Disjunctive association rule. See **Hierarchical classes model**.

Display mode. See **Joint biplot**.

Distance form. The form of the coordinates in a principal component analysis in which the Euclidean distances between subjects are preserved in the full decomposition and are represented as well as possible in the reduced space. The subject coefficients are in principal coordinates and the variable coefficients in standard coordinates. This form follows naturally from conceptualizing the components in PCA as linear combinations of the original variables. See also **Covariance form**.

Double centering. See **Centering**.

EM-algorithm. Expectation–maximization algorithm. An algorithm in which sets of parameters are estimated in turn. One step consists of calculating the expectations of a data point given the model if it is missing, the other of estimating the parameters of the model given the (imputed) data via maximization of the loss function. Alternating least squares algorithms can be seen as EM-algorithms.

Equilibrium circle. The circle drawn in a plot of normalized coordinates, which helps to assess the fit of the levels of a mode in low-dimensional space.

Equivalence relation. See **Hierarchical classes model**.

Euclidean distances. The straight-line distance between two points in a standard Cartesian coordinate system: $d(x, y) = \sqrt{\sum_s^S (x_s - y_s)^2}$, where S is the dimensionality of the space.

Exchangeable. Two entities are exchangeable if they are drawn from the same population and it is irrelevant to the validity of generalizations made from the sample which of the entities is included in the sample. Entities in random samples are automatically exchangeable. In multiway analysis random samples are rare, but one hopes that the entities included in a sample are at least exchangeable with nonsampled entities from the population.

Extended core array. A core array, most commonly three-way, in which one of the ways is not condensed into its components. An extended three-way core array is slice-diagonal if the slices featuring the two condensed ways are diagonal.

Extended multiway models. Multiway models that contain both additive and multiplicative terms, such as models with additive main effects and multiplicative terms for the interactions. Alternative terms are "biadditive model" (multi = two) and "triadditive models" (multi = three).

Factorial data. Within the context of multiway analyses, multiway factorial data are assumed to originate from a factorial design with a single observation per cell. The contents of the cells may be the scores or means of a single dependent variable, for instance, yield, with varieties, locations, and years as the three factors.

Fibers. A fiber is a one-dimensional subarray or vector indexed by all but one of the indices of the full array. For three-way data the fibers are vectors with two subscripts, in particular, rows ($\mathbf{x}_{ik}$), columns ($\mathbf{x}_{jk}$) and tubes ($\mathbf{x}_{ij}$). In the four-way case the fibers are: $\mathbf{x}_{ikl}$, $\mathbf{x}_{jkl}$, $\mathbf{x}_{ijl}$, and $\mathbf{x}_{ijk}$; the latter are called *pipes* in this book.

Flattening. Making a two-way matrix out of a three-way array by removing one of the ways, for instance, by averaging.

Free parameters. The number of free parameters in a model is the number of parameters in a model minus the number of restrictions on those parameters, such as nonnegativity constraints and orthogonality constraints.

Fully crossed multiway data. Multiway data for which in principle all possible combinations of levels from all ways exist. Thus, multiset data do not fall into this class.

Global dependence. See **Dependence, total**.

Hierarchical classes model. A hierarchical classes (or HICLAS) model is a model for binary data in which the parameters are also binary. These models create classes of levels in each mode, and in principle such classes can have hierarchical relations with each other. Because of the Boolean algebra defined for these models, the relations follow a disjunctive association rule, that is, a subject belongs to a class if the person has at least one but not necessarily all qualifications specified for a class. Models following conjunctive association rules have also been formulated, that is, a subject only belongs to a class if the person has all qualifications specified for a class.

Hierarchical relation. See **Hierarchical classes model**.

Indirect fitting. Fitting a model to data that have been nonlinearly derived from raw data, for instance, by first transforming raw data into cross products or correlations, or by changing distances into scalar products. When data have been transformed linearly (e.g., centered and/or normalized) they can be fitted directly. See also **Direct fitting**.

Individual differences scaling. INDSCAL. A model for sets of symmetric dissimilarity or similarity matrices. The space of the stimuli, which constitute the first and second way, is shared by the levels of the third mode (judges, subjects, etc.). The axes of their individual spaces are arbitrarily stretched or a shrunken versions of the common axes.

Individual differences in orientation scaling. IDIOSCAL. A model for sets of symmetric dissimilarity or similarity matrices similar to INDSCAL. In this model, however, the common space may also be rotated before the shrinking or stretching of the axes.

Inner product. The inner product of two vectors produces a scalar: if $\mathbf{y}$ and $\mathbf{x}$ are two vectors, then their inner product in matrix notation is $z = \mathbf{y}'\mathbf{x}$. It is equal to the sum of the products of the corresponding elements: $\sum y_i x_i$. Geometrically, it is the product of the lengths of the vectors times the cosine of the angle between them. This product is frequently used in interpreting biplots.

Interactive biplot. See **Nested-mode biplot**.

Interactive coding. Variables $\mathbf{y}$ with I levels and $\mathbf{x}$ with J levels are interactively coded into a new variable $\mathbf{z}$ with IJ levels if $\mathbf{z}$ is a fully crossed version of $\mathbf{y}$ and $\mathbf{x}$.

Interstructure. In STATIS, this is the component solution or structure based on the conditions covariance matrix. For each condition, the matrix with similarities between subjects is vectorized. And the interstructure covariance matrix consists of the covariances between these vectorized matrices.

Intrinsic axis property. Components or axes of a model have the intrinsic axis property if, under "reasonable" conditions, the orientations of the axes are determined by the specification of the model. The components of the Parafac model have the intrinsic axis property if the data structure is trilinear and the components show distinct variation in at least three modes.

Intrastructure. In STATIS, the structure within each condition's similarity matrix.

Jackknife. A jackknife estimator is based on systematically recomputing a statistical estimate leaving out one or more observations at a time from the sample. From the set of values for the statistic, an estimate for the bias of the statistic can be calculated, as well as an estimate for the variance of the statistic over repeated sampling. It is closely related to, and often used for the same purposes as, the bootstrap.

Joint biplot. In three-mode analysis, a joint biplot is a biplot for two of the modes (the *display modes*) conditional on a component of the third mode (the *reference mode*).

Kronecker product. The Kronecker product of an $I \times P$ matrix $\mathbf{A} = (a_{ip})$ and a $J \times Q$ matrix $\mathbf{B}$ is the $IJ \times PQ$ matrix composed of $J \times Q$ submatrices of the form $a_{ip}\mathbf{B}$.

Latent covariance matrix. For the Tucker2 model, the latent covariance matrix contains the (co)variances of the scores of the subjects on the latent-variable–prototype-condition combinations. Depending on the scaling, the covariances may be cross products, "real" covariances, or correlations. For the Tucker3 and Parafac models, this matrix contains the variances and covariances between the scores of the 'idealized subjects' on the latent-variable–prototype-condition combinations.

Levels. Levels are the units into which a mode is divided, each represented by a distinct value of the mode's subscript (e.g., rows are the levels of Mode A). "Level" is often used as a generic term for the entities in a way or mode, such as a variable, a subject, a crop variety, and so on.

Level-fit plot. See **Sums-of-squares plot**.

Line plot. See **Per-component plot**. The term *line plot* is also used for what is here called here a **All-components plot**.

Loadings. In two-way analysis loadings are regression coefficients relating the levels of a given mode to the components extracted from that mode. Some people restrict the definition to coefficients relating levels of the "variables" mode to its components; or even further to regression of orthogonal components on standard score variables, i.e., to variable–component correlations. Such conventions are less common in multiway contexts and/or outside of the social and behavioral sciences.

Marginal dependence. See **Dependence, global**.

Matricization. Matricization is the stringing out of an $I \times J \times K$ array into a two-way matrix, mostly of the order I by $(J \times K)$. Also referred to as *stringing-out*, *juxtapositioning*, and, especially in chemometrics, *unfolding*. The use of this last term, however, is now discouraged, both in chemometrics and elsewhere.

Maximum-product rule. The maximum-product rule states that if the size of one of the ways in a three-way data set is larger than the product of the other two, the largest way can be reduced without loss of information essential for the components of the two other ways and the core. Specifically, when $I > JK$, the data can be reduced to a $JK \times J \times K$ data set. Moreover, ignoring restrictions on the parameters, the number of free parameters in the largest mode is reduced from $I \times P$ to $JK \times P$.

Means plot. A means plot is a plot in which the values of a single variable (second mode) are indicated on the vertical axis and the conditions (the third mode) are marked on the horizontal axis. In the plot lines connect the condition means of the groups that are constituted on the basis of a cluster analysis.

Minimum-product rule. The minimum-product rule states that for a Tucker$n$ model the product of the numbers of components of $n - 1$ modes must always be equal or larger than that of the remaining mode, so that for the Tucker3 model $P \times Q \geq R$, $P \times R \geq Q$, and $Q \times R \geq P$.

Mode. The term "mode" refers to any one of the indexed directions of elements of an array, such as in "Mode A corresponds to rows" or "Mode B is the Subject mode". "Mode" is typically used as a synonym for "way", but the term has fewer alternative meanings. There is also a more specialized definition, by which mode means "type of classification". However, the restricted definition is still less common. See also **Way**.

Model. A model is a simplified description of a system or complex entity that retains its essential features; a mathematical expression that specifies the structural form of the patterns to be recovered from a data set, as in the "Tucker3 model". A statistical model has a low-rank structural part plus a "stochastic" part to represent disturbances or random error. See also **Decomposition**.

Model array. Or *Model matrix*; see **Structural image**.

Monopolar scale. A rating scale referring to a single adjective that is anchored by 0 at the lower end (for instance, not pleasant at all) and, say, 10 (as pleasant as can be) at the other end. An alternative is a single-adjective scale anchored by adverbs such as "extremely" (from extremely unpleasant [−5] to extremely pleasant [5]).

Multiblock models. Models for data that consist of more than one multiway data array (or block of data). Often one of the arrays is the dependent array to be predicted from the other arrays. Thus, the models fit into the generalized linear model framework.

Multidimensional point of view. In the multidimensional point of view data primarily considered as a population; stochastic considerations such as distributions of variables play only a limited role. High-dimensional spaces defined by the levels of the modes are generally projected into maximum-variance low-dimensional subspaces, which are interpreted as subspace; thus using a *subspace interpretation*.

Multilinear model. A model that is linear in each of its sets of parameters. Basic PCA is bilinear, the original Parafac model is *trilinear*, and the Tucker3 model is *quadrilinear*.

Multimode covariance matrix. A patterned covariance matrix in which one or more modes are nested in another mode, and the modes are fully crossed. Multitrait–multimethod matrices are two-mode covariance matrices in which the traits and methods are fully crossed. When no normalization has been carried out, the matrix is also referred to as *multimode cross product matrix*.

Multiple imputation. A procedure to substitute missing data estimates into a data set such that standard analysis procedures can be carried out. The essence of the method is to create a number of parallel data sets, each with different values for the same missing data (imputation), and estimate the variabilities on the basis of the different outcomes of the analyses on these imputed data sets.

Multiset data. Multiway data in which not all possible combinations of levels from all ways exist, but the data have at least one way in common. Examples are sets of cross-sectional data, data in which the same variables have been measured on different sets of individuals, or the same individuals have responded to different sets of variables. Examples of analyses of such data are not included in this book.

Multivariate point of view. The multivariate point of view concerns systems of random samples. It is based on the asymmetric roles of the modes. For instance, we may have J random variables measured on K samples of I individuals each who may or may not be the same, or alternatively that there are $J \times K$ random variables with a sample of I subjects. The results form multiway component analyses are generally interpreted via a direct interpretation of the components, i.e. via a *component interpretation*

Multiway scree plot. A plot of the deviance or residual sum of squares against the sum of the numbers of components of the modes. This plot is the multiway variant of Cattell's scree plot. Inspecting the models on the convex hull may aid model and dimensionality selection.

Murakami form. The form of the Tucker2 model $(\mathbf{A}\mathbf{H}_k\mathbf{B}')$ in terms of first-order component loadings $\mathbf{B}$, second-order component loadings $\mathbf{H}_k$, and second-order component scores $\mathbf{A}$. A Tucker3 variant has been formulated as well.

Murakami core. When in a Tucker3 model the product of the number of components of two modes -1 is equal to the number of components of the third mode, the associated core array is called a Murakami core. When $QR - 1 = P$ (or $PR - 1 = Q$ or $PQ - 1 = R$), and $Q \geq R$ (as can be done without loss of generality) the core array has exactly $R(Q + R - 2)$ nonzero elements, provided the mode with the largest number of components is in principal coordinates.

Natural weights. A term used for the relative weights of the modes in simultaneous rotations of components and (three-way) core arrays.

Nested-mode biplot. A biplot in which the row markers are fully crossed combinations of two modes, so that one mode is nested within the other (the index of the nested or inner mode runs fastest, and that of the nesting or outer mode slowest). The column markers are the levels of the of the remaining or reference mode. The plotting space is that of the reference mode. These plots are also called *interactive biplots*.

Nested-mode per-component plot. A plot in which a single component of a nested-mode component matrix is plotted. The vertical axis indicates the values on the (reference) component, the horizontal axis the levels of one of the nested modes (mostly an ordered inner mode one). The levels of the other nesting mode are plotted separately and connected in accordance with the levels of the inner mode.

Nesting of models. A model's solutions with different numbers of components are nested when the components from a solution with a smaller number of components are the same as the corresponding components from a solution with a larger number of components. The solutions of multiway (multi > 2) models with different numbers of components are rarely if ever nested.

Normalization. The process of equalizing the sums of squares of a subarray of a multiway array. For three-way profile data the normalization is mostly done per variable slice, so that the sum of squares of the values in a slice is equal to 1.

Normalized coordinates. A component is in normalized coordinates if it has a length equal to 1. See also **Standard coordinates**.

Object variation. Unlike system variation, with object variation the component structure is not the same from one occasion to the next. For instance, a set of variables can have different correlational structures under different conditions. See also **System variation**.

Offset term. The scalar that functions as the "real" zero point of an interval scale. For example, the intercept term in a regression model.

Optimal scaling. Optimal scaling is the process of finding a transformation for values, mostly categories, of a variable, such that the new quantifications or numerical values are in accordance with the specified measurement level of the variable and fit a specific model optimally in a least squares sense. For instance, if the measurement level is ordinal, the numerical values assigned by the transformation must have the same (partial) order as the original values.

Orthogonal. Two vectors are orthogonal if they are perpendicular to each other irrespective of their lengths. An $I \times J$ matrix $\mathbf{X}$ is (column-wise) orthogonal if $\mathbf{X}'\mathbf{X} = \mathbf{\Lambda}_J$, where $\mathbf{\Lambda}$ is a $J \times J$ diagonal matrix with arbitrary nonnegative numbers.

Orthonormal. Two vectors are orthonormal if they are orthogonal and have length 1. An $I \times J$ matrix $\mathbf{X}$ is (column-wise) orthonormal if $\mathbf{X}'\mathbf{X} = \mathbf{I}_J$, where $\mathbf{I}_J$ is a $J \times J$ diagonal matrix with ones and possibly some zeroes on the diagonal. In mathematics the name "orthogonal matrix" is sometimes restricted to mean a square orthonormal matrix, with non-square matrices described as "columnwise" orthogonal or orthonormal.

Outer product. The outer product of two vectors produces a matrix: if $\mathbf{y}$ and $\mathbf{x}$ are vectors, then their outer product is the matrix $\mathbf{Z} = \mathbf{yx}'$

Paired-components plot. A plot in which the components of a mode are plotted against one another. They are especially useful to examine the spatial arrangement of the plotted points. Such plots need to have an aspect ratio of one.

Parafac core array. The three-way core array that is calculated using the components of a Parafac solution.

Parafac model. The Parafac (parallel proportional profiles) model is a multilinear model with components for each of its modes and a superdiagonal core array. It specifies a parallel weighting of the components of one way by those of the other ways. The Parafac model is sometimes contrasted with the "Parafac decomposition" which refers to the mathematical decomposition of an array into a sum of rank

1 subarrays, analogous to the singular value decomposition. In this latter sense it is equivalent to the canonical decomposition CANDECOMP. See also **Parallel proportional profiles**.

Parallel proportional profiles. Components are parallel proportional across another mode, say, samples, if samples multiply the components with a constant so that from one sample to the next the complete component is larger or smaller, but there are no changes within the component.

Parallel solution. A solution of a Parafac model with two proportional columns in any of the component matrices

Partial dependence. See **Dependence, global**.

Per-component plot. A plot that contains one component of each mode. Such a plot is especially useful for the Parafac models, in which components are linked exclusively to each other. Another term for such plots is *across-modes plot*.

Principal coordinates. If a component is in principal coordinates, its length is equal to the singular value, which is equal to the square root of its eigenvalue.

Profile data. Data in which each subject is considered to have scores for a set of variables; these scores are called the profile of a subject.

Preprocessing. The application of procedures to a multiway data set before a multiway model is fitted; especially centering, normalization, and standardization.

Profile preprocessing. Profile preprocessing is the standard way to preprocess three-way profile data: variable–condition (fiber) centering and variable (lateral slice) normalization.

Projection. The projection of a vector $\mathbf{y}$ onto another vector $\mathbf{x}$ is the vector $\mathbf{y}'$ collinear with $\mathbf{x}$ that can be found by dropping a perpendicular line from $\mathbf{y}$ onto $\mathbf{x}$.

Quadrilinear. See **Multilinear**.

Postprocessing. The application of procedures to the estimated parameters of a multiway model to enhance interpretation, such as transformations, rotations, and scalings of components and/or core array.

Rao distance. The length d_i of a subject i's perpendicular to the component space defined by the S fitted components. The length indicates the extent to which the structural image fails to fit its data, and it is thus a measure for the extent that the subject is an outlier.

Rank of an array. The rank of an array $\mathcal{A}$ is the minimal number of rank-one arrays necessary to yield $\mathcal{A}$ as a linear combination of rank-one arrays.

Rank-one array. A multiway array of rank 1. It is generally the outer product of one component of each way. For instance, each term in the Parafac model for a three-way array, $\mathbf{a}_s \otimes \mathbf{b}_s \otimes \mathbf{c}_s$, is a rank-one array.

Rating scale data. Data consisting of subjects' judgments of a concept on a number of rating scales with the same measurement properties. An example of a rating scale is a variable with five answer categories running from "extremely disagree" to "extremely agree". When several concepts have to be judged the data become three-way rating scale data.

Reference mode. The column mode for the nested-mode biplot, or the mode conditional on which joint biplots are made. See also **Joint biplot** and **Nested-mode biplot**.

Relative fit. 'Relative fit' is an abbreviation for "relative fitted sum of squares" and is the fitted sum of squares divided by the total sum of squares. The term applies to specific parts of a model such as the levels of a mode. It is equal to 1-Relative residual sum of squares.

Relative residual sum of squares. Residual sum of squares divided by the total sum of squares. The term applies to specific parts of a model such as the levels of a mode.

Replicated PCA. A Tucker3 model with only one component in one of the modes. It results in a single diagonal core slice, given the orthogonality of the component matrices. Also called *Weighted PCA*

Row-isometric. A biplot is row-isometric if the rows are in principal coordinates and the columns in normalized or standard coordinates.

Scores. A name given to component coefficients for subjects, objects, etc.

Simultaneous factor analysis. A set of structural equation models defined for sets of covariance matrices. These models are characterized by factor matrices comparable for all covariance matrices. The models can be restricted to have common loading patterns, identical factor loadings, identical factor correlations, identical unique variances, and so on.

Singular value decomposition. The decomposition of a matrix $\mathbf{X}$ into a left set of orthonormal vectors $\mathbf{U}$, a right set of orthonormal vectors $\mathbf{V}$, and a diagonal matrix $\mathbf{\Lambda}$ of singular values, which are the square roots of the eigenvalues of both $\mathbf{X}'\mathbf{X}$ and $\mathbf{X}\mathbf{X}'$. Thus $\mathbf{X} = \mathbf{U}\mathbf{\Lambda}\mathbf{V}'$. The SVD represents the basic structure of a matrix and forms the basis of the generalization to methods for multiway arrays.

Slice diagonality. See **Extended core array**.

Slices of an array. For three-way data a slice is matrix of values with a single subscript: *frontal slices* $\mathbf{X}_k$, *lateral slices* $\mathbf{X}_j$, and *horizontal slices* $\mathbf{X}_i$. For N-way data a more general definition is: two-way subarrays of an array, that is, the set of elements defined by keeping all but two subscripts fixed and allowing the two that vary to take on all the values in their range. Slices are sometimes also called *slabs*.

st-**criterion.** A criterion for selecting an adequate multiway model on the basis of the sharpest angle in the convex hull in a deviance plot.

Standardization. Combination of centering and normalization.

Standard coordinates. A component is in standard coordinates if it has a mean of 0 and a length or variance equal to 1.

Standardized (component) weights. The eigenvalue of a component divided by the total sum of squares. Thus for PCA, SS(Fit)$_s$/SS(Total) is the standardized weight of component s. In the Parafac model these quantities are g_{sss}/SS(Total). In the Tucker models the weights are SS(Fit)$_\ell$/SS(Total), where $\ell = p, q,$ or r.

STATIS. A method for analyzing three-way arrays that consists of three phases, the computation of (1) the *interstructure* between conditions, (2) *compromise structure* for the subjects, and (3) the *intrastructure* within each condition.

Stringing-out. See **Matricization**.

Structural image of the data. The structural image of a data set is the collection of values calculated on the basis of a model; they are supposed to approximate the data to a satisfactory degree. Thus, in the equation $y = \hat{y} + error$, $\hat{y}$ is the structural image. In a simple regression equation, $\hat{y}$ is calculated on the basis of $b_0 + b_1 * x$. Other terms in use: *model matrix, implied data (by a model), fitted data, data estimated on the basis of a model.*

Structured component scores (loadings). Component scores (loadings) that consist of the product of the scores (loadings) of two modes, that is, $u_{\ell s} = u_{(jk)s} = a_{js}b_{ks}$. The term "structured loading matrix" is also used for $\mathbf{L} = (\mathbf{C} \otimes \mathbf{B})\mathbf{G}_a$, in which the core array is part of the definition.

Subspace. A low-dimensional space that falls completely within a higher-dimensional space.

Subspace uniqueness. A uniqueness referring to an entire subspace without assuming uniqueness for the orientation of the axes in this subspace. The subspaces derived by Tucker models usually have subspace uniqueness but not component uniqueness. The solution of a singular value decomposition usually also has subspace uniqueness. See also **Component uniqueness** and **Uniqueness**.

Subspace interpretation. See **Multidimensional point of view**.

Successive dyadic fitting. A procedure to estimate the components of a two-mode component analysis by alternating regression procedures on single components at a time.

Sums-of-squares plot. A plot with the residual sum of squares on the vertical axis and the fitted sum of squares on the horizontal axis. These plots are used to assess the extent to which the data of the levels of a mode are fitted by the multiway model.

Superdiagonality. A core array is superdiagonal if it is a (hyper)cube and only the elements g_{sss} or $g_{sssss..}$ are nonzero. The term *body diagonality* has also been used in the past. See also **Core array**.

Supermatrix. See **Tall combination-mode matrix**.

Symmetric scaling. As used in optimal scaling or biplot construction, the coordinates of row and column markers are symmetrically scaled when neither of the marker sets has normalized or principal coordinates, but the scale factor is equally divided between them. The emphasis is on assessing the inner products between row and column markers, rather than the relationships among the row or column markers. The inner products are the approximations of the values in the data table. See also **Asymmetric scaling**.

System variation. A component shows system variation across levels of a given mode (e.g., across levels of mode C) if its pattern of relative influence across the levels of the other modes (e.g., A, and B) does not change, but its magnitude of influence at different levels is stepped up or down proportionally (e.g., its effect changes proportionally across all elements from one level of C to the next). See also **Object variation**.

Tall combination-mode matrix. A two-way matrix resulting from matricizing a three-way array such that the matrix has $K \times I$ rows and J columns. Thus, it is a partitioned matrix whose elements are themselves matrices.

Tensor. Sometimes used in the multiway literature as an alternative term for array, for example, in the expression "higher-order tensor". In this book, the term "array" is always used, because in mathematics the term "tensor" has a deeper meaning than simply "array".

Three-mode factor analysis. A term initially used for three-mode component analysis. Now primarily used for *three-mode common factor analysis* and three-mode exploratory factor analysis. In three-mode common factor analysis, multimode covariance matrices are modeled using structural equation models that include separate factors for the variables and occasions and sometimes a form of a core array as well. Direct-product models form a limiting case. See also **Simultaneous factor analysis**.

Three-mode scaling. The application of the Tucker2 model to sets of similarity or dissimilarity matrices.

Three-mode scree plot. See **Multiway scree plot**.

Total dependence. See **Dependence, total**.

Trajectories. A trajectory refers to the collection of points in a component plot (mostly a nested-mode biplot), where for the level of one mode the levels of another ordered mode are connected in their proper order. For instance, if a subject i has coordinate values on a component q for time points $k = 1, \ldots, K$, then a trajectory is the connected set of coordinates $\mathbf{a}_{i1}, \ldots, \mathbf{a}_{iK}$ in the Q-dimensional component space of B, where $\mathbf{a}_{ik}$ is a Q-dimensional vector.

Triadditive models. Models for three-way data that contain both additive and multiplicative terms, such as models with additive main effects and multiplicative terms for one or more interactions.

Trilinear. See **Multilinear**.

Tucker variation. A term used in the context of the Parafac model. In the Parafac model, the components of a mode are only linked to one component of another mode. If, however, links between different numbered components exist, then the components show Tucker variation.

Tucker models. Multiway models in which component matrices are specified for at least one of the ways, and in which the relationships between components from the ways are represented in a core array. A Tucker1 model is an ordinary PCA model but applied to a tall combination-mode matrix; a Tucker2 model has components for two of its ways with an extended core array, a Tucker$n$ model has components for n of its ways. Generally, Tucker$n$ models are applied to n-way data arrays.

Uniqueness. A (Parafac) model is unique if the set of components or axes cannot be rotated or linearly transformed without affecting fit to the data (apart from trivial rescalings and permutations). One refers to *weak uniqueness* when there are nonunique models with almost the same fit, and to *strong uniqueness* when there are no such models. *Partial uniqueness* refers to the situation when some but not all of the components are uniquely determined. See also **Component uniqueness** and **Subspace uniqueness**.

Validation, external. Evaluation and verification of the results of an analysis by means of information that was not used in the analysis itself.

Validation, internal. Evaluation of the solution from an analysis using fit statistics, residual analyses, bootstrap analysis, and so on.

Vectorization. The rewriting of a matrix as a vector by stacking the columns of the matrix underneath each other. The operation is indicated by the vec operator. Thus if a column of $\mathbf{X}$ is written as $\mathbf{x}'_j$, then $vec(\mathbf{X}) = \mathbf{x}'$ where $\mathbf{x}' = (\mathbf{x}'_1, \ldots, \mathbf{x}'_J)'$. The vectorization of a multiway array is a generalization of this process and can be realized by multiple applications of the vec operator. However, vec is then written only once.

Way. The *ways* of an array correspond to the individual indices used to reference elements in the array. The term 'way' is adopted from similar applications in analysis of variance. Carroll and Arabie have proposed a distinction between *way* and *mode*. In particular, their proposal entails that if the contents of the ways are different, say, subjects, variables, and conditions, the (three-way) array is said to have three modes. If it has two ways with the same content, the array is said to have two modes and three ways. See also **Mode**.

Weighted PCA. See **Replicated PCA**.

Wide combination-mode matrix. A two-way matrix resulting from matricizing a three-way array, such that the matrix has I rows and $K \times J$ columns.

ACRONYMS

ALS   Alternating least squares

ALSCOMP3   Alternating least-squares component analysis for three-way data

AMMI   Additive main effects and multiplicative interaction model

ANOVA   Analysis of variance

ANACOR   Correspondence analysis

CANDECOMP   Canonical decomposition

CANDELINC   Canonical decomposition with linear constraints

CORCONDIA   Core consistency diagnostic

DIFFIT   Differential fit statistic

EM   Expectation-maximization [algorithm]

FANOVA   Factor analysis of variance

GEMANOVA   General multiplicative analysis of variance

GEPCAM   Generalized principal component analysis with missing data

GSVD   Generalized singular value decomposition

HICLAS	Hierarchical classes model
HOSVD	Higher-order singular value decomposition
IDIOSCAL	Individual differences in orientation scaling
INDCLASS	Individual differences classification
INDCLUS	Individual differences clustering
INDSCAL	Individual differences scaling
MCD	Minimum covariance determinant (estimator)
MICE	Multivariate imputation by chained equations
MIXCLUS3	Mixture method of clustering for three-mode data
MLPCA	Maximum likelihood principal component analysis
OLS	Ordinary least squares
PARAFAC	Parallel factor analysis
PCA	Principal component analysis
PLS	Partial least squares
PRESS	Predictive residual error sum of squares
RobPCA	Robust principal component analysis
SCA	Simultaneous component analysis
SEM	Structural equation modeling
SVD	Singular value decomposition
STATIS	Structuration des tableaux à trois indices de la statistique
TUCKALS2	Tucker2 model alternating least-squares algorithm
TUCKALS3	Tucker3 model alternating least-squares algorithm

AUTHOR INDEX

---

[1] Entries refer to publications, and are ordered by first author irrespective the presence of co-authors.

545

SUBJECT INDEX

WILEY SERIES IN PROBABILITY AND STATISTICS
ESTABLISHED BY WALTER A. SHEWHART AND SAMUEL S. WILKS

The *Wiley Series in Probability and Statistics* is well established and authoritative. It covers many topics of current research interest in both pure and applied statistics and probability theory. Written by leading statisticians and institutions, the titles span both state-of-the-art developments in the field and classical methods.

Reflecting the wide range of current research in statistics, the series encompasses applied, methodological and theoretical statistics, ranging from applications and new techniques made possible by advances in computerized practice to rigorous treatment of theoretical approaches.

This series provides essential and invaluable reading for all statisticians, whether in academia, industry, government, or research.

BECHHOFER, SANTNER, and GOLDSMAN · Design and Analysis of Experiments for Statistical Selection, Screening, and Multiple Comparisons

BELSLEY · Conditioning Diagnostics: Collinearity and Weak Data in Regression

† BELSLEY, KUH, and WELSCH · Regression Diagnostics: Identifying Influential Data and Sources of Collinearity

BENDAT and PIERSOL · Random Data: Analysis and Measurement Procedures, *Third Edition*

BERRY, CHALONER, and GEWEKE · Bayesian Analysis in Statistics and Econometrics: Essays in Honor of Arnold Zellner

BERNARDO and SMITH · Bayesian Theory

BHAT and MILLER · Elements of Applied Stochastic Processes, *Third Edition*

BHATTACHARYA and WAYMIRE · Stochastic Processes with Applications

BILLINGSLEY · Convergence of Probability Measures, *Second Edition*

BILLINGSLEY · Probability and Measure, *Third Edition*

BIRKES and DODGE · Alternative Methods of Regression

BISWAS, DATTA, FINE, and SEGAL · Statistical Advances in the Biomedical Sciences: Clinical Trials, Epidemiology, Survival Analysis, and Bioinformatics

BLISCHKE AND MURTHY (editors) · Case Studies in Reliability and Maintenance

BLISCHKE AND MURTHY · Reliability: Modeling, Prediction, and Optimization

BLOOMFIELD · Fourier Analysis of Time Series: An Introduction, *Second Edition*

BOLLEN · Structural Equations with Latent Variables

BOLLEN and CURRAN · Latent Curve Models: A Structural Equation Perspective

BOROVKOV · Ergodicity and Stability of Stochastic Processes

BOULEAU · Numerical Methods for Stochastic Processes

BOX · Bayesian Inference in Statistical Analysis

BOX · R. A. Fisher, the Life of a Scientist

BOX and DRAPER · Response Surfaces, Mixtures, and Ridge Analyses, *Second Edition*

* BOX and DRAPER · Evolutionary Operation: A Statistical Method for Process Improvement

BOX and FRIENDS · Improving Almost Anything, *Revised Edition*

BOX, HUNTER, and HUNTER · Statistics for Experimenters: Design, Innovation, and Discovery, *Second Editon*

BOX and LUCEÑO · Statistical Control by Monitoring and Feedback Adjustment

BRANDIMARTE · Numerical Methods in Finance: A MATLAB-Based Introduction

† BROWN and HOLLANDER · Statistics: A Biomedical Introduction

BRUNNER, DOMHOF, and LANGER · Nonparametric Analysis of Longitudinal Data in Factorial Experiments

BUCKLEW · Large Deviation Techniques in Decision, Simulation, and Estimation

CAIROLI and DALANG · Sequential Stochastic Optimization

CASTILLO, HADI, BALAKRISHNAN, and SARABIA · Extreme Value and Related Models with Applications in Engineering and Science

CHAN · Time Series: Applications to Finance

CHARALAMBIDES · Combinatorial Methods in Discrete Distributions

CHATTERJEE and HADI · Regression Analysis by Example, *Fourth Edition*

CHATTERJEE and HADI · Sensitivity Analysis in Linear Regression

CHERNICK · Bootstrap Methods: A Guide for Practitioners and Researchers, *Second Edition*

CHERNICK and FRIIS · Introductory Biostatistics for the Health Sciences

CHILÈS and DELFINER · Geostatistics: Modeling Spatial Uncertainty

CHOW and LIU · Design and Analysis of Clinical Trials: Concepts and Methodologies, *Second Edition*

CLARKE and DISNEY · Probability and Random Processes: A First Course with Applications, *Second Edition*

* COCHRAN and COX · Experimental Designs, *Second Edition*

*Now available in a lower priced paperback edition in the Wiley Classics Library.
†Now available in a lower priced paperback edition in the Wiley–Interscience Paperback Series.

---

*Now available in a lower priced paperback edition in the Wiley Classics Library.
†Now available in a lower priced paperback edition in the Wiley–Interscience Paperback Series.

† FLEMING and HARRINGTON · Counting Processes and Survival Analysis
  FULLER · Introduction to Statistical Time Series, *Second Edition*
† FULLER · Measurement Error Models
  GALLANT · Nonlinear Statistical Models
  GEISSER · Modes of Parametric Statistical Inference
  GELMAN and MENG · Applied Bayesian Modeling and Causal Inference from
    Incomplete-Data Perspectives
  GEWEKE · Contemporary Bayesian Econometrics and Statistics
  GHOSH, MUKHOPADHYAY, and SEN · Sequential Estimation
  GIESBRECHT and GUMPERTZ · Planning, Construction, and Statistical Analysis of
    Comparative Experiments
  GIFI · Nonlinear Multivariate Analysis
  GIVENS and HOETING · Computational Statistics
  GLASSERMAN and YAO · Monotone Structure in Discrete-Event Systems
  GNANADESIKAN · Methods for Statistical Data Analysis of Multivariate Observations,
    Second Edition
  GOLDSTEIN and LEWIS · Assessment: Problems, Development, and Statistical Issues
  GREENWOOD and NIKULIN · A Guide to Chi-Squared Testing
  GROSS and HARRIS · Fundamentals of Queueing Theory, *Third Edition*
* HAHN and SHAPIRO · Statistical Models in Engineering
  HAHN and MEEKER · Statistical Intervals: A Guide for Practitioners
  HALD · A History of Probability and Statistics and their Applications Before 1750
  HALD · A History of Mathematical Statistics from 1750 to 1930
† HAMPEL · Robust Statistics: The Approach Based on Influence Functions
  HANNAN and DEISTLER · The Statistical Theory of Linear Systems
  HEIBERGER · Computation for the Analysis of Designed Experiments
  HEDAYAT and SINHA · Design and Inference in Finite Population Sampling
  HEDEKER and GIBBONS · Longitudinal Data Analysis
  HELLER · MACSYMA for Statisticians
  HINKELMANN and KEMPTHORNE · Design and Analysis of Experiments, Volume 1:
    Introduction to Experimental Design, *Second Edition*
  HINKELMANN and KEMPTHORNE · Design and Analysis of Experiments, Volume 2:
    Advanced Experimental Design
  HOAGLIN, MOSTELLER, and TUKEY · Exploratory Approach to Analysis
    of Variance
* HOAGLIN, MOSTELLER, and TUKEY · Exploring Data Tables, Trends and Shapes
* HOAGLIN, MOSTELLER, and TUKEY · Understanding Robust and Exploratory
    Data Analysis
  HOCHBERG and TAMHANE · Multiple Comparison Procedures
  HOCKING · Methods and Applications of Linear Models: Regression and the Analysis
    of Variance, *Second Edition*
  HOEL · Introduction to Mathematical Statistics, *Fifth Edition*
  HOGG and KLUGMAN · Loss Distributions
  HOLLANDER and WOLFE · Nonparametric Statistical Methods, *Second Edition*
  HOSMER and LEMESHOW · Applied Logistic Regression, *Second Edition*
  HOSMER, LEMESHOW, and MAY · Applied Survival Analysis: Regression Modeling
    of Time-to-Event Data, *Second Edition*
† HUBER · Robust Statistics
  HUBERTY · Applied Discriminant Analysis
  HUBERTY and OLEJNIK · Applied MANOVA and Discriminant Analysis,
    Second Edition
  HUNT and KENNEDY · Financial Derivatives in Theory and Practice, *Revised Edition*

*Now available in a lower priced paperback edition in the Wiley Classics Library.
†Now available in a lower priced paperback edition in the Wiley–Interscience Paperback Series.

*Now available in a lower priced paperback edition in the Wiley Classics Library.
†Now available in a lower priced paperback edition in the Wiley–Interscience Paperback Series.

*Now available in a lower priced paperback edition in the Wiley Classics Library.
†Now available in a lower priced paperback edition in the Wiley–Interscience Paperback Series.

*Now available in a lower priced paperback edition in the Wiley Classics Library.
†Now available in a lower priced paperback edition in the Wiley–Interscience Paperback Series.

SCHUSS · Theory and Applications of Stochastic Differential Equations
SCOTT · Multivariate Density Estimation: Theory, Practice, and Visualization
† SEARLE · Linear Models for Unbalanced Data
† SEARLE · Matrix Algebra Useful for Statistics
† SEARLE, CASELLA, and McCULLOCH · Variance Components
SEARLE and WILLETT · Matrix Algebra for Applied Economics
SEBER · A Matrix Handbook For Statisticians
† SEBER · Multivariate Observations
SEBER and LEE · Linear Regression Analysis, *Second Edition*
† SEBER and WILD · Nonlinear Regression
SENNOTT · Stochastic Dynamic Programming and the Control of Queueing Systems
* SERFLING · Approximation Theorems of Mathematical Statistics
SHAFER and VOVK · Probability and Finance: It's Only a Game!
SILVAPULLE and SEN · Constrained Statistical Inference: Inequality, Order, and Shape
    Restrictions
SMALL and McLEISH · Hilbert Space Methods in Probability and Statistical Inference
SRIVASTAVA · Methods of Multivariate Statistics
STAPLETON · Linear Statistical Models
STAPLETON · Models for Probability and Statistical Inference: Theory and Applications
STAUDTE and SHEATHER · Robust Estimation and Testing
STOYAN, KENDALL, and MECKE · Stochastic Geometry and Its Applications, *Second
    Edition*
STOYAN and STOYAN · Fractals, Random Shapes and Point Fields: Methods of
    Geometrical Statistics
STREET and BURGESS · The Construction of Optimal Stated Choice Experiments:
    Theory and Methods
STYAN · The Collected Papers of T. W. Anderson: 1943–1985
SUTTON, ABRAMS, JONES, SHELDON, and SONG · Methods for Meta-Analysis in
    Medical Research
TAKEZAWA · Introduction to Nonparametric Regression
TANAKA · Time Series Analysis: Nonstationary and Noninvertible Distribution Theory
THOMPSON · Empirical Model Building
THOMPSON · Sampling, *Second Edition*
THOMPSON · Simulation: A Modeler's Approach
THOMPSON and SEBER · Adaptive Sampling
THOMPSON, WILLIAMS, and FINDLAY · Models for Investors in Real World Markets
TIAO, BISGAARD, HILL, PEÑA, and STIGLER (editors) · Box on Quality and
    Discovery: with Design, Control, and Robustness
TIERNEY · LISP-STAT: An Object-Oriented Environment for Statistical Computing
    and Dynamic Graphics
TSAY · Analysis of Financial Time Series, *Second Edition*
UPTON and FINGLETON · Spatial Data Analysis by Example, Volume II:
    Categorical and Directional Data
VAN BELLE · Statistical Rules of Thumb
VAN BELLE, FISHER, HEAGERTY, and LUMLEY · Biostatistics: A Methodology for
    the Health Sciences, *Second Edition*
VESTRUP · The Theory of Measures and Integration
VIDAKOVIC · Statistical Modeling by Wavelets
VINOD and REAGLE · Preparing for the Worst: Incorporating Downside Risk in Stock
    Market Investments
WALLER and GOTWAY · Applied Spatial Statistics for Public Health Data
WEERAHANDI · Generalized Inference in Repeated Measures: Exact Methods in
    MANOVA and Mixed Models
WEISBERG · Applied Linear Regression, *Third Edition*

*Now available in a lower priced paperback edition in the Wiley Classics Library.
†Now available in a lower priced paperback edition in the Wiley–Interscience Paperback Series.

WELSH · Aspects of Statistical Inference

WESTFALL and YOUNG · Resampling-Based Multiple Testing: Examples and Methods for *p*-Value Adjustment

WHITTAKER · Graphical Models in Applied Multivariate Statistics

WINKER · Optimization Heuristics in Economics: Applications of Threshold Accepting

WONNACOTT and WONNACOTT · Econometrics, *Second Edition*

WOODING · Planning Pharmaceutical Clinical Trials: Basic Statistical Principles

WOODWORTH · Biostatistics: A Bayesian Introduction

WOOLSON and CLARKE · Statistical Methods for the Analysis of Biomedical Data, *Second Edition*

WU and HAMADA · Experiments: Planning, Analysis, and Parameter Design Optimization

WU and ZHANG · Nonparametric Regression Methods for Longitudinal Data Analysis

YANG · The Construction Theory of Denumerable Markov Processes

YOUNG, VALERO-MORA, and FRIENDLY · Visual Statistics: Seeing Data with Dynamic Interactive Graphics

ZELTERMAN · Discrete Distributions—Applications in the Health Sciences

* ZELLNER · An Introduction to Bayesian Inference in Econometrics

ZHOU, OBUCHOWSKI, and McCLISH · Statistical Methods in Diagnostic Medicine

*Now available in a lower priced paperback edition in the Wiley Classics Library.

†Now available in a lower priced paperback edition in the Wiley–Interscience Paperback Series.